A-PLUS NOTES
FOR
BEGINNING ALGEBRA
(Pre-Algebra and Algebra 1)

US Common Core State Standards for High School Mathematics
(Based on the Common Core Standards at www.corestandards.org)

Since the official standards cover all math subjects and topics, the main topics for Beginning Algebra are picked up and organized clearly in the categories below.

A-Plus Notes Learning Center

Library of Congress Control Number: 2002096952
Washington, DC 20540-4320.

ISBN–13: 978-0-9654352-2-2
ISBN–10: 0-9654352-2-9
Printed in the United States of America

A-Plus Notes Learning Center
A publisher
Torrance, CA 90503
Tel: (310) 870-7093

WHERE TO BUY:

Available at Barnes & Noble, Amazon.com., and local educational bookstores across the United States, such as Teacher Supply, School Supply,····, etc.

Revised and Updated, 13th printing

Preface

Teaching math in an urban school of Los Angeles for several years, I am aware of significant differences between math students here and the students I taught for many years in Asia. The students in my classroom here come from different countries all over the world with widespread differences in their academic achievement. Many have difficulty in comprehending simple math concepts and are confused about the problem-solving steps necessary for simplifying math expressions or solving word problems.

In order to provide a clear and easy way for my students to learn algebra in my classes, I designed notes which simplify and outline the concepts and skills in the course textbooks. The notes focus on sequential, easy steps in the problem-solving process. Each topic is organized in one or two pages for quick understanding. For the past few years, I have distributed the notes to my students and encouraged them to use the notes together with their course textbooks. My students have told me that my notes *do* help facilitate their learning process in algebra. They always ask for my notes whenever they enter my classroom. Some teachers and students from other schools have also contacted me and requested copies of my notes. This is the reason why I have decided to print and publicly publish my notes for use as a reference book for all math students.

This reference book provides a clear summary and outline of all topics in algebra. Each topic begins with concepts, definitions, formulas, theorems, and problem-solving steps, followed by well-designed examples which illustrate the summary. The book covers most topics introduced and developed in high school and college textbooks, **ranging from basic math skills through Pre-Algebra and Algebra 1**. It provides students with a useful guide for review and further study.

There are 18,000 examples and exercises in this reference book–7,000 are real-life word problems modeled after typical questions on standardized tests that are given across the United States.

I wrote this book following the same pattern of the math books used in the schools in Asia. This pattern has been proved as excellent for student-learning.
I am trying to help the students here to catch up and meet the math-level in Asia.
Because I know the benefits from how I was taught, I want you to learn it as well.

Rong Yang
Los Angeles, California.

HOW TO USE THIS BOOK

1. Understand the summary and outline in each section.
2. Study the examples and be familiar with the problem-solving steps.
3. Practice the exercises and check your answers in the back of the book.
4. Pre-Algebra Students: Follow Page 8 "Lesson Plan for Pre-Algebra".
5. Algebra 1 Students: The students should study the whole book.
6. If you are a top math student, read and study "additional examples" on the last part of each chapter to succeed in getting high scores in all kinds of achievement tests, and win nationwide brainpower competition such as Super Quiz and Academic Decathlon.

ATTENTION: MATH TEACHERS AND STUDENTS

This book will make the teachers' job and the students' learning process easier if the students have it on hand in the class.

It makes a teacher's job easier because the teacher need not " write notes " on the board in the class. It makes a student's learning process easier because the student need not spend time to " take notes " in the class. This book is a complete set of lecture " notes " for the students. They need only to read it.

Most schools in the United States take back the textbook from the student after the course is completed. The student has no material on hand for future review. This book is simple and easy like a dictionary for the student to review when needed in the future.

This book is strongly recommended for math teachers and students. You may use it as a reference book together with the textbook, or use it as the textbook in your class.

For more information, visit this book and readers' comments, or write your comments at :

www.amazon.com and **www.barnes & noble.com**

The Publisher

Table of Contents

Lesson Plan for Pre-Algebra Only

Notes: 1. Read the instructions and examples of each section in the book.
2. Work on the following exercises. If your time is limited, choose only the exercises that are numbered odd or even.

Chapter	Pages	Exercises	Time Needed (hours)
1	25~26	1~92, 113~122, 131~136	3.0
2	39~40	1~125	3.0
3	55~56	1~90, 106~113, 126~138	3.0
4	77~78	1~72, 81~95, 111~118	3.0
5	95~96	1~129	4.0
6	129~130	1~90	3.0
7	165~166	1~47, 61~111	4.0
8	189~190	1~32, 41~50, 71~76, 81~89 101~107, 121~130, 136~144	4.0
9	209~210	1~35, 61~63, 88~91	3.0
10	225~226	1~18, 25~30, 52~57	3.0
11	241~242	1~20, 29~32, 65~70, 86~91, 101~104	3.0
12	255~256	1~24, 56~64, 71~82, 90, 91	3.0
13	271~272	1~24, 48~58, 76~80	3.0
14	274 276 282 285~286 287~288	1~6, 16~20, 26~27 1~16 1~8, 26~28 1~32, 41~55 1~21	2.0 2.0 2.0 2.0 2.0

Total: 52 hours

Numbers

1-1 Real Numbers

Understanding the classification of numbers is a great help in the study of mathematics. In mathematics, the following numbers are called **real numbers**:

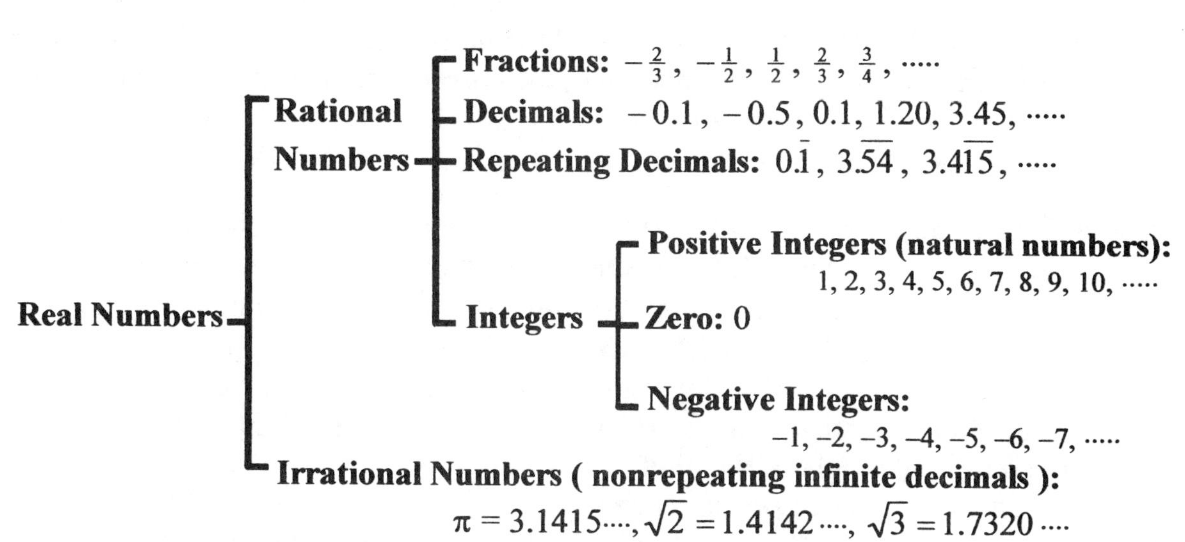

We can graph all of the above real numbers on the **number line** :

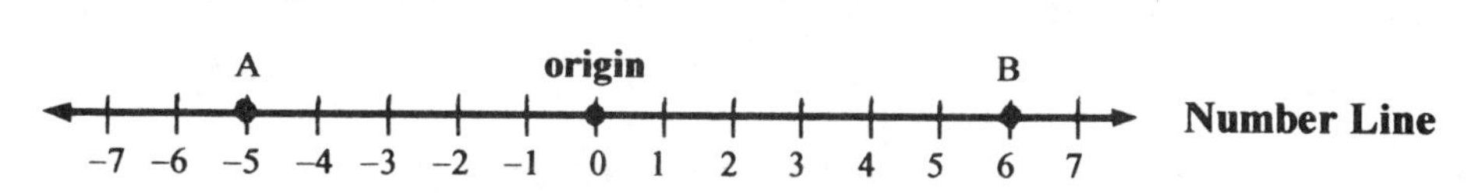

Definition of Real Numbers: Numbers which can be located on the number line.

Each point on the number line is called the **graph** of the number on the line. The number is called the **coordinate** of the point. The coordinate of point A is –5. The graph of –5 is point A. The graph of 0 is the point called **origin**. The coordinate of the origin is 0.

Real numbers include **rational numbers** and **irrational numbers**.

Rational Numbers: Numbers which can be written as the ratio of two integers.

$0.1 = \frac{1}{10}, \quad 0.5 = \frac{5}{10} = \frac{1}{2}, \quad 1.2 = \frac{12}{10} = \frac{6}{5}, \quad 0.\overline{1} = \frac{1}{9}, \quad 0.\overline{54} = \frac{54}{99} = \frac{6}{11}$.

Every rational number can be expressed as either a terminating decimal or as a repeating decimal.

Irrational Numbers: Numbers which cannot be written as the ratio of two integers. They are nonterminating (infinite), nonrepeating decimals.

Integers: They are positive integers, negative integers and zero.
Zero is an integer, neither positive nor negative.

Positive Integers: Numbers which we use to count objects. (1, 2, 3, 4, 5, 6, 7, 8, 9, 10, ·····.)
They are also called **natural numbers** or **counting numbers**.

Whole Numbers: They are 0 and positive integers. (0, 1, 2, 3, 4, 5, 6, 7, 8, 9, 10, ·····.)

Prime Number: A whole number other than 0 and 1 which is divisible only by 1 and itself.
(2, 3, 5, 7, 11, 13, 17, 19, 23, 29, 31, 37, 41, ·····.) The only even prime is 2.

Composite Number: A whole number other than 0 and 1 which is not a prime number.
(4, 6, 8, 9, 10, 12, 14, 15, 16, 18, 20, 21, 22, 24, 25, 26, 27, 28, 30, ·····.)

On a number line, the numbers to the right of the origin are positive, and the numbers to the left of the origin are negative. On a number line, the number on the right is greater than the number on the left. $>$ means "is greater than". $5 > 2$, $2 > -1$, $-3 > -6$.
$<$ means "is less than". $2 < 5$, $-1 < 2$, $-6 < -3$.

Rule of Rounding : When an exact computation is not needed for the answer of a problem, we use the **rules of rounding** to estimate numbers. We can round each number to the nearest 10, 100, 1000, ····. It depends on how accurate the answer we need. We use the symbol "$\approx$" which means "approximately equal to".

1. If the digit to the right of the digit to be rounded is 5 or more, we **round up** by increasing the digit to be rounded by **1** and replacing the digits to the right with zeros.
2. If the digit to the right of the digit to be rounded is less than 5, we **round down** by leaving the digit to be rounded **the same** and replacing the digits to the right with zeros.

$928{,}426 \approx 928{,}430 \approx 928{,}400 \approx 928{,}000 \approx 930{,}000$.

Place –Value of Numbers: Every digit that makes up a number has a **place-value** name.

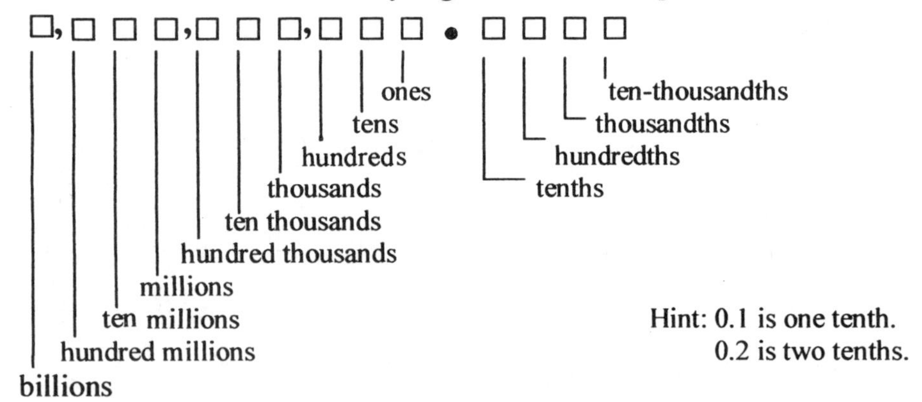

The short word name of 21, 234, 567, 809.1234 is:
" 21 billion, 234 million, 567 thousand, 809 and 1 thousand 234 ten-thousandths."

Read or write in **the number-word name**:
"twenty-one billion, two hundred thirty-four million, five hundred sixty-seven thousand, eight hundred nine and one thousand two hundred thirty-four ten-thousandths."

To write a number in words, the comma "," can be omitted.
To write a number in words, we must use a hyphen "-" between 20 and 100.
Never use the word "and" when we write the whole-number part.
To write the decimal point, we write the word "**and**".

EXERCISES

Classify each of the following numbers according to the categories. Indicate with a check mark.

1.

	Real number	Rational number	Irrational number	Whole number	Integer	Fraction	Decimal	Composite number	Prime number
31	✓	✓		✓	✓				✓
0									
0.25									
−3/7									
2.4···									
28									
1									
π									
3.12									
$3.\overline{12}$									
5									
−2.5									

Refer to the number lines below, write the coordinate of each point described.

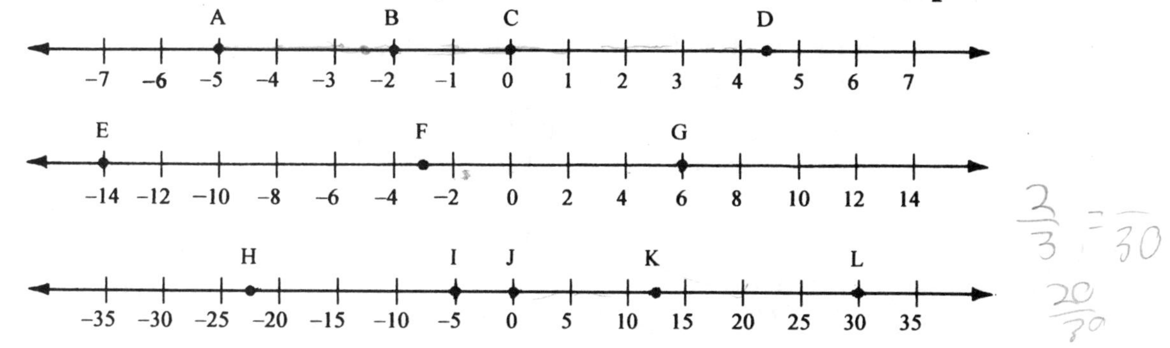

2. A **3.** B **4.** C **5.** D **6.** E **7.** F **8.** G **9.** H **10.** I **11.** J **12.** K **13.** L

14. The point whose distance is 6 points to the right of A.

15. The point whose distance is 3 points to the left of D.

16. The midpoint between A and C.

17. The point halfway between E and G.

18. The point two third of the way from J to L.

19. The point to the right of A is twice as far as the distance from A to B.

20. The point to the left of J is 1.5 times the distance from J to I.

-----Continued-----

State the distance between two points on a number line.

21. 2 and 6 **22.** 6 and 2 **23.** –2 and 6 **24.** 2 and –6 **25.** –2 and –6
26. –3 and –9 **27.** –5 and 12 **28.** 2.5 and 5.5 **29.** –5.2 and 5.5 **30.** –2.5 and –0.5

31. In the diagram, D is the midpoint between C and E.
C is the midpoint between A and E.
D is 6 points to the left of E
C is 5 points to the right of B.
The coordinate of A is 0. Find the coordinates of B, C, D, and E.

A B C D E
0

32. In the diagram, C is the midpoint of A and E.
A is 5 points to the left of B.
D is 22 points to the right of A.
The coordinate of A is –2.
The coordinate of E is 28. Find the coordinates of B, C, D.

A B C D E
–2 28

Compare each pair of numbers, using the symbols <, >, or =.

33. 2 and 5 **34.** 8 and 3 **35.** –2 and 5 **36.** –2 and –5 **37.** 2 and –5 **38.** –8 and –3
39. 2.5 and 3.5 **40.** –5.3 and –5.4 **41.** –4.8 and –4.7 **42.** –38 and –60 **43.** –9 and –9.

Round each number to the nearest ten.

44. 3 **45.** 8 **46.** 65 **47.** 64 **48.** 344 **49.** 348 **50.** 3,409 **51.** 5,801 **52.** 6,555

Round each number to the nearest hundred, thousand, and ten thousand.

53. 37,212 **54.** 24,360 **55.** 31,402 **56.** 60,305 **57.** 67,485 **58.** 369,998 **59.** 499,999

Round each decimal to the nearest tenth and hundredth.

60. 2.641 **61.** 3.027 **62.** 32.374 **63.** 323.909 **64.** 245.949 **65.** 247.989 **66.** 258.999

Round each decimal to the nearest cent.

67. $0.985 **68.** $4.423 **69.** $5.399 **70.** $19.989 **71.** $56.893 **72.** $125.998

73. Round 1,274,557,829 **to the nearest hundred million.**

Write each number in the short word name.

74. 1,234 **75.** 2,347 **76.** 45,786 **77.** 345,625 **78.** 0.6 **79.** 0.627 **80.** 204.005

Write each number in the word name.

81. 1,234 **82.** 2,347 **83.** 45,786 **84.** 345,625 **85.** 2,425,256 **86.** 23,435,253
87. 0.01 **88.** 0.02 **89.** 0.278 **90.** 63.05 **91.** 25.124 **92.** 58.3674
93. 406.046 **94.** 905.078 **95.** 4,000.0012 **96.** 1,000.00001 **97.** 5,000.00024

1-2 Adding Integers

Rules for adding any two integers a and b :

1) The numbers to be added are called the **addends**. The answer is called the **sum**.
2) **Commutative** and **associative** hold for addition.
   Commutative: $8+4=4+8$. Associative: $(8+4)+2=8+(4+2)$.
3) No sign on a number means it is positive. $1=+1$, $2=+2$.
4) If a and b are both positive, then the sum is positive. $2+3=5$.
5) If a and b are both negative, then the sum is negative. $-2+(-3)=-5$.
6) **Absolute Value**: The positive of any number "a". It is denoted by $|a|$.

   $$|3|=3,\quad |-3|=3,\quad |-2|=2,\quad -|-3|=-3$$

   If a is positive and b is negative and $|a|>|b|$, then $a+b$ is positive.
   $3+(-2)=1$. (3 has the larger absolute value, the sum is positive.)
   If a is positive and b is negative and $|a|<|b|$, then $a+b$ is negative.
   $2+(-3)=-1$. (-3 has the larger absolute value, the sum is negative.)
7) **The Property of Opposites**: $a+(-a)=0$. $3+(-3)=0$.
8) **The Property of Additive Identity**: $a+0=a$. $3+0=3$.
9) Addition of numbers can be written in **horizontal form** or **vertical form**.
10) To add numbers in vertical form, we line up like values.

You may consider a positive number as **"receiving"**, a negative number as **"giving away"**. The sum is **"the net result"**.

$-5+3=-2$ → Give away 5 and receive 3, -2 is the net result.
$-3+5=2$ → Give away 3 and receive 5, 2 is the net result.
$(-3)+(-5)=-8$ → Give away 3 and give away 5, -8 is the net result.

Examples

1. $8+4=12$, $4+8=12$, $-8+(-4)=-12$, $8+(-4)=4$, $4+(-8)=-4$.
2. $8+(-8)=0$, $-8+8=0$, $8+0=8$, $0+8=8$, $0+(-8)=-8$.
3. $|18|=18$, $|-18|=18$, $-|-18|=-18$. **4.** $|15+(-24)|+6=|-9|+6=9+6=15$.

Evaluate each expression if $a=2$, $b=-3$, $c=-4$.

5. $a+b=2+(-3)=-1$. **6.** $a+|b|+(-|c|)=2+|-3|+(-|-4|)=2+3+(-4)=1$.

7. John weighed 126 pounds. He gains 25 pounds this month. How much does John weigh this month ?
Solution: $126+25=151$ pounds.

8. Jorge lost 10 pounds in January, lost 9 pounds in February, and lost 4 pounds in March. What was his total weight loss for these three months ?
Solution: $(-10)+(-9)+(-4)=-23$ pounds. He lost 23 pounds.

EXERCISES

Find the sum.

1. $2+(-6)$ **2.** $-2+(-6)$ **3.** $-6+2$ **4.** $-6+(-2)$
5. $31+54$ **6.** $-31+54$ **7.** $-31+(-54)$ **8.** $31+(-54)$
9. $-31+54+(-13)$ **10.** $-42+40+20$ **11.** $-32+(-24)+(-3)$ **12.** $-20+(-20)$
13. $-89+12+70$ **14.** $-89+12+(-70)$ **15.** $-89+(-12)+(-70)$ **16.** $-20+20$
17. $-42+42+(-42)$ **18.** $-42+(-42)+24$ **19.** $320+(-120)+100$ **20.** $-120+245$
21. $-2+4+(-6)+(-8)+10+(-12)+14$ **22.** $24+(-26)+15+(-112)+120+70$
23. $2+|-6|$ **24.** $-2+|6|$ **25.** $|-2|+|-6|$ **26.** $-|-2|+|-6|$
27. $-24+32+|-28|$ **28.** $|-45|+36+|-15|$ **29.** $102+|-29|+|102|-|-245|+(-25)$

Evaluate each expression if $a=2$, $b=-1$, $c=5$, $d=-3$.

30. $a+c$ **31.** $b+c$ **32.** $a+b+c$ **33.** $b+c+(-d)$ **34.** $a+b+c+d$
35. $a+|c|$ **36.** $b+|d|$ **37.** $-|a+b|+d$ **38.** $b+|c|-|d|$ **39.** $|a+b+d|+|d+a|+a$

40. A-Plus Notes Center sold 250 books last month and 272 books this month to Barnes & Noble Books. How many books were sold to Barnes & Noble Books for these two months ?
41. Cindy weighed 130 pounds two months ago. She gained 7 pounds last month. She lost 3 pounds this month. How much does Cindy weigh now ?
42. An atom of a chemical element contains neutrons, protons, and electrons. The neutron is a neutral particle. The proton carries one unit of positive charge. The electron carries one unit of negative charge. An atom of a certain chemical element has 20 protons, 20 neutrons, and 25 electrons. What is the total charge of this atom ?
43. The base price of a new car is \$22,500. The destination charge is \$250 and the cost of the optional equipment is \$2,100. What is the sticker price of this new car ?
44. The temperature was $-15°C$ in the morning. It increased $12°C$ by 10:00 A.M. It then decreased $8°C$ by 10:00 P.M. What was the temperature reading at 10:00 P.M. ?
45. A submarine is located 200 feet below sea level. It ascends 120 feet . What is the final location of the submarine ?
46. What is the distance between 125 feet (below sea level) and 125 feet (above sea level) ?
47. Danny purchased stock on Amazon Books. The price per share of stock fell by \$6 in the first month, rose \$5 in the second month, and rose by \$3 in the third month. How much did he gain (or lose) per share of stock for the past three months ?

A kid said to an old lady sitting on a park chair.
kid: How are your teeth ?
old lady: No good. They are all gone.
kid: Please hold my apple. I want to play ball with my friends for a while.

1-3 Subtracting Integers

Rules for subtracting any two integers a and b :

1) The number to be subtracted is called the **subtrahend**. The number we subtract from in a subtraction is called the **minuend**. The answer is called the **difference**.

Minuend – Subtrahend = Difference.

$8-2=6$.

2) **Commutative** and **associative** do not hold for subtraction.

$a-b \neq b-a$. **Examples:** $8-4 \neq 4-8$.

$(a-b)-c \neq a-(b-c)$. $(8-4)-2 \neq 8-(4-2)$.

3) Subtraction is the **inverse** of addition. To subtract a number, we add its **opposite**.

$-(-a)=a$, $a-b=a+(-b)$, $a-(-b)=a+b$.

Examples: $-(-8)=8$, $8-2=8+(-2)$, $8-(-2)=8+2$.

4) $+(-a)$ and $-(+a)$ are usually simplified to "$-a$". $+(-8)=-8$, $-(+8)=-8$.
$a+(-b)$ is usually simplified to "$a-b$". $8+(-4)=8-4$.

5) $-(a+b)=-a-b$. $-(a-b)=-a+b$. A negative sign in front of a parenthesis means " the **opposite** of the expression in the parenthesis ". $-(a+b-3)=-a-b+3$.

6) If the signs are the **same**, we combine the numbers and keep the sign.

$3+2=5$; $-3-2=-5$.

If the signs are different, we take the **difference** and use the sign of the number having the larger absolute value. $3-1=2$; $-3+2=-1$.

Examples

1. $8+(-4)=8-4=4$.
$4+(-8)=4-8=-4$.
$-4-8=-12$.
$8-(-4)=8+4=12$.
$-8-(-4)=-8+4=-4$.
$-8-8=-16$.

2. $-56-(-24)=-56+24=-32$.
$56-(-24)=56+24=80$.

3. $-15-12=-27$.
$-(15-12)=-15+12=-3$.
$-(15+12)=-15-12=-27$.

4. Evaluate each expression if $x=-4$, $y=3$, and $z=-2$.

a) $x-y-z=-4-3-(-2)=-7+2=-5$.

b) $-x+|z|-|x|=-(-4)+|-2|-|-4|=4+2-4=6-4=2$.

5. John weighed 196 pounds. He lost 25 pounds. How much does John weigh then ?
Solution: 196 – 25 = 171 pounds.

6. Find the distance between two points –5 and 7 on the number line.
Solution: $d=|7-(-5)|=|7+5|=|12|=12$. Or $d=|-5-7|=|-12|=12$.

7. What is the distance between 125 feet (below sea level) and 102 feet (below sea level) ?
Solution: $d=|-125-(-102)|=|-125+102|=|-23|=23$ feet.

EXERCISES

Evaluate each expression.

1. $-(-1)$ **2.** $-(-2)$ **3.** $-(-3)$ **4.** $-(-4)$ **5.** $-(-5)$ **6.** $-(-99)$ **7.** $-(-100)$

Find the difference.

8. $6-2$ **9.** $2-6$ **10.** $6-(-2)$ **11.** $2-(-6)$ **12.** $-6-(-2)$ **13.** $-2-(-6)$

14. $31-54$ **15.** $-31-54$ **16.** $-31-(-54)$ **17.** $31-(+54)$ **18.** $54-(-31)$ **19.** $-54-(+31)$

Simplify each expression.

20. $-31+54-(+13)$ **21.** $-42+40-20$ **22.** $-32-(-24)+(-3)$ **23.** $-20-(-20)$

24. $-89+12-70$ **25.** $-89+12-(-70)$ **26.** $89-(-12)-70$ **27.** $-20-20$

28. $-42-42+(-42)$ **29.** $42-(-42)-24$ **30.** $320-(-120)-100$ **31.** $-120-245$

32. $-2-4+(-6)-(-8)-10+(-42)+14$ **33.** $24-(-26)+15-(-112)+120-70$

34. $2+|-6|-|-8|$ **35.** $-2+|6|-8$ **36.** $|-2|-|-6|+8$ **37.** $-|-2|-|-20|$

38. $-24+32-|-28|$ **39.** $|-45|-36-|15|$ **40.** $102-|-29|-|102|+|-245|-(-25)$

Evaluate each expression if $a=2$, $b=-1$, $c=5$, $d=-3$.

41. $a-c$ **42.** $b-c$ **43.** $a-b+c$ **44.** $b+c-d$ **45.** $a-b-c-d$

46. $a-|b|$ **47.** $b-|d|$ **48.** $|b-a|+d$ **49.** $b+|c|-|d|$ **50.** $|a-b+d|-|d+a|+a$

51. A company sold 250 books to Barnes & Noble Books last month. 12 books were returned due to shipping damages. How many books were sold to Barnes & Noble last month ?

52. Cindy weighed 130 pounds two months ago. She lost 5 pounds last month and gained 7 pounds this month. How much does Cindy weigh now ?

53. The temperature was $-15°C$ in the morning. It decreased $10°C$ by 12:00 A.M. It then increased $8°C$ by 10:00 P.M. What was the temperature reading at 10:00 P.M. ?

54. A submarine is located 200 feet below sea level. It ascends 120 feet and then descends 50 feet . What is the final location of the submarine ?

55. What is the distance between 125 feet (below sea level) and 250 feet (above sea level) ?

56. What is the distance between –12 and –25 on the number line ?

57. The sale price is $18 less than the regular price. If the regular price is $99, what is the sale price ?

58. A number is the sum of 17 and 28, decreased by 6. What is the number ?

59. What is the change in temperature from $-13°C$ to $12°C$?

60. What is the net change in voltage across an electric circuit from 124 volts to –115 volts ?

1-4 Multiplying Integers

Rules for multiplying any two integers a and b :

1) The number to be multiplied is called the **multiplicand.** The multiplicand is multiplied by the **multiplier**. The answer is called the **product.**
   The multiplicand and the multiplier are called the **factors** of the product.
   Multiplicand × multiplier = product
   $2 \times 4 = 8 \rightarrow$ 2 and 4 are the **factors** of product 8.
2) Different ways to indicate multiplication: $2 \times 4 = 2 \cdot 4 = 2(4) = (2)4 = (2)(4) = 8$.
3) If a and b have the same sign, then the product is positive.
   $(a)(b) = ab$, $(-a)(-b) = ab$. **Examples:** $(8)(4) = 32$, $(-8)(-4) = 32$.
   If a and b have opposite signs, then the product is negative.
   $(a)(-b) = -ab$, $(-a)(b) = -ab$. **Examples:** $(8)(-4) = -32$, $(-8)(4) = -32$.
4) **Distributive property** : $a(b + c) = ab + ac$, $a(b - c) = ab - ac$.
5) **Multiplication** is **commutative** and **associative** : $ab = ba$, $(ab)c = a(bc)$.
6) The product of any integer and -1 is the additive inverse of the integer.
   $(-1)\ a = -a$, $(-1)(-1) = -(-1) = 1$.
   $(-a)(-b) = (-1 \cdot a)(-1 \cdot b) = (-1)(-1)(ab) = 1(ab) = ab$.
7) $a \cdot 1 = 1 \cdot a = a$ and $a \cdot 0 = 0 \cdot a = 0$.

Examples

1. $(5)(4) = 20$, $(-5)(-4) = 20$.
   $(5)(-4) = -20$, $(-5)(4) = -20$.
2. $-8(0) + 8(0) = 0 + 0 = 0$.
   $-8 \cdot 0 \cdot 6 = 0$.
3. $4(-1)(-2) = -4(-2) = 8$
4. $-2 \cdot -3 \cdot -4 = 6 \cdot -4 = -24$.
5. $4(-2)(-3)(-4) = -8(-3)(-4)$
   $= 24(-4) = -96$.
6. $-2(3-4) = -2(3) - 2(-4) = -6 + 8 = 2$.
   Or $-2(3-4) = -2(-1) = 2$.
7. Harvey sold 110 books last month. Each book earned a \$3 profit. How much was the profit last month ?
   Solution: \$3 × 110 = \$ 330.
8. There are 12 eggs in one box. You bought 5 boxes. How many eggs did you buy ?
   Solution: $12 \times 5 = 60$ eggs.
9. A number is the product of 12 and 8, minus 15. What is the number ?
   Solution: $12 \times 8 - 15 = 96 - 15 = 81$.
10. Light travels at the rate of about 300 million meters per second. It takes 8 minutes and 20 seconds for the light from the Sun to reach the Earth. Find the distance from the Sun to the Earth.
    Solution: $8 \times 60 + 20 = 500$ seconds.
    $d = 300$ million × 500 = 150,000 million = 150 billion meters.
11. If your heart beats 75 times per minute, how many times does your heart beat in one hour ?
    Solution: $75 \times 60 = 4{,}500$ beats.

EXERCISES

Find the product.

1. $5(4)$ **2.** $5(-4)$ **3.** $-5 \cdot -4$ **4.** $-5(4)$ **5.** $-1(4)$ **6.** $-1 \cdot (-4)$
7. $-1(0)$ **8.** $4(0)(9)$ **9.** $2(-3)(-4)$ **10.** $-5(-1)(5)$ **11.** $-1(-2)(-3)$ **12.** $-10(25)$

Simplify each expression.

13. $2(-3)+3$ **14.** $2(-3)-3$ **15.** $-2 \cdot -3+6$ **16.** $12 \cdot -10 - 5$ **17.** $7(-4)-7$
18. $5(-13)-3$ **19.** $-4(-8)+30$ **20.** $-6-(-5)$ **21.** $-12-(-24)$ **22.** $-(-1-2-3)$

Evaluate each expression if $a=-3$, $b=2$, $c=4$, $d=-2$.

23. $4a$ **24.** $5c$ **25.** $12b$ **26.** $-15d$ **27.** $-24c$
28. $3a+4$ **29.** $4c-2$ **30.** $5a-6$ **31.** $-3d+7$ **32.** $-12b-10$
33. $3ab$ **34.** $5ad$ **35.** $7ac-b$ **36.** $-2ad+c$ **37.** $-ab-d$
38. The product of a and c. **39.** a times d, decreased by c.
40. The product of a and b, plus c. **41.** c times d, plus a.

42. Mary made 4 deposits of \$120 each this month in her savings account. What is the total amount that Mary deposited this month ?
43. There are 9 balls in each bag. There are 25 bags. How many balls are there total ?
44. A high-school track is 400 meters around. A race is 4 times around the track. How many meters are in the race ?
45. A bus was filled with 35 passengers on each trip to a national zoo. The bus made 22 trips last week. How many passengers were transported by the bus last week ?
46. There are 50 nails in each package. John bought 12 packages. How many nails did John buy ?
47. Gorge purchased 16 gallons of gasoline for \$2 per gallon. What was the total cost he paid ?
48. A plane travels at an average speed of 485 miles per hour. It takes 14 hours for the plane to travel nonstop from Los Angeles to Taipei. What is the distance from Los Angeles to Taipei ?
49. A-Plus Company sold at an average of 250 books per month to Barnes & Noble Books. How many books did A-Plus Company sell to Barnes & Noble Books in the last 5 years ?
50. Maria works 8 hours a day and 5 days a week at MacDonald. She earns \$6 per hour. How much does she earn per week ?
51. The telephone rate for calls made from Los Angeles to China is \$2 for the first minute and \$1 for each additional minute. Mark made a call for 9 minutes. What was the total cost ?
52. The area of a rectangle is given by the formula $A = l \cdot w$. Find the area of a rectangle when $l = 14$ in. and $w = 9$ in.

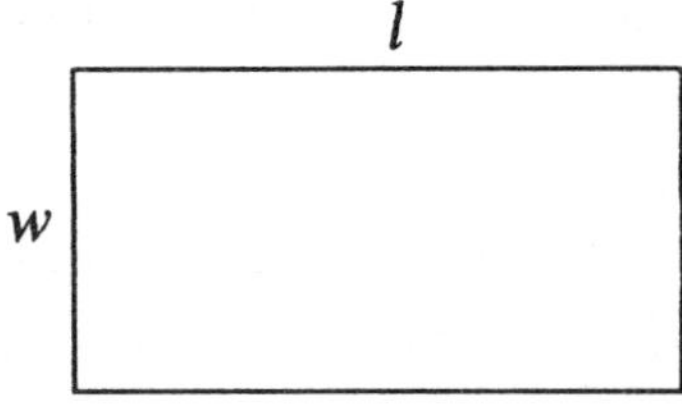

1-5 Dividing Integers

Rules for dividing any two integers a and b:

1) The number to be divided is called the **dividend.** The dividend is divided by the **divisor**. The answer is called the **quotient**.

   Dividend ÷ divisor = quotient

   $8 \div 4 = 2 \rightarrow$ 2 is the quotient of $(8 \div 4)$.

2) Different ways to indicate division: $8 \div 4 = 4\overline{)8} = \frac{8}{4} = 8/4 = 8 \cdot \frac{1}{4} = \frac{1}{4} \cdot 8 = 2$.

3) If a and b have the same sign, then the quotient is positive.

   If $b \neq 0$, $\frac{a}{b} = \frac{a}{b}$ and $\frac{-a}{-b} = \frac{a}{b}$. **Examples:** $\frac{8}{4} = 2$, $\frac{-8}{-2} = 4$.

   If a and b have opposite signs, then the quotient is negative.

   If $b \neq 0$, $\frac{a}{-b} = -\frac{a}{b}$ and $\frac{-a}{b} = -\frac{a}{b}$. **Examples:** $\frac{8}{-2} = -4$, $\frac{-8}{2} = -4$.

4) **Distributive property** : If $c \neq 0$, $\frac{a+b}{c} = \frac{a}{c} + \frac{b}{c}$ and $\frac{a-b}{c} = \frac{a}{c} - \frac{b}{c}$.

5) **Division** is not **commutative** and **associative** :

   $\frac{a}{b} \neq \frac{b}{a}$; $a \div b \div c \neq a \div (b \div c)$. **Examples:** $\frac{8}{4} \neq \frac{4}{8}$, $8 \div 4 \div 2 \neq 8 \div (4 \div 2)$.

6) $\frac{0}{a} = 0$; $\frac{a}{0}$ is **undefined** (it means " Division by zero is not allowed. ").

Examples

1. $8 \div 2 = 4$.
2. $-8 \div (-2) = 4$.
3. $8 \div (-2) = -4$.
4. $-8 \div 2 = -4$.
5. $\frac{0}{8} = 0$, $\frac{8}{0}$ is undefined.
6. $8 \div (-4) \cdot 10 = -2 \cdot 10 = -20$.
7. You need 60 eggs. Each box contains 12 eggs. How many boxes should you buy ?
   Solution: $60 \div 12 = 5$ boxes.
8. Roger worked 40 hours last week. He earned \$240. How much does he earn per hour ?
   Solution: $240 \div 40 = \$6$.
9. 900 yearbooks are packaged in 60 cartons. How many yearbooks are in each carton ?
   Solution: $900 \div 60 = 15$ yearbooks.
10. In a fund-raising program, a class has planned to sell 720 Christmas cards. There are 12 volunteers to sell the cards. How many cards must each student sell ?
    Solution: $720 \div 12 = 60$ cards.
11. A player scored a total of 1,692 points in 47 games. What is his average points per game ?
    Solution: $1{,}692 \div 47 = 36$ points.
12. If your heart beats 25 times in 20 seconds, how many times does your heart beat in one minute ?
    Solution: $(60 \div 20) \times 25 = 3 \times 25 = 75$ times. Or $25 \times (60 \div 20) = 25 \times 3 = 75$ times.

EXERCISES

Find the quotient.

1. $15 \div (-3)$ **2.** $-15 \div (-3)$ **3.** $-15 \div 3$ **4.** $0 \div (-5)$ **5.** $(-5) \div 0$

6. $-98 \div 2$ **7.** $-98 \div (-2)$ **8.** $121 \div (-11)$ **9.** $-144 \div (-12)$ **10.** $\frac{100}{0}$

Simplify each expression.

11. $2(-3) \div 2$ **12.** $4 \div (-1) \cdot 2$ **13.** $-16 \div (-4) \cdot (-6)$ **14.** $100(-10) \div (-5)$

15. $-12 \div (-2) - 9$ **16.** $54(-3) \div (-9)$ **17.** $26 \div 13 \times (-2) + 8$ **18.** $-42 \div 6 \times 4 - 5 - 4$

Evaluate each expression if $a = -3,\ b = 2,\ c = 4,\ d = -2$.

19. $4a \div b$ **20.** $5c \div (-2)$ **21.** $-15d \div b$ **22.** $-24c \div d$

23. $ac \div b$ **24.** $ab \div d$ **25.** $-2ad \div c$ **26.** $-ac \div d$

27. $3ab \div d$ **28.** $-2ac \div b + (-7)$ **29.** $3ad \div b \times c$ **30.** $4bd \div c - a + d$

31. The quotient of b and d. **32.** The quotient of ac and b.

33. The quotient of ab and d. **34.** b times d, divided by c, minus a.

35. Mary made 4 deposits of the same amount this month. The total amount she deposited is \$480. What is the amount of each deposit this month ?

36. There are 225 balls in 25 bags. How many balls are there in each bag ?

37. John works 38 hours at McDonald Burger each week. He earns \$228 per week. How much does he earn per hour ?

38. A-Plus Company sold \$1,080 worth of books to Book Star last week. Each book is \$9. How many book were sold to Book Star last week ?

39. A-Plus Company sold 5,760 books to Barnes & Noble Books last year. What is the average amount of books sold to Barnes & Noble Books each month ?

40. A rope is 468 inches long. If we cut it in 26 pieces of equal length. How long is each piece ?

41. A rope is 468 inches long. If we cut it to 18 inches in length for each piece. How many pieces do we have ?

42. If your heart beats 18 times in 15 seconds, how many times does your heart beat in one minute ?

43. If your pulse rates were 70, 74, 76, and 72. What was your average pulse rate ?

44. Gary's math test scores are 85, 76, 83, 92, and 99. What is his average score ?

45. The daily temperatures for last five days were $70°$, $69°$, $74°$, $67°$, and $75°$. What was the average temperature ?

46. If you are dividing a large 50-pound bag of dog food into smaller bags to sell, how many 5-pound bags are there ?

47. 825 yearbooks are packaged in boxes. Each box contains 25 books. How many boxes are there ?

48. Nelson is a lazy boy. He slept for 50,400 seconds yesterday. How many hours did he sleep ?

1-6 Powers and Exponents

Exponent (power): In the expression 5^3, 3 is called the **exponent** (or **power**) of 5.
5 is called the **base.** 5^3 is the exponential (or power) form of $5 \times 5 \times 5$.
The exponent 3 in the expression 5^3 means that there are three **repeated factors** of 5.
Exponential form or **power form** a^n is a short way to write a **multiplication (product) of** n **repeated factors of** a: $a^n = a \cdot a \cdot a \cdots \cdot a \to n$ repeated factors of a

$5^1 \to$ read " five to the first power".
$5^2 \to$ read " five to the second power " or " five squared ".
$5^3 \to$ read " five to the third power " or " five cubed ".
$5^n \to$ read " five to the nth power ".

Any number can be written as " **the number to the first power** ".
$1 = 1^1$, $2 = 2^1$, $(-3) = (-3)^1$, $4 = 4^1$, $-5 = (-5)^1$, $99 = 99^1$, $100 = 100^1$, ...

Examples: Write each product using exponents.
$2 \cdot 2 = 2^2$, $2 \cdot 2 \cdot 2 = 2^3$, $2 \cdot 2 \cdot 2 \cdot 2 = 2^4$, $-2 \cdot -2 \cdot -2 \cdot -2 \cdot -2 = (-2)^5$.
$3 \cdot 3 = 3^2$, $3 \cdot 3 \cdot 3 = 3^3$, 7 squared $= 7^2$, 7 cubed $= 7^3$.

The expression $3a^2$ means that a is squared, but not 3. $3a^2 = 3 \cdot a \cdot a$.
The expression $(3a)^2$ means that 3 and a are both squared. $(3a)^2 = 3a \cdot 3a = 9a^2$.

Notice that: $-3^2 \neq (-3)^2$. $-3^2 = -3 \cdot 3 = -9$, $(-3)^2 = (-3) \cdot (-3) = 9$.
$-2^4 = -2 \cdot 2 \cdot 2 \cdot 2 = -16$, $(-2)^4 = (-2) \cdot (-2) \cdot (-2) \cdot (-2) = 16$.

$(-a)^n$ is positive if n is an even number. $(-1)^2 = 1$, $(-1)^{50} = 1$, $(-1)^{100} = 1$.
$(-a)^n$ is negative if n is an odd number. $(-1)^3 = -1$, $(-1)^{49} = -1$, $(-1)^{99} = -1$.
$(-2)^{19} = -2^{19}$, $(-2)^{20} = 2^{20}$.

General rules of exponents (powers): If a is any nonzero real number, then:

1. $a^m \cdot a^n = a^{m+n}$. **2.** $\dfrac{a^m}{a^n} = a^{m-n}$. **3.** $a^0 = 1$ if $a \neq 0$

Rules: 1. To multiply two powers of the same base, we add the powers.
2. To divide two powers of the same base, we subtract the powers.
3. If a is any nonzero number, then $a^0 = 1$. $1^0 = 1$, $2^0 = 1$, $3^0 = 1$, ...,$100^0 = 1$.
0^0 is undefined (it means that the power "0" does not apply to 0 itself).

Examples:

1. $2^{30} \cdot 2^{10} = 2^{30+10} = 2^{40}$. **2.** $\dfrac{2^{30}}{2^{10}} = 2^{30-10} = 2^{20}$. **3.** $\dfrac{\cancel{2^{10}}}{\cancel{2^{10}}} = 1$, or $\dfrac{2^{10}}{2^{10}} = 2^{10-10} = 2^0 = 1$.

EXERCISES

Write each product (expression) using exponents.

1. $10 \cdot 10 \cdot 10$ **2.** $9 \cdot 9$ **3.** $7 \cdot 7 \cdot 7 \cdot 7$ **4.** $4 \cdot 4 \cdot 4 \cdot 4$ **5.** $1 \cdot 1 \cdot 1 \cdot 1 \cdot 1 \cdot 1 \cdot 1$

6. $a \cdot a$ **7.** $p \cdot p \cdot p$ **8.** $x \cdot x \cdot x \cdot x$ **9.** $x \cdot x \cdot y \cdot y \cdot y$ **10.** $x \cdot y \cdot x \cdot y \cdot x$

11. 8 cubed **12.** 8 squared **13.** $a \cdot a \cdot 5 \cdot 5$ **14.** $y \cdot y \cdot -a \cdot -a$ **15.** 9 to the fifth power

16. $-3 \cdot -3$ **17.** $-3 \cdot 3$ **18.** $-6 \cdot -6 \cdot -6$ **19.** $-x \cdot x \cdot -y \cdot -y$ **20.** $-a \cdot -a \cdot -a \cdot 2y$

Write each power (expression) as a product of the same factor.

21. 1^5 **22.** 3^4 **23.** 4^3 **24.** 6^4 **25.** 4^6

26. 10 squared **27.** 10 cubed **28.** a^3 **29.** p^4 **30.** 3 to the sixth power

31. $(-4)^3$ **32.** -4^3 **33.** -3^4 **34.** $(-3)^4$ **35.** -10^4

36. $-a^3$ **37.** $(-a)^3$ **38.** $3x^3$ **39.** $(3x)^3$ **40.** $-(3x)^2$

Evaluate each exponent (power).

41. 5^3 **42.** 3^5 **43.** 2^5 **44.** 5^2 **45.** 4^3

46. 9 squared **47.** 9 cubed **48.** 1^5 **49.** 1^9 **50.** 3^4

51. $(-2)^3$ **52.** $(-2)^6$ **53.** -2^3 **54.** $(-6)^3$ **55.** -6^2

56. $(-1)^{20}$ **57.** $(-1)^{25}$ **58.** $(-1)^{15}$ **59.** $(-a)^{16}$ **60.** $(-a)^{73}$

61. 7^0 **62.** 10^0 **63.** 99^0 **64.** 0^0 **65.** 1^0

Evaluate each expression if $a=-3$, $b=2$, $c=4$, $d=-2$.

66. a^3 **67.** d^2 **68.** b^4 **69.** $2b^3$ **70.** $4a^2$

71. abc **72.** $4d^3$ **73.** $6a^4d$ **74.** $3ad^3$ **75.** $(4a)^2$

76. $2d^3$ **77.** $(2d)^3$ **78.** $-4c^2$ **79.** $-(4c)^2$ **80.** $(-4c)^2$

81. The area of a square is given by the formula $A = s^2$ (where s is the length of one side). Find the area of a square with one side 12 meters.

82. The volume of a cube is given by the formula $V = s^3$ (where s is the length of one side). Find the volume of a cube with one side 12 meters.

83. The distance in feet that an object falls in t seconds is given by the formula $d = \frac{1}{2}gt^2$ (where $g = 32$). Find the distance of an object falls in 5 seconds.

84. Light travels at the rate of about 200,000 miles per second. Write this number as the product of 2 and a power of 10.

85. The distance from the Earth to the Moon is about 400 million meters. Write this number as the product of 4 and a power of 10.

86. Light travels at the rate of about 200,000 miles per second. It takes 500 seconds for the light from the Sun to reach the Earth. Find the distance from the Sun to the Earth in the form as a the product of 1 and a power of 10.

87. Evaluate: **a)** $a^7 \cdot a^5$ **b)** $a^7 \div a^5$ **c)** $a^7 \div a^7$ **d)** $3^{30} \cdot 3^{12}$ **e)** $3^{30} \div 3^{12}$ **f)** $3^{30} \div 3^{30}$

1-7 Order of Operations

To simplify an expression with numbers, we use the following **order of operations**:

1) Do all operations within the grouping symbols first, start with the innermost grouping symbols.
2) Do all operations with exponents.
3) Do all multiplications and divisions in order from left to right.
4) Do all additions and subtractions in order from left to right.

In mathematics, we have agreed the above order of operations to ensure that there is exactly one correct answer.

It is important that we always multiply and divide before adding and subtracting.

Grouping Symbols: A grouping symbol is used to enclose an expression that should be simplified first. First simplify the expression in the innermost grouping symbol. Then work toward the outmost grouping symbol. Grouping symbols are also called **inclusion symbols**.

Parentheses ( ), Brackets [ ], Braces { }, Fraction Bar ——.

Examples

1. $1+2\cdot 3=1+6=7$.
2. $(1+2)\cdot 3=3\cdot 3=9$.
3. $2\cdot 3-4=6-4=2$.
4. $4-6\div 2=4-3=1$
5. $(4-6)\div 2=-2\div 2=-1$.
6. $2(3-4)=2(-1)=-2$.
7. $3+2^3=3+8=11$.
8. $(3+2)^3=5^3=125$.
9. $3-3^3=3-27=-24$.
10. $2^3\cdot 3^2=8\cdot 9=72$.
11. $2^3-3^2=8-9=-1$.
12. $2^3+3^2=8+9=17$.
13. $1^5+2\cdot 3^4=1+2\cdot 81=1+162=163$.
14. $12-35\div 7+2=12-5+2=7+2=9$.
15. $7-\frac{72-30\div 5}{3}+2=7-\frac{72-6}{3}+2=7-\frac{66}{3}+2=7-22+2=-15+2=-13$.

16.
$$\begin{aligned}&2+3\{4-[6-2(3-1)]\div 2\}\\&=2+3\{4-[6-2(2)]\div 2\}\\&=2+3\{4-[6-4]\div 2\}\\&=2+3\{4-2\div 2\}\\&=2+3(4-1)\\&=2+3(3)=2+9=11.\end{aligned}$$

17.
$$\begin{aligned}&3-2\{5-4^2\cdot[2-3(1-3)]\}\\&=3-2\{5-16\cdot[2-3(-2)]\}\\&=3-2\{5-16\cdot[2+6]\}\\&=3-2\{5-16\cdot 8\}\\&=3-2(5-128)\\&=3-2(-123)=3+246=249.\end{aligned}$$

18. Joe finished 15 assignments and 3 tests (which he received 2 A's and 1 B) in math class. Each assignment earns him 2 points; for each test he can earn 10 points for an A and 7 points for a B. What is his total point ?
Solution: $15\times 2+10\times 2+7\times 1=30+20+7=57$ points.

EXERCISES

Evaluate each expression.

1. $12-10+6$. **2.** $12+10-6$. **3.** $36\div6\times4$. **4.** $36-6\times3$. **5.** $36+6\div3$.
6. $16-(1-5)$. **7.** $8\times4\div16$. **8.** $(5-1)\div2\cdot3$. **9.** $7-8\div4\cdot3$. **10.** $24\div(8-5)$.
11. $8-3\cdot2$. **12.** $8-4\div2$. **13.** $(8-4)\div2$. **14.** $8\div4+4\div2$. **15.** $6+3\cdot4+3^2$.
16. $2^3\times3^2$. **17.** $1^5\times5^1$. **18.** $1^4\times0^4$. **19.** $4^1\times4^0$. **20.** $5^0\times0^5$.
21. $2^4\times10^2$. **22.** $10^3\times10^0$. **23.** $2^5\times5^2$. **24.** $3^4\times10^5$. **25.** $5^2\times3^3$.
26. 2^3+3^2. **27.** 1^5-5^1. **28.** 2^4+0^4. **29.** 4^2-5^0. **30.** 5^2+3^3.
31. $\dfrac{30-5}{6-1}$. **32.** $\dfrac{30+5}{6+1}$. **33.** $\dfrac{75-30}{3\times3}$. **34.** $\dfrac{20\times30}{20-30}$. **35.** $\dfrac{36-(4\times5)}{4}$.
36. $\dfrac{36\cdot20}{5\cdot6}$. **37.** $\dfrac{360\div12}{30\div5}$. **38.** $\dfrac{120\div3}{2\times5}$. **39.** $\dfrac{5\times6}{40\div4}$. **40.** $\dfrac{32+(15\div5)}{11-(2\times3)}$.
41. $7-6\div3+4$. **42.** $15-2(6)+3$. **43.** $5+6\div3\cdot4$. **44.** $10\div2\cdot4-12$. **45.** $16\div2^3-2$.
46. $82\div2-40$. **47.** $94-3^3\div9$. **48.** $3^2-2^4+6\div2$. **49.** $2(4-1)-7$. **50.** $(4+3\cdot2)\div5$.
51. $(4+6\div3)\times(4\cdot6\div3)$. **52.** $(7-6\div3-4)\div(4-3)$. **53.** $26-4^2-5\cdot2+4$.
54. $2^4-[7-3(10)]$. **55.** $[12(5)+10(2)]\div40$. **56.** $3[(2+4)\cdot3-5]\div3$.
57. $4^3-6\cdot3-2\cdot8$. **58.** $20\cdot4\div5+3-4(1^5-5)$. **59.** $3[(2+4\cdot3)\div7-5]+2$.
60. $4-\{2[3-(2+5)]-5\}$. **61.** $1+\{2-[3-(4+5)+6]\}$. **62.** $4\{[8-(4^3-24)]-22\}$.
63. $2\{3[4(5+6)]-7\}$. **64.** $5\{5-5[5-5(5+5)+5]\}-5$. **65.** $2+2\{2-[2+2(2)]\}+2^2$.
66. $32-[6(3)+(4-|-2-4|)]$. **67.** $6-3\{4-[2-4(3^2-4)]\}$. **68.** $6+3\{-4[2-(2^3+4)]\}$.
69. $6-3^2\cdot2-7-\dfrac{-4+2}{3-1}\cdot8\div2$. **70.** $9-\dfrac{9-9\div9}{8}+9\div9-9^2$.

Evaluate each expression if $a=-3$, $b=2$, $c=4$, $d=-2$.

71. $a+bc$. **72.** $a(b+c)$. **73.** $ab+ac$. **74.** a^2-c^2. **75.** a^2-d^2.
76. c^2+a^2. **77.** $a-(b-c)$. **78.** $b+(a^3-d)$. **79.** $a-b-c^2$. **80.** $d^2-(a+c)$.
81. $b+d^3-3a$. **82.** $d-bc+a$. **83.** $a^2+b^2-c^2$. **84.** $3ab-c^3$. **85.** $2a+3b-4c$.
86. $2ad(c+d)$. **87.** $ab(3c-4d)$. **88.** $2ac(a-b)$. **89.** $a[b-(c+d)]$. **90.** $d-[c-(b+a)]$.

91. Maria bought two movie tickets for adults and three for children. Each adult ticket is \$8. Each child ticket is \$5. How much did Maria pay for the tickets ?

92. The perimeter of a rectangle is given by the formula $p=2(a+b)$. Find the perimeter of a rectangle when $a=6\ cm$ and $b=4\ cm$.

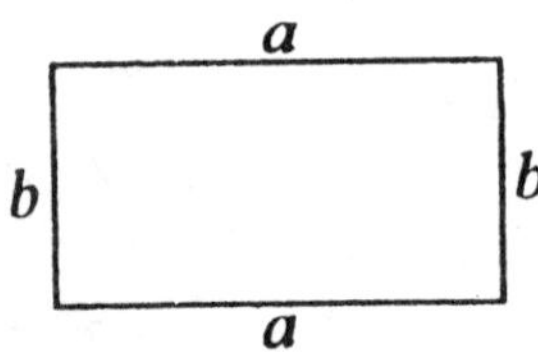

CHAPTER 1 EXERCISES

Classify each of the following numbers according to the categories.

1. 20 **2.** 21 **3.** 0 **4.** 2.5 **5.** 0.20 **6.** 29 **7.** 35 **8.** −8 **9.** $-\frac{1}{2}$ **10.** 2.15

11. 37 **12.** 38 **13.** −3 **14.** $2.\overline{15}$ **15.** $\frac{4}{5}$ **16.** π **17.** 11 **18.** −11 **19.** 1 **20.** 1.732···

Answer the following questions regarding the number line.

21. What is the number whose distance is 5 to the left of 12 ?
22. What is the number whose distance is 5 to the right of −12 ?
23. What is the number whose distance is $\frac{1}{2}$ to the left of 13 ?
24. What is the number whose distance is $\frac{1}{2}$ to the right of 13 ?
25. What is the number whose distance is $\frac{1}{2}$ to the left of −13 ?
26. What is the midpoint between 6 and 12 ? **27.** What is the distance between −6 and 12 ?
28. What is the distance between −2 and −8 ? **29.** What is the midpoint between −6 and 12 ?
30. What is the distance between 2 and −8 ? **31.** What is the midpoint between −1 and −8 ?
32. What is the number to the right of 4 that is 3 times the distance from 4 to 7 ?
33. What is the number halfway between −10 and 13 ?

Compare each pair of numbers, using the symbols $<$, $>$, or $=$.

34. −7 and 5 **35.** 7 and −5 **36.** −7 and −5 **37.** 4.5 and 4.9 **38.** −4.5 and −4.9
39. −1 and −2 **40.** −2.5 and 3 **41.** −2 and −2 **42.** 4.1 and 3.9 **43.** 0.5 and $\frac{1}{2}$

Round each number to the nearest ten, hundred, thousand.

44. 12,345 **45.** 54,521 **46.** 23,546 **47.** 44,454 **48.** 555,545

Round each decimal to the nearest tenth and hundredth.

49. 3.545 **50.** 3.525 **51.** 45.362 **52.** 235.327 **53.** 245.363

Write each number in "short word name" and "word name".

54. 12,345 **55.** 325,423 **56.** 0.24 **57.** 214.252 **58.** 31,324,534

59. Round 2,365,467,939 to the nearest million.
60. Round 2,365,467,939 to the nearest ten million.
61. Round 2,365,467,939 to the nearest hundred million.
62. Round \$0.954 to the nearest cent.
63. Round \$19.887 to the nearest cent.
64. Round \$42.794 to the nearest cent.

-----Continued-----

Evaluate each expression.

65. $-(+6)$ **66.** $-(-6)$ **67.** $-(-20)$ **68.** $+(-20)$

69. $3-7$ **70.** $-3-7$ **71.** $4+(-6)$ **72.** $4-(-6)$

73. $10-(-20)$ **74.** $10+(-20)$ **75.** $-3+(-7)$ **76.** $-3-(-7)+2$

77. $42-56+30$ **78.** $35-24-40$ **79.** $-22+24-15$ **80.** $-12+(-30)-(-13)$

81. $4+|-6|-(-2)$ **82.** $7-|-5|+(-2)$ **83.** $-|-8|-|6|-(-5)$ **84.** $|-9|-|8|+|-30|$

85. $(-2)(6)+4$ **86.** $(-2)(-6)-4$ **87.** $-6\div 3\times(-4)$ **88.** $-2(-3+1-4)$

89. $8\div(-4)-6\times 2$ **90.** $-36\div(-6)-8$ **91.** $32\times 2\div(-8)\times 3$ **92.** $8\times(-2)-8\div(-2)$

93. $(-3)^2+(-3)^3$ **94.** $(-2)^3-(-2)^4$ **95.** $9^0-0^9-9^1$ **96.** $(-1)^{42}+(-1)^{33}+1^{21}$

97. $25^0\times 4^2+(-4)$ **98.** $0^{12}-12^0\times 2$ **99.** $48\div 2^4-3(1-2)$ **100.** $10^4\div 10^2\cdot 10^1$

101. $6-3\cdot 4+8\div 2$ **102.** $4+6\div 2-(-4)$ **103.** $15-6^2-3\cdot 4+4\div 2-1$

104. $2-[4-(3-2)\cdot 6]$ **105.** $[(2-6)-(6-2)\times 4]\div(-5)$ **106.** $3[6-(3^4+4)\div 5]\div 11$

107. $\dfrac{2(4-1)+34}{5-3(2-7)}$ **108.** $\dfrac{(4^2-4\cdot 5)^3}{2^5}$ **109.** $5-\dfrac{5-5\cdot 5}{5+5}\cdot 5-5^2$

110. $2+\{4[6-3(5-2)]-1\}$ **111.** $7-\{5[4-(3-1)\cdot 2^4]-30\}$ **112.** $4\{4-[4(4+4)\cdot 4]\div 4\}$

Evaluate each expression if $a=2$, $b=-4$, $c=-3$, $d=5$.

113. $a+c$ **114.** $a-c$ **115.** $b-c$ **116.** $a+b-c$ **117.** $d+|c|-|b|$

118. $|a+b|-c$ **119.** $|b-a|+c$ **120.** $3ab$ **121.** $-2ac+7$ **122.** $3ac-b$

123. $5ab\div(2d)$ **124.** $-bc\div a$ **125.** $2ac-d$ **126.** The quotient of bc and a, minus d.

127. b^3-c^2 **128.** $2ab^2+d$ **129.** $3c^2-(2b)^2$ **130.** The product of b and c, plus a.

131. The perimeter of the following figure is $2a+4b+2c$. Find the perimeter when $a=10$, $b=4$ and $c=2$.

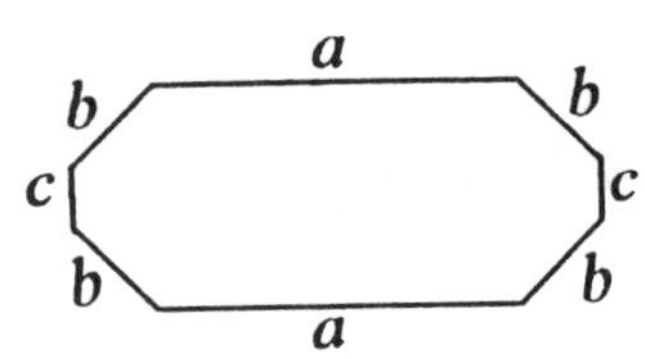

132. The area of a trapezoid is $\frac{1}{2}(a+b)h$. Find the area when $a=10$, $b=16$ and $h=6$.

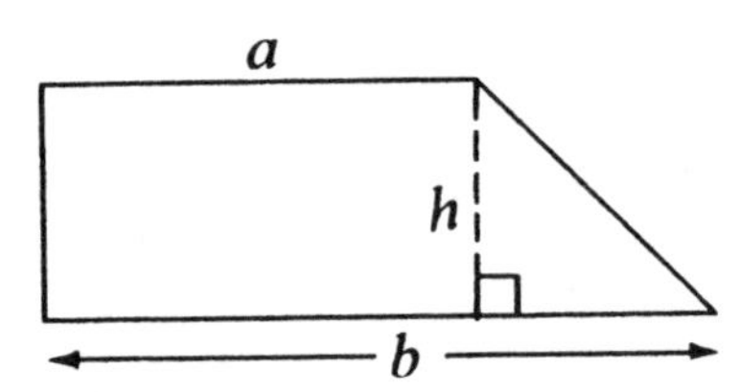

133. You bought 24 tennis balls for \$72. What is the price per ball ?

134. The average cost per card is 15 cents. You bought 18 cards. What was the total cost you spent in cents ?

135. The average cost per exposure for a 24-exposure roll of film is 18 cents. What is the cost of the roll of film in cents ?

136. John bought 15 folders for \$45. The price of the folders on sale is \$2 each. How much could John save per folder by buying at the price on sale ?

Number Properties

Axiom (Postulate): It is a statement that we assume to be true without proof.
The following basic axioms (postulates) about the properties of real numbers are used very frequently in the field of algebra.

For all real numbers *a*, *b*, and *c*, we have :

1. **Reflexive Property**: $a = a$
2. **Symmetric Property**: If $a = b$, then $b = a$.
3. **Transitive Property:** If $a = b$ and $b = c$, then $a = c$.
4. **Substitution Property:** If $a = b$, then a can be replaced by b.
5. **Commutative Property:** $a + b = b + a$ and $ab = ba$
6. **Associative Property:** $(a + b) + c = a + (b + c)$ and $(ab)c = a(bc)$
7. **Distributive Property:** $a(b + c) = ab + ac$ and $(b + c)a = ba + ca = ab + ac$
8. **Multiplicative Property of Zero:** $a \times 0 = 0$ and $0 \times a = 0$
9. **Additive Identity Property:** $a + 0 = a$ and $0 + a = a$
10. **Multiplicative Identity Property:** $a \times 1 = a$ and $1 \times a = a$
11. **Addition Property of Equality:** If $a = b$, then $a + c = b + c$.
12. **Subtraction Property of Equality:** If $a = b$, then $a - c = b - c$.
13. **Multiplication Property of Equality:** If $a = b$, then $ac = bc$.
14. **Division Property of Equality:** If $a = b$ and $c \neq 0$, then $\frac{a}{c} = \frac{b}{c}$.
15. **Additive Inverse (Axiom of Opposites):** $a + (-a) = 0$
16. **Multiplication Inverse (Axiom of Reciprocals):** If $a \neq 0$, then $a \times \frac{1}{a} = 1$.
17. **Property of the Opposite of a Sum:** $-(a + b) = (-a) + (-b)$
18. **Property of the Reciprocal of a Product:** If $a \neq 0$ and $b \neq 0$, then $\dfrac{1}{ab} = \dfrac{1}{a} \times \dfrac{1}{b}$.
19. **Property of Proportions:** If $\dfrac{a}{b} = \dfrac{c}{d}$, then $ad = bc$. (The cross products are equal.)

Additional Examples

1. Find the sum $1 + 2 + 3 + 4 + \cdots + 100$.
Solution:

$$\overbrace{1 + \overbrace{2 + 3 + 4 + \cdots + 98 + 99}^{101} + 100}^{101}$$

Pair the numbers.
Each pair has a sum of 101.
There are 50 pairs from 1 to 100.
We have: Sum $= 101 \times 50 = 5{,}050$.

2. Find the sum $1 + 2 + 3 + 4 + \cdots + 1000$.
Solution:

$$\overbrace{1 + \overbrace{2 + 3 + 4 + \cdots + 988 + 999}^{1001} + 1000}^{1001}$$

Pair the numbers.
Each pair has a sum of 1001.
There are 500 pairs from 1 to 1000.
We have: Sum $= 1001 \times 500 = 500{,}500$.

3. $111{,}111 \times 111{,}111 = 12{,}345{,}654{,}321$

4. $111{,}111{,}111 \times 111{,}111{,}111 = 12{,}345{,}678{,}987{,}654{,}321$

Roman Numerals

There are only seven symbols (letters in capitals) in Roman numerals. These symbols are:

Roman Numerals	I	V	X	L	C	D	M
Values	1	5	10	50	100	500	1000

The Roman numerals from 1 to 10 are:

Roman Numerals	I	II	III	IV	V	VI	VII	VIII	IX	X
Values	1	2	3	4	5	6	7	8	9	10

Other numbers are formed from the above symbols by adding or subtracting.
The value of a symbol following another of the same or greater value is added. XV=15
The value of a symbol preceding one of greater value is subtracted. IX = 9
The value of a symbol standing between two of greater values is subtracted from that of the second, the remainder being added to that of the first. XIX = 19
No symbol appears more than 3 times in a row.
A bar over a symbol indicates multiplication by 1,000. $\overline{V} = 5{,}000$
To write a Roman number, no symbol appears more than 3 times in a row.
To write 4, we write IV, which means 5 – 1.
To write 9, we write IX, which means 10 – 1.
To write 40, we write XL, which means 50 – 10.
To write 90, we write XC, which means 100 – 10.
The following table shows how to write Roman numerals between 11 and 50:

11	XI	21	XXI	31	XXXI	41	XLI
12	XII	22	XXII	32	XXXII	42	XLII
13	XIII	23	XXIII	33	XXXIII	43	XLIII
14	XIV	24	XXIV	34	XXXIV	44	XLIV
15	XV	25	XXV	35	XXXV	45	XLV
16	XVI	26	XXVI	36	XXXVI	46	XLVI
17	XVII	27	XXVII	37	XXXVII	47	XLVII
18	XVIII	28	XXVIII	38	XXXVIII	48	XLVIII
19	XIX	29	XXIX	39	XXXIX	49	XLIX
20	XX	30	XXX	40	XL	50	L

Examples:

LX = 60 LXX = 70 LXXX = 80 XC = 90 C = 100
CX = 110 CXX = 120 CCC = 300 CCCX = 310 CCCXX = 320
CD = 400 D = 500 DC = 600 DCC = 700 DCCC = 800
CM = 900 M = 1,000 MMM = 3,000 $\overline{X} = 10{,}000$ $\overline{M} = 1$ million
DCCLXVII = 767 $\overline{DCCLXVII} = 767{,}000$ $\overline{MCMXLVIII} = 1{,}948{,}000$

CHAPTER TWO

Decimals

2-1 Decimals

In Section 1-1, we have learned how to read, write and round decimals.
A decimal has a **whole-number** part and a **decimal** part separated by a **decimal point**.
To compare decimals, we write one decimal under the other and line up the decimal points and like place-values. Starting from left to right, compare the first digits that are **not alike**. Add zeros after the last digit when necessary.

4.125	3.4**6**2	5.430**4**
2.346	3.4**7**3	5.430**0**

Compare the digits 4 and 2, $4 > 2$, we have $4.125 > 2.346$. $-4.125 < -2.346$.
Compare the digits 6 and 7, $6 < 7$, we have $3.462 < 3.473$. $-3.462 > -3.473$.
Compare the digits 4 and 0, $4 > 0$, we have $5.4304 > 5.43$. $-5.4304 < -5.43$.

Every integer can be written as an equal decimal by adding one or more zeros following the decimal point. $1 = 1.0 = 1.00 = 1.000 = \cdots$; $10 = 10.0 = 10.00 = 10.000 = \cdots$

Decimals include:

1. **Terminating** (or **finite**) **decimal**. It ends in a given place. 0.1, 0.3, 2.54.
2. **Nonterminating** (or **infinite**) **decimal**. It does not end in any given place.
   3.1415····. 1.4142····. 1.7320····. 2.71828····.
3. **Repeating decimal**. It is a nonterminating decimal that repeats a pattern.
   To write a repeating decimal, we place a bar above the digit that repeats.
   $0.111\cdots = 0.\bar{1}$. $0.333\cdots = 0.\bar{3}$. $2.125454\cdots = 2.12\overline{54}$.

Examples

1. Write the following decimals in order from the least to the greatest.

0.12, 0.034, 1.245, 5.134, 1.354, 5.012

Solution:

0.034, 0.12, 1.245, 1.354, 5.012, 5.134

2. Order the following decimals from the largest to the smallest.

3.45, –2.325, 3.452, 2.423, –2.335, 4.12, 2.43

Solution:

4.12, 3.452, 3.45, 2.43, 2.423, –2.325, –2.335

EXERCISES

Write the place-value of the underlined digit. (review Section 1-1)

1. 2.$\underline{5}$ **2.** 3.4$\underline{5}$ **3.** 1$\underline{4}$.124 **4.** 7.25$\underline{8}$ **5.** $\underline{3}$6.45 **6.** $\underline{6}$32.277 **7.** 32.492$\underline{8}$3 **8.** 4$\underline{7}$,134.23

Write each decimal in the short word name. (review Section 1-1)

9. 3.05 **10.** 0.005 **11.** 2.054 **12.** 0.0235 **13.** 24.0015 **14.** 324.12 **15.** 3,452.8246

Write each decimal in the word name. (review Section 1-1)

16. 2.01 **17.** 23.02 **18.** 0.325 **19.** 354.048 **20.** 9,782.0046 **21.** 6,000.1423

Round each decimal to the nearest tenth, hundredth, and thousandth. (review Section 1-1)

22. 3.0151 **23.** 32.1436 **24.** 0.25447 **25.** 7.3509 **26.** 72.14382 **27.** 0.10206

Write each decimal as an equal integer.

28. 1.000 **29.** 3.0 **30.** 6.00 **31.** 12.0 **32.** 0.0 **33.** 0.000 **34.** 423.00 **35.** 243.0000

Compare each pair of decimals with “ < ” or “ > ”.

36. 0.3 $\underline{?}$ 0.5 **37.** –0.3 $\underline{?}$ –0.5 **38.** 0.7 $\underline{?}$ 0.09 **39.** 0.07 $\underline{?}$ 0.9
40. –0.02 $\underline{?}$ –0.01 **41.** 7.01 $\underline{?}$ 7.05 **42.** –7.01 $\underline{?}$ –7.05 **43.** 2.01 $\underline{?}$ 2.011
44. 23.68 $\underline{?}$ 23.67 **45.** –28.34 $\underline{?}$ –29.43 **46.** 34.431 $\underline{?}$ 34.432 **47.** 0.00010 $\underline{?}$ 0.00012

Identify each decimal as terminating, nonterminating or repeating decimal.

48. 1.02 **49.** 3.115 **50.** 0.0555···· **51.** 0.050505···· **52.** 2.82847···· **53.** 3.31662····
54. 23.40 **55.** 0.11···· **56.** 3.873 **57.** 3.8729···· **58.** 5.12323···· **59.** 9.125454····

Write each repeating decimal using bar notation in the simplest form.

60. 0.111···· **61.** 0.0555···· **62.** 0.0202···· **63.** 3.545454···· **64.** 0.01212···· **65.** 0.234234····

Write the following decimals from the greatest to the least.

66. 0.2, 0.25, 0.02, 0.12, 0.025, 0.22 **67.** 0.2, –0.25, –0.02, 0.12, –0.025, 0.22
68. 1.6, 0.16, 3.43, 3.34, 3.30, 1.61 **69.** –1.6, 0.16, 3.43, –3.34, –3.30, –1.61

Write the following decimals from the smallest to the largest.

70. –0.12, –0.012, 0.12, –0.10, 0.21 **71.** 8.54, 8.45, –8.45, –8.54, 9.54, –6.45
72. 4.04, 3.04, –4.39, 4.039, –3.039 **73.** –2.04, –2.039, 2.40, 2.39, –2.39, 2.04

74. Using the digits 0,2,6,8,1,3 and write the smallest decimal possible that is larger than 2.
75. The grade-point averages (GPA) of five students are 3.091, 3.270, 3.910, 3.712, 3.207.
a) What is the highest score ? *b*) What is the lowest score ?

2-2 Adding and Subtracting Decimals

To add or subtract with decimals:

1. We write one decimal under the other and line up the decimal points. This automatically line up the place-values.
2. Add zeros after the last digit when necessary.
3. Add or subtract as with whole numbers.

To add or subtract decimals and whole numbers, we rewrite each whole number as an equal decimal with a decimal point and zeros after the last digit.

$$\begin{array}{r} 33.050 \\ +\ 24.573 \\ \hline 57.623 \end{array} \qquad \begin{array}{r} 33.050 \\ -\ 24.573 \\ \hline 8.477 \end{array} \qquad \begin{array}{r} 12.00 \\ +\ 34.45 \\ \hline 46.45 \end{array} \qquad \begin{array}{r} 34.00 \\ -\ 12.83 \\ \hline 21.17 \end{array}$$

General rules for adding and subtracting two decimals:

1. If the signs are the same, we combine the decimals and keep the sign.
$2.3 + 3.6 = 5.9$; $-9.3 - 6.8 = -16.1$

2. If the signs are different, we take the difference and use the sign of the decimal having the larger absolute value. $-8.4 + 5.2 = -3.2$; $8.4 - 5.2 = 3.2$

Examples

Add or subtract as indicated.

1. $1.1 + 0.11 + 1.143 + 10$ **2.** $0.702 + 2.06 + 100.8 + 24$ **3.** $25.12 + 7.34 + 95 + 0.3$
4. $0.89 - 0.5$ **5.** $326 - 28.24$ **6.** $343.7 - 125.325$

Solution: **1.**
$$\begin{array}{r} 1.100 \\ 0.110 \\ 1.143 \\ +\ 10.000 \\ \hline 12.353 \end{array}$$
2.
$$\begin{array}{r} 0.702 \\ 2.060 \\ 100.800 \\ +\ 24.000 \\ \hline 127.562 \end{array}$$
3.
$$\begin{array}{r} 25.12 \\ 7.34 \\ 95.00 \\ +\ 0.30 \\ \hline 127.76 \end{array}$$

4.
$$\begin{array}{r} 0.89 \\ -\ 0.50 \\ \hline 0.39 \end{array}$$
5.
$$\begin{array}{r} 326.00 \\ -\ 28.24 \\ \hline 297.76 \end{array}$$
6.
$$\begin{array}{r} 343.700 \\ -\ 125.325 \\ \hline 218.375 \end{array}$$

7. In a 800–meter relay, the time ran in seconds by four runners are 28.43, 29.2, 28.52 and 30. What is the total time ?
Solution: $28.43 + 29.2 + 28.52 + 30 = 116.15$ seconds.

8. John has \$284.56 in his checking account. What will his balance be after writing two checks for \$102.30 and \$46.52 ?
Solution: \$284.56 – \$102.30 – \$46.52 = \$135.74.

EXERCISES

Evaluate each expression

1. 1.5 + 2.7 **2.** 2.5 + 4.3 **3.** 0.02 + 2.13 **4.** 2.3 + 4.56 **5.** 3.128 + 12.2
6. 4.8 –1.7 **7.** 12.3 – 8.4 **8.** 1.78 – 0.26 **9.** 15 – 4.12 **10.** 9.54 – 6.807
11. 8.51 – 7.82 **12.** 12.20 – 1.45 **13.** 5.3 – 8.5 **14.** 3.5 – 8.7 **15.** 4.82 – 7.124
16. –5.2 – 3.8 **17.** –11.4 – 1.2 **18.** –11.4 + 1.2 **19.** 1.2 – 11.4 **20.** –1.2 + 11.4
21. 100 + 0.123 **22.** 100 – 0.123 **23.** 200 – 0.138 **24.** –200 + 0.342 **25.** –200 – 0.342
26. 100 – 23.6 **27.** –300 + 2.14 **28.** –500 – 200.45 **29.** –400 + 0.99 **30.** –500 + 0.239
31. 1.2 – (–2.3) **32.** 0.32 – (–0.2) **33.** –2.4 – (–4.5) **34.** 0 – (–5.9) **35.** –8.98 – (–4)

36. 0.01 + 1.23 + 0.45 **37.** 1.8 + 3.4 + 6.84 **38.** 22.45 + 4.2 + 16.06 + 12
39. 24.105 + 0.34 + 125.4 **40.** 0.92 + 12.405 + 112.1 **41.** 4.521 + 4.9 + 102.34 + 6
42. 3.46 – 2.43 – 4.78 **43.** –3.54 + 2.24 – 5.89 **44.** 524.3 – 324.12 + 120.231

Evaluate each expression if $a = 2.4$, $b = 3.5$, $c = -12.3$, $d = 8$.

45. $a + b$ **46.** $a - b$ **47.** $b - c$ **48.** $b + c - a$ **49.** $b - a + d$ **50.** $a - b + c$
51. $a + a + c$ **52.** $c + c - a$ **53.** $b + d - d$ **54.** $c + c + c$ **55.** $b - c - c$ **56.** $d - b - c$
57. $|c| - c$ **58.** $b - |c|$ **59.** $|b - d| - c$ **60.** $|a - b - c| + d$

61. You have \$412.86 in your checking account. What will your balance be after writing one check for \$123.4.
62. It rained for the last five days with rainfall of 1.34 inches, 2 inches, 0.92 inches, 0.6 inches and 1.57 inches. What was the total rainfall for the last five days ?
63. Marry weighed 124 pounds. She gained 7.5 pounds this month. How much does Marry weigh this month ?
64. John weighed 140.5 pounds two months ago. He gained 7.8 pounds last month. He lost 3.4 pounds this month. How much does John weigh now ?
65. Roger purchased stock on Cisco Company. The price per share of stock fell by \$1.12 in the first month, rose \$0.75 in the second month, and rose \$0.50 in the third month. How much did he gain (or lose) per share of stock for the past three months ?
66. Find the distance between two points -5.65 and 7.32.
67. What is the distance between 126.8 feet (below sea level) and 103.4 feet (below sea level) ?
68. What is the distance between 115.25 feet (below sea level) and 95.24 feet (above sea level) ?
69. In a 88-meter relay, the time ran in seconds by four runners are 27.54, 28.43, 29.4 and 29. What is the total time ?
70. The sale price is \$19.50 less than the regular price. If the regular price is \$78.90, what is the sale price ?
71. A number is the sum of 12.4 and 23.79, decreased by 5.27. What is the number ?
72. What is the net change in voltage across an electric circuit from 23.45 volts to –17.26 volts ?

2-3 Multiplying Decimals

To multiply with decimals:

1. We multiply as with two whole numbers.
2. Count the total number of decimal places in both numbers being multiplied.
3. Put the same number of decimal places in the product.
4. Add zeros in the product when necessary.

To multiply a decimal by a power of 10 (10, 100, 1000,·····), we move the decimal point one decimal place to the right for each zero in the power of 10.

$$\begin{array}{r} 2.5 \\ \times\ 0.37 \\ \hline 175 \\ +\ 75\ \ \\ \hline 0.925 \end{array} \qquad \begin{array}{r} 8.12 \\ \times\ 3.45 \\ \hline 4060 \\ 3248\ \ \\ +\ 2436\ \ \ \ \\ \hline 28.0140 \end{array} \qquad \begin{array}{r} 0.04 \\ \times\ 0.02 \\ \hline 0.0008 \end{array} \qquad \begin{array}{r} 0.015 \\ \times\ 0.25 \\ \hline 75 \\ +\ 30\ \ \\ \hline 0.00375 \end{array}$$

$1.3 \times 10 = 13$ $\quad 1.3 \times 100 = 130$ $\quad 1.3 \times 1000 = 1{,}300$

$0.1234 \times 10 = 1.234$ $\quad 0.1234 \times 100 = 12.34$ $\quad 0.1234 \times 1000 = 123.4$

$5.125 \times 10 = 51.25$ $\quad 5.125 \times 100 = 512.5$ $\quad 5.125 \times 1000 = 5{,}125$

$0.00123 \times 10 = 0.0123$ $\quad 0.00123 \times 100 = 0.123$ $\quad 0.00123 \times 1000 = 1.23$

$1.3 \times 10^1 = 13$ $\quad 1.3 \times 10^3 = 1{,}300$ $\quad 1.3 \times 10^5 = 130{,}000$

$1.234 \times 10^2 = 123.4$ $\quad 1.234 \times 10^4 = 12{,}340$ $\quad 1.234 \times 10^6 = 1{,}234{,}000$

Examples

1. A publisher sold 200 books last month. Each book earned a \$3.25 profit. How much was the profit last month ?
   Solution: $\$3.25 \times 200 = \$650$.
2. A number is the product of 1.25 and 8.4, minus 2.56. What is the number ?
   Solution: $1.25 \times 8.4 - 2.56 = 10.5 - 2.56 = 7.94$.
3. Maria purchased 18 gallons of gasoline for \$3.79 per gallon. What was the total cost she paid ?
   Solution: $\$3.79 \times 18 = \$68.22$.
4. John works 8 hours a day and 6 days a week at Macy Company. He earns \$9.45 per hour. How much does he earn per week ?
   Solution: $\$9.45 \times 8 \times 6 = \$453.60$.
5. Find the perimeter and the area of a rectangle when the length is 12.5 meters and the width is 8.25 meters.
   Solution: Perimeter $= (12.5 + 8.25) \times 2 = 20.75 \times 2 = 41.5$ meters.
   Area $= 12.5 \times 8.25 = 103.125\ m^2$ (square meter).
6. There are 365.242199 days in a year. Write it in days, hours, minutes, and seconds.
   Solution: 24 hours $\times\ 0.242199 = 5.812776$ hours
   60 minutes $\times\ 0.812776 = 48.76656$ minutes
   60 seconds $\times\ 0.76656 = 45.9936$ seconds $\approx$ 46 seconds.
   There are 365 days, 5 hours, 48 minutes, and 46 seconds in a year.

EXERCISES

Find the product.

1. 0.1×0.2 **2.** $(-0.3)(0.4)$ **3.** $(-0.5)(-0.6)$ **4.** $0.3(2.4)$ **5.** 3.8×0.7
6. 3.45×1.2 **7.** 0.05×0.007 **8.** 0.15×0.06 **9.** 12×0.125 **10.** 0.26×3.23
11. 1.2×0.036 **12.** 24.6×15 **13.** 12.56×0.04 **14.** 214×0.028 **15.** 1.243×3.12
16. $1.2(3.4)$ **17.** $2.5(-4.6)$ **18.** $-3.2(-1.5)$ **19.** $-1 \cdot (-4.9)$ **20.** $-1.7(-2.7)$
21. 2.3×10 **22.** 3.2×100 **23.** 100×0.12 **24.** 0.004×100 **25.** $1{,}000 \times 0.01$
26. 3.2×10^2 **27.** 3.2×10^3 **28.** 2.13×10^2 **29.** 0.04×10^3 **30.** 0.0035×10^2

Simplify each expression.

31. $3.6 - 4(1.25)$ **32.** $5.2(3.6) + 3.27$ **33.** $-4.1(3.6) - 7.2$ **34.** $6(7.4) + 4(5.3)$
35. $-3.2(-1.5) - 4$ **36.** $-1.2 - (-4.5)$ **37.** $-(-4.5) + 1.2$ **38.** $5(-2.3) - 3(4.5)$
39. $-(1.2 \times 2.5) + 8$ **40.** $2.3(3.5) - 4.23$ **41.** $9 + (-1.6)(-3.5)$ **42.** $7 - (3.4)(4.3)$

Evaluate each expression if $x = 3$, $y = 2.1$, $z = -3.2$, $w = 5.6$.

43. xy **44.** yz **45.** $5y$ **46.** $7z$ **47.** $-12y$ **48.** $2.4w$
49. $3x + y$ **50.** $3x - y$ **51.** $3y - w$ **52.** $-3z + y$ **53.** $2z + w$ **54.** $-2y - z$
55. $5xy$ **56.** $3xz$ **57.** $2xy - w$ **58.** $-yz + x$ **59.** $3wz - y$ **60.** wyz
61. x^2 **62.** y^2 **63.** z^2 **64.** $x^3 + z$ **65.** $w^2 - x$ **66.** $10 - y^3$
67. The product of x and z. **68.** y times z, plus w.
69. The product of w and y, decreased by x. **70.** w times x, increased by y.

71. Adrian works 40 hours a week. He earns $5.75 per hour. How much does he earn per week ?
72. Find the area of a rectangle when the length is 13.5 inches and the width is 7.12 inches.
73. Carlos bought 16 gallons of gasoline for $2.59 per gallon. What was the total cost he paid ?
74. A case of apples costs $7.99. What is the cost of 5 cases ?
75. A number is the product of 12.32 and 7.35, plus 3.47. What is the number ?
76. Roger bought 2.4 pounds of cheese for $2.99 per pound and 3.5 pounds of beef for $5.99 per pound. How much did he pay ?
77. Larry bought 4.5 pounds of pork for $1.99 per pound. How much change should he receive from a $10 bill ?
78. Tony works 8 hours a day and 6 days a week. He earns $12.75 per hour. How much does he earn for a month (four weeks) ?
79. The cost of electricity is $0.019 per hour to operate a refrigerator. What is the total cost to operate a refrigerator for four weeks ? Round the answer to the nearest cent.
80. Find the perimeter and the area of a rectangle when the length is 10.24 inches and the width is 6.48 inches.
81. John ordered 1,000 cards for $1.98 per card. How much did he pay ?
82. Home Depot ordered 10,000 nails for $0.02 per nail. What was the total cost ?

2-4 Dividing Decimals

To divide with decimals:

1. We move both decimal points to the right by the same number of places in order to get a whole-number divisor.
2. Place a decimal point in the quotient directly above the decimal point in the dividend.
3. Add zeros after the last digit on the dividend when necessary.
4. Divide as with whole numbers.
5. Round the quotient as needed.

To divide a decimal by a whole-number divisor, we start from step 2.

Examples

1. $5.16 \div 12 = ?$

```
     0.43
12 )5.16
    4 8
      36
      36
       0
```

$5.16 \div 12 = 0.43$

2. $-5.16 \div 1.2 = ?$

```
        4.3
1.2 )5.1 6
     4 8
       3 6
       3 6
         0
```

$-5.16 \div 1.2 = -4.3$

3. $5.16 \div 0.0012 = ?$

```
             4300.
0.0012 )5.1600
        4 8
          36
          36
           0
```

$5.16 \div 0.0012 = 4300$

4. Find $58 \div 2.4$ and round the solution to the nearest tenth.
Solution:

```
       2 4.16
2.4 )58.0 00
     48
     10 0
      9 6
        4 0
        2 4
        1 60
        1 44
          16
```

$58 \div 2.4 \approx 24.2$

5. Find $56.4 \div (-2.43)$ and round the solution to the nearest hundredth.
Solution:

```
         23.209
2.43 )56.40 000
      48 6
       7 80
       7 29
         51 0
         48 6
          2 400
          2 187
            213
```

$56.4 \div (-2.43) \approx -23.21$

6. Find the average (mean) of the numbers 31.6, 24.7, 56 and 34.2.
Solution:

$$\frac{31.6 + 24.7 + 56 + 34.2}{4} = \frac{146.5}{4} = 36.625.$$

EXERCISES

Find the quotient.

1. $2\overline{)2.5}$ **2.** $4\overline{)2.5}$ **3.** $8\overline{)2.5}$ **4.** $0.2\overline{)2.5}$ **5.** $0.4\overline{)2.5}$

6. $0.8\overline{)2.5}$ **7.** $0.02\overline{)2.5}$ **8.** $0.04\overline{)2.5}$ **9.** $0.08\overline{)2.5}$ **10.** $0.02\overline{)0.25}$

11. $0.0002\overline{)2.5}$ **12.** $15\overline{)78.75}$ **13.** $1.5\overline{)78.75}$ **14.** $0.015\overline{)78.75}$ **15.** $0.15\overline{)78.75}$

16. $15\overline{)76.2}$ **17.** $0.15\overline{)76.2}$ **18.** $0.15\overline{)0.762}$ **19.** $21.6\overline{)524.88}$ **20.** $0.216\overline{)524.88}$

21. $0.2\overline{)25}$ **22.** $0.02\overline{)25}$ **23.** $1.5\overline{)762}$ **24.** $0.15\overline{)762}$ **25.** $0.0015\overline{)762}$

Find the quotient. Round to the nearest tenth.

26. $25 \div 0.3$ **27.** $-25 \div 0.03$ **28.** $2.5 \div (-0.03)$ **29.** $-2.5 \div (-0.7)$ **30.** $2.5 \div 0.07$

31. $0.25 \div 0.9$ **32.** $25 \div (-0.11)$ **33.** $-24.45 \div 1.3$ **34.** $24.45 \div 0.13$ **35.** $-2445 \div 1.3$

Find the quotient. Round to the nearest hundredth.

36. $29 \div 0.3$ **37.** $29 \div (-0.03)$ **38.** $2.9 \div 0.03$ **39.** $-2.9 \div (-0.7)$ **40.** $-2.9 \div 0.07$

41. $0.29 \div 0.9$ **42.** $-29 \div (-0.11)$ **43.** $-0.029 \div 1.1$ **44.** $29.35 \div (-4.3)$ **45.** $2935 \div 4.3$

Simplify each expression. Round the nonterminating decimal to the nearest hundredth.

46. $1 + 5 \div 2$ **47.** $1 - 5 \div 2$ **48.** $5 \times 2 \div 3$ **49.** $5 \div 2 \times 3$ **50.** $7 \times 4 \div 0.6$

51. $5\,(2.6) \div 4$ **52.** $9 \div 2 - 8 \div 4$ **53.** $3(-6.2) + 2$ **54.** $8 - 6.5 \div (-4)$ **55.** $7 + (-5) \div 3$

56. $-5 \div 2 + 3$ **57.** $5.5 \div 2 - 4.3$ **58.** $9 - 5.5 \div 2.4$ **59.** $-9 \div (-4) + 4$ **60.** $7 - 6 \div (-7)$

Evaluate each expression if $x = 3$, $y = 2.1$, $z = -3.2$, $w = 5.6$. Round the nonterminating decimal to the nearest hundredth.

61. $y \div x$ **62.** $x \div y$ **63.** $2y \div z$ **64.** $2w \div x$ **65.** $-2.4x \div y$

66. $3y \div (-x)$ **67.** $5 - w \div z$ **68.** $x - w \div (-y)$ **69.** $w \div (-z) - x$ **70.** $w + y \div x$

71. $x^2 \div w$ **72.** $y^2 \div x$ **73.** $x - z^2 \div 2y$ **74.** $z^3 \div (-x)$ **75.** $2y^2 - w^2$

76. The quotient of $2y$ and x. **77.** The quotient of $2x$ and y.

78. It takes 3.5 hours for a plane to travel nonstop from city A to city B. The distance from city A to city B is 960 miles. Find the average speed. Round it to the nearest hundredth.

79. A number is the quotient of 5.52 and 11.5, decreased by 2.79. What is the number ?

80. John bought 12.5 gallons of gas at a total cost of $26.75. What is the cost per gallon of gas ?

81. John drove 120 miles on 12.5 gallons of gas. What is the mileage for a gallon of gas ?

82. Four students go to a restaurant. The waiter brings a bill totaling $42.50. How much, including tip $5, does each student pay ?

2-5 Scientific Notations

To write very large numbers or very small numbers, we use the **scientific notations.**

The distance from the Sun to the Earth = 93 million miles = 9.3×10^7 miles = 1.5×10^{11} meters.

The distance from the Earth to the Moon = 250,000 miles = 2.5×10^5 miles = 4×10^8 meters.

The mass of one atom of hydrogen = 1.674×10^{-24} gram.

Scientific Notation: **(A number)** = $m \times 10^n$, $1 \le m < 10$.

m is a number greater or equal to 1, but less than 10. n is an integer.

The following powers of 10 are useful in scientific notations.

$10 = 10^1$, $100 = 10^2$, $1{,}000 = 10^3$, $10{,}000 = 10^4$, $100{,}000 = 10^5$,........

$0.1 = 10^{-1}$, $0.01 = 10^{-2}$, $0.001 = 10^{-3}$, $0.0001 = 10^{-4}$, $0.00001 = 10^{-5}$,........

To write a number in scientific notation, we move the decimal point to get a number that is at least 1 but less than 10. Count the number of decimal places it moves and use it as the power of 10.

$123{,}000{,}000 = 1.23 \times 10^8$. (Moved 8 digits to the left. The power is positive.)

$0.0000000123 = 1.23 \times 10^{-8}$. (Moved 8 digits to the right. The power is negative.)

The Milky Way is the home galaxy of our solar system and contains about 200 billion stars ($200 \times 10^9 = 2 \times 10^{11}$ stars).

A **light-year** is the distance light travels in one year in a vacuum. A light-year is about 5.9 trillion miles (5.9×10^{12} miles).

The nearest star to our Sun, Alpha Centauri, is 4.25 light-years (2.51×10^{13} miles) away. The diameter of our solar system is 1.18×10^{11} miles. The size of the universe is estimated at least 2 million light-years (2×10^6 light-years $= 1.18 \times 10^{19}$ miles) in diameter.

Examples (Write in scientific notations.)

1. $100 = 1 \times 100 = 1 \times 10^2$.

2. $5{,}000 = 5 \times 1{,}000 = 5 \times 10^3$.

3. $0.02 = 2 \times 0.01 = 2 \times 10^{-2}$.

4. $0.005 = 5 \times 0.001 = 5 \times 10^{-3}$.

5. $12{,}000 = 1.2 \times 10{,}000 = 1.2 \times 10^4$.

6. $0.00012 = 1.2 \times 0.0001 = 1.2 \times 10^{-4}$.

7. $235 = 2.35 \times 100 = 2.35 \times 10^2$.

8. $0.235 = 2.35 \times 0.1 = 2.35 \times 10^{-1}$.

9. $2{,}350{,}000 = 2.35 \times 10^6$.

10. $0.00000235 = 2.35 \times 10^{-6}$.

11. $2 \times 10^5 \times 6 \times 10^7 = 12 \times 10^{12} = 1.2 \times 10 \times 10^{12} = 1.2 \times 10^{13}$.

12. $2.35 \times 10^2 \times 5.2 \times 10^4 = (2.35)(5.2) \times 10^6 = 12.22 \times 10^6 = 1.222 \times 10 \times 10^6 = 1.222 \times 10^7$.

13. An atom of oxygen has a mass of 2.658×10^{-23} gram. Write it in decimal numeral.

Solution: $2.658 \times 10^{-23} = 0.00000000000000000000002658$ gram.

14. Light travels at the rate of 186,000 miles per second. It takes 8 minutes and 20 seconds for the light from the Sun to reach the Earth. Find the distance from the Sun to the Earth in scientific notation.

Solution: $d = 186{,}000 \times (8 \times 60 + 20) = 93{,}000{,}000 = 9.3 \times 10^7$ miles.

EXERCISES

Write each number in scientific notation.

1. 320 **2.** 342 **3.** 1,000 **4.** 3,000 **5.** 3,580
6. 9,000 **7.** 10,000 **8.** 32,000 **9.** 76,500 **10.** 50,123
11. 234,000 **12.** 2,345,000 **13.** 305,000 **14.** 2,220,000 **15.** 324,432
16. 777,000 **17.** 777,700 **18.** 40,200,000 **19.** 20,220,000 **20.** 19,300,000
21. 700,000,000 **22.** 712,000,000 **23.** 901,000,000 **24.** 801,500,000 **25.** 40,253,000
26. 320.45 **27.** 4,582.67 **28.** 56,346.12 **29.** 32,254.123 **30.** 243,142.18
31. 0.000001 **32.** 0.000123 **33.** 0.0000007 **34.** 0.00000042 **35.** 0.00000436
36. 0.14258 **37.** 0.05328 **38.** 0.004567 **39.** 0.000368 **40.** 0.283456

Write each number in decimal numeral.

41. 1×10^2 **42.** 1.2×10^2 **43.** 2.4×10^3 **44.** 4×10^7 **45.** 4.6×10^5
46. 4.06×10^7 **47.** 3.52×10^6 **48.** 1.101×10^3 **49.** 4.02×10^8 **50.** 5.0×10^5
51. 1.48×10^1 **52.** 4.243×10^5 **53.** 9.99×10^5 **54.** 2.345×10^2 **55.** 2.034×10^1
56. 1×10^{-1} **57.** 1×10^{-2} **58.** 2×10^{-3} **59.** 2.4×10^{-2} **60.** 5×10^{-6}
61. 7.6×10^{-5} **62.** 2.145×10^{-2} **63.** 5.24×10^{-7} **64.** 8.1×10^{-6} **65.** 2.05×10^{-2}
66. 7×10^{-7} **67.** 3.2345×10^{-3} **68.** 8.02×10^{-6} **69.** 1.111×10^{-5} **70.** 1×10^{-9}

Evaluate each of the following expressions in scientific notation.

71. $4\times10^5+2\times10^5$ **72.** $4\times10^5+2\times10^4$ **73.** $4\times10^5-2\times10^5$
74. $4\times10^5-2\times10^4$ **75.** $4\times10^5\times2\times10^4$ **76.** $4\times10^5\times2\times10^{-4}$
77. $5\times10^6\times3\times10^5\times2\times10^4$ **78.** $6\times10^4\times7\times10^8\times2\times10^{-5}$ **79.** $6\times10^4\times7\times10^{-8}$
80. $2\times10^{-7}\times4\times10^{-6}$ **81.** $2.5\times10^9\times3\times10^6$ **82.** $5.3\times10^{10}\times2.6\times10^4$
83. $\dfrac{8\times10^6}{4\times10^2}$ **84.** $\dfrac{15\times10^2}{3\times10^7}$ **85.** $\dfrac{14\times10^7}{4\times10^2}$

86. The distance from the Earth to the Moon is about 399,000,000 meters. Write this number in scientific notation.

87. A light-year is about 5.9×10^{12} miles. Write this number in decimal numeral.

88. The speed of light is about 982,000,000 feet per second. Write this number in scientific notation.

89. The diameter of the solar system is about 118,000,000,000 miles. Write this number in scientific notation.

90. A light-year is about 5.9×10^{12} miles. The nearest star to the Earth, Alpha Centauri, is 4.25 light-years away. Find the distance from the Earth to Alpha Centauri.

91. The size of the universe is estimated 2 million light-years in diameter. A light-year is about 5.9×10^{12} miles. What is the diameter of the universe?

CHAPTER 2 EXERCISES

Write the place-value of the underlined digit.

1. $2.\underline{0}5$ **2.** $3.40\underline{5}$ **3.** $12\underline{4}.12$ **4.** $70.125\underline{8}$ **5.** $\underline{3}01.47$ **6.** $42{,}1\underline{2}3.14$

Write each decimal in the word name.

7. 3.21 **8.** 4.01 **9.** 542.005 **10.** 42.012 **11.** 5,023.045 **12.** 3,000.1243

Compare each pair of decimal with "<" or ">".

13. 0.4 ? 0.3 **14.** −0.4 ? 0.3 **15.** −0.4 ? −0.3 **16.** 0.05 ? 0.5
17. 3.02 ? 3.01 **18.** 3.02 ? 3.021 **19.** 54.453 ? 54.54 **20.** −53.4 ? −54.3

Identify each decimal as terminating, nonterminating or repeating decimal.

21. 1.0202···· **22.** 3.121 **23.** 0.011···· **24.** $0.\overline{3}1$ **25.** 0.31 **26.** 4.3521····
27. $4.\overline{3541}$ **28.** 4.3541 **29.** 6.24343···· **30.** 7.653 **31.** $7.\overline{656}$ **32.** 9.097····

Write the following decimals from the smallest to the largest.

33. 0.32, 0.23, 3.23, 3.32, 0.023 **34.** 0.2, −0.3, −0.03, 3.01, −2.32

Write the following decimals from the greatest to the least.

35. 5.02, 6.13, 0.61, 0.50, 1.78, 5.03 **36.** −4.24, 2.2, −4.2, 3.25, 3.52, −2.2

Evaluate each expression.

37. $0.3 + 0.2$ **38.** $0.3 - 0.2$ **39.** $0.2 - 0.3$ **40.** $0.13 + 2.35$ **41.** $3.26 + 0.28$
42. $-2.8 - 8.2$ **43.** $1.94 - 0.28$ **44.** $8.26 - 6.124$ **45.** $-2.7 + 7.2$ **46.** $-7.2 + 2.7$
47. $6.34 - 8.522$ **48.** $-0.45 - 3.69$ **49.** $200 - 0.01$ **50.** $-400 + 1.324$ **51.** $-100 + 1.2$
52. $2.7 - (-3.4)$ **53.** $0.4 - (-0.6)$ **54.** $-1.4 - (-2.8)$ **55.** $1.2 + 3.5 - 4.9$ **56.** $3.8 - 7.2 + 3$

Find the product.

57. 0.01×0.02 **58.** -1.2×0.5 **59.** 0.12×0.006 **60.** 2.62×3.5 **61.** 23.5×40
62. $-3.2 \times (-4)$ **63.** $2.4 \times (-56.2)$ **64.** 1.05×10^3 **65.** 0.26×10^4 **66.** 0.034×10^5

Find the quotient. Round the nonterminating decimal to the nearest hundredth.

67. $46 \div 3$ **68.** $-4.6 \div 4$ **69.** $-6.4 \div (-3.2)$ **70.** $4.5 \div 0.003$ **71.** $4.35 \div 1.24$
72. $-0.24 \div (-4)$ **73.** $24.57 \div (-2)$ **74.** $0.045 \div 0.009$ **75.** $24.56 \div 0.24$ **76.** $1{,}428 \div 0.34$

Simplify each expression.

77. $4.2 + 3(2.5)$ **78.** $5.6 - (3.2)(4.8)$ **79.** $5 \div 2 + 2 \div 5$ **80.** $24 \div 1.2 \times 2.3$
81. $-2.5 - 3(-4.4)$ **82.** $3.8 - 4.2 \times 2 \div 6$ **83.** $7.2 \div 6 + 3.2 \times 4 - 10$

-----Continued-----

Evaluate each expression if $x = 2$, $y = 2.5$, $z = -2.1$, $w = 4$. Round the nonterminating decimal to the nearest hundredth.

84. $xy - z$ **85.** $x \div y + w$ **86.** $7y - 2z$ **87.** $-2w \div y$ **88.** $6 - xyz$

89. $x + y \div x$ **90.** $y \div z - z \div y$ **91.** $x^2 + 10y$ **92.** $-y^2 - z$ **93.** $z^2 - y$

94. $w^2 \div y$ **95.** $y^2 - z^2$ **96.** $5y \div 4z$ **97.** $y^3 - z$ **98.** $z^3 \div y$

Write each number in scientific notation.

99. 5,620,000 **100.** 0.000562 **101.** 0.000012 **102.** 76,800,000 **103.** 204,000

104. 0.004532 **105.** 0.4532 **106.** 14,532.72 **107.** 124,123.43 **108.** 0.00014

Write each number in decimal numeral.

109. 1.2×10^1 **110.** 3.1×10^2 **111.** 2.3×10^2 **112.** 4.5×10^3 **113.** 1×10^4

114. 5.03×10^5 **115.** 1.2×10^{-1} **116.** 4×10^{-2} **117.** 4.2×10^{-3} **118.** 3.42×10^{-5}

119. What is the perimeter ?

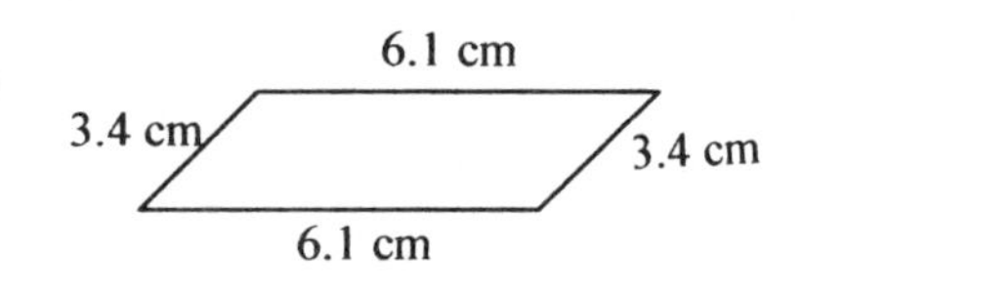

120. What is the area ?

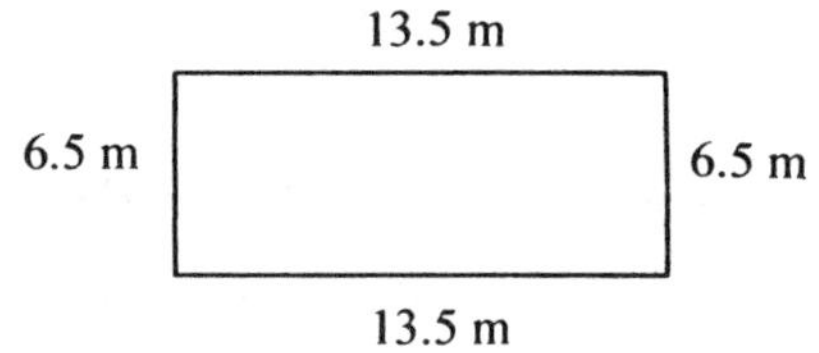

121. You bought 24 tennis balls for \$70.90. What is the price per ball ?

122. John bought 15 folders for \$37.50. The price of the folder on sale is \$2.15 each. How much could John save per folder by buying at the price on sale ?

123. The volume of a cube is given by the formula $V = s^3$ (where s is the length of one side). Find the volume of a cube with one side being 11.4 meters long.

124. The area of a circle is given by the formula $A = \pi r^2$ (where $\pi \approx 3.14$ and r is the radius). Find the area of a circle with a radius of $4\,cm$.

125. In a 800–meter relay, the total time ran by four runners is 116.16 seconds. What is the average time in seconds ran by each runner ?

126. In our Milky Way Galaxy, there are 200 billion stars banded together by gravity in an enormous spiral disk. Write this number in scientific notation.

127. The diameter of the solar system is about 118,000,000,000 miles. Write this number in scientific notation.

128. The nearest star to our Earth, Alpha Centauri, is 272,000 AU (astronomical unit) away. 1 AU = 9.3×10^7 miles . Find the distance from Alpha Centauri to our Earth.

129. An electron has a mass of $5.49 \times 10^{-4}\ u$ (where $u = 1.661 \times 10^{-24}$ g). Find the mass of an electon. (Hint: u is one atomic mass unit. See next page.)

130. A proton has a mass of $1.0073\,u$ (where $u = 1.661 \times 10^{-24}$ g). Find the mass of a proton.

131. A neutron has a mass of $1.0087\,u$ (where $u = 1.661 \times 10^{-24}$ g). Find the mass of a neutron.

132. The speed of light is 186,000 miles per second. The Sun is 9.3×10^7 miles from the Earth. How long (in seconds) does it take a beam of light to reach the Earth from the Sun ?

Additional Examples

1. The astronomical distance is very large. Scientists use the **astronomical unit** (AU) as a standard of measurement. An astronomical unit $\text{AU} = 9.3 \times 10^7$ miles (it is the distance from the Earth to the Sun).

2. The atomic mass is very small. Scientists use the atomic mass unit (u) as a standard of measurement. An atomic mass unit $u = 1.661 \times 10^{-24}$ gram (it is the mass of a carbon atom).

3. The gravitational force attracted each other between any two objects in the universe is given by the **Newton's Law of Universal Gravitation** (it is an inverse square law):

$$F = G\frac{m_1 m_2}{r^2} \quad \text{(Force in newtons)}$$

$G = 6.67 \times 10^{-11}$ (it is a universal constant).
m_1 and m_2 are their masses (in kilograms).
r is the distance apart (in meters).

Examples

1. Find the gravitational force between two objects at a distance 30,000 meters apart Their masses are 6.4×10^5 kg and 5×10^8 kg.

Solution: $F = G\frac{m_1 m_2}{r^2} = 6.67 \times 10^{-11} \times \frac{(6.4 \times 10^5)(5 \times 10^8)}{(3 \times 10^4)^2}$

$= \frac{6.67 \times 6.4 \times 5}{9} \times 10^{-11+5+8-8} = 23.7 \times 10^{-6} = 2.37 \times 10^{-5}$ newton. Ans.

2. Find the gravitational force between a 70 kg boy and a 50 kg girl who are 2 meters apart.

Solution: $F = G\frac{m_1 m_2}{r^2} = 6.67 \times 10^{-11} \times \frac{(70)(50)}{2^2} = \frac{6.67 \times 70 \times 50}{4} \times 10^{-11}$

$= 5836.25 \times 10^{-11} = 5.84 \times 10^{-8}$ newton. Ans.
(It is too small to attract each other.)

4. According to the famous **Einstein's Mass-Energy Relation** (**Theory of Relativity**), the mass of a particle can be converted into energy by the formula: $E = mc^2$, where E is the energy in joules, $c = 3 \times 10^8$ (speed of light in meters per second), m is the mass in kg.

Example: What is the energy related from 1 gram of water if the water is completely converted to energy ?
Solution:

$$m = 1 \text{ gram } = 1 \times 10^{-3} \text{ kg}$$

$$E = mc^2 = 1 \times 10^{-3} \times (3 \times 10^8)^2 = 9 \times 10^{-3} \times 10^{16} = 9 \times 10^{13} \text{ joules. Ans.}$$

(A small loss of mass produces a huge energy.)

Additional Example

5. Newton's Three Laws of Motion

Newton's Laws of Motion are first found by Newton to explain the results of stars and planets observed in the universe by Galileo Galilei and Johannes Kepler. These Laws are applied for all objects of nature to obey.
The Laws discuss the effects of forces on objects. These forces are often represented with vectors. We can add and subtract these forces to obtain components, unit vectors, and resultants.

1. **Newton's First Law of Motion**

   An object continues in its state of rest, or move with a constant velocity in a straight line, unless a force is applied on it to change that state.

2. **Newton's Second Law of Motion**

   The change of motion of an object is proportional to the applied force and in the direction which that force acts.
   Mathematically, if a net force (F) acts on an object of mass (m), the object will have acceleration (a).
   Therefore, the second law is sometimes expressed as:

   $$F = ma$$

   F is measured in newtons.

   a is measured in meters per second squared (m/s^2).

   m is measured in kilograms.

3. **Newton's Third Law of Motion**

   For every action force on an object, there is an equal but opposite reaction force.
   In real life, we can observe action and reaction forces when air is released from a balloon.

> Using a phone, a four-year old boy calls his mother who is working in a foreign country.
> Son: Mom, are you coming home today ?
> Mother: No, I am very far away at the other side of the earth.
> Son: Where is the other side of the earth ?
> Mother: Earth is like a basketball. You are on top of the ball. I am at the bottom.
> Son: It must be very hard for you with your head upside-down when you drink water.

CHAPTER THREE

Number Theory

3-1 Factors and Multiples

Before we study the operations with fractions, we need to learn some of the properties of whole numbers. **Number Theory** is the study of the properties of whole numbers. In this chapter, some of the properties of whole numbers are discussed.

Whole numbers: 0 and positive integers. They are 0, 1, 2, 3, 4, 5, 6, 7, 8, and so on.
Divisible: A whole number is said to be divisible by another number if the remainder is 0.
Even numbers: Numbers which are divisible by 2 (such as 2, 4, 6, 8, 10, ...).
Odd numbers: Numbers which are not divisible by 2 (such as 1, 3, 5, 7, 9, ...).
0 is considered to be an even number, neither positive nor negative.

Factor: The whole numbers we multiply are called **factors** of their product.

Examples: $1 \times 12 = 12.$ 1 and 12 are factors of 12.
$2 \times 6 = 12.$ 2 and 6 are factors of 12.

A number is said to be **divisible** by each of its **factors**.
To find the factors of a number, we divide that number by each whole number, starting with 1. If the remainder is 0 (divisible), then the whole number is a factor. If the remainder is not 0, then the whole number is not a factor.
1 is the smallest factor of every whole number. The only factor of 1 is 1.
A number divided by 0 is undefined. Therefore, 0 is not a factor of any number.
0 divided by any nonzero integer is 0. Therefore, every nonzero integers is a factor of 0. ($0 \div 1 = 0$, $0 \div 2 = 0$, $0 \div 3 = 0$, ...).

Multiple: A **multiple** of a number is the product of that number and any whole number.

Examples: The multiples of 3 are 0, 3, 6, 9, 12, 15, 18, 21, ...
The multiples of 4 are 0, 4, 8, 12, 16, 20, 24, 28, ...

Any nonzero integer multiplied by 0 is 0. Therefore, 0 is a multiple of any nonzero integer. ($1 \times 0 = 0$, $2 \times 0 = 0$, $3 \times 0 = 0$, ...)

Conventionally, we apply factors and multiples on whole numbers only.
(± 3 are the factors of 12 or -12. But, negative numbers are excluded.)

Examples

1. Find all the whole–number factors of 12.
Solution: 12 is divisible by 1, 2, 3, 4, 6, and 12.
Therefore, the factors are 1, 2, 3, 4, 6, and 12.
(We do not have to try the factors which begin to repeat. Stop at $12 \div 4 = 3$.)

2. Write the first five multiples of 6.
Solution: Multiplying 6 by 0,1,2,3, and 4. Therefore, the multiples are 0, 6, 12, 18, and 24.

EXERCISES

Write True or False for each statement.

1. 10 is divisible by 2.
2. 9 is divisible by 4.
3. 3 is a factor of 12.
4. 15 is a multiple of 5.
5. 45 is a multiple of 13.
6. 4 is a factor of 25.
7. 13 is a multiple of 13.
8. 0 is a factor of 5.
9. 0 is a multiple of 5.
10. 121 is divisible by 11.
11. 1 is a factor of 10.
12. 144 is divisible by 12.
13. 169 is divisible by 12.
14. 13 is a multiple of 169.
15. 13 is a factor of 169.

Find all the whole–number factors of each number.

16. 10 **17.** 9 **18.** 8 **19.** 7 **20.** 6 **21.** 5 **22.** 4 **23.** 3 **24.** 2 **25.** 1
26. 15 **27.** 18 **28.** 20 **29.** 48 **30.** 75 **31.** 80 **32.** 90 **33.** 95 **34.** 99 **35.** 112

Write the first five multiples of each number.

36. 1 **37.** 2. **38.** 3 **39.** 4 **40.** 5 **41.** 6 **42.** 7 **43.** 8 **44.** 9 **45.** 10
46. 11 **47.** 13 **48.** 15 **49.** 21 **50.** 25 **51.** 30 **52.** 40 **53.** 75 **54.** 92 **55.** 101

Identify each number as Odd or Even.

56. 7 **57.** 9 **58.** 12 **59.** 17 **60.** 0 **61.** 1 **62.** 29 **63.** 391 **64.** 408 **65.** 105

66. Write the remainder when 24 is divided by 6.
67. Write the remainder when 24 is divided by 5.
68. Is 100 a factor of 1,200 ?
69. Is 1000 a factor of 20,300 ?
70. Zero is a factor of what number ?
71. Zero is a multiple of what number ?
72. Is every nonzero integer a factor of itself ?
73. Is every nonzero integer a multiple of itself ?
74. What number is the smallest factor of every whole number ?
75. What number is a factor of every whole number ?
76. What number is a multiple of every nonzero integer ?
77. What number has all nonzero integers as its factors ?
78. What is the smallest whole number having 3 and 4 as factors ?
79. What is the smallest number which is the sum of all its factors except itself ?
80. Find the missing digit of the number | 5 | 2 | | if we know that it has 4 and 6 as factors.
81. Write all the possible quantities to separate 100 toys equally, including 1 each and 100 each.
82. Factor 16 in as many different ways as possible.
83. If x is a nonnegative integer, What is the total number of factors of 2^x ?
84. If x and y are nonnegative integers, what is the total number of factors of $2^x 3^y$?
85. If x, y, and z are nonnegative integers, what is the total number of factors of $2^x 3^y 5^z$?

3-2 Divisibility Tests

A whole number is divisible (the remainder is 0) by each of its factors.
To find the factors of a number, sometimes it is helpful to test the divisibility by inspecting the patterns of the digits without actually dividing.

Divisibility Tests: A whole number is:

divisible by **2** : if its last digit is 0, 2, 4, 6, or 8.
divisible by **3** : if the sum of its digits is divisible by 3.
divisible by **4** : if its last two digits are divisible by 4. (**See proof on example 2**)
divisible by **5** : if its last digit is 0 or 5.
divisible by **6** : if it is divisible by both 2 and 3.
divisible by **8** : if its last three digits are divisible by 8.
divisible by **9** : if the sum of its digits is divisible by 9. (**See proof on page 58**)
divisible by **10** : if its last digit is 0.
divisible by **11** : if the difference between the sum of the odd digits and the sum of the even digits is 0 or 11. (**See proof on page 58**)

25894 is divisible by 11 because $(2+8+4)-(5+9)=0$.
86801 is divisible by 11 because $(8+8+1)-(6+0)=11$.

Examples

1. Test each number for divisibility by 2, 3, 4, 5, 6, 8, 9, 10, or 11.

a. 105 *b.* 1248 *c.* 9170 *d.* 78903

Solution:

a. 105 is divisible:
by 3 because the sum of its digits $(1+0+5=6)$ is divisible by 3.
by 5 because its last digit is 5.

b. 1248 is divisible:
by 2 because its last digit is 6.
by 3 because the sum of its digits $(1+2+4+8=15)$ is divisible by 3.
by 4 because its last two digits 48 is divisible by 4.
by 6 because it is divisible by both 2 and 3.
by 8 because the last three digits 248 is divisible by 8.

c. 9170 is divisible:
by 2, 5, and 10 because its last digit is 0.

d. 78903 is divisible:
by 3 and 9 because the sum of its digits is divisible by both 3 and 9.
by 11 because $(7+9+3)-(8+0)=11$.

2. Find a test for divisibility by 4.

Solution: Any multiple of 100 is divisible by 4. We simple check the last two digits of a number which is "any multiple of 100 + two-digit whole number".
Therefore, a number is divisible by 4 if its last two digits is divisible by 4.

EXERCISES

Test each number for divisibility by 2.

1. 251 **2.** 346 **3.** 1,414 **4.** 7,250 **5.** 3,645 **6.** 6,322 **7.** 32,498

Test each number for divisibility by 3.

8. 252 **9.** 320 **10.** 1,411 **11.** 2,748 **12.** 6,724 **13.** 8,220 **14.** 42,912

Test each number for divisibility by 4.

15. 258 **16.** 348 **17.** 2,210 **18.** 3,420 **19.** 5,234 **20.** 7,456 **21.** 23,718

Test each number for divisibility by 5.

22. 620 **23.** 728 **24.** 1,535 **25.** 4,310 **26.** 6,665 **27.** 8,552 **28.** 31,520

Test each number for divisibility by 6.

29. 681 **30.** 563 **31.** 1,542 **32.** 1,520 **33.** 3,426 **34.** 9,628 **35.** 46,324

Test each number for divisibility by 8.

36. 1,016 **37.** 2,808 **38.** 2,240 **39.** 4,823 **40.** 8,832 **41.** 9,871 **42.** 96,423

Test each number for divisibility by 9.

43. 333 **44.** 3,333 **45.** 3,339 **46.** 5,225 **47.** 5,229 **48.** 8,244 **49.** 78,327

Test each number for divisibility by 10.

50. 590 **51.** 5,955 **52.** 5,900 **53.** 6,810 **54.** 7,352 **55.** 8,200 **56.** 89,059

Test each number for divisibility by 11.

57. 990 **58.** 2,349 **59.** 2,398 **60.** 3,718 **61.** 6,919 **62.** 7,235 **63.** 58,278

64. Find the missing digits in $\boxed{1}\boxed{4}\boxed{} = 2 \times \boxed{}\boxed{}$.

65. Find the missing digits in $\boxed{2}\boxed{}\boxed{5} = 3 \times \boxed{}\boxed{}$.

66. Find the missing digit in $\boxed{5}\boxed{3}\boxed{}$ if it has the factors 2 and 3.
67. Develop a test for divisibility by 50.
68. If a number is divisible by 6, is it also divisible by 2 ?
69. If a number is divisible by 5, is it also divisible by 10 ?
70. What is the greatest 4-digit number that is divisible by 4 ?
71. What is the smallest 5-digit number that is divisible by 9 ?
72. There are 7,650 pounds of potatoes. How many additional pounds are needed to completely fill the bags that hold 8 pounds each ?

3-3 Prime Factorization

Prime Number: A whole number other than 0 and 1 which is divisible only by 1 and itself.
Here are the prime numbers: 2, 3, 5, 7, 11, 13, 17, 19, 23, 29, 31, 37,·····.
A prime number has exactly two factors, 1 and itself.

0 is divisible by every nonzero integer (0 has many factors). Therefore, 0 is not a prime number.

1 is divisible only by itself (1 has only one factor, itself). Therefore, 1 is not a prime number (see the other reason at the bottom of this page).

All even numbers greater than 2 are not prime numbers. 2 is the only even prime number.

Therefore, to identify whether or not an odd number is a prime number, we test the divisibility by the prime numbers 2, 3, 5, 7, 11, 13, 17, 19,·····.

Example: 121 is divisible by 11. Therefore, 121 is not a prime number.

Composite Number: A whole number other than 0 and 1 which is not a prime number.
Here are the composite numbers: 4, 6, 8, 9, 10, 12, 14, 15, 16, ·····.
A composite number has more than two factors.

The numbers 0 and 1 are neither prime nor composite.

Prime Factorization: It is to write a composite number as a product of prime numbers.
We can factor a number into prime factors by using either of the following two methods: **Shortcut division** and **Factor tree.**

In prime factorization, we divide the number by a prime number (from small to large) until the bottom row is a prime number. Then, we write the prime factors in order from least to greatest.

Example: Find the prime factorization of 24.
Solution:
Method 1: **Shortcut division** Method 2: **Factor tree**

$$\begin{array}{r|l} 2 & 24 \\ \hline 2 & 12 \\ \hline 2 & 6 \\ \hline & 3 \end{array}$$

```
24
/  \
2 × 12
    /  \
   2 × 6
       /  \
      2 × 3
```

Answer: $24 = 2 \times 2 \times 2 \times 3 = 2^3 \times 3$.
(Exponents are often used to express the prime factorization.)

Fundamental Theorem of Arithmetic:
Every composite number can be factored as a product of prime factors in exactly one way, except for the order of the prime factors.

1 is not considered as a prime number to avoid $24 = 1 \times 1 \times 1 \times \cdots \times 2^3 \times 3$.
The Fundamental Theorem of Arithmetic would be false if 1 was defined as a prime number.

EXERCISES

Identify each number as prime or composite.

1. 19 **2.** 51 **3.** 21 **4.** 23 **5.** 41 **6.** 24 **7.** 50 **8.** 59 **9.** 57 **10.** 29
11. 13 **12.** 49 **13.** 39 **14.** 33 **15.** 91 **16.** 144 **17.** 89 **18.** 125 **19.** 519 **20.** 781

Find all the whole–number factors of each number.

21. 8 **22.** 9 **23.** 15 **24.** 18 **25.** 20 **26.** 48. **27.** 90 **28.** 99 **29.** 112 **30.** 169

Find all the prime factors of each number.

31. 8 **32.** 9 **33.** 15 **34.** 18 **35.** 20 **36.** 48 **37.** 90 **38.** 99 **39.** 112 **40.** 169
41. 35 **42.** 25 **43.** 30 **44.** 36 **45.** 47 **46.** 80 **47.** 55 **48.** 46 **49.** 76 **50.** 57

Write the prime factorization of each number using exponents.

51. 8 **52.** 9 **53.** 15 **54.** 18 **55.** 20 **56.** 48 **57.** 90 **58.** 99 **59.** 112 **60.** 169
61. 12 **62.** 50 **63.** 28 **64.** 39 **65.** 40 **66.** 63 **67.** 84 **68.** 51 **69.** 70 **70.** 196

71. What is the smallest prime factor of 16 ?
72. What is the smallest prime factor of 119 ?
73. What number is the only prime even number ?
74. Is the number 0 a prime number ? Explain.
75. Is the number 1 a prime number ? Explain.
76. Choose the prime factorization of 12 as an example to explain why the number 1 is not considered as a prime number.
77. Explain why the sum of two prime numbers greater than 2 can never be a prime number.
78. Find the prime factorization of 1,575.
79. Find the prime factorization of 3,213.
80. Find the prime factorization of 6,825.
81. Find the prime factorization of 10,780.
82. Find the prime factorization of 40,425.

A kid said to an old lady sitting on a park chair.
Kid: How are your teeth ?
Old lady: No good. They are all gone.
Kid: Please hold my apple. I want to play ball with my friends for a while.

3-4 Greatest Common Factor and Least Common Multiple

Greatest Common Factor (GCF): The greatest number that is a factor of two or more given numbers.

Factors of 12: 1, 2, 3, 4, **6**, 12 ; Factors of 18: 1, 2, 3, **6**, 9, 18

The common factors of 12 and 18 are 1, 2, 3, **6**. The GCF of 12 and 18 is **6**.

To find the GCF of two or more given numbers, we write the prime factorizations of the given numbers, and find the **lowest power** of each factor that occurs in **all** of the factorizations. The product of these lowest powers is the GCF.

Example 1: Find the GCF of 12 and 18.
Solution:
$$12 = 2^2 \times 3 \ ; \ 18 = 2 \times 3^2$$
$$\therefore \ \text{GCF} = 2 \times 3 = 6.$$

Example 2: Find the GCF of 54 and 90.
Solution:
$$54 = 2 \times 3^3 \ ; \ 90 = 2 \times 3^2 \times 5$$
$$\therefore \ \text{GCF} = 2 \times 3^2 = 18.$$

Least Common Multiple (LCM): The smallest number that is a multiple of two or more given numbers.

Multiples of 12: 0, 12, 24, **36**, 48, 60, **72**, ···· ; Multiples of 18: 0,18, **36**, 54, **72**, ····

The common multiples of 12 and 18 are **36**, **72**, ····. The LCM of 12 and 18 is **36**.

To find the LCM of two or more given numbers, we write the prime factorizations of the given numbers, and find the **highest power** of each factor that occurs in **any** of the factorizations. The product of these highest powers is the LCM.

Example 3: Find the LCM of 12 and 18.
Solution:
$$12 = 2^2 \times 3 \ ; \ 18 = 2 \times 3^2$$
$$\therefore \ \text{LCM} = 2^2 \times 3^2 = 36.$$

Example 4: Find the LCM of 54 and 90.
Solution:
$$54 = 2 \times 3^3 \ ; \ 90 = 2 \times 3^2 \times 5$$
$$\therefore \ \text{LCM} = 2 \times 3^3 \times 5 = 270.$$

Example 5: Find the GCF and the LCM of $24x^2y^2$ and $84x^3y$.
Solution:
$$24x^2y^2 = 2^3 \cdot 3 \cdot x^2 \cdot y^2 \ ; \ 84x^3y = 2^2 \cdot 3 \cdot 7 \cdot x^3 \cdot y$$
$$\therefore \ \text{GCF} = 2^2 \cdot 3 \cdot x^2 \cdot y = 12x^2y. \quad \text{LCM} = 2^3 \cdot 3 \cdot 7 \cdot x^3 \cdot y^2 = 168x^3y^2.$$

The above examples show the following relationship between GCF and LCM.

Relationship between GCF and LCM

Let (a, b) represents the GCF, $[a, b]$ represents the LCM of two numbers a and b.

The product of two numbers *a* and *b* is equal to the product of their GCF and LCM.

Formula: $a \times b = (a, b) \times [a, b]$

Example 6. If the product of two numbers is 216 and their GCF is 6, what is their LCM?
Solution: $216 = 6 \times \text{LCM}$ $\therefore \text{LCM} = 36$. Ans.

Example 7. If the product of two numbers is 4860 and their LCM is 270,What is their GCF?
Solution: $4860 = 270 \times \text{GCF}$ $\therefore \text{GCF} = 18$. Ans.

-----Continued-----

The Shortcut method

1. To find the GCF of two or more given numbers, we choose a prime number (from small to large) that can divide into **all** of the given numbers. Continue to choose the divisors until no divisor can be used. The product of all the divisors is the GCF.
   Or, find the largest number that can divide into all of the given numbers.

2. To find the LCM of two or more given numbers, we choose a prime number (from small to large) that can divide into **any two or more** of the given numbers. Write the quotients and the **remaining** numbers on the next row. Continue to choose the divisors until no divisor can be used. The product of all the divisors and the numbers on the bottom row is the LCM.

3. To find GCF and LCM of two very large numbers, we use the **Rollabout Divisions** to find the largest number that can divide into all of the given numbers (see page 345).

Example 8: Find the GCF and the LCM of 54 and 90.

Solution:

2	54	90
3	27	45
3	9	15
	3	5

Or:

18	54	90
	3	5

$\therefore$ GCF $= 2\times3\times3=18$

LCM $= 2\times3\times3\times3\times5=270$.

Example 9: Find the GCF and the LCM of 12, 30 and 45.

Solution:

3	12	30	45	
2	4	10	15	
5	2	5	**15**	← 15 is the remaining number.
	2	1	3	← 2 is the remaining number.

$\therefore$ GCF $=3$. (3 is the only common divisor.)

LCM $= 3\times2\times5\times2\times1\times3=180$.

Example 10: Find the GCF and the LCM of 15, 21 and 35.

Solution:

3	15	21	35	← no common divisor except 1.
5	5	7	**35**	← 35 is the remaining number.
7	1	**7**	7	← 7 is the remaining number.
	1	1	1	← 1 is the remaining number.

$\therefore$ GCF $=1$. (There is no common divisor except 1.)

LCM $= 3\times5\times7\times1\times1\times1=105$.

Example 11: If a and b are integers and $a<b$, GCF = 18, LCM = 270, find a and b.

Solution: (See Example 8)

$270\div18=15$, 15 is the product of the numbers on the bottom row.

$15=1\times15$ or $15=3\times5$

$a=1\times18=18$ and $b=15\times18=270$. Ans.

Or: $a=3\times18=54$ and $b=5\times18=90$. Ans.

EXERCISES

Find the GCF for each set of numbers or algebraic expressions.

1. 1, 2 **2.** 2, 3 **3.** 2, 4 **4.** 2, 5 **5.** 4, 8 **6.** 3, 9
7. 5, 9 **8.** 6, 9 **9.** 3, 10 **10.** 12, 20 **11.** 15, 17 **12.** 20, 25
13. 24, 30 **14.** 16, 32 **15.** 28, 32 **16.** 15, 50 **17.** 24, 40 **18.** 19, 57
19. 42, 56 **20.** 69, 93 **21.** 30, 99 **22.** 70, 140 **23.** 72, 144 **24.** 61, 244
25. 1, 2, 3 **26.** 2, 3, 4 **27.** 4, 6, 8 **28.** 6, 8, 12 **29.** 12, 18, 30 **30.** 24, 56, 64
31. $2, 6x$ **32.** $2x, 6x$ **33.** $2x, 6x^2$ **34.** $6ab, 10b$ **35.** $4a^2b, 20ab$ **36.** $7y^3, 28y$
37. $5cd^2, 9cd$ **38.** $12uv^3, 3v^2$ **39.** $21mn^2, 35m^2n$ **40.** $15r^3s, 45r^2s^2$

Find the LCM for each set of numbers or algebraic expressions.

41. 1, 3 **42.** 2, 6 **43.** 3, 5 **44.** 4, 6 **45.** 4, 9 **46.** 6, 10
47. 3, 20 **48.** 6, 18 **49.** 7, 12 **50.** 5, 14 **51.** 5, 20 **52.** 7, 28
53. 12, 20 **54.** 16, 18 **55.** 12, 21 **56.** 15, 60 **57.** 19, 57 **58.** 30, 99
59. 1, 2, 3 **60.** 4, 5, 6 **61.** 6, 8, 12 **62.** 3, 10, 15 **63.** 12, 18, 30 **64.** 7, 14, 42
65. $2, 6x$ **66.** $2, 7x$ **67.** $4x, 6x^2$ **68.** $5a^2b, 7ab$ **69.** $12uv, 36u^2$ **70.** $13x^2y, 9y^3$

71. Find the GCF and the LCM of $42x^3y^2$ and $180xy^3$.
72. Find the GCF and the LCM of $3ab^2$, $12a^3b$ and $14ab^2$.
73. Find the GCF and the LCM of $2\times3^3\times5^2$ and $2^2\times3^2\times5^2\times7$.
74. Find the GCF and the LCM of 130 and 221.
75. Find the GCF and the LCM of 4,389 and 5,320.
76. Find the GCF and the LCM of 126, 300 and 330.
77. There are two numbers. Their GCF is 5. Their LCM is 150. The first number is 25. What is the second number ?
78. There are three classes with student numbers 28, 35 and 42. The teacher wants to divide each class into groups with the same number of students in each group without having any student left over. What is the greatest number of students each group has ?
79. Two runners race in a high school circular track. The first runner can circle the track in 50 seconds, and the second runner can circle the track in 60 seconds. From the start point of the race, how many seconds will it be before the first runner pass the second runner ?
80. Two satellites make one complete orbit around the earth every 60 and 90 minutes, respectively. If both satellites are directly over Hawthorne City at 11:00 AM, what time will they pass over Hawthorne City again ?
81. Three pieces of wood with length 72 inches, 60 inches and 42 inches. John wants to cut each wood into pieces with the same length without having any wood left over. What is the greatest length of each piece he can cut ? How many pieces does he has ?
82. A 1890–feet bridge has a light every 30 feet, including two end sides. In order to save the electricity, the lights will be moved every 42 feet. How many lights will not be moved ?
83. If the product of two numbers is 200 and their GCF is 5, what is their LCM ?
84. If the product of two numbers is equal to their GCF, what is their LCM ?
85. If a and b are integers and $a < b$, GCF = 5, LCM = 105, find a and b.

3-5 Sequences and Number Patterns

Patterns appear in nature and our daily life. Studying sequences and number patterns helps us to identify, understand patterns and make predictions.

Suppose a new car is worth $30,000. The value of the car decreases $2,000 for each year. The following numbers show how the value of the car decreases for the next five years.

30000, 28000, 26000, 24000, 22000

The pattern of numbers shown above is a **sequence**. Each number in the sequence is 2,000 less than the preceding. We can predict its value of 15 or 20 years.

Suppose the chart below shows the seat numbers of economy class in an airplane. There are 42 seats in the economy class. The chart shows six sequences. Each sequence has a pattern with a common difference (d = 6) between any two successive numbers.

1	3	5		6	4	2
7	9	11		12	10	8
13	15	17		18	16	14
.	.	.		.	.	.
.	.	.		.	.	.

1, 7, 13, 19, 25, 31, 37.
3, 9, 15, 21, 27, 33, 39.
5, 11, 17, 23, 29, 35, 41.
6, 12, 18, 24, 30, 36, 42.
4, 10, 16, 22, 28, 34, 40.
2, 8, 14, 20, 26, 32, 38.

Sequence: A consecutive nature numbers in a certain order. Each number is called a term of the sequence.

Arithmetic Sequence: A sequence which has a common difference between any two successive terms.

1, 2, 3, 4, 5, 6, 7, ·····. Common difference: d = 1
2, 4, 6, 8, 9, 10, 11, ·····. Common difference: d = 2
62, 51, 40, 29, 18, 7, ·····. Common difference: d = −11

Fibonacci Sequence: Each number in the sequence is the sum of the two preceding numbers.

1, 1, 2, 3, 5, 8, 13, ·····.
2, 3, 5, 8, 13, 21,34, ·····.
3, 4, 7, 11, 18, 29, 47, ·····.

Triangular Sequence: Each number in the sequence can be arranged to form a triangle.

1 , 3 , 6 , 10 , 15 , 21, ·····.
2 3 4 5 6
(differences)

1 3 6 10

----Continued----

Rectangular Sequence: Each number in the sequence can be arranged to form a rectangle.

2 , 6 , 12 , 20 , 30, ·····.
1×2   2×3   3×4   4×5   5×6

2   6   12   20

Studying and discovering the pattern of a number sequence, we make charts, tables or diagrams to organize the information into a pattern to show how the numbers relate to each other. Then, we continue the pattern to extend the numbers in the sequence.
To find the pattern of a sequence, we find the **differences** between any two successive terms.

Example 1: Find the next five terms of the sequence 2, 4, 6, 8, 10, ·····.
Solution: The terms increase by 2. The differences are constant.
The next five terms are 12, 14, 16, 18, and 20.

Example 2: Find the next five terms of the sequence 106, 95, 84, 73, 62, ····.
Solution: The terms decrease by 11. The differences are constant.
The next five terms are 51, 40, 29, 18, and 7.

Example 3: Find the next five terms of the sequence 1, 3, 6, 10, 15, ·····.
Solution: The differences are not constant.
The next five terms are continued by adding 6, 7, 8, 9, and 10.
The next five terms are 21, 28, 36, 45, and 55.

1 , 3 , 6 , 10 , 15 , **21** , **28** , **36** , **45** , **55**
2   3   4   5   **6**   **7**   **8**   **9**   **10**   (differences)

Example 4: Find the next five terms of the sequence 2, 6, 12, 20, 30, ·····.
Solution: The differences are not constant.
The next five terms are continued by adding 12, 14, 16, 18, and 20.
The next five terms are 42, 56, 72, 90, and 110.

2 , 6 , 12 , 20 , 30 , **42** , **56** , **72** , **90** , **110**
4   6   8   10   **12**   **14**   **16**   **18**   **20**   (differences)

Example 5: Each of the five teams play each of the other teams just once. How many games will there be ? How many games will there be for 8 teams ?
Solution:

Number of teams	1	2	3	4	5	6	7	8
Number of games	0	1	3	6	10	15	21	28

There will be 10 games for 5 teams.
There will be 28 games for 8 teams.

EXERCISES

Find the next five terms of each arithmetic sequence.

1. 4, 7, 10, 13, 16, ·····

2. 5, 8, 11, 14, 17·····

3. 2, 6, 10, 14, 18, ·····

4. 14, 22, 30, 38, 46, ·····

5. 21, 19, 17, 15, 13, ·····

6. 29, 26, 23, 20, 17, ·····

7. 1, 5, 9, 13, 17, ·····

8. 56, 52, 48, 44, 40, ·····

9. 3, 8, 13, 18, 23, ·····

10. 16, 23, 30, 37, 44, ·····

11. 1.5, 2.0, 2.5, 3.0, 3.5, ·····

12. 1.5, 3, 4.5, 6, 7.5, ·····

13. 18, 15, 12, 9, 6, ·····

14. –3, –7, –11, –15, –19, ·····

15. –21, –18, –15, –12, –9, ·····

Find the next five terms of each sequence.

16. 2, 4, 7, 11, 16, ·····

17. 1, 2, 5, 10, 17, ·····

18. 0, 2, 6, 12, 20, ·····

19. 2, 3, 6, 11, 18, ·····

20. 1, 1, 2, 3, 5, 8, ·····

21. 1, 5, 6, 11, 17, ·····

22. 1, 2, 4, 8, 16, ·····

23. 3, 4, 7, 11, 18, ·····

24. 2, 3, 5, 8, 13, ·····

25. 1, 4, 9, 16, 25, ·····

26. 9, 10, 13, 18, 25, ·····

27. 0, 0.5, 2, 4.5, 8, ·····

28. 4, 5.5, 8, 11.5, 16, ·····

29. –9, –8, –6, –3, 1, ·····

30. –9, –10, –12, –15, –19, ·····

State whether each sequence is an arithmetic sequence.

31. 6, 9, 12, 15, 18, ·····

32. 7, 9, 12, 16, 21, ·····

33. Is the sequence $1, \frac{1}{2}, \frac{1}{3}, \frac{1}{4}, \frac{1}{5}$ an arithmetic sequence ?

34. John plans to jog for 10 minutes the first day and increases 5 minutes every day. At which day will John be able to jog for 1 hour ?

35. The bus schedule at a station is at these times: 8:00, 8:05, 8:15, 8:30. If this pattern continues, what is the time of the next bus ?

36. How many games have to be played to determine a final winner from 16 teams in a single-elimination competition ?

37. There are 9 students in the party. Each student shakes hands exactly once with each of the others. How many handshakes take place ?

38. What is the greatest number of pieces we can get from a round pizza with 1, 2, 3, 4 or 5 cuts ?

39. If the sequence a, b, c and d is an arithmetic sequence, is the sequence a^2, b^2, c^2 and d^2 an arithmetic sequence ?

40. If the sequence a, b, c and d is an arithmetic sequence, is the sequence $2a$, $2b$, $2c$ and $2d$ an arithmetic sequence ?

CHAPTER 3 EXERCISES

Write all the whole–number factors of each number.

1. 12 **2.** 14 **3.** 17 **4.** 19 **5.** 25 **6.** 35 **7.** 49 **8.** 81 **9.** 93 **10.** 124

Write the first five multiples of each number.

11. 1 **12.** 4 **13.** 12 **14.** 16 **15.** 20 **16.** 24 **17.** 31 **18.** 42 **19.** 60 **20.** 91

Test whether each number is divisible by 2, 3, 4, or 5.

21. 44 **22.** 45 **23.** 48 **24.** 69 **25.** 75 **26.** 80 **27.** 84 **28.** 92 **29.** 150 **30.** 225

Test whether each number is divisible by 6, 8, 9, 10, or 11.

31. 48 **32.** 54 **33.** 72 **34.** 88 **35.** 90 **36.** 96 **37.** 110 **38.** 210 **39.** 720 **40.** 2,024

Identify each number as prime or composite.

41. 0 **42.** 1 **43.** 2 **44.** 3 **45.** 11 **46.** 14 **47.** 33 **48.** 37 **49.** 135 **50.** 671

Find all the prime factors of each number.

51. 12 **52.** 14 **53.** 17 **54.** 19 **55.** 25 **56.** 35 **57.** 49 **58.** 81 **59.** 93 **60.** 124

Write the prime factorization of each number using exponents.

61. 10 **62.** 14 **63.** 24 **64.** 44 **65.** 45 **66.** 72 **67.** 92 **68.** 121 **69.** 144 **70.** 225

Find the GCF and the LCM.

71. 1, 5 **72.** 2, 8 **73.** 4, 10 **74.** 5, 15 **75.** 6, 12
76. 3, 18 **77.** 4, 19 **78.** 8, 24 **79.** 10, 20 **80.** 15, 19
81. 20, 35 **82.** 17, 51 **83.** 30, 66 **84.** 42, 54 **85.** 64, 75
86. 65, 91 **87.** 60, 120 **88.** 2, 4, 6 **89.** 5, 8, 12 **90.** 2, 6, 24
91. 6, 15, 60 **92.** 20, 30, 45 **93.** 13, 18, 52 **94.** 20, 40, 75 **95.** 16, 18, 80
96. $3,\ 4x$ **97.** $4x,\ 6x^2$ **98.** $2a,\ 8ab$ **99.** $5y,\ 10y^3$ **100.** $6uv^2,\ 8u^3v$
101. $2^3\times3^2,\ 2\times3^2$ **102.** $3^2\times2^4,\ 2^2\times3$ **103.** $4\times5^2,\ 5\times4^2$
104. $2\times5^2\times9^2,\ 3^2\times5^3\times9$ **105.** $2^3\times3^2\times4,\ 2\times3^3\times4^2\times5$

-----Continued-----

In a biology class, a student said to his teacher.
Student: I just killed five mosquitoes, three were males and two were females.
Teacher: How did you know which were males and which were females ?
Student: Three were killed on a beer bottle. Two were killed on a mirror.

Find the next five terms of each sequence.

106. 5, 7, 9, 11, 13, ...

107. 3, 8, 13, 18, 23, ...

108. 2, 9, 16, 23, 30, ...

109. 25, 23, 21, 19, 17, ...

110. 35, 32, 29, 26, 23, ...

111. 0.5, 3, 5.5, 8, 10.5, ...

112. 14, 12.5, 11, 9.5, 8, ...

113. 15, 14.5, 14, 13.5, 13, ...

114. 5, 5.5, 6, 6.5, 7, 7.5, ...

115. −14, −12, −10, −8, −6,

116. 5, 3, 1, −1, −3, ...

117. 1, 2, 4, 7, 11, ...

118. 0, 3, 9, 18, 30, 45, ...

119. 0, 1, 4, 9, 16, ...

120. 4, 5, 9, 14, 23, ...

121. 2, 2, 4, 6, 10, ...

122. 3, 5, 9, 15, 23, ...

123. −11, −10, −7, −2, 5, ...

124. 3, 7, 12, 16, 21, 25, ...

125. 1, 2.5, 5, 8.5, 13, ...

126. Zero is a factor of what number ?

127. Zero is a multiple of what number ?

128. What number is a factor of every whole number ?

129. What number is a multiple of every nonzero integer ?

130. Find the missing digit of the number | 9 | 3 | | if we know that it has 6 and 8 as factors ?

131. Write all the possible number of students to group 120 students equally, including 1 each and 100 each.

132. What is the smallest 4-digit number that is divisible by 8 ?

133. What is the greatest 5-digit number that is divisible by 11 ?

134. There are 354 students. How many additional students are needed to completely fill the groups that hold 12 students each ?

135. Find the prime factorization of 4,860.

136. Find the prime factorization of 5,544.

137. Find the GCF and the LCM of 140 and 220.

138. Find the GCF and the LCM of 70, 140 and 220.

139. There are two numbers. Their GCF is 7. Their LCM is 315. The first number is 35. What is the second number ?

140. A 2200-meter bridge has a light every 25 meters, including two end sides. In order to save electricity, the lights will be moved every 40 meters. How many lights will not be moved ?

141. To divide three classes with students 24, 30 and 36 into groups with the same number of students in each group without having any student left over. What is the greatest number of students each group has ?

142. How many games have to be played to determine a final winner from 32 teams in a single-elimination competition ?

143. There are 11 students in the party. Each student shakes hands exactly once with each of the others. How many handshakes take place ?

144. What is the greatest number of pieces we can get from a round pizza with 6 cuts ?

145. If the sequence a, b, c and d is an arithmetic sequence, is the sequence $2a+5$, $2b+5$, $2c+5$ and $2d+5$ an arithmetic sequence ?

Additional Examples

1. What is the smallest number which is divided by 3, divided by 5, or divided by 8 ? All divisions have the same remainder of 7.
Solution: $3 \times 5 \times 8 + 7 = 127$. Ans. (Hint: $3 \times 5 \times 8 = 120$ is the LCM.)

2. Roger spent all \$9 to buy tennis balls and baseballs. If tennis balls are 60 cents each and baseballs are 90 cents each. What is the greatest possible number of tennis balls that Roger could have bought ? How many baseballs he bought ?
Solution:

LCM of 60 cents and 90 cents is 540 cents.
$\frac{540}{60} = 9$ tennis balls (greatest possible number), $\frac{900-540}{90} = 4$ baseballs. Ans.

3. What is the last digit of 3^{71} ?
Solution:
$3^1 = 3$, $3^2 = 9$, $3^3 = 27$, $3^4 = 81$
$3^5 = 243$, $3^6 = 729$, $3^7 = 2187$, $3^8 = 6561$
3^{69} 3^{70} 3^{71}
The last digit repeats by " 3, 9, 7, 1 ".
Therefore, the last digit of 3^{71} is 7. Ans.
Hint: The exponent of 1st number of last row is $1 + 4n$, where $n = 17$.

4. What is the remainder of $(4^{110} \div 7)$?
Solution: The remainder of
$4^2 \div 7$ is 2, $4^3 \div 7$ is 1, $4^4 \div 7$ is 4
$4^5 \div 7$ is 2, $4^6 \div 7$ is 1, $4^7 \div 7$ is 4
$4^{110} \div 7$ $4^{111} \div 7$ $4^{112} \div 7$
The remainder repeats by "2, 1, 4".
Therefore, the remainder of $4^{110} \div 7$ is 2. Ans.
Hint: The exponent of 1st number of last row is $2 + 3n$, where $n = 36$.

5. The unit price and the demand (number of products sold) of a product are shown in the following table. If the pattern continues, what is the unit price of the product which makes the highest sale ?

Unit price	Demand (Number of product sold)	Sales
\$1	19	\$19
2	17	34
3	15	45

Solution:

\$4	13	\$52
5	11	**55**
6	9	54

The unit price of the product to make the highest sales is \$5. Ans.

6. There are 7 students in a game. Each student plays exactly one game with each of the others. How many games are played ?
Solution:

Number of students:	1	2	3	4	5	6	7
Number of games:	0	1	3	6	10	15	**21**

21 games are played. Ans.

7. The annual sales of A-plus Company are shown below. If the pattern shown continues, what will the company's sales be in the 6th year ?

Year	1	2	3	4	5
Sales	\$8,000	\$22,000	\$36,000	\$50,000	\$64,000

Solution:

The pattern shown continues to increase \$14,000 annually.
Therefore, the sales in the 6th year will be \$78,000.

8. Prove the following statement

"A whole number is divisible by 9 if the sum of its digits is divisible by 9."

Proof:

1. A **two-digit** number with digits a and b has the value of $10a + b = 9a + (a + b)$.
   $9a$ is divisible by 9. Therefore, a two-digit number is divisible by 9 if $(a + b)$ is divisible by 9.
2. A **three-digit** number with digits a, b, and c has the value of
   $$100a + 10b + c = 99a + 9b + (a + b + c)$$
   $99a + 9b$ is divisible by 9. Therefore, a three-digit number is divisible by 9 if $(a + b + c)$ is divisible by 9.

Similarly, the statement is true for four, five, six, ······-digit number.

9. Prove the following statement

"A whole number is divisible by 11 if the difference between the sum of the odd digits and the sum of the even digits is 0 or 11."

Proof:

1. A **two-digit** number with digits a and b has the value of $10a + b = 11a - (a - b)$.
   $11a$ is divisible by 11. Therefore, a two-digit number is divisible by 11 if $(a - b) = 0$.
2. A **three-digit** number with digits a, b, and c has the value of
   $$100a + 10b + c = 99a + 11b + (a + c - b)$$
   $99a + 11b$ is divisible by 11. Therefore, a three-digit number is divisible by 11 if $(a + c - b)$ is 0 or 11.
3. A **four-digit** number with digits a, b, c, and d has the value of
   $$1000a + 100b + 10c + d = 1001a + 99b + 11c - [(a + c) - (b + d)]$$
   $1001a + 99b + 11c$ is divisible by 11. Therefore, a four-digit number is divisible by 11 if $[(a + c) - (b + d)]$ is 0 or 11.

Similarly, the statement is true for five, six, seven,·····-digit number.

10. Find the twinprimes of less than 100.

Solution: (5, 7), (11, 13), (17, 19), (29, 31), (41, 43), (59, 61), (71, 73). Ans.

Hint: **Twinprimes**: Two numbers which are both prime numbers and only two units apart.

CHAPTER FOUR

Fractions

4-1 Introduction to Fractions

Fractions are commonly used in our daily life.

1. A fraction is used to represent each part of a whole number that is divided into smaller and equal parts.
   Suppose we divide the number 1 into 5 equal parts. Each part is represented by the fraction $\frac{1}{5}$. Two parts are represented by the fraction $\frac{2}{5}$, and so on.
   If we divide \$1 into 4 equal parts. Each part is $\frac{1}{4}$ of \$1, or a quarter (25 cents).
2. A fraction is also used to represent a division of two whole numbers.
   2 and 5 are whole numbers. 2 divided by 5 is written as $\frac{2}{5}$ or $2/5$.

In the fraction (division) $\frac{2}{5}$, 2 is called the **numerator**, 5 is called the **denominator**, and the horizontal bar is called the **fraction bar**.
In a fraction, the denominator cannot be 0. We call a fraction with zero denominator as "**undefined**". $\frac{2}{0}$ is undefined (it means that "division by 0 is not allowed.").

Equivalent fractions (equal fractions): Fractions that have the same value.

If we divide the number 1 into 5 equal parts, 2 parts are represented by $\frac{2}{5}$.

If we divide the number 1 into 10 equal parts, 4 parts are represented by $\frac{4}{10}$.

$\frac{2}{5}$ and $\frac{4}{10}$ are equivalent because they represent the same part of the number 1.

To check if two fractions are equivalent (equal), we may use the following ways:

1. Two fractions are equal if their **cross products** are equal.
   $\frac{2}{5} = \frac{4}{10}$ because $2 \times 10 = 5 \times 4$.
   $\frac{2}{5} \neq \frac{3}{10}$ because $2 \times 10 \neq 5 \times 3$.
2. **Multiply** or **divide** both the numerator and the denominator by the same number.
   $\frac{2}{5} = \frac{2 \times 2}{5 \times 2} = \frac{4}{10}$, or $\frac{4}{10} = \frac{4 \div 2}{10 \div 2} = \frac{2}{5}$

Proper fraction: The numerator is less than the denominator. Such as $\frac{1}{2}, \frac{2}{3}, \frac{4}{5}, \ldots$

Improper fraction: The numerator is greater than or equal to the denominator.
Such as $\frac{2}{1}, \frac{3}{2}, \frac{5}{4}, \frac{1}{1}, \frac{2}{2}, \frac{5}{1}, \ldots$

Mixed numbers: A whole number plus a fraction. Such as $1\frac{1}{2}, 1\frac{1}{4}, 3\frac{2}{5}, \ldots$

-----Continued-----

A whole number can be rewritten as a improper fraction with a denominator of 1.

$1=\frac{1}{1}$, $2=\frac{2}{1}$, $3=\frac{3}{1}$, and so on.

To change an improper fraction as an equal mixed number, we divide the numerator by the denominator.

$\frac{2}{1}=2\div 1=2$, $\frac{5}{2}=5\div 2=2\frac{1}{2}$, $\frac{14}{3}=14\div 3=4\frac{2}{3}$

To change a mixed number as an equal improper fraction, we multiply the whole number by the denominator and then add the numerator. The result is the numerator of improper fraction. The denominator remains the same.

$4\frac{2}{3}=\frac{4\times 3+2}{3}=\frac{14}{3}$

A proper fraction is in **lowest terms** when the numerator and the denominator have no common factor other than 1.

To simplify a proper fraction in lowest terms, we divide both the numerator and the denominator by the same nonzero number (the greatest common factor " GCF ").

$\frac{2}{4}=\frac{2\div 2}{4\div 2}=\frac{1}{2}$, $\frac{12}{20}=\frac{12\div 4}{20\div 4}=\frac{3}{5}$

A mixed number is in **simplest form** when its fractional part is in lowest terms.
When we write a fraction for an answer, we always write it in **simplest form.**
Conventionally, the **simplest form** of a fraction can be:

1. a proper fraction in lowest terms.
2. a mixed number when its fractional part is in lowest terms.

The simplest form of $5\frac{4}{8}$ is $5\frac{1}{2}$.

The simplest form of $5\frac{9}{8}$ is $6\frac{1}{8}$.

Read and write the word name of a fraction:

$\frac{2}{1}$ as "two-wholes", $\frac{1}{2}$ as "one-half", $\frac{3}{2}$ as "three-halves"
$\frac{1}{3}$ as "one-third' , $\frac{2}{3}$ as " two-thirds", $\frac{3}{4}$ as " three-fourths"
$\frac{1}{5}$ as "one-fifth" , $\frac{5}{6}$ as "five-sixths" , $\frac{1}{7}$ as "one-seventh"
$\frac{3}{8}$ as "three-eighths" , $\frac{4}{9}$ as "four-ninths" , $\frac{7}{10}$ as "seven-tenths"
$2\frac{1}{9}$ as "two and one-ninth" , $3\frac{2}{5}$ as "three and two-fifths"

To compare fractions, we express each fraction with the **least common denominator (LCD).** LCD is the least common multiple (LCM) of the denominators (see Section 3-4).

Example : Compare $\frac{2}{3}$ and $\frac{3}{5}$, using the symbols > , <, or =.

Solution: The cross product $2\times 5\neq 3\times 3$. They are not equal.
The LCD is 15.

$\frac{2}{3}=\frac{2\times 5}{3\times 5}=\frac{10}{15}$; $\frac{3}{5}=\frac{3\times 3}{5\times 3}=\frac{9}{15}$

$\frac{10}{15}>\frac{9}{15}$. Therefore, $\frac{2}{3}>\frac{3}{5}$. Ans.

Unit Fraction: It is a fraction that has a numerator of 1. $\frac{1}{2}$, $\frac{1}{3}$, $\frac{1}{4}$, ... are unit fractions.

Example: Express $\frac{3}{4}$ as the sum of two unit fractions.

Solution: $\frac{3}{4}=\frac{1}{4}+\frac{2}{4}=\frac{1}{4}+\frac{1}{2}$. Ans.

EXERCISES

Classify each of the following fractions as proper, improper, or undefined.

1. $\frac{1}{4}$ 2. $-\frac{3}{4}$ 3. $\frac{7}{4}$ 4. $\frac{2}{3}$ 5. $-\frac{5}{1}$ 6. $\frac{5}{0}$ 7. $\frac{8}{7}$ 8. $\frac{0}{8}$ 9. $-\frac{1}{50}$ 10. $\frac{9}{9}$

Tell whether the fractions are equal or not .

11. $\frac{1}{2}$ and $\frac{5}{10}$ 12. $\frac{2}{3}$ and $\frac{8}{12}$ 13. $\frac{3}{4}$ and $\frac{2}{5}$ 14. $\frac{0}{3}$ and $\frac{0}{4}$ 15. $\frac{5}{8}$ and $\frac{3}{4}$
16. $\frac{3}{5}$ and $\frac{9}{15}$ 17. $\frac{3}{7}$ and $\frac{4}{8}$ 18. $\frac{3}{4}$ and $\frac{9}{12}$ 19. $\frac{2}{9}$ and $\frac{10}{54}$ 20. $\frac{5}{8}$ and $\frac{15}{24}$

Simplify each fraction in simplest form.

21. $\frac{5}{10}$ 22. $\frac{4}{8}$ 23. $-\frac{4}{10}$ 24. $\frac{3}{21}$ 25. $\frac{3}{24}$ 26. $\frac{12}{24}$ 27. $5\frac{18}{24}$ 28. $-4\frac{12}{18}$
29. $5\frac{24}{18}$ 30. $-4\frac{18}{12}$

Write each improper fraction as simplest form.

31. $\frac{4}{3}$ 32. $\frac{5}{1}$ 33. $\frac{8}{3}$ 34. $\frac{8}{8}$ 35. $-\frac{10}{3}$ 36. $\frac{15}{2}$ 37. $\frac{16}{4}$ 38. $-\frac{20}{11}$ 39. $\frac{43}{9}$ 40. $\frac{56}{22}$

Write each mixed number as improper fraction.

41. $2\frac{2}{3}$ 42. $3\frac{1}{3}$ 43. $-4\frac{2}{5}$ 44. $7\frac{3}{4}$ 45. $5\frac{7}{3}$ 46. $-10\frac{2}{3}$ 47. $9\frac{3}{4}$ 48. $6\frac{5}{8}$
49. $8\frac{2}{3}$ 50. $-1\frac{48}{49}$

Compare the fractions, using the symbols >, <, or =.

51. $\frac{1}{2}$ and $\frac{4}{8}$ 52. $\frac{1}{2}$ and $\frac{3}{8}$ 53. $\frac{2}{3}$ and $\frac{4}{7}$ 54. $\frac{3}{5}$ and $\frac{4}{7}$ 55. $\frac{1}{3}$ and $\frac{1}{6}$
56. $\frac{1}{2}$ and $\frac{6}{10}$ 57. $\frac{4}{5}$ and $\frac{16}{20}$ 58. $\frac{3}{14}$ and $\frac{5}{28}$ 59. $\frac{20}{30}$ and $\frac{2}{3}$ 60. $\frac{15}{45}$ and $\frac{7}{15}$

Write the word name for each fraction.

61. $\frac{1}{4}$ 62. $\frac{3}{4}$ 63. $\frac{4}{5}$ 64. $\frac{1}{8}$ 65. $\frac{7}{8}$ 66. $\frac{5}{2}$ 67. $\frac{7}{11}$ 68. $\frac{11}{7}$
69. $\frac{10}{12}$ 70. $\frac{24}{50}$ 71. $\frac{5}{1}$ 72. $1\frac{2}{3}$ 73. $4\frac{1}{5}$ 74. $6\frac{5}{9}$ 75. $10\frac{10}{13}$

Write the fractions in order from least to greatest.

76. $\frac{1}{2}, \frac{1}{3}, \frac{2}{3}, \frac{3}{5}, \frac{5}{6}$ 77. $\frac{3}{4}, \frac{11}{16}, \frac{3}{8}, \frac{1}{2}, \frac{5}{8}$ 78. $\frac{5}{6}, \frac{3}{4}, \frac{7}{12}, \frac{2}{3}, \frac{19}{24}$
79. $\frac{2}{5}, \frac{2}{3}, -\frac{7}{10}, \frac{8}{15}, -\frac{17}{30}$ 80. $-\frac{1}{2}, -\frac{1}{3}, \frac{5}{9}, \frac{5}{6}, \frac{13}{18}$

Write each expression in simplest form.

81. $\frac{6x}{3}$ 82. $\frac{12x^2}{4}$ 83. $\frac{18x^3y}{3x}$ 84. $\frac{16a^2b}{ab}$ 85. $\frac{12ab^3}{8ab}$ 86. $\frac{20c^3d^2}{15c^2d}$
87. $\frac{24m^2}{36mn}$ 88. $\frac{12yz^2}{18y^2z}$ 89. $\frac{21xy}{14y^2}$ 90. $\frac{32mn^3}{24m^2n}$

-----Continued-----

Rewrite each of the fractions with the least common denominator (LCD).

91. $\frac{1}{3}, \frac{1}{9}$ **92.** $\frac{1}{4}, \frac{1}{8}$ **93.** $\frac{2}{3}, \frac{1}{12}$ **94.** $\frac{3}{4}, \frac{5}{8}$ **95.** $\frac{2}{5}, \frac{3}{4}$

96. $\frac{3}{8}, \frac{5}{16}$ **97.** $\frac{2}{9}, \frac{3}{18}$ **98.** $\frac{5}{6}, \frac{4}{7}$ **99.** $-\frac{2}{7}, \frac{5}{14}$ **100.** $\frac{7}{12}, \frac{7}{18}$

101. $-\frac{9}{11}, \frac{1}{22}$ **102.** $\frac{4}{15}, \frac{13}{20}$ **103.** $\frac{11}{30}, -\frac{2}{3}$ **104.** $\frac{4}{9}, \frac{8}{11}$ **105.** $\frac{6}{25}, \frac{13}{75}$

106. $\frac{3}{12}, \frac{5}{11}$ **107.** $-\frac{7}{48}, \frac{11}{12}$ **108.** $\frac{13}{15}, \frac{3}{13}$ **109.** $\frac{9}{42}, \frac{7}{56}$ **110.** $\frac{10}{63}, \frac{3}{56}$

111. $\frac{1}{2}, \frac{1}{3}, \frac{1}{4}$ **112.** $-\frac{1}{4}, \frac{1}{5}, \frac{1}{8}$ **113.** $\frac{2}{3}, -\frac{3}{4}, \frac{4}{5}$

114. $\frac{5}{12}, \frac{3}{8}, \frac{1}{4}$ **115.** $-\frac{1}{15}, \frac{3}{18}, -\frac{19}{24}$

Evaluate

116. $\frac{3x}{8}$ if $x = 2$ **117.** $\frac{4x}{5}$ if $x = 3$ **118.** $\frac{2x-3}{15}$ if $x = 3$ **119.** $\frac{3x+2}{12}$ if $x = 4$

120. $\frac{5a+1}{28}$ if $a = 5$ **121.** $\frac{4-3a}{10}$ if $a = 2$ **122.** $\frac{6}{4-3x}$ if $x = 1$ **123.** $\frac{18}{5x+4}$ if $x = 2$

124. $\frac{3x+5}{4y-6}$ if $x = 1$, $y = 3$ **125.** $\frac{4a-3b}{2a}$ if $a = 3$, $b = 2$

Rewrite each of the fractional expressions with a common denominator.

126. $\frac{1}{x}, \frac{1}{y}$ **127.** $\frac{2}{a}, \frac{1}{b}$ **128.** $\frac{1}{2a}, \frac{2}{a}$ **129.** $\frac{2}{3x}, \frac{3}{2y}$ **130.** $\frac{1}{x}, \frac{2}{y}, \frac{3}{z}$

On graduation day, a high school graduate said to his math teacher.
Student: Thank you very much for giving a "D" in your class.
For your kindness, do you want me to do anything for you ?
Teacher: Yes. Please do me a favor.
Just don't tell anyone that I was your math teacher.

4-2 Fractions and Decimals

To change a fraction to a decimal, we divide the numerator by the denominator.

Examples: **1.** $\frac{1}{2} = 1 \div 2 = 0.5$ **2.** $\frac{1}{3} = 0.333\cdots = 0.\overline{3}$

3. $-\frac{1}{8} = -(1 \div 8) = -0.125$ **4.** $\frac{3}{4} = 3 \div 4 = 0.75$

5. $\frac{5}{12} = 5 \div 12 = 0.41666\cdots = 0.41\overline{6}$ **6.** $\frac{3}{11} = 3 \div 11 = 0.2727\cdots = 0.\overline{27}$

To change a decimal to a fraction, we rewrite the decimal as a fraction whose denominator is a power of 10 and then write the fraction in simplest form .

Examples: **7.** $0.5 = \frac{5}{10} = \frac{1}{2}$ **8.** $0.25 = \frac{25}{100} = \frac{1}{4}$

9. $-0.125 = -\frac{125}{1000} = -\frac{1}{8}$ **10.** $3.75 = 3\frac{75}{100} = 3\frac{3}{4}$

The following example shows how to change a repeating decimal to a fraction.

Example 11. Change $0.\overline{27}$ as a fraction in lowest terms.

Solution: Let $n = 0.\overline{27} = 0.2727\cdots$

$$100n = 27.\overline{27}$$
$$-)\quad n = 0.\overline{27}$$
$$99n = 27\ ,\quad n = \frac{27}{99} = \frac{3}{11} \qquad \therefore 0.\overline{27} = \frac{3}{11}$$

QUICK METHOD: The following examples show an easier way to change a repeating decimal to a fraction. For each repeating digit, we write the denominator with a number of 9, and 0 for each nonrepeating digit. Subtract the nonrepeating digits from the numerator.

Examples: (You may use a calculator to verify the answers.)

12. $0.\overline{1} = \frac{1}{9}$ **13.** $0.\overline{2} = \frac{2}{9}$ **14.** $0.\overline{3} = \frac{3}{9} = \frac{1}{3}$ **15.** $0.\overline{9} = \frac{9}{9} = 1$

16. $0.\overline{27} = \frac{27}{99} = \frac{3}{11}$ **17.** $0.\overline{123} = \frac{123}{999} = \frac{41}{333}$ **18.** $0.\overline{1234} = \frac{1234}{9999}$

19. $0.0\overline{1} = \frac{1}{90}$ **20.** $0.4\overline{6} = \frac{46-4}{90} = \frac{42}{90} = \frac{7}{15}$ **21.** $0.12\overline{34} = \frac{1234-12}{9900} = \frac{1222}{9900} = \frac{611}{4950}$

EXERCISES

Change each fraction to a decimal.

1. $\frac{1}{4}$ **2.** $\frac{1}{5}$ **3.** $\frac{1}{6}$ **4.** $-\frac{2}{5}$ **5.** $\frac{3}{8}$ **6.** $-\frac{2}{9}$

7. $\frac{5}{6}$ **8.** $\frac{4}{9}$ **9.** $-\frac{2}{3}$ **10.** $\frac{9}{10}$ **11.** $\frac{8}{9}$ **12.** $-\frac{5}{18}$

13. $\frac{4}{3}$ **14.** $\frac{5}{12}$ **15.** $\frac{8}{5}$ **16.** $\frac{5}{8}$ **17.** $\frac{7}{3}$ **18.** $\frac{17}{36}$

19. $\frac{17}{45}$ **20.** $\frac{1}{99}$ **21.** $-\frac{13}{99}$ **22.** $\frac{13}{90}$ **23.** $\frac{11}{99}$ **24.** $\frac{11}{90}$

25. $1\frac{1}{2}$ **26.** $-2\frac{1}{9}$ **27.** $4\frac{5}{6}$ **28.** $5\frac{7}{10}$ **29.** $5\frac{7}{9}$ **30.** $2\frac{5}{7}$

31. $1\frac{2}{3}$ **32.** $1\frac{1}{12}$ **33.** $3\frac{6}{25}$ **34.** $4\frac{1}{20}$ **35.** $7\frac{6}{125}$

Change each decimal to a fraction in simplest form.

36. 0.02 **37.** 0.25 **38.** –0.04 **39.** 0.005 **40.** 0.125 **41.** 0.625

42. –0.75 **43.** 0.45 **44.** –0.62 **45.** 0.225 **46.** 0.375 **47.** 0.05

48. 0.001 **49.** –0.36 **50.** 0.016 **51.** 0.875 **52.** 0.008 **53.** 2.35

54. –4.45 **55.** 10.01 **56.** –9.25 **57.** 1.004 **58.** 6.025 **59.** 5.125

60. 8.016

Change each repeating decimal to a fraction in simplest form.

61. $0.\overline{13}$ **62.** $0.\overline{12}$ **63.** $0.\overline{21}$ **64.** $0.\overline{234}$ **65.** $0.0\overline{12}$ **66.** $0.0\overline{21}$

67. $0.0\overline{6}$ **68.** $0.0\overline{06}$ **69.** $-0.00\overline{6}$ **70.** $0.00\overline{12}$ **71.** $0.\overline{10}$ **72.** $0.1\overline{2}$

73. $0.1\overline{6}$ **74.** $0.12\overline{3}$ **75.** $0.1\overline{23}$ **76.** $0.01\overline{2}$ **77.** $0.00\overline{3}$ **78.** $0.01\overline{3}$

79. $0.5\overline{42}$ **80.** $0.54\overline{2}$ **81.** $-0.12\overline{345}$ **82.** $0.0012\overline{3}$ **83.** $1.\overline{12}$ **84.** $1.4\overline{9}$

85. $2.1\overline{8}$ **86.** $-2.\overline{25}$ **87.** $3.3\overline{9}$ **88.** $4.0\overline{12}$ **89.** $5.24\overline{8}$ **90.** $1.08\overline{3}$

4-3 Adding and Subtracting Fractions

Rules for adding or subtracting fractions:

1. If the fractions have a common denominator, we combine (add or subtract) the numerators and write the result with the same denominator.
2. If the fractions have different denominators, we rewrite the fractions with a common denominator before adding and subtracting.
3. Convert mixed numbers to improper fractions before adding and subtracting. Or, use vertical form to add or subtract mixed numbers.

Examples: **1.** $\frac{3}{5}+\frac{1}{5}=\frac{3+1}{5}=\frac{4}{5}$

2. $\frac{3}{5}-\frac{1}{5}=\frac{3-1}{5}=\frac{2}{5}$

3. $\frac{3}{12}-\frac{7}{12}=\frac{3-7}{12}=\frac{-4}{12}=-\frac{1}{3}$

4. $\frac{1}{2}+\frac{1}{3}=\frac{3}{6}+\frac{2}{6}=\frac{3+2}{6}=\frac{5}{6}$

5. $\frac{3}{4}-\frac{11}{12}=\frac{9}{12}-\frac{11}{12}=\frac{9-11}{12}=\frac{-2}{12}=-\frac{1}{6}$

6. $2\frac{2}{3}+3\frac{1}{3}=\frac{8}{3}+\frac{10}{3}=\frac{18}{3}=6$ Or: $2\frac{2}{3}+3\frac{1}{3}=5\frac{2+1}{3}=5\frac{3}{3}=6$

7. $2\frac{2}{3}+4\frac{5}{6}=\frac{8}{3}+\frac{29}{6}=\frac{16}{6}+\frac{29}{6}=\frac{16+29}{6}=\frac{45}{6}=7\frac{1}{2}$

8. $6\frac{1}{4}-3\frac{3}{5}=\frac{25}{4}-\frac{18}{5}=\frac{125}{20}-\frac{72}{20}=\frac{125-72}{20}=\frac{53}{20}=2\frac{13}{20}$

(Vertical Form) 9.

$$
\begin{array}{rl}
 & 2\frac{2}{3} = 2\frac{4}{6} \\
+) & 4\frac{5}{6} = 4\frac{5}{6} \\
\hline
 & 6\frac{9}{6} = 7\frac{3}{6} = 7\frac{1}{2}. \text{ Ans.}
\end{array}
$$

10.

$$
\begin{array}{rl}
 & 6\frac{1}{4} = 6\frac{5}{20} = 5\frac{25}{20} \\
-) & 3\frac{3}{5} = 3\frac{12}{20} \\
\hline
 & 2\frac{13}{20}. \text{ Ans.}
\end{array}
$$

11. $\frac{1}{x}+\frac{2}{x}=\frac{1+2}{x}=\frac{3}{x}$

12. $\frac{6}{a}-\frac{9}{a}=\frac{6-9}{a}=\frac{-3}{a}=-\frac{3}{a}$

13. $\frac{4d}{7}-\frac{3d}{14}=\frac{8d}{14}-\frac{3d}{14}=\frac{8d-3d}{14}=\frac{5d}{14}$

----Continued-----

Examples

1. John is 5 ft tall. He grew $1\frac{1}{3}$ in. during the last year. How tall in inches was he one year ago ? (Hint: 1 foot = 12 inches)
   Solution:
   5 feet = 60 inches

   $60-1\frac{1}{3}=59\frac{3}{3}-1\frac{1}{3}=58\frac{2}{3}$ in. Ans.

2. The price per share of stock on A-Plus Company was $\$54\frac{3}{4}$. It rose by $\$2\frac{5}{8}$. What is its price per share of stock now ?
   Solution:
   $54\frac{3}{4}+2\frac{5}{8}=54\frac{6}{8}+2\frac{5}{8}=56\frac{11}{8}=\$57\frac{3}{8}$. Ans.

3. What is the perimeter ?

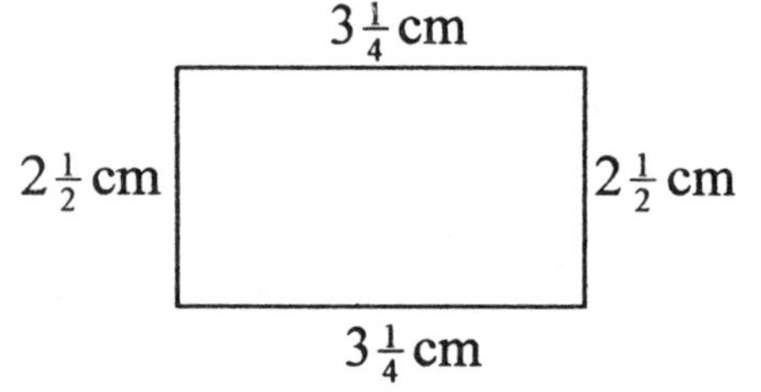

Solution:

$3\frac{1}{4}+3\frac{1}{4}+2\frac{1}{2}+2\frac{1}{2}$
$=3\frac{1}{4}+3\frac{1}{4}+2\frac{2}{4}+2\frac{2}{4}$
$=10\frac{6}{4}=11\frac{1}{2}$ cm. Ans.

4. What is the perimeter ?

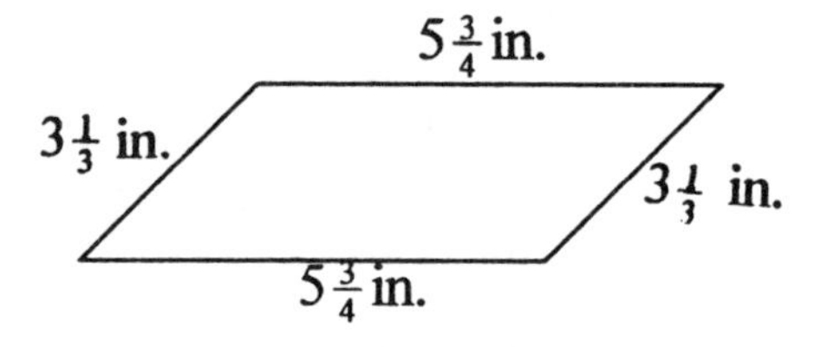

Solution:

$5\frac{3}{4}+5\frac{3}{4}+3\frac{1}{3}+3\frac{1}{3}$
$=5\frac{9}{12}+5\frac{9}{12}+3\frac{4}{12}+3\frac{4}{12}$
$=16\frac{26}{12}=18\frac{1}{6}$ in. Ans.

4. Evaluate $\frac{1}{7}-(\frac{1}{7}-\frac{1}{15})$
   Solution:
   $\frac{1}{7}-(\frac{1}{7}-\frac{1}{15})=\frac{1}{7}-\frac{1}{7}+\frac{1}{15}=\frac{1}{15}$.

5. Evaluate $\frac{2}{3}-15\frac{1}{4}-4\frac{1}{2}+3\frac{5}{6}$
   Solution:

$\frac{2}{3}-15\frac{1}{4}-4\frac{1}{2}+3\frac{5}{6}$
$=(-15-4+3)+(\frac{2}{3}-\frac{1}{4}-\frac{1}{2}+\frac{5}{6})$
$=(-16)+\frac{8-3-6+10}{12}$
$=-16+\frac{9}{12}$
$=-16+\frac{3}{4}$
$=\frac{-64+3}{4}=-\frac{61}{4}=-15\frac{1}{4}$.

OR: $\frac{2}{3}-15\frac{1}{4}-4\frac{1}{2}+3\frac{5}{6}$
$=\frac{8}{12}-15\frac{3}{12}-4\frac{6}{12}+3\frac{10}{12}$
$=(\frac{8}{12}+3\frac{10}{12})+(-15\frac{3}{12}-4\frac{6}{12})$
$=3\frac{18}{12}+(-19\frac{9}{12})$
$=3\frac{18}{12}+(-18\frac{21}{12})$
$=-15\frac{3}{12}=-15\frac{1}{4}$.

EXERCISES

Add or subtract. Write the answer in simplest form.

1. $\frac{1}{5}+\frac{2}{5}$
2. $\frac{4}{5}+\frac{3}{5}$
3. $\frac{4}{5}-\frac{1}{5}$
4. $-\frac{4}{5}+\frac{2}{5}$
5. $-\frac{1}{5}-\frac{3}{5}$
6. $\frac{2}{5}-\frac{4}{5}$
7. $\frac{5}{7}+\frac{6}{7}$
8. $\frac{3}{8}-\frac{5}{8}$
9. $\frac{3}{10}+\frac{2}{10}$
10. $-\frac{9}{16}-\frac{12}{16}$
11. $\frac{13}{7}+\frac{1}{7}$
12. $\frac{10}{3}-\frac{2}{3}$
13. $\frac{7}{11}-\frac{7}{11}$
14. $\frac{11}{12}+\frac{7}{12}$
15. $-\frac{7}{15}+\frac{2}{15}$
16. $\frac{7}{9}-(-\frac{5}{9})$
17. $-\frac{7}{16}+(-\frac{3}{16})$
18. $-\frac{15}{16}-(-\frac{1}{16})$
19. $\frac{7}{5}+(-\frac{12}{5})$
20. $\frac{11}{12}-(-\frac{13}{12})$
21. $3\frac{1}{3}+2\frac{2}{3}$
22. $7\frac{2}{5}+5\frac{4}{5}$
23. $7\frac{4}{5}-5\frac{2}{5}$
24. $8\frac{2}{7}-4\frac{1}{7}$
25. $8\frac{1}{7}-2\frac{3}{7}$
26. $4\frac{6}{7}+6\frac{5}{7}$
27. $-6\frac{1}{5}+2\frac{2}{5}$
28. $5\frac{2}{9}-2\frac{7}{9}$
29. $-\frac{2}{6}-4\frac{5}{6}$
30. $6\frac{3}{8}-2\frac{5}{8}$
31. $\frac{8}{x}-\frac{3}{x}$
32. $\frac{9}{a}-\frac{14}{a}$
33. $\frac{x}{x^2}-\frac{2y}{x^2}$
34. $\frac{2a}{c}+\frac{b}{c}$
35. $\frac{4}{2a}-\frac{6}{2a}$

Add or subtract. Write the answer in simplest form.

36. $\frac{1}{3}+\frac{1}{2}$
37. $\frac{2}{3}+\frac{1}{4}$
38. $\frac{3}{4}-\frac{2}{3}$
39. $\frac{5}{12}-\frac{3}{4}$
40. $\frac{7}{8}+\frac{1}{2}$
41. $-\frac{2}{5}-\frac{3}{4}$
42. $-\frac{5}{7}+\frac{2}{3}$
43. $\frac{4}{9}-\frac{4}{5}$
44. $\frac{5}{6}+\frac{7}{9}$
45. $\frac{3}{16}-\frac{1}{2}$
46. $2\frac{1}{3}+3\frac{1}{2}$
47. $5\frac{5}{6}+4\frac{2}{3}$
48. $6\frac{2}{7}-3\frac{1}{5}$
49. $7\frac{5}{8}-4\frac{2}{8}$
50. $8\frac{4}{5}-5\frac{3}{4}$
51. $12\frac{2}{7}-6\frac{1}{2}$
52. $9\frac{3}{5}-7\frac{3}{10}$
53. $4\frac{1}{4}-3\frac{1}{2}$
54. $10\frac{7}{10}-5\frac{2}{5}$
55. $5\frac{5}{8}-2\frac{3}{4}$
56. $10+\frac{7}{8}$
57. $10-\frac{7}{8}$
58. $12-4\frac{3}{5}$
59. $15-6\frac{5}{9}$
60. $15+6\frac{5}{9}$
61. $\frac{x}{4}+\frac{y}{5}$
62. $\frac{a}{3}-\frac{b}{2}$
63. $\frac{2a}{7}-\frac{b}{2}$
64. $\frac{m}{5}+\frac{2n}{8}$
65. $\frac{2x}{5}-\frac{y}{10}$

-----Continued-----

66. Jorge purchased stock on A-Plus Company. The price per share of stock fell by $\$1\frac{5}{8}$ in the first month, rose $\$2\frac{3}{4}$ in the second month. How much did he gain (or lose) per share of stock for the past two months ?

67. Find the distance between two points $-4\frac{1}{2}$ and $3\frac{3}{5}$ on the number line.

68. A number is the sum of $2\frac{1}{4}$ and $5\frac{2}{3}$, decreased by $4\frac{3}{5}$. What is the number ?

69. The perimeter of the following figure is $a+b+c+d$. Find the perimeter when $a = 6\frac{1}{2}$, $b = 5$, $c = 4\frac{1}{3}$, $d = 8\frac{5}{6}$.

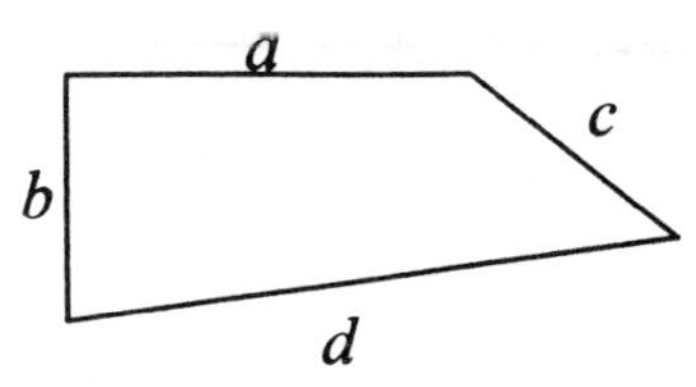

70. Roger is 6 ft tall. He grew $1\frac{2}{7}$ in. during the last year. How tall in inches was he one year ago ? (Hint: 1 foot = 12 inches)

71. In the figure shown below, what fractional part of the circle is shaded ?

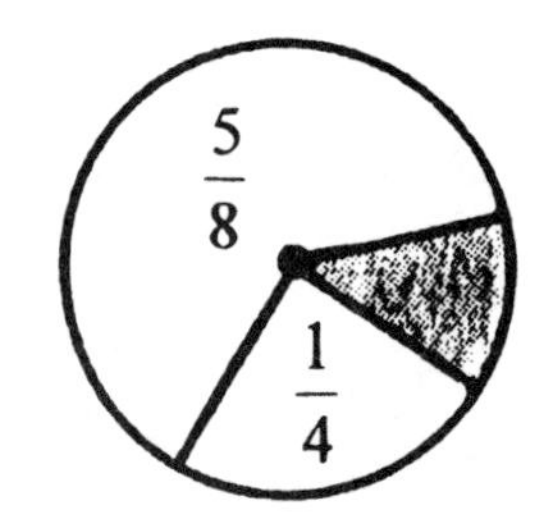

72. Evaluate $1\frac{3}{4} - 12 + 3\frac{4}{5} - 6\frac{7}{10}$.

73. Evaluate $-\frac{5}{6} - \left(2\frac{3}{4} - \frac{1}{6}\right)$.

74. The difference in height of two buildings is $6\frac{3}{4}$ feet. The higher building is at a height of $60\frac{7}{12}$ feet. What is the height of the lower building ?

75. Maria lost $9\frac{1}{2}$ pounds in January, lost 7 pounds in February, and lost $5\frac{3}{4}$ pounds in March. What was his total weight loss for these three months ?

76. Peter lost $2\frac{1}{3}$ pounds the last month and lost $4\frac{2}{5}$ pounds this month. He now weighs $152\frac{3}{10}$ pounds. How much did he weigh before ?

77. The rainfall was $2\frac{3}{8}$ inches on Monday, $3\frac{9}{16}$ inches on Tuesday, and $1\frac{3}{4}$ inches on Wednesday. What was the total rainfall for the three days ?

78. Carl spent $1\frac{5}{7}$ hours studying math, $\frac{7}{8}$ hour studying science, and $1\frac{1}{4}$ hours studying English. How much time did he spend altogether ?

79. Maria and George were driving on the freeway. Maria used $\frac{1}{4}$ of the gas. George used $\frac{2}{3}$ of the gas. How much (in fractional part) of the gas was left ?

80. Evaluate $\frac{c}{a-b} + \frac{a}{b-c} + \frac{b}{c-a}$ if $a = 1$, $b = 2$, $c = -2$.

4-4 Multiplying Fractions

Rules for multiplying fractions:

1. To multiply with fractions, we multiply their numerators and multiply their denominators. Then write the answer in simplest form.
2. Write each mixed number as an improper fraction before multiplying.
3. It is easier if we can "simplify first when possible" and then multiply.
4. To multiply fractions and whole numbers, we rename each whole number as an equal fraction (use "1" as its denominator). $2 = \frac{2}{1}$, $3 = \frac{3}{1}$, $4 = \frac{4}{1}$, ...

Examples

1. $\frac{1}{3} \times \frac{2}{3} = \frac{1 \times 2}{3 \times 3} = \frac{2}{9}$.

2. $-\frac{4}{5} \times 7 = -\frac{4}{5} \times \frac{7}{1} = -\frac{4 \times 7}{5 \times 1} = -\frac{28}{5} = -5\frac{3}{5}$.

3. $\frac{2}{\cancel{5}_1} \times \frac{\cancel{10}^2}{17} = \frac{2 \times 2}{1 \times 17} = \frac{4}{17}$. (Simplify first and then multiply.)

OR: $\frac{2}{5} \times \frac{10}{17} = \frac{2 \times 10}{5 \times 17} = \frac{20}{85} = \frac{4}{17}$. (Multiply first and then simplify.)

4. $\frac{7}{3} \times (-2\frac{2}{3}) = \frac{7}{3} \times (-\frac{8}{3}) = -\frac{7 \times 8}{3 \times 3} = -\frac{56}{9} = -6\frac{2}{9}$.

5. $2\frac{1}{5} \times 3\frac{3}{4} = \frac{11}{\cancel{5}_1} \times \frac{\cancel{15}^3}{4} = \frac{11 \times 3}{1 \times 4} = \frac{33}{4} = 8\frac{1}{4}$.

6. $5 \times \frac{4}{7} = \frac{5}{1} \times \frac{4}{7} = \frac{5 \times 4}{1 \times 7} = \frac{20}{7} = 2\frac{6}{7}$.

7. $-4\frac{2}{3} \times (-7) = \frac{14}{3} \times \frac{7}{1} = \frac{14 \times 7}{3 \times 1} = \frac{98}{3} = 32\frac{2}{3}$.

8. $2\frac{2}{3} \times 4 \times \frac{2}{7} = \frac{8}{3} \times \frac{4}{1} \times \frac{2}{7} = \frac{8 \times 4 \times 2}{3 \times 1 \times 7} = \frac{64}{21} = 3\frac{1}{21}$.

9. $\frac{3}{\cancel{x}} \times \frac{\cancel{x}}{2} = \frac{3}{2} = 1\frac{1}{2}$, $x \neq 0$.

10. $\frac{a}{4} \times \frac{2b}{3} = \frac{2ab}{12} = \frac{ab}{6}$.

11. Find $\frac{1}{2}$ of $\frac{1}{3}$.
Solution:
$\frac{1}{3} \times \frac{1}{2} = \frac{1}{6}$. Ans.

12. Find $\frac{2}{3}$ of $\frac{7}{10}$.
Solution:
$\frac{7}{10} \times \frac{2}{3} = \frac{14}{30} = \frac{7}{15}$. Ans.

-----Continued-----

Examples

13. Tom ate one-half of the pizza. Roger ate three-fifths of the remaining part. What fractional part of the whole pizza did Roger eat ? How much the fractional part of the pizza was left ?

Solution:

$$\frac{1}{2}\times\frac{3}{5}=\frac{3}{10} \text{ of the pizza (Roger ate). Ans.}$$

$$1-\frac{1}{2}-\frac{3}{10}=\frac{10}{10}-\frac{5}{10}-\frac{3}{10}=\frac{10-5-3}{10}=\frac{2}{10}=\frac{1}{5} \text{ of the pizza was left. Ans.}$$

14. John picked up one-half of the book on the shelf. Maria picked up one-half of the remaining books. Janet picked up 6 books that were left. How many books were on the shelf to begin with ?

Solution:

We work backward from the end.
There are 6 books left for Janet.
There are 12 books left for Maria (she picked up 6 books).
There are 24 books in the beginning for John (He picked up 12 books).
Ans: 24 books

(Hint: We will learn to use "equation" to solve it. See Section 6-8. Example 9)

15. The formula to find the Celsius temperature for each Fahrenheit temperature is $C=\frac{5}{9}(F-32)$. Find the Celsius temperature for the temperature $98^o F$.

Solution: $C=\frac{5}{9}(98-32)=\frac{5}{9}\times 66=\frac{5\times 66}{9\times 1}=\frac{330}{9}=36.7^o$. Ans.

16. There are 42 students in class. We expect one-third of the students to get "A" in the final test. How many students can be expected to get "A" in the final test ?

Solution:

$$42\times\frac{1}{3}=\frac{42}{1}\times\frac{1}{3}=\frac{42}{3}=14 \text{ students. Ans.}$$

17. Find the area.

$3\frac{1}{3}$ in.

$2\frac{1}{4}$ in.

Solution: Area $=3\frac{1}{3}\times 2\frac{1}{4}=\frac{10}{\cancel{3}_{1}}\times\frac{\cancel{9}^{3}}{4}=\frac{30}{4}=7\frac{1}{2}\ in.^2$ or $7.5\ in.^2$ Ans.

EXERCISES

Multiply. Write the answer in simplest form.

1. $\frac{3}{4}\times\frac{5}{7}$
2. $\frac{2}{5}\times\frac{1}{7}$
3. $\frac{3}{7}\times\frac{6}{5}$
4. $\frac{2}{9}\times\frac{4}{5}$
5. $-\frac{4}{9}\times\frac{1}{2}$
6. $\frac{6}{7}\times(-\frac{2}{3})$
7. $-\frac{12}{17}\times\frac{17}{6}$
8. $\frac{5}{6}\times(-\frac{3}{10})$
9. $\frac{12}{25}\times\frac{5}{6}$
10. $-\frac{14}{27}\times(-\frac{9}{7})$
11. $-\frac{5}{9}\times2\frac{4}{5}$
12. $\frac{4}{13}\times3\frac{3}{4}$
13. $4\frac{3}{5}\times\frac{2}{3}$
14. $3\frac{2}{5}\times\frac{5}{4}$
15. $-\frac{5}{16}\times(-2\frac{4}{5})$
16. $4\frac{1}{5}\times(-\frac{10}{21})$
17. $5\frac{2}{5}\times\frac{5}{9}$
18. $\frac{11}{20}\times1\frac{3}{22}$
19. $-\frac{2}{9}\times(-4\frac{3}{4})$
20. $-3\frac{2}{3}\times\frac{6}{11}$
21. $2\frac{1}{2}\times3\frac{2}{5}$
22. $4\frac{1}{5}\times5\frac{2}{3}$
23. $-3\frac{2}{5}\times4\frac{5}{17}$
24. $1\frac{3}{5}\times(-6\frac{4}{5})$
25. $7\frac{1}{2}\times3\frac{4}{15}$
26. $-1\frac{1}{2}\times(-2\frac{4}{5})$
27. $9\frac{1}{8}\times2\frac{16}{73}$
28. $10\frac{2}{3}\times5\frac{9}{16}$
29. $-15\frac{3}{4}\times3\frac{8}{21}$
30. $7\frac{2}{13}\times4\frac{1}{3}$
31. $\frac{5}{6}\times7$
32. $-\frac{4}{9}\times9$
33. $\frac{9}{4}\times9$
34. $\frac{6}{5}\times7$
35. $-2\frac{5}{8}\times16$
36. $12\times\frac{5}{24}$
37. $-15\times(-\frac{3}{5})$
38. $24\times4\frac{5}{6}$
39. $21\frac{4}{5}\times2$
40. $-2\frac{7}{9}\times(-5)$
41. $\frac{x}{4}\times\frac{2}{5}$
42. $\frac{y}{6}\times\frac{3}{5}$
43. $\frac{m}{7}\times\frac{n}{2}$
44. $\frac{a}{4}\times\frac{2b}{5}$
45. $3\frac{2}{3}\times\frac{6d}{7}$
46. $1\frac{3}{4}\times\frac{6t}{11}$
47. $\frac{x}{5}\times\frac{6y}{5}$
48. $\frac{5}{2x}\times\frac{4x}{3}$, $x\neq0$
49. $\frac{7}{6y}\times\frac{2y}{21}$, $y\neq0$
50. $4\frac{2}{3}\times\frac{6p}{7}$

Evaluate. Write the answer in simplest form.

51. $\frac{2}{3}-\frac{1}{3}\times\frac{2}{5}$
52. $\frac{2}{7}+\frac{3}{4}\times\frac{1}{2}$
53. $\frac{4}{5}\times\frac{3}{8}-\frac{5}{6}$
54. $\frac{5}{7}\times\frac{7}{10}-5$
55. $5-1\frac{1}{4}\times3$
56. $3\frac{1}{2}-\frac{1}{2}\times8$
57. $5\frac{1}{4}-1\frac{1}{3}\times\frac{3}{5}$
58. $3\times(4\frac{2}{3}-2)$
59. $\frac{1}{2}\times(\frac{1}{3}+\frac{1}{4})$
60. $\frac{1}{2}\times\frac{1}{3}-\frac{1}{3}\times\frac{1}{4}$

-----Continued-----

61. Find $\frac{3}{4}$ of $\frac{1}{2}$. **62.** Find $\frac{1}{2}$ of $\frac{3}{4}$. **63.** Find $\frac{3}{8}$ of 16. **64.** Find $\frac{2}{3}$ of $5\frac{3}{8}$

65. Maria ate $\frac{1}{3}$ of the pizza. David ate $\frac{2}{5}$ of the remaining. What fractional part of the whole pizza did David eat ?

66. Maria ate $\frac{1}{3}$ of the pizza. David ate $\frac{2}{5}$ of the remaining. How much the fractional part of the pizza was left ?

67. John picked up $\frac{2}{3}$ of the books on the shelf. Mario picked up 18 books that were left. How many books were on the shelf to begin with ?

68. John picked up $\frac{2}{3}$ of the books on the shelf. Mario picked up $\frac{1}{2}$ of the remaining books. David picked up 12 books that were left. How many books were on the shelf to begin with ?

69. Find the area. **70.** Find the area.

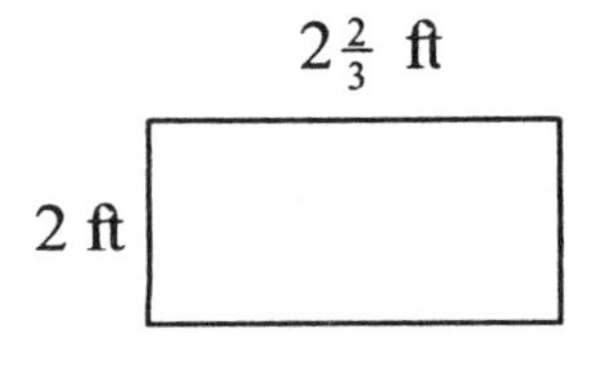

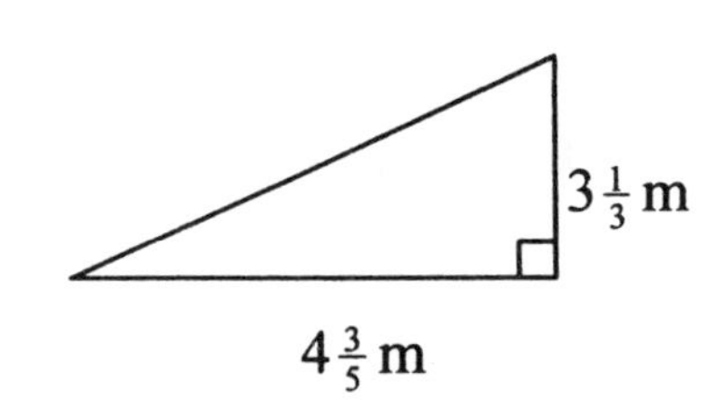

71. The volume of a rectangular box is the product of its length (l) of the base, the width (w) of the base, and the altitude h. That is the formula $V = l\ w\ h$.
Find the volume of a rectangular box if $l = 9$ in., $w = 5\frac{1}{2}$ in., and $h = 4\frac{2}{3}$ in..

72. There are 39 students in class. We expect two-third of the students to pass the test. How many students can be expected to pass to test.

73. There are 3,195 students in Hawthorne High. Two-third of the students are boys. How many boys and how many girls are there in Hawthorne High.

74. The formula to find the Fahrenheit temperature for each Celsius temperature is $F = \frac{9}{5}C + 32$. Find the Fahrenheit temperature for the temperature $36.7^o C$.

75. A car averages 11 miles by one gallon of gasoline. The tank holds a full tank of $12\frac{1}{2}$ gall-llons. How far will the car travel on a full tank of gasoline ?

76. The size of the floor is $12\frac{3}{4}$ feet long and $8\frac{4}{5}$ feet wide. How many square feet of carpet needed to cover the floor ?

77. Jason works $2\frac{1}{2}$ hours each night. How many hours does he work in 4 weeks if he works 5 nights a week ? How much will he earn if he makes $6.50 per hour ?

78. Each "A-plus notes for algebra" book weighs $2\frac{1}{4}$ pounds. There are 20 copies per box. How much will 8 boxes of books weigh ?

79. Evaluate $-\frac{3}{5} \times (-\frac{4}{27}) \times (-2\frac{16}{19})$.

80. Evaluate $-2\frac{3}{11} \times 12 \times (-3\frac{24}{25})$.

4-5 Dividing Fractions

Rules for dividing fractions:

1. To divide with fractions, we **multiply** the reciprocal of the divisor.
2. Multiply as the rules in Section 4-4.
3. Division by 0 is undefined (it means "Division by zero is not allowed.").

Reciprocal: Two numbers are reciprocals of each other if their product is 1.

$\frac{2}{3}\times\frac{3}{2}=1.$ $\frac{2}{3}$ and $\frac{3}{2}$ are reciprocals of each other.

$\frac{7}{10}\times\frac{10}{7}=1.$ $\frac{7}{10}$ and $\frac{10}{7}$ are reciprocals of each other.

$2\times\frac{1}{2}=1.$ 2 and $\frac{1}{2}$ are reciprocals of each other.

0 times any nonzero number is 0. Therefore, 0 has no reciprocal.
The reciprocal of 1 is 1.

Dividing by a number gives the same result as multiplying by its reciprocal.

$$15\div 3=5 \rightarrow 15\times\frac{1}{3}=5 \;;\; 15\div\frac{2}{3}=15\times\frac{3}{2}=\frac{15}{1}\times\frac{3}{2}=\frac{45}{2}=22\frac{1}{2}$$

Examples

1. $\frac{1}{3}\div\frac{2}{3}=\frac{1}{\cancel{3}}\times\frac{\cancel{3}}{2}=\frac{1}{2}$

2. $\frac{-\frac{4}{5}}{\frac{7}{3}}=-\frac{4}{5}\times\frac{3}{7}=-\frac{12}{35}$

3. $\frac{7}{3}\div(-2\frac{2}{3})=\frac{7}{3}\div(-\frac{8}{3})=-\frac{7}{\cancel{3}}\times\frac{\cancel{3}}{8}=-\frac{7}{8}$

4. $0\div\frac{2}{5}=0$; $\frac{2}{5}\div 0$ is undefined.

5. $2\frac{1}{5}\div 3\frac{3}{4}=\frac{11}{5}\div\frac{15}{4}=\frac{11}{5}\times\frac{4}{15}=\frac{44}{75}$

6. $\frac{4}{7}\div 5=\frac{4}{7}\div\frac{5}{1}=\frac{4}{7}\times\frac{1}{5}=\frac{4}{35}$

7. $-7\div(-4\frac{2}{3})=\frac{7}{1}\div\frac{14}{3}=\frac{\overset{1}{\cancel{7}}}{1}\times\frac{3}{\underset{2}{\cancel{14}}}=\frac{3}{2}=1\frac{1}{2}$

8. $4\frac{2}{3}\div 7=\frac{14}{3}\div\frac{7}{1}=\frac{\overset{2}{\cancel{14}}}{3}\times\frac{1}{\underset{1}{\cancel{7}}}=\frac{2}{3}$

9. $\frac{3}{x}\div\frac{x}{2}=\frac{3}{x}\times\frac{2}{x}=\frac{6}{x^2},\ x\neq 0$

10. $\frac{a}{4}\div\frac{2b}{3}=\frac{a}{4}\times\frac{3}{2b}=\frac{3a}{8b},\ b\neq 0$

-----Continued-----

If a number is given by a fractional part of an unknown number, we can find that unknown number by division.

Examples

11. $\frac{1}{6}$ is $\frac{1}{2}$ of what number ?

Solution:

$$\frac{1}{6} \div \frac{1}{2} = \frac{1}{6} \times \frac{2}{1} = \frac{1}{3}. \quad \text{Ans.}$$

Check: $\frac{1}{2}$ of $\frac{1}{3} = \frac{1}{2} \times \frac{1}{3} = \frac{1}{6}$.

12. $\frac{7}{15}$ is $\frac{2}{3}$ of what number ?

Solution:

$$\frac{7}{15} \div \frac{2}{3} = \frac{7}{15} \times \frac{3}{2} = \frac{7}{10}. \quad \text{Ans.}$$

Check: $\frac{2}{3}$ of $\frac{7}{10} = \frac{2}{3} \times \frac{7}{10} = \frac{7}{15}$.

13. David bought one-half of the pizza. He cut it into 5 equal pieces. What the fractional part of the whole pizza does each piece has ?

Solution:

$$\frac{1}{2} \div 5 = \frac{1}{2} \times \frac{1}{5} = \frac{1}{10} \text{ of the whole pizza. Ans.}$$

14. A car travels 150 miles on $8\frac{1}{3}$ gallons of fuel. How far does it travel on one gallon of fuel ?

Solution: $150 \div 8\frac{1}{3} = 150 \div \frac{25}{3} = 150 \times \frac{3}{25} = 18$ miles. Ans.

15. Richard stayed at home $\frac{4}{9}$ of the day. The time he studied math was $\frac{3}{4}$ of the time he stayed at home. What fractional part of the day did Richard study math ?

Solution:

The problem is similar to " Find $\frac{3}{4}$ of $\frac{4}{9}$.".

$$\frac{4}{9} \times \frac{3}{4} = \frac{1}{3} \text{ of the day to study math. Ans.}$$

16. Richard studied math $\frac{1}{3}$ of the day. The time he studied math was $\frac{3}{4}$ of the time he stayed at home. What fractional part of the day did Richard stay at home ?

Solution:

The problem is similar to "$\frac{1}{3}$ is $\frac{3}{4}$ of what number ? ".

$$\frac{1}{3} \div \frac{3}{4} = \frac{1}{3} \times \frac{4}{3} = \frac{4}{9} \text{ of the day he stayed at home. Ans.}$$

17. How many pieces of wires $2\frac{1}{2}$ feet long can be cut from a wire $27\frac{1}{2}$ feet long ?

Solution:

$$27\frac{1}{2} \div 2\frac{1}{2} = \frac{55}{2} \div \frac{5}{2} = \frac{55}{2} \times \frac{2}{5} = 11 \text{ pieces. Ans.}$$

EXERCISES

Write the reciprocal of each number.

1. $\frac{2}{7}$ **2.** $\frac{7}{2}$ **3.** $\frac{3}{8}$ **4.** $\frac{9}{5}$ **5.** $\frac{13}{4}$ **6.** $\frac{3}{16}$ **7.** $3\frac{1}{4}$ **8.** $5\frac{2}{3}$ **9.** 0 **10.** 1

Divide. Write the answer in simplest form.

11. $\frac{3}{4} \div \frac{5}{7}$ **12.** $\frac{2}{5} \div \frac{1}{7}$ **13.** $\frac{3}{7} \Big/ \frac{6}{5}$ **14.** $\frac{2}{9} \div \frac{4}{5}$ **15.** $-\frac{4}{9} \div \frac{1}{2}$

16. $\frac{6}{7} \Big/ -\frac{2}{3}$ **17.** $-\frac{12}{17} \div \frac{6}{17}$ **18.** $\frac{5}{6} \div (-\frac{10}{3})$ **19.** $\frac{\frac{12}{25}}{\frac{6}{5}}$ **20.** $-\frac{14}{27} \div (-\frac{7}{9})$

21. $-\frac{5}{9} \div 2\frac{1}{3}$ **22.** $\frac{\frac{4}{13}}{3\frac{4}{13}}$ **23.** $4\frac{2}{5} \div \frac{11}{15}$ **24.** $3\frac{1}{5} \div \frac{8}{11}$ **25.** $-\frac{5}{16} \div (-2\frac{3}{4})$

26. $2\frac{1}{2} \div 3\frac{1}{2}$ **27.** $4\frac{2}{5} \div 5\frac{1}{5}$ **28.** $-3\frac{2}{5} \div 3\frac{1}{10}$ **29.** $\frac{-1\frac{3}{5}}{2\frac{2}{15}}$ **30.** $7\frac{2}{13} \div 4\frac{2}{13}$

31. $\frac{5}{6} \div 7$ **32.** $-\frac{4}{9} \div 9$ **33.** $\frac{9}{4} \div 9$ **34.** $\frac{6}{5} \div 7$ **35.** $-2\frac{5}{8} \div 21$

36. $12 \div \frac{5}{24}$ **37.** $-15 \div (-\frac{5}{6})$ **38.** $24 \div 4\frac{4}{5}$ **39.** $21\frac{4}{5} \div 2$ **40.** $-2\frac{7}{9} \div (-5)$

41. $\frac{x}{4} \div \frac{2}{5}$ **42.** $\frac{y}{6} \div \frac{3}{5}$ **43.** $\frac{m}{7} \div \frac{1}{2}$ **44.** $\frac{a}{4} \div \frac{5}{2}$ **45.** $\frac{6d}{7} \div 3\frac{2}{3}$

46. $\frac{6t}{11} \div 1\frac{3}{4}$ **47.** $\frac{x}{5} \div 1\frac{3}{5}$ **48.** $\frac{5}{2x} \div \frac{3x}{4},\ x \neq 0$ **49.** $\frac{6y^2}{7} \div \frac{2y}{21},\ y \neq 0$ **50.** $\frac{6p}{7} \div 1\frac{5}{7}$

Evaluate. Write the answer in simplest form.

51. $\frac{2}{3} - \frac{1}{3} \div \frac{2}{3}$ **52.** $\frac{2}{7} + \frac{3}{4} \div \frac{1}{2}$ **53.** $\frac{4}{5} \div \frac{8}{3} - \frac{5}{6}$ **54.** $\frac{5}{7} \div \frac{10}{7} + 5$ **55.** $5 + 1\frac{4}{5} \div 3$

56. $3\frac{1}{2} - \frac{1}{2} \div \frac{1}{8}$ **57.** $5\frac{1}{4} - 1\frac{1}{3} \div 4$ **58.** $3 \div (4\frac{2}{3} - 2)$ **59.** $\frac{1}{2} \div (\frac{1}{3} + \frac{1}{4})$ **60.** $\frac{1}{2} \div \frac{1}{3} - \frac{1}{3} \div \frac{1}{4}$

-----Continued-----

61. $\frac{3}{8}$ is $\frac{3}{4}$ of what number ? **62.** 6 is $\frac{3}{8}$ of what number ? **63.** $3\frac{7}{12}$ is $\frac{2}{3}$ of what number ?

64. A car travels 125 miles on $12\frac{1}{2}$ gallons of fuel. How far does it travel on one gallon of fuel ?

65. Norma bought two-third of the pizza. He cut it into 6 equal pieces. What the fractional part of the whole pizza does each piece has ?

66. How many pieces of wires $1\frac{1}{2}$ meters long can be cut from a wire $40\frac{1}{2}$ meters long ?

67. Connie studied science $\frac{1}{4}$ of the day. The time he studied science was $\frac{3}{8}$ of the time he stayed at home. What fractional part of the day did Connie stayed at home ?

68. The area of a rectangle is $5\frac{1}{3}$ $ft.^2$. Its length is $2\frac{2}{3}$ $ft.$. What is the width ?

69. One bar is made from $15\frac{1}{2}$ meters of iron. How many bars can we make from 310 meters of iron ?

70. How many $2\frac{1}{2}$-pound bags of dog food can we make from a 80-pound bag of dog food ?

71. A gasoline tank is $\frac{3}{5}$ full with 15 gallons of gasoline. How many gallons will it take to fill the full tank ?

72. Anna bought $52\frac{1}{2}$ pounds of beef. She divided them into 15 equal packages. How many pounds were in each package ?

73. Mark can run $3\frac{3}{5}$ miles in 9 minutes. How many miles can he run in one minute ?

74. John can finish the job in 5 hours. Steve can finish the same job in 4 hours. How many hours do they need to finish the job if they work together ?

75. Three-fifths of the students are boys in class. There are 24 boys. How many students are there in class ?

76. John spent $\frac{1}{5}$ of his money to buy books, and $\frac{1}{4}$ of his money to buy clothes. He spent a total of $180. How much money did he have before he bought these items ?

77. The drawing shows the length and the width of a rectangular field on a map. What is the actual width if the actual length is 200 meters ?

$1\frac{1}{4}$ m

$\frac{3}{5}$ m

78. Evaluate $14\times(-\frac{8}{21})-8\div(-\frac{16}{9})$. **79.** Evaluate $7\times(-\frac{3}{14})+(-8)\div(-\frac{16}{9})$.

80. Evaluate $\frac{1}{2}-\frac{1}{2}\times1\frac{1}{4}+(-6)\div2$. **81.** If $\frac{5}{8}$ of $\frac{4}{9}=\frac{2}{9}$ of $\frac{x}{8}$, what is the value of x ?

82. Erica can paint a wall in 6 hours. John can paint the same wall in 4 hours. If they work together, how long will it take them to paint the wall ?

PUZZLE

A farmer wants to give his 17 lambs to his three sons. Each son will have $\frac{1}{2}$, $\frac{1}{3}$, and $\frac{1}{9}$ of the lambs. How could the farmer give the lambs to his sons perfectly ?

CHAPTER 4 EXERCISES

Tell whether the fractions are equal or not.

1. $\frac{1}{3}$ and $\frac{2}{5}$ **2.** $\frac{1}{3}$ and $\frac{2}{6}$ **3.** $\frac{3}{4}$ and $\frac{9}{12}$ **4.** $\frac{3}{4}$ and $\frac{5}{8}$ **5.** $\frac{5}{7}$ and $\frac{6}{8}$

6. $\frac{0}{5}$ and $\frac{0}{6}$ **7.** $\frac{5}{9}$ and $\frac{15}{27}$ **8.** $\frac{7}{12}$ and $\frac{14}{24}$ **9.** $\frac{13}{25}$ and $\frac{27}{50}$ **10.** $\frac{2}{9}$ and $\frac{18}{72}$

Write each fraction or expression as simplest form.

11. $\frac{9}{12}$ **12.** $\frac{3}{5}$ **13.** $\frac{21}{35}$ **14.** $\frac{35}{21}$ **15.** $\frac{12}{12}$ **16.** $\frac{0}{12}$ **17.** $-\frac{18}{6}$ **18.** $\frac{27}{4}$

19. $-\frac{53}{13}$ **20.** $\frac{75}{10}$ **21.** $-\frac{7}{21}$ **22.** $\frac{12}{36}$ **23.** $7\frac{15}{20}$ **24.** $4\frac{5}{4}$ **25.** $-2\frac{10}{3}$ **26.** $7\frac{15}{7}$

27. $\frac{3x}{15}$ **28.** $\frac{4y^2}{2y}$, $y \neq 0$ **29.** $\frac{12a^2b}{8ab}$, $a \neq 0$ and $b \neq 0$ **30.** $\frac{36mn^2}{48m^2n}$, $m \neq 0$ and $n \neq 0$

Write the fractions as equivalent fractions with the least common denominator (LCD).

31. $\frac{3}{4}, \frac{1}{8}$ **32.** $\frac{4}{5}, \frac{5}{6}$ **33.** $-\frac{4}{7}, \frac{2}{3}$ **34.** $\frac{3}{7}, \frac{5}{21}$ **35.** $\frac{1}{11}, -\frac{4}{33}$

36. $\frac{7}{9}, \frac{1}{4}$ **37.** $\frac{5}{6}, \frac{9}{11}$ **38.** $\frac{5}{6}, -\frac{3}{8}$ **39.** $\frac{5}{12}, \frac{7}{8}$ **40.** $\frac{5}{12}, \frac{5}{18}$

41. $-\frac{3}{42}, -\frac{2}{48}$ **42.** $\frac{1}{3}, \frac{1}{4}, \frac{2}{5}$ **43.** $-\frac{4}{5}, \frac{5}{6}, -\frac{2}{15}$ **44.** $\frac{7}{8}, \frac{5}{12}, \frac{1}{6}$ **45.** $\frac{1}{18}, \frac{7}{12}, \frac{3}{10}$

Change each fraction to a decimal.

46. $\frac{3}{5}$ **47.** $\frac{4}{5}$ **48.** $\frac{5}{9}$ **49.** $\frac{7}{9}$ **50.** $\frac{3}{10}$ **51.** $-\frac{5}{8}$

52. $-\frac{9}{4}$ **53.** $\frac{18}{5}$ **54.** $-\frac{13}{99}$ **55.** $\frac{11}{90}$ **56.** $\frac{7}{30}$ **57.** $\frac{29}{90}$

58. $-3\frac{2}{3}$ **59.** $6\frac{1}{4}$ **60.** $-12\frac{7}{8}$

Change each decimal to a fraction in simplest form.

61. 0.05 **62.** 0.15 **63.** –0.04 **64.** 0.45 **65.** 0.65 **66.** 1.1

67. 2.01 **68.** –9.5 **69.** 12.55 **70.** 4.005 **71.** –10.001 **72.** 8.125

73. $0.\overline{05}$ **74.** $0.\overline{15}$ **75.** $-0.0\overline{4}$ **76.** $0.4\overline{5}$ **77.** $0.3\overline{68}$ **78.** $1.\overline{1}$

79. $2.0\overline{1}$ **80.** $-4.1\overline{23}$

-----Continued-----

Evaluate each of the following expression. Write the answer in simplest form.

81. $\frac{2}{3}+\frac{3}{4}$ **82.** $\frac{2}{3}-\frac{3}{4}$ **83.** $-\frac{2}{3}\times\frac{3}{4}$ **84.** $\frac{2}{3}\div\frac{3}{4}$ **85.** $\frac{4}{7}+\frac{5}{7}$

86. $\frac{7}{5}\times\frac{4}{7}$ **87.** $-\frac{4}{7}-\frac{5}{7}$ **88.** $\frac{4}{7}\div\frac{5}{7}$ **89.** $\frac{3}{8}-0.6$ **90.** $0.3+\frac{2}{5}$

91. $4\frac{1}{2}+3\frac{5}{8}$ **92.** $4\frac{5}{8}-3\frac{1}{2}$ **93.** $3\frac{2}{3}\times 6$ **94.** $8\div 5\frac{1}{3}$ **95.** $\frac{2}{5}\Big/3\frac{3}{10}$

96. $\frac{4}{11}\times(-4\frac{1}{8})$ **97.** $-\frac{4}{11}\Big/4\frac{1}{8}$ **98.** $\dfrac{2\frac{1}{5}}{3\frac{1}{10}}$ **99.** $7\frac{4}{5}\div 13$ **100.** $\frac{7}{8}\times 18$

101. $\frac{1}{4}+\frac{1}{3}-\frac{1}{2}$ **102.** $\frac{1}{3}-\frac{1}{4}+\frac{1}{2}$ **103.** $\frac{2}{5}+\frac{2}{3}\times\frac{3}{5}$ **104.** $\frac{2}{7}\div\frac{14}{21}-\frac{1}{14}$ **105.** $1-\frac{1}{2}\times\frac{1}{3}$

106. $2\frac{1}{2}-\frac{3}{4}\div\frac{5}{8}$ **107.** $\frac{1}{2}\times\frac{1}{3}\div\frac{1}{4}$ **108.** $4\frac{1}{5}\div(\frac{1}{2}-\frac{2}{3})$ **109.** $\frac{7}{8}\div\frac{7}{18}-5$ **110.** $6\div\frac{3}{5}\times\frac{2}{5}$

111. Evaluate $7-\frac{2}{3}\times\frac{6}{11}+\frac{6}{11}\div\frac{2}{3}$.

112. Find the perimeter and the area.

$4\frac{1}{2}$ in.

$1\frac{2}{3}$ in.

113. The volume of a cube is given by the formula $V=s^3$ (where s is the length of one side). Find the volume of a cube with one side $4\frac{2}{3}$ in..

114. In the figure shown below, what fractional part of the circle is shaded ?

115. Find $\frac{4}{5}$ of $\frac{2}{3}$.

116. $\frac{8}{15}$ is $\frac{4}{5}$ of what number ?

117. David ate one-third of the pizza. Donna ate two-fifths of the remaining part. What fractional part of the whole pizza did Donna eat ? How much the fractional part of the pizza was left ?

118. How many pieces of wires $5\frac{2}{3}$ feet long can be cut from a wire 340 feet long ?

Scale Drawings and Scale Factors

Scale Drawing: A reduced or enlarged figure that is similar to the actual object.
Scale Factor: A ratio of the measurement of a scale drawing to the actual measurement.

Examples

1. The scale of the drawing on a map is 1 inch = 10 inches. What is the actual length of a line if the length on the map is 8 inches ?
Solution:

Method 1:
Actual length $= 8 \times 10 = 80$ inches. Ans.

Method 2:
Let $x =$ the actual length of the line

$$\frac{1}{10} = \frac{8}{x} \quad \therefore x = 80 \text{ inches. Ans.}$$

2. The scale of the drawing on a map is $\frac{1}{2}$ inch = 10 inches. What is the actual length of a line if the length on the map is $5\frac{1}{2}$ inches ?
Solution:

Method 1: $10 \div \frac{1}{2} = 20$
The scale is 1 in. = 20 in.
The actual length $= 5\frac{1}{2} \times 20 = 110$ in. Ans.

Method 2:
Let $x =$ the actual length of the line

$$\frac{\frac{1}{2}}{10} = \frac{5\frac{1}{2}}{x} \quad \therefore \tfrac{1}{2}x = 10(5\tfrac{1}{2})$$

$$\tfrac{1}{2}x = 55$$

$$x = 110 \text{ inches.}$$

Ans.

3. The scale of the drawing on a map is $1\frac{1}{2}$ inches = 18 miles. Los Angeles and San Diego are 10 inches apart on the map. What is the actual distance between the two cities ?
Solution:
$18 \div 1\frac{1}{2} = 12$, the scale is 1 inch = 12 miles
Actual distance $= 12 \times 10 = 120$ miles. Ans.

4. The scale of the drawing on a map is 1 inch = 1 foot. How tall would the drawing be if you are 5 feet 8 inches tall ?

Solution:

The scale is 1 inch = 12 inches
5 feet 8 inches = 68 inches
$68 \div 12 = 5\frac{2}{3}$ inches. Ans.

5. The measurement on the scale drawing is 4 inches. The actual measurement is 5 yards. What is the scale factor ? (1 yard = 36 inches)
Solution:
5 yards = 180 inches
Scale factor $= \frac{4}{180} = \frac{1}{45}$. Ans.

6. The scale of the drawing of a house map is 1 inch = 20 feet. Find the scale factor ?

Solution: 20 feet = 240 inches

Scale factor $= \frac{1}{240}$. Ans.

Converting Repeating Decimals into Fractions

We can convert a repeating decimal into a common fraction. Most textbooks use the following method to convert a repeating decimal into a common fraction.

Example: Convert $0.5\overline{42}$ into fraction.

Solution:

$$\text{Let} \quad N = 0.5\overline{42}$$

$$100N = 54.2\overline{42}$$
$$-) \quad N = 0.5\overline{42}$$
$$99N = 53.7$$

$$N = \frac{53.7}{99} = \frac{537}{990} = \frac{179}{330} \qquad \therefore \; 0.5\overline{42} = \frac{179}{330}. \text{ Ans.}$$

Quick Method : For each repeating digit, we write the denominator with a number of 9, and 0 for each nonrepeating digit. Subtract the nonrepeating digits from the numerator. This method was first introduced into schools in U.S.A by Rong Yang (1948 ~ 2068 A.D.).

$$0.\bar{1} = \frac{1}{9}, \quad 0.\bar{2} = \frac{2}{9}, \quad 0.\bar{3} = \frac{3}{9} = \frac{1}{3}, \quad 0.\bar{9} = \frac{9}{9} = 1, \quad 0.\overline{12} = \frac{12}{99} = \frac{4}{33}, \quad 0.\overline{123} = \frac{123}{999} = \frac{41}{333}$$

$$0.0\bar{1} = \frac{1}{90}, \quad 0.0\bar{2} = \frac{2}{90} = \frac{1}{45}, \quad 0.0\bar{3} = \frac{3}{90} = \frac{1}{30}, \quad 0.0\overline{12} = \frac{12}{990} = \frac{4}{330}, \quad 0.00\overline{12} = \frac{12}{9900} = \frac{1}{825}$$

$$0.1\bar{2} = \frac{12-1}{90} = \frac{11}{90}, \quad 0.1\overline{23} = \frac{123-1}{990} = \frac{122}{990} = \frac{61}{495}, \quad 0.12\bar{3} = \frac{123-12}{900} = \frac{111}{900} = \frac{37}{300}$$

$$0.12\overline{345} = \frac{12345-12}{99900} = \frac{12333}{99900} = \frac{4111}{33300}, \quad 0.0012\bar{3} = \frac{123-12}{90000} = \frac{111}{90000} = \frac{37}{30000}$$

$$1.\overline{12} = 1\frac{12}{99} = 1\frac{4}{33}, \qquad 0.\overline{73} + 0.\overline{53} = \frac{73}{99} + \frac{53}{99} = \frac{126}{99} = \frac{99}{99} + \frac{27}{99} = 1.\overline{27}$$

$$0.\overline{314} + 0.\overline{26} = \frac{314314}{999999} - \frac{262626}{999999} = \frac{51688}{999999} = 0.\overline{051688}, \quad (\textbf{Hint: } \text{LCM} = 6 \text{ digits})$$

$$0.\overline{624} + 0.\overline{1432} = \frac{624624624624}{999999999999} + \frac{143214321432}{999999999999} = \frac{767838946056}{999999999999} = 0.\overline{767838946056}$$

Prove: $0.\overline{ab} = \dfrac{ab}{99}$

Proof: $0.\overline{ab} = 0.ab + 0.00ab + 0.0000ab + \cdots = ab(0.01 + 0.0001 + 0.000001 + \cdots)$

$$= ab\left(\frac{0.01}{1-0.01}\right) = ab\left(\frac{0.01}{0.99}\right) = ab\left(\frac{1}{99}\right) = \frac{ab}{99}.$$

It is the sum of an infinite geometric series.

Read the book "A-Plus Notes for Algebra, page 361".

CHAPTER FIVE

Ratios, Proportions, and Percents

5-1 Ratios and Rates

A **ratio** is used to compare two numbers. We write a ratio as a fraction in **lowest terms**. If an improper fraction represents a ratio, we do not change it to a mixed number. Note that the lowest terms of $\frac{18}{10}$ is $\frac{9}{5}$, the simplest form of $\frac{18}{10}$ is $1\frac{4}{5}$.
If there are 15 boys and 25 girls in a class, we can compare the number of boys to the number of girls by writing a ratio (fraction). A comparison of 15 boys to 25 girls is $\frac{15}{25}$. It is read as " 15 to 25 ". A ratio can be written in three ways: 15 to 25, 15:25, or $\frac{15}{25}$.
Just as with fractions, a ratio can be written in lowest terms: $\frac{15}{25} = \frac{3}{5}$. We say that the ratio of the number of the boys to the number of the girls is " 15 to 25 ", or " 3 to 5 ".
A ratio $\frac{15}{25}$ can be also represented to the statement " 15 out of 25 ". It is equivalent to the statement " 3 out of 5 ".
If a ratio is used to compare two measurements, we must use the measurements in the same unit (like measures). (see example 4).

A **rate** is a ratio of two unlike measures. A rate is usually simplified to a **per unit form**, or called a **unit rate**. The rate of 120 miles to 8 gallon is $\frac{120}{8} = 15$ miles per gallon.
To find a rate in a problem, it always means to find the unit rate.

Examples

1. Write the ratio " 6 to 9 " as a fraction in lowest terms.

Solution: $\frac{6}{9} = \frac{2}{3}$. Ans.

2. Write the ratio " 18 : 10 " as a fraction in lowest terms.

Solution: $\frac{18}{10} = \frac{9}{5}$. Ans.

3. Write the ratio $\frac{1.8}{1.2}$ in lowest terms.

Solution: $\frac{1.8}{1.2} = \frac{18}{12} = \frac{3}{2}$. Ans.

4. Write the ratio $\frac{2h}{150\,\text{min}}$ in lowest terms.

Solution: $\frac{2h}{150\,\text{min}} = \frac{120\,\text{min}}{150\,\text{min}} = \frac{4}{5}$. Ans.

5. You drive 240 miles in 4 hrs. At that rate, how many miles could you drive in 7 hrs. ?
Solution:

$$\text{unit rate (average speed)} = \frac{240}{4} = 60 \text{ mph (miles per hour).}$$

$$\therefore \text{distance} = 60 \times 7 = 420 \text{ miles in 7 hrs.}$$

EXERCISES

Write each ratio as a fraction in lowest terms.

1. 5 to 8 **2.** 8 to 5 **3.** 6 to 9 **4.** 9 to 6 **5.** 12 : 10 **6.** 10 : 12

7. 18 : 24 **8.** 27 : 15 **9.** 8 to 2 **10.** 2 to 8 **11.** 5 : 25 **12.** 25 : 5

13. 3.5 : 4.5 **14.** 4.5 : 3.5 **15.** 2.4 to 3.2 **16.** 3.2 to 2.4 **17.** 4.2 : 7 **18.** 7 to 4.2

19. 3.12 : 1.2 **20.** 3.15 : 10 **21.** 2.25 : 20 **22.** 10 to 2.4 **23.** $\frac{1}{5}$ to 0.4 **24.** 4.2 . $\frac{2}{5}$

25. $\frac{1}{3} : \frac{1}{4}$ **26.** $\frac{1}{4} : \frac{1}{3}$ **27.** $\frac{3}{4}$ to $\frac{3}{8}$ **28.** $\frac{3}{8}$ to $\frac{8}{3}$ **29.** $3\frac{3}{4} : 2\frac{1}{7}$ **30.** $2\frac{2}{3}$ to 6

31. 60 minutes to 20 minutes **32.** 20 minutes to one hour **33.** 8 days to 2 weeks
34. 3 minutes to 90 seconds **35.** 1 year : 5 months **36** 16 feet to 6 feet
37. 2 dollars to 60 cents **38.** 3 days : 48 hours **39.** 4 meters : 150 centimeters
40. 12 out of 48 boys **41.** 25 out of 60 students **42.** 12 out of 20 free throws

43. A line segment is shown below. Find each ratio as a fraction in lowest terms.

a. $\overline{AB} : \overline{BC}$ *b.* $\overline{BC} : \overline{AB}$ *c.* $\overline{AB} : \overline{AC}$

A —— 10 —— B —— 4 —— C

44. A baseball team won 15 games and lost 6 games. Find each ratio as a fraction in lowest terms *a.* wins to losses *b.* losses to wins *c.* wins to total *d.* losses to total

45. The population of a city is 150,000 in 1990 and 250,000 in 2000. Find the ratio (in lowest terms) of the growth of the population to its population in 2000.

46. Ronald can finish the job in 6 hours. What part of the job can he finish in 2 hours ?

47. The cost of a first-class letter is 35 cents. The cost of an express letter is $1.25. Find the ratio (in lowest terms) to compare the cost of a first-class letter to the cost of an express letter.

48. There are 80 apples in a bag. 5 of them are overripe. How many apples on average are there to have one overripe apple ?

49. There are 42 students in a class. 14 of them wear glasses. How many students on average are there to have one student wearing glasses ?

50. A 12-liter salt-water solution contains 2 liters of pure salt. What part (ratio) of the solution is pure salt ?

51. You drive 250 miles on 4 gallons of gasoline. Find the rate of fuel consumption in miles per gallon.

52. You drive 313 miles in 5 hours. What is your average speed ? At that rate, how many miles could you drive in 7 hours ?

53. What is the unit price if 12 eggs sell for $2.16 ? How many eggs could you buy for $18 ?

54. A can of 3 lb of A-brand ground beef costs $9.75. A can of 4 lb of B-brand ground beef costs $12.60. Based on the unit price only, which brand is the better buy ?

55. You spent $8.10 for 6 gallons of gasoline. How much would you spend for 22 gallons ?

5-2 Proportions

A **proportion** is formed by two equivalent ratios. In a proportion, we say that the given two ratios are equal or **proportional**.

$\frac{1}{2} = \frac{2}{4}$ is a proportion.

If two ratios are not equal, we say that the two given ratios are **not proportional**.
$\frac{1}{2}$ and $\frac{3}{4}$ are not equal (not proportional). We can not form them as a proportion.

$$\frac{1}{2} \neq \frac{3}{4}$$

To determine if two given ratios are equal (proportional), we use the following statement:

Property of Proportions: In a proportion, the cross products are equal.

If $\frac{a}{b} = \frac{c}{d}$, then $ad = bc$. (b and d are nonzero numbers)

Example: $\frac{1}{2} = \frac{2}{4}$ is a proportion because $1 \times 4 = 2 \times 2$.

Sometimes one of the 4 terms in a proportion is a variable. An equation is formed.

$$\frac{1}{2} = \frac{n}{4}$$

To find n in this proportion, we use equivalent fractions with a common denominator.

$$\frac{1 \times 2}{2 \times 2} = \frac{n}{4} \rightarrow \frac{2}{4} = \frac{n}{4} \qquad \therefore n = 2$$

We can find n also by **cross-multiplying** in the proportion.

$$\frac{1}{2} = \frac{n}{4} \rightarrow 1 \times 4 = 2 \times n, \quad 4 = 2n, \quad n = \frac{4}{2} = 2. \text{ Ans.}$$

Examples

1. Are the ratios $\frac{5}{8}$ and $\frac{15}{24}$ proportional ?
Solution:
Cross multiply: $5 \times 24 = 8 \times 15$
$(120 = 120)$
They are proportional.

2. Are the ratios $\frac{5}{6}$ and $\frac{7}{8}$ proportional ?
Solution:
Cross multiply: $5 \times 8 \neq 6 \times 7$
$(40 \neq 42)$
They are not proportional.

3. Solve $\frac{n}{24} = \frac{5}{8}$.
Solution:
$n \times 8 = 24 \times 5$
$8n = 120 \quad \therefore n = \frac{120}{8} = 15$. Ans.

4. Solve $\frac{3}{5} = \frac{x}{20}$.
Solution:
$3 \times 20 = 5 \times x$
$60 = 5x \quad \therefore x = \frac{60}{5} = 12$. Ans.

5. Two pounds of cheese cost $7. How much do five pounds cost ?
Solution: Write a proportion and use cross-products to solve it.

$$\frac{7}{2} = \frac{x}{5}, \quad 7 \times 5 = 2 \times x, \quad 35 = 2x, \quad \therefore x = \frac{35}{2} = \$17.50 \quad \text{Ans.}$$

EXERCISES

Determine whether the ratios are proportional (equivalent).

1. $\frac{2}{3}, \frac{8}{12}$ **2.** $\frac{4}{5}, \frac{18}{24}$ **3.** $\frac{4}{15}, \frac{8}{30}$ **4.** $\frac{7}{9}, \frac{14}{18}$ **5.** $\frac{3}{2}, \frac{12}{7}$ **6.** $\frac{11}{4}, \frac{5}{2}$

7. $\frac{2}{9}, \frac{10}{45}$ **8.** $\frac{9}{15}, \frac{4}{5}$ **9.** $\frac{7}{8}, \frac{14}{17}$ **10.** $\frac{14}{9}, \frac{5}{3}$ **11.** $\frac{12}{15}, \frac{4}{5}$ **12.** $\frac{12}{120}, \frac{2}{20}$

13. $\frac{2}{1.5}, \frac{8}{6}$ **14.** $\frac{1.2}{5}, \frac{6}{25}$ **15.** $\frac{2.4}{7}, \frac{3}{5}$ **16.** $\frac{3}{4}, \frac{3.6}{4.8}$ **17.** $\frac{1.2}{3.5}, \frac{3.6}{9.5}$ **18.** $\frac{7}{8.5}, \frac{2.8}{3.4}$

19. $\frac{3}{1.2}, \frac{9}{3.5}$ **20.** $\frac{2.1}{5}, \frac{8.4}{20}$

Solve each proportion.

21. $\frac{n}{5} = \frac{3}{15}$ **22.** $\frac{2}{15} = \frac{n}{45}$ **23.** $\frac{5}{7} = \frac{x}{21}$ **24.** $\frac{x}{9} = \frac{12}{27}$ **25.** $\frac{y}{12} = \frac{1}{2}$ **26.** $\frac{7}{8} = \frac{y}{24}$

27. $\frac{k}{4} = \frac{12}{16}$ **28.** $\frac{4}{k} = \frac{16}{12}$ **29.** $\frac{1}{3} = \frac{14}{n}$ **30.** $\frac{7}{n} = \frac{21}{15}$ **31.** $\frac{6}{7} = \frac{p}{21}$ **32.** $\frac{r}{18} = \frac{5}{6}$

33. $\frac{x}{2} = \frac{7}{5}$ **34.** $\frac{4}{x} = \frac{2}{3}$ **35.** $\frac{4}{5} = \frac{y}{7}$ **36.** $\frac{m}{4} = \frac{8}{5}$ **37.** $\frac{4}{25} = \frac{n}{26}$ **38.** $\frac{5}{p} = \frac{2}{9}$

39. $\frac{n}{1.4} = \frac{6}{7}$ **40.** $\frac{1.2}{x} = \frac{9}{1.5}$

41. If 2 pounds of beef cost $5.60, how much do 5 pounds cost ?
42. If 2 pounds of beef cost $5.60, how many pounds can you buy for $14 ?
43. Bob earns $25 for 3 hours of cleaning floor. How many hours must he work to earn $125 ?
44. A picture 12 by 15 cm is enlarged to 20 by x cm. What is the length of x ?
45. A paper 11 by 8.5 in. is reduced by a copying machine. If the reduced paper is 8.8 in. long, what is its width ?
46. The length and the width of a rectangular field are 1.2 by 0.5 meters on a map. What is the actual width if the actual length is 120 meters ?
47. A bus traveled 320 miles in 5 hours. How far would it travel at the same speed in 8 hours ?
48. A bus uses 4.5 gallons of gasoline to travel 75 miles. How many gallons would the bus use to travel 90 miles ?
49. Bob saves $5 out of every $20 he earns. He earns $820 a month. How much does he save a year.
50. A map of 2.2 inches represents 250 miles. How many inches would represent 450 miles ?

5-3 Percents

A **percent** is a ratio of a number to 100. The symbol for percent is " %".
Percent means: **1. per hundred (or hundredths).**

$$20\% = \frac{20}{100}$$

2. out of 100.

20% means " 20 out of 100 ". Therefore, $20\% = \frac{20}{100}$.

Then we have: 20% of 100 is 20. We multiply: $100 \times 20\% = 100 \times \frac{20}{100} = 20$.
20% of 200 is 40. $200 \times 20\% = 200 \times \frac{20}{100} = 40$.

$\frac{20}{100} = \frac{2}{10} = \frac{1}{5}$. " 20 out of 100 " is equivalent to " 2 out of 10 ".
" 20 out of 100 " is equivalent to " 1 out of 5 ".
Therefore, $\frac{20}{100}$, $\frac{2}{10}$, and $\frac{1}{5}$ are all equivalent to 20%.

Examples: 1. Find 20% of 300.
Solution:
$$300 \times 20\%$$
$$= 300 \times \frac{20}{100}$$
$$= 60. \quad \text{Ans.}$$

2. Find $\frac{1}{5}$ of 300.
Solution:
$$300 \times \frac{1}{5} = 60. \quad \text{Ans.}$$

3. What number is 5% 0f 40 ?
Solution:
$$n = 40 \times 5\%$$
$$= 40 \times \frac{5}{100}$$
$$= 2. \quad \text{Ans.}$$

4. What number is 8% of 50 ?
Solution:
$$n = 50 \times 8\%$$
$$= 50 \times \frac{8}{100}$$
$$= 4. \quad \text{Ans.}$$

5. 60 out of 300 students are boys. What percent of the students is boy ?
Solution:
$$p = \frac{60}{300} = \frac{20}{100} = 20\%. \quad \text{Ans.}$$

We can also use a proportion to change a fraction to a percent.

6. Write $\frac{1}{5}$ as a percent.
Solution:
$$\frac{1}{5} = \frac{n}{100}$$
$$5n = 100$$
$$n = \frac{100}{5} = 20$$
$$\therefore \frac{1}{5} = 20\%. \quad \text{Ans.}$$

7. What percent is 60 out of 300 ?
Solution:
$$\frac{60}{300} = \frac{n}{100}$$
$$300n = 6{,}000$$
$$n = \frac{6{,}000}{300} = 20$$
$$\therefore \frac{60}{300} = 20\%. \quad \text{Ans.}$$

EXERCISES

Write each fraction as a percent (use a proportion).

1. $\frac{1}{4}$ **2.** $\frac{2}{4}$ **3.** $\frac{3}{4}$ **4.** $\frac{4}{4}$ **5.** $\frac{5}{4}$ **6.** $\frac{8}{4}$ **7.** $\frac{1}{2}$ **8.** $\frac{2}{5}$

9. $\frac{5}{5}$ **10.** $\frac{3}{10}$ **11.** $\frac{9}{10}$ **12.** $\frac{10}{10}$ **13.** $\frac{1}{25}$ **14.** $\frac{4}{25}$ **15.** $\frac{10}{25}$ **16.** $\frac{15}{25}$

17. $\frac{1}{3}$ **18.** $\frac{2}{3}$ **19.** $\frac{4}{3}$ **20.** $\frac{15}{4}$ **21.** $\frac{5}{5}$ **22.** $\frac{47}{50}$ **23.** $\frac{23}{50}$ **24.** $\frac{9}{5}$

25. $\frac{1}{100}$ **26.** $\frac{37}{100}$ **27.** $\frac{2}{8}$ **28.** $\frac{12}{5}$ **29.** $\frac{7}{4}$ **30.** $\frac{7}{20}$

Write each percent as a proper fraction in lowest terms or as a mixed number in simplest form.

31. 2% **32.** 4% **33.** 5% **34.** 9% **35.** 10% **36.** 50% **37.** 75% **38.** 90%

39. 12% **40.** 15% **41.** 18% **42.** 25% **43.** 55% **44.** 98% **45.** 99% **46.** 102%

47. 120% **48.** 125% **49.** 130% **50.** 375%

51. Find 4% of 100.
52. Find 5% of 100.
53. Find 5% of 40.
54. Find 20% of 40.
55. What number is 15% of 100 ?
56. What number is 25% of 200 ?
57. What number is 40% of 150 ?
58. What number is 45% of 25 ?
59. What number is 100% of 10 ?
60. What number is 150% of 100 ?
61. What number is 180% of 20 ?
62. What number is 180% of 200 ?
63. What number is 200% of 15 ?
64. What number is 300% of 150 ?
65. What percent is 10 out of 50 ?
66. What percent is 25 out of 50 ?
67. What percent is 40 of 50 ?
68. What percent is 40 of 200 ?
69. What number is 15% of 45 ?
70. What number is 25% of 55 ?
71. What number is 45% of 90 ?
72. What number is 22% of 40 ?
73. What percent is 3 of 5 ?
74. What percent is 5 of 200 ?
75. What percent is 9 of 9 ?
76. What percent is 5.5 of 50 ?
77. What percent is 1.2 of 20 ?
78. What percent is 2.5 of 40 ?
79. What percent is 1 of 1 ?
80. What percent is 2.5 of 2.5 ?

5-4 Percents and Decimals

The following examples show us a general relationship between percents and decimals.

$$45\% = \frac{45}{100} = 0.45, \qquad 120\% = \frac{120}{100} = 1.20, \qquad 25.4\% = \frac{25.4}{100} = 0.254$$

Therefore, we have the following rules:

A) To change a percent to a decimal, we move the decimal point 2 places to the left and remove the % sign.

Examples: **1.** 20% = 0.2 ; 2% = 0.02 ; 0.2% = 0.002 ; 0.02% = 0.0002.

2. 25% = 0.25 ; 2.5% = 0.025 ; 0.25% = 0.0025.

3. 125% = 1.25 ; 225% = 2.25 ; 925% = 9.25.

4. 25.7% = 0.257 ; 125.7% = 1.257 ; 1325.7% = 13.257.

B) To change a decimal to a percent, we move the decimal point 2 places to the right and add the % sign.

Examples: **1.** 0.2 = 20% ; 0.02 = 2% ; 0.002 = 0.2% ; 0.0002 = 0.02%.

2. 0.25 = 25% ; 0.025 = 2.5% ; 0.0025 = 0.25%.

3. 1.25 = 125% ; 2.25 = 225% ; 9.25 = 925%.

4. 0.257 = 25.7% ; 1.257 = 125.7% ; 13.257 = 1325.7%

Examples

1. What number is 5% of 40 ?

Solution:

$$\begin{aligned} n &= 40 \times 5\% \\ &= 40 \times 0.05 \\ &= 2. \quad \text{Ans.} \end{aligned}$$

2. Find 80% of 50.

Solution:

$$\begin{aligned} &50 \times 80\% \\ &= 50 \times 0.8 \\ &= 40. \quad \text{Ans.} \end{aligned}$$

3. What percent is 15% of 80% ?

Solution:

$$\begin{aligned} &80\% \times 15\% \\ &= 0.80 \times 0.15 \\ &= 0.12 = 12\%. \quad \text{Ans.} \end{aligned}$$

4. Find the sales tax on \$50 if the tax rate is 8.25 %.

Solution:

$$50 \times 8.25\% = 50 \times 0.0825 = \$4.13.$$

Ans.

EXERCISES

Write each percent as a decimal.

1. 2% **2.** 4% **3.** 5% **4.** 9% **5.** 10% **6.** 50% **7.** 75%

8. 90% **9.** 12% **10.** 15% **11.** 18% **12.** 25% **13.** 55% **14.** 98%

15. 99% **16.** 102% **17.** 0.5% **18.** 0.05% **19.** 0.005% **20.** 0.45% **21.** 0.045%

22. 82.4% **23.** 79.8% **24.** 2.35% **25.** 150% **26.** 1120% **27.** 9.99% **28.** 99.9%

29. 1001% **30.** 249%

Write each decimal as a percent.

31. 0.2 **32.** 0.02 **33.** 0.002 **34.** 0.9 **35.** 0.1 **36.** 1.0 **37.** 0.01

38. 0.99 **39.** 1.01 **40.** 0.099 **41.** 0.125 **42.** 1.25 **43.** 4.5 **44.** 0.835

45. 0.0035 **46.** 0.0001 **47.** 0.071 **48.** 7.05 **49.** 14.12 **50.** 0.0029 **51.** 0.0219

52. 21.2 **53.** 10.25 **54.** 2.1 **55.** 0.333 **56.** 3.33 **57.** 0.033 **58.** 1.001

59. 0.875 **60.** 0.0875

61. Find 4% of 100.
62. Find 5% of 100.
63. Find 5% of 80.
64. Find 25% of 40.
65. What number is 15% Of 90 ?
66. What number is 20% of 200 ?
67. What number is 15% of 150 ?
68. What number is 45% of 25 ?
69. What number is 120% of 100 ?
70. What number is 120% of 80 ?
71. What percent is 22% of 40% ?
72. What percent is 18% of 55% ?

73. What is the sale tax on $90 if the tax rate is 8.25% ?
74. What is the sales tax on $1.99 if the tax rate is 8% ?
75. What is the sales tax on $2.99 if the tax rate is 7.25% ?
76. What is the income tax on $1,200 if the tax rate is 30% ?
77. What is the income tax on $2,500 if the tax rate is 15% ?
78. What is the income tax on $25,000 if the tax rate is 28% ?
79. What is the income tax on $120,000 if the tax rate is 39.2% ?
80. 30% of teens do not consider themselves to be risk for sexuality-related diseases and 4% of this group accounts for new sexuality-related disease cases each year. How many new sexuality-related disease cases are there each year at a high school with 3,000 students ?

5-5 Percents and Fractions

We have learned how to use a proportion to change a fraction to a percent in Section 5-3. In this Section, we introduce an easier way to change a fraction to percent.

To change a fraction to a percent, we first change it to a decimal and then to a percent.

Examples

1. $\frac{1}{2} = 0.5 = 50\%$, $\frac{1}{5} = 0.2 = 20\%$, $\frac{2}{5} = 0.4 = 40\%$, $\frac{3}{4} = 0.75 = 75\%$

$\frac{3}{8} = 0.375 = 37.5\%$, $\frac{5}{8} = 0.625 = 62.5\%$

2. $\frac{1}{3} = 0.333\cdots = 33.\bar{3}\,\% = 33.3\,\%$ (**It is the nearest tenth of a percent.**)

$\frac{1}{3} = 0.33\frac{1}{3} = 33\frac{1}{3}\,\%$ (**It is an exact percent.**)

3. $\frac{3}{7} = 0.42\overline{857142} = 0.42\frac{857142}{999999} = 0.42\frac{857142 \div 142857}{999999 \div 142857} = 0.42\frac{6}{7} = 42\frac{6}{7}\%$

(Hint: To find the GCF = 142857, read "Rollabout Divisions" on Page 353.)

To change a percent to a fraction, we change it to an equivalent fraction with a denominator of 100.

Examples

1. $50\,\% = \frac{50}{100} = \frac{1}{2}$, $20\,\% = \frac{20}{100} = \frac{1}{5}$, $40\,\% = \frac{40}{100} = \frac{2}{5}$, $125\,\% = \frac{125}{100} = 1\frac{1}{4}$

$12.5\,\% = \frac{12.5}{100} = \frac{125}{1000} = \frac{1}{8}$, $62.5\,\% = \frac{62.5}{100} = \frac{625}{1000} = \frac{5}{8}$

2. $\frac{1}{2}\% = \frac{1}{2} \times \frac{1}{100} = \frac{1}{200}$, $2\frac{1}{2} = \frac{5}{2}\% = \frac{5}{2} \times \frac{1}{100} = \frac{5}{200} = \frac{1}{40}$

$33\frac{1}{3}\% = \frac{100}{3}\% = \frac{100}{3} \times \frac{1}{100} = \frac{1}{3}$

Examples

1. Find $5\frac{1}{3}\%$ of 33. Solution: $33 \times 5\frac{1}{3}\% = 33 \times \frac{16}{3}\% = 33 \times \frac{16}{3} \times \frac{1}{100} = \frac{528}{300} = 1.76$. Ans.

2. What percent is 12 of 30 ? Solution: $.\frac{12}{30} = 0.4 = 40\%$. Ans.

EXERCISES

Write each fraction as a percent.

1. $\frac{1}{4}$ **2.** $\frac{3}{4}$ **3.** $\frac{4}{5}$ **4.** $\frac{7}{8}$ **5.** $\frac{1}{10}$ **6.** $\frac{3}{10}$ **7.** $\frac{1}{25}$

8. $\frac{9}{5}$ **9.** $\frac{12}{5}$ **10.** $2\frac{3}{5}$ **11.** $6\frac{2}{5}$ **12.** $2\frac{3}{8}$ **13.** $1\frac{7}{8}$ **14.** $3\frac{1}{8}$

15. $\frac{1}{100}$ **16.** $\frac{3}{50}$ **17.** $\frac{23}{100}$ **18.** $\frac{3}{200}$ **19.** $\frac{5}{4}$ **20.** $\frac{8}{5}$

Write each fraction as a percent. Round to the nearest tenth of a percent.

21. $\frac{5}{12}$ **22.** $\frac{5}{9}$ **23.** $\frac{5}{6}$ **24.** $\frac{3}{7}$ **25.** $\frac{1}{15}$ **26.** $\frac{2}{9}$. **27.** $\frac{4}{13}$

28. $1\frac{1}{3}$ **29.** $2\frac{2}{3}$ **30.** $\frac{13}{12}$

Write each fraction as an exact percent.

31. $\frac{5}{12}$ **32.** $\frac{5}{9}$ **33.** $\frac{5}{6}$ **34.** $\frac{3}{7}$ **35.** $\frac{1}{15}$ **36.** $\frac{2}{9}$ **37.** $\frac{4}{13}$

38. $1\frac{1}{3}$ **39.** $2\frac{2}{3}$ **40.** $\frac{13}{12}$

Write each percent as a proper fraction in lowest terms or as a mixed number in simplest form.

41. 2% **42.** $1\frac{1}{4}$% **43.** 5% **44.** 9% **45.** 10% **46.** $5\frac{1}{2}$% **47.** 75%

48. $1\frac{1}{2}$% **49.** 12% **50.** $3\frac{1}{2}$% **51.** $4\frac{1}{3}$% **52.** 25% **53.** 90% **54.** $66\frac{2}{3}$%

55. 99% **56.** 102% **57.** 0.5% **58.** 0.05% **59.** 0.45% **60.** 1001%

61. Find $1\frac{1}{2}$% of 100.
62. Find $4\frac{2}{3}$% of 30.
63. Find $2\frac{1}{3}$% of 60.
64. Find $20\frac{1}{4}$% of 80.
65. What number is $3\frac{1}{5}$% of 100 ?
66. What number is $10\frac{1}{4}$% of 200 ?
67. What number is $\frac{4}{5}$% of 250 ?
68. What number is $5\frac{4}{5}$% of 55 ?
69. What percent is 1 of 4 ?
70. What percent is 4 of 5 ?
71. What percent is 4 out of 20 ?
72. What percent is 2.25 out of 15 ?

5-6 Percent and Applications

In this section, we apply the ideas we have learned about percent to solve real consumer problems.

A) Finding a percent of one number of another, we write a " p " for the percent.

Examples:

1. What percent of 40 is 5 ? (5 is what percent of 40 ?)
   Solution:

$$40 \times p = 5 \qquad \therefore p = \frac{5}{40} = \frac{1}{8} = 0.125 = 12.5\% . \quad \text{Ans.}$$

2. Joe spent \$50 and sales tax \$4 to buy a pair of shoes. What percent is the sales tax ?
   Solution:

$$50 \times p = 4 \qquad \therefore p = \frac{4}{50} = \frac{2}{25} = 0.08 = 8\% . \quad \text{Ans.}$$

B) Finding a number when both the percent and another number are known, we write a " n " for the number.

Examples:

3. 40 is 5% of what number ? (5% of what number is 40 ?)
   Solution:

$$40 = n \times 5\%$$

$$40 = n \times 0.05 \qquad \therefore n = \frac{40}{0.05} = 800 . \quad \text{Ans.}$$

4. $5\frac{1}{3}\%$ of what number is 1.76 ?
   Solution:

$$n \times 5\tfrac{1}{3}\% = 1.76 \ , \quad n \times \frac{16}{3}\% = 1.76 \ , \quad n \times \frac{16}{300} = 1.76$$

$$\therefore n = 1.76 \times \frac{300}{16} = \frac{528}{16} = 33 . \quad \text{Ans.}$$

5. Joe paid \$4 sales tax for a pair of shoes he bought. The sales-tax rate is 8%. What is the price of the shoes ?
   Solution:

Let $p =$ the price

$$p \times 8\% = 4 \ , \quad p \times 0.08 = 4$$

$$\therefore p = \frac{4}{0.08} = \$50 . \quad \text{Ans.}$$

-----Continued-----

C) Finding the percent of increase (or decrease), we write the ratio of the amount of increase (or decrease) to the original amount.

Examples

6. What is the percent of increase from 4 to 6 ?

Solution: percent of increase $= \dfrac{6-4}{4} = \dfrac{1}{2} = 0.5 = 50\%$. Ans.

7. What is the percent of decrease from 6 to 4 ?

Solution: percent of decrease $= \dfrac{6-4}{6} = \dfrac{2}{6} = \dfrac{1}{3} = 0.33\tfrac{1}{3} = 33\tfrac{1}{3}\%$. Ans.

8. The bill of electricity was \$60 last month. It decreased to \$45 this month. Find the percent of decrease.
Solution:

percent of decrease $= \dfrac{60-45}{60} = \dfrac{15}{60} = \dfrac{1}{4} = 0.25 = 25\%$. Ans.

9. The population of a city increased from 125,000 to 180,000 in 10 years. Find the percent of increase.
Solution:

percent of increase $= \dfrac{180{,}000 - 125{,}000}{125{,}000} = \dfrac{55{,}000}{125{,}000} = \dfrac{11}{25} = 0.44 = 44\%$.

D) Discount and Commission

A discount is a decrease in the price of a product. A discount can be expressed as a percent of the original price.
A commission is a payment to the salesman for selling products. A commission can be expressed as a percent of the total sales.

Examples

10. A suit that is regularly \$45.90 is on sale for 30% discount. What is the sale price of the suit ?
Solution:

the discount: $\$45.90 \times 30\% = 45.90 \times 0.30 = \$13.77$.
the sale price: $\$45.90 - \$13.77 = \$32.13$. Ans.
(Or: $\$45.90 \times 70\% = 45.90 \times 0.70 = \$32.13$)

11. Mary sold sport products worth \$32,000 this week. She receives 3% commission on her sales. How much is her commission this week ?
Solution:

commission: $\$32{,}000 \times 3\% = 32{,}000 \times 0.03 = \$960$. Ans.

-----Continued-----

E) Simple Interest

Interest is the amount of money charged for the use of borrowed money. Interest is also the amount of money earned for the money deposited in a bank.
The interest rate is usually expressed as a percent of the principal over a period of time.
If the interest rate is computed on a yearly basis, it is called a yearly rate or an annual rate.
When interest is computed on the principal only (no interest on the interest previously earned), it is called **simple interest**.

The formula to compute simple interest is:

$$\text{Interest} = \text{principal} \times \text{rate} \times \text{time}$$

$$\mathbf{I = p\,r\,t}$$

Examples:

12. Find the simple interest you pay if you borrow \$500 for 3 years at a yearly rate 7%.
Solution:

$$I = p\,r\,t, \quad I = \$500 \times 7\% \times 3 = \$500 \times 0.07 \times 3 = \$105. \quad \text{Ans.}$$

13. You have a saving account with \$500. The annual interest rate is 3.5%. How much simple interest will your account earn in 3 months ? How much do you have in your account after 3 months ?
Solution:

$$I = p\,r\,t, \quad I = 500 \times 3.5\% \times \frac{3}{12} = 500 \times 0.035 \times \frac{1}{4} = \$4.38. \quad \text{Ans.}$$

$$\text{Total} = \$500 + \$4.38 = \$504.38. \quad \text{Ans}$$

F) Compound Interest

Compound interest is computed on "the principal plus the interest previously earned". The principal increases when the interest is compounded. The compound interest is usually given in daily, monthly, quarterly, semiannually, or yearly.

The formula to compute the compound amount (principal plus compound interest) is:

$$A = p(1+r)^n\,, \quad r: \text{annual interest rate} \quad n: \text{number of years}$$

Examples

14. You have a savings account with \$500. The annual interest rate is 7%, compounded yearly. How much do you have in your account after 3 years ?
Solution:

$$A = p(1+r)^n = 500(1+0.07)^3 = 500(1.07)^3 = 500(1.225043) = \$612.52 \quad \text{Ans.}$$

15. \$500 is deposited in a bank at an annual interest rate 6%, compounded monthly. Find the new principal after 3 months ?
Solution: monthly rate $r = 0.06 \div 12 = 0.005$

$$A = p(1+r)^n = 500(1+0.005)^3 = 500(1.005)^3$$

$$= 500(1.01508) = \$507.54. \quad \text{Ans.}$$

-----Continued-----

EXERCISES

Find each answer.

1. What percent of 70 is 14 ?
2. What percent of 90 is 27 ?
3. What percent of 70 is 17.5 ?
4. What percent of 90 is 31.5 ?
5. What percent of 64 is 16 ?
6. What percent of 45 is 27 ?
7. What percent of 64 is 35.2 ?
8. What percent of 45 is 38.25 ?
9. 21 is what percent of 60 ?
10. 3 is what percent of 60 ?
11. 12.6 is what percent of 30 ?
12. 20 is what percent of 60 ?
13. 30 is 5% of what number ?
14. 40 is 25% of what number ?
15. 32 is 8% of what number ?
16. 96 is 40% of what number ?
17. 20 is 100% of what number ?
18. 15 is 200% of what number ?
19. 15 % of what number is 48 ?
20. 160% of what number is 40 ?
21. 0.5% of what number is 15 ?
22. 0.2% of what number is 2 ?
23. $2\frac{1}{3}$% of what number is 2.1 ?
24. $3\frac{2}{3}$% of what number is 2.2 ?
25. You spent $90 and sales tax $6.75 to buy a suit. What percent is the sales tax ?
26. You paid $6.75 sales tax for a suit you bought. The sales-tax rate is 7.5%. What is the price of the suit ?
27. Ronald bought a computer table that costs $180 has a down payment of $54. What percent of the total cost is the down payment ?
28. Ronald paid a down payment of $54 for a computer table. The down payment is 30% of the price. What is the price of the computer table ?
29. You got 35 questions right in a high school math exit test with 50 questions. What percent of the questions did you get right ?
30. You got 35 questions right in a high school math exit test. It is 70% of the questions. How many questions were on the test ?
31. The price of a gallon of gasoline was $1.65 last month. It increased to $1.98 this month. Find the percent of increase.
32. The annual budget for medical research at a local hospital was $1,200,000 last year. The budget is $900,000 this year. What is the percent of decrease ?
33. The number of students at a high school increased 130% from the original total of 1,250 students ten years ago. What is the number of students now ?
34. A chair that is regularly $19.90 is on sale for 20% discount. What is the new price ?
35. David sold a new house worth $180,000. He receives 2.5% commission on the selling price of the house. How much did he earn in commission ?
36. David earned $4,500 in commission on the new house he sold. The commission is 2.5% of the selling price of the house. What is the selling price of the house ?
37. Find the simple interest you paid if you borrowed $2,000 for 5 years at an annual rate 12% ?
38. If you deposit $2,000 in a bank at a simple-interest yearly rate 4.8%, how much do you have in the bank after 2 years ?
39. If you deposit $2,000 in a bank at a yearly rate 4.8%, compounded monthly, how much do you have in the bank after 2 months ?
40. Find the new principal amount after 3 months in problem 39.

CHAPTER 5 EXERCISES

Write each ratio as a fraction in lowest terms.

1. 7 to 4 **2.** 9 to 3 **3.** 15 : 3 **4.** 3 : 12 **5.** 2.5 to 1.5 **6.** 7 : 2.8
7. $\frac{1}{4}$ to 0.4 **8.** 2.1 to 0.7 **9.** $\frac{3}{5}:\frac{6}{7}$ **10.** $\frac{2}{7}:\frac{3}{7}$ **11.** $1\frac{2}{3}:2\frac{1}{6}$ **12.** $1\frac{1}{2}$ to 9
13. 7 to $1\frac{5}{9}$ **14.** $3\frac{1}{3}:2\frac{2}{9}$ **15.** $12x:6x$ **16.** $6xy:15y$ **17.** 10 out of 25 students
18. 2 minutes to 45 seconds **19.** 3 weeks to 9 days **20.** 20 wins to 12 losses

Determine whether the ratios are proportional (equivalent).

21. $\frac{1}{2}$ and $\frac{4}{8}$ **22.** $\frac{1}{2}$ and $\frac{8}{4}$ **23.** $\frac{3}{4}$ and $\frac{9}{11}$ **24.** $\frac{13}{15}$ and $\frac{26}{30}$

25. $\frac{6}{1.3}$ and $\frac{12}{2.6}$ **26.** $\frac{1.2}{4}$ and $\frac{4.5}{16}$ **27.** $\frac{9}{1.2}$ and $\frac{63}{8.4}$ **28.** $\frac{4}{5}$ and $\frac{4.8}{6.5}$

29. $\frac{1.2}{3.1}$ and $\frac{3.9}{9.3}$ **30.** $\frac{2.3}{4.2}$ and $\frac{2.76}{5.04}$

Solve each proportion.

31. $\frac{n}{2}=\frac{4}{8}$ **32.** $\frac{1}{2}=\frac{8}{n}$ **33.** $\frac{3}{4}=\frac{n}{11}$ **34.** $\frac{13}{x}=\frac{26}{30}$ **35.** $\frac{y}{16}=\frac{3}{48}$

36. $\frac{m}{1.3}=\frac{12}{2.6}$ **37.** $\frac{1.2}{k}=\frac{4.8}{16}$ **38.** $\frac{9}{1.2}=\frac{x}{8.4}$ **39.** $\frac{4}{p}=\frac{4.8}{3}$ **40.** $\frac{y}{3.1}=\frac{3.9}{9.3}$

Use a proportion to write each fraction as a percent.

41. $\frac{1}{5}$ **42.** $\frac{5}{5}$ **43.** $\frac{4}{5}$ **44.** $\frac{1}{50}$ **45.** $\frac{11}{50}$ **46.** $\frac{3}{25}$ **47.** $\frac{25}{25}$ **48.** $\frac{7}{8}$

49. $\frac{7}{100}$ **50.** $\frac{13}{40}$

Write each percent as a proper fraction in lowest terms or as a mixed number in simplest form.

51. 20% **52.** 8% **53.** 40% **54.** 99% **55.** 1% **56.** 101% **57.** 199% **58.** 55%
59. 120% **60.** 325%

Write each percent as a decimal.

61. 8% **62.** 11% **63.** 4.5% **64.** 99% **65.** 1% **66.** 1.09% **67.** 0.1% **68.** 0.02%
69. 225% **70.** 22.5%

Write each decimal as a percent.

71. 0.08 **72.** 0.11 **73.** 0.045 **74.** 0.99 **75.** 0.01 **76.** 0.0109 **77.** 0.001 **78.** 0.0002
79. 2.25 **80.** 0.225

-----Continued-----

Write each fraction as a percent. If the fraction is a repeating decimal, write it as an exact percent.

81. $\frac{1}{4}$ **82.** $\frac{3}{5}$ **83.** $\frac{8}{5}$ **84.** $4\frac{1}{2}$ **85.** $\frac{7}{8}$ **86.** $\frac{23}{40}$

87. $\frac{2}{3}$ **88.** $\frac{1}{30}$ **89.** $\frac{5}{6}$ **90.** $\frac{5}{9}$

Write each percent as a fraction in simplest form.

91. 25% **92.** 60% **93.** 160% **94.** 450% **95.** 87.5% **96.** 57.5%

97. $66\frac{2}{3}$% **98.** $3\frac{1}{3}$% **99.** $83\frac{1}{3}$% **100.** $55\frac{5}{9}$%

101. Find 2% of 60

102. Find 28% of 25 ?

103. What number is $\frac{1}{4}$ of 150 ?

104. What number is 120% of 15 ?

105. What number is 1.5% of 45 ?

106. What number is 12.5% of 70 ?

107. What percent is 5 of 40 ?

108. What percent is 3.5 of 40 ?

109. What percent is 0.72 of 4.5 ?

110. What percent is 30 of 30 ?

111. What percent is 12% of 60% ?

112. What percent is 2% of 65% ?

113. Find $4\frac{1}{4}$% of 80.

114. Find $4\frac{1}{3}$% of 900.

115. What number is $5\frac{2}{3}$% of 600 ?

116. What number is $5\frac{1}{5}$% of 60 ?

117. There are 32 students in a class. 24 of them are girls. Find the ratio (in lowest terms) of the number of the girls to the number of the boys.

118. There are 32 students in a class. 24 of them are girls. What is the percentage of the girls in this class ?

119. A 25-liter alcohol-water solution contains 18 liters of pure alcohol. What is the percentage of alcohol in the solution ?

120. A picture 3 by 6 cm is enlarged to 8 by x cm. What is the length of x ?

121. A bus traveled 370 miles in 6 hours. How far would it travel at the same speed in 9 hours ?

122. What is the sales tax on $19.50 if the tax rate is 7.5% ?

123. You spent $120 and sales tax $9 to buy a table. What is the sales-tax rate ?

124. You paid $18 sales tax to buy a table. The sales-tax rate is 7.5%. What is the price of the table ?

125. You got 28 questions right in a math test with 50 questions. What percent of the questions did you get right ?

126. The population of a city increased from 120,000 to 150,000 in 10 years. Find the percent of increase.

127. The price of a gallon of gasoline was $1.98 last month. It decreased to $1.65. Find the percent of decrease.

128. A table that is regularly $120 is on sale for 30% discount. What is the new price ?

129. Find the simple interest you pay if you borrow $5,000 for 5 years at an annual rate 9.5% ?

130. If you deposit $5,000 in a bank at a simple-interest yearly rate 4.5%, how much do you have in the bank after 5 years ?

131. If you deposit $5,000 in a bank at a yearly rate 7.2%, compounded monthly, how much do you have in the bank after 4 months ?

Additional Examples

1. The ratio of the number of boys to the number of girls is $4:3$ in a class. There are 24 boys. How many girls are there in class ?
Solution:

Let x = number of girls

$$4:3 = 24:x \quad , \quad \frac{4}{3} = \frac{24}{x} \quad , \quad 4x = 3(24)$$

$\therefore x = \frac{3(24)}{4} = \frac{72}{4} = 18$. Ans: There are 18 girls in class.

2. There are 42 students in a class. The ratio of the number of boys to the number of girls is $4:3$. Find the number of boys and the number of girls in class.
Solution:

$4 + 3 = 7$.
4 out of 7 are boys. 3 out of 7 are girls.
boys: $42 \times \frac{4}{7} = 24$, girls: $42 \times \frac{3}{7} = 18$
Ans: There are 24 boys and 18 girls in class.

3. The ratio of sugar to water in a 200-gram sugar-water solution is $4:21$. How many grams of sugar are there in the solution ?
Solution:

$4 + 21 = 25$.
4 out of 25 are sugar.
sugar: $200 \times \frac{4}{25} = \frac{800}{25} = 32$ grams. Ans.

4. The ratio of the lengths of three sides of a triangle is 3 : 4 : 5. What is the length of each of the three sides if the perimeter is 96 meters ?
Solution:

$3 + 4 + 5 = 12$
We divide the perimeter of the triangle into 12 equal parts.

Three sides are: $96 \times \frac{3}{12} = 24$, $96 \times \frac{4}{12} = 32$, $96 \times \frac{5}{12} = 40$

Ans: 24 meters, 32 meters, and 40 meters.

We can check the answers (divided by 8): $24:32:40 = 3:4:5$

-----Continued-----

Additional Examples

5. A 16-gallon salt-water solution contains 4 gallons of pure salt. What is the percentage of salt in the solution ?
Solution:

$$p = \frac{4}{16} = 0.25 = 25\%. \quad \text{Ans.}$$

6. A 16-gallon salt-water solution contains 25% pure salt. How much water should be added to produce the solution to 20% salt ?
Solution:

Original solution: $16 \times 25\% = 16 \times 0.25 = 4$ gallons of pure salt
Let x = gallons of water added

$$\frac{4}{16+x} = 20\%, \quad \frac{4}{16+x} = \frac{20}{100}$$

$$4(100) = 20(16+x), \quad 400 = 320 + 20x, \quad 80 = 20x$$

$$\therefore x = \frac{80}{20} = 4 \text{ gallons of water.} \quad \text{Ans.}$$

7. A 20-gallon salt-water solution contains 10% pure salt and 90% water. If 5 gallons of water are added to the solution, what is the new concentration of pure salt ?
Solution:

Original solution: $20 \times 10\% = 20 \times 0.10 = 2$ gallons of pure salt

$$\frac{2}{20+5} = 0.08 = 8\% \text{ of pure salt.} \quad \text{Ans.}$$

8. In a 10-gram alcohol-water solution, the ratio of water to alcohol is 4 to 1. If a 30-gram alcohol-water solution containing 10 gram pure alcohol is added to the 10-gram solution, what fraction of the new solution is pure alcohol ?
Solution:

$$10 \times \frac{1}{5} = 2 \text{ grams of pure alcohol in 10-gram solution}$$

$$\frac{2+10}{10+30} = \frac{12}{40} = \frac{3}{10} \text{ of the new solution is pure alcohol.} \quad \text{Ans.}$$

9. If 32 students donated to a fund and the remaining 8 did not, what percent of the students did not donated ?
Solution:

$$\frac{8}{32+8} = \frac{8}{40} = \frac{1}{5} = 0.20 = 20\%. \quad \text{Ans.}$$

Scale Factors in similar figures

If two geometric figures are similar (the same shape), the ratios of all corresponding dimensions (length, width, height, perimeter, radius, circumference) are equal. These ratios in lowest terms are called the **scale factors**, or **similarity ratios**.
If the scale factor of two similar figures is $a : b$ (or $\frac{a}{b}$) , we have the following formulas:

The ratio of the **surface areas**: $\frac{A_1}{A_2} = \left(\frac{a}{b}\right)^2$; The ratio of the **volumes**: $\frac{V_1}{V_2} = \left(\frac{a}{b}\right)^3$

Examples:

1. The following two rectangular solids are similar,

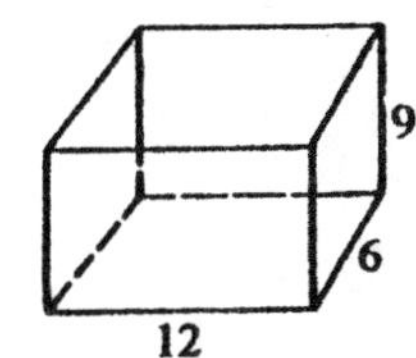

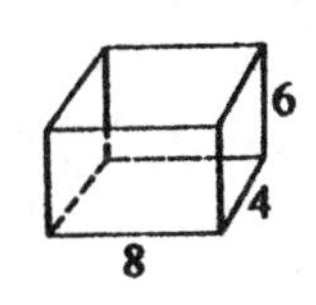

a) what is the ratio of their corresponding sides ?
b) what is the ratio of their surface areas ?
c) what is the ratio of their volumes ?

Solution: a) $\frac{a}{b} = \frac{6}{4} = \frac{9}{6} = \frac{12}{8} = \frac{3}{2}$. It is the **scale factor**.

b) $\frac{A_1}{A_2} = \frac{2(12 \cdot 6 + 12 \cdot 9 + 9 \cdot 6)}{2(8 \cdot 4 + 8 \cdot 6 + 6 \cdot 4)} = \frac{2(234)}{2(104)} = \frac{468}{208} = \frac{9}{4}$. **OR:** $\frac{A_1}{A_2} = \left(\frac{3}{2}\right)^2 = \frac{9}{4}$.

c) $\frac{V_1}{V_2} = \frac{12 \cdot 6 \cdot 9}{8 \cdot 4 \cdot 6} = \frac{648}{192} = \frac{27}{8}$. **OR:** $\frac{V_1}{V_2} = \left(\frac{3}{2}\right)^3 = \frac{27}{8}$.

2. The following two cylinders are similar,

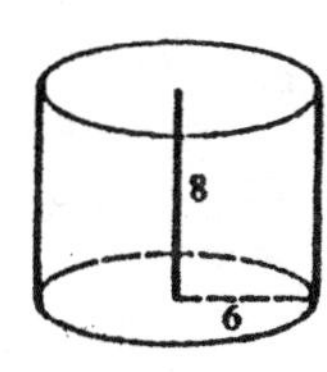

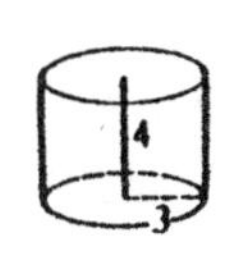

a) what is the ratio of their radii ?
b) what is the ratio of their total surface areas ?
c) what is the ratio of their volumes ?

Solution: a) $\frac{r_1}{r_2} = \frac{6}{3} = \frac{2}{1}$. It is the **scale factor**.

b) $\frac{A_1}{A_2} = \frac{2(\pi \cdot 6^2) + \pi \cdot 12 \cdot 8}{2(\pi \cdot 3^2) + \pi \cdot 6 \cdot 4} = \frac{168\pi}{42\pi} = \frac{4}{1}$. **OR:** $\frac{A_1}{A_2} = \left(\frac{r_1}{r_2}\right)^2 = \left(\frac{2}{1}\right)^2 = \frac{4}{1}$.

c) $\frac{V_1}{V_2} = \frac{\pi \cdot 6^2 \cdot 8}{\pi \cdot 3^2 \cdot 4} = \frac{288\pi}{36\pi} = \frac{8}{1}$. **OR:** $\frac{V_1}{V_2} = \left(\frac{r_1}{r_2}\right)^3 = \left(\frac{2}{1}\right)^3 = \frac{8}{1}$.

-----Continued-----

Examples

3. The following two cones are similar,

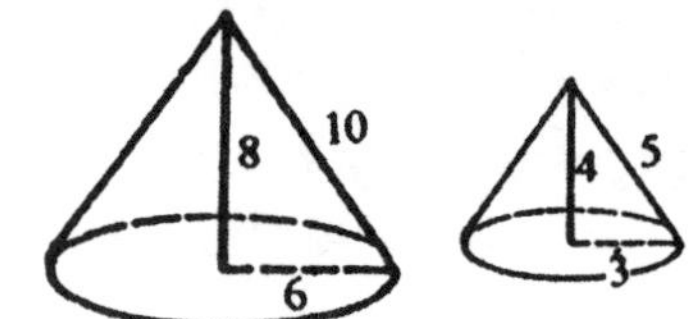

a) what is the ratio of their radii ?
b) what is the ratio of their total surface areas ?
c) what is the ratio of their volumes ?

Solution: a) $\dfrac{r_1}{r_2} = \dfrac{6}{3} = \dfrac{2}{1}$. It is the **scale factor**.

b) $\dfrac{A_1}{A_2} = \dfrac{\pi \cdot 6^2 + \pi \cdot 6 \cdot 10}{\pi \cdot 3^2 + \pi \cdot 3 \cdot 5} = \dfrac{96\pi}{24\pi} = \dfrac{4}{1}$. **OR:** $\dfrac{A_1}{A_2} = \left(\dfrac{2}{1}\right)^2 = \dfrac{4}{1}$.

c) $\dfrac{V_1}{V_2} = \dfrac{\frac{1}{3} \cdot \pi \cdot 6^2 \cdot 8}{\frac{1}{3} \cdot \pi \cdot 3^2 \cdot 4} = \dfrac{96\pi}{12\pi} = \dfrac{8}{1}$. **OR:** $\dfrac{V_1}{V_2} = \left(\dfrac{2}{1}\right)^3 = \dfrac{8}{1}$.

4. The scale factor of two similar pyramids is 5 : 1. If the surface area of the smaller pyramid is 35 square feet, what is the surface area of the larger pyramid ?

Solution: Let A_1 = the surface area of the larger pyramid.
A_2 = the surface area of the smaller pyramid.

$$\frac{A_1}{A_2} = \left(\frac{5}{1}\right)^2 = \frac{25}{1} \qquad \therefore A_1 = A_2 \cdot 25 = 35 \cdot 25 = 875 \text{ square feet.}$$

5. The scale factor of two similar spheres is 2 : 5. If the volume of the larger sphere is 375 cubic inches, what is the volume of the smaller sphere ?

Solution: Let V_1 = the volume of the smaller sphere.
V_2 = the volume of the larger sphere.

$$\frac{V_1}{V_2} = \left(\frac{2}{5}\right)^3 = \frac{8}{125} \qquad \therefore V_1 = V_2 \cdot \frac{8}{125} = 375 \cdot \frac{8}{125} = 24 \text{ cubic inches.}$$

6. If the ratio of the total surface areas of two similar cylinders is 9 : 4,
a) what is the ratio of their radii ? b) what is the ratio of their volumes ?

Solution: a) $\dfrac{r_1}{r_2} = \sqrt{\dfrac{9}{4}} = \dfrac{3}{2}$. b) $\dfrac{V_1}{V_2} = \left(\dfrac{3}{2}\right)^3 = \dfrac{27}{8}$.

CHAPTER SIX

Equations

6-1 Variables and Expressions

Variable: A letter that can be used to represent one or more numbers.

x is a variable. If $x = 1$, then $2x = 2(1) = 2$. If $x = 2$, then $2x = 2(2) = 4$.

We use variables to form algebraic expressions and equations.

Numerical expression: An expression that includes only numbers.

Algebraic expression: An expression that combines numbers and variables by operations (add, subtract, multiply or divide).

$xy + x + 2y - 3$ is an algebraic expression.

Algebraic equation: A statement by placing an equal sign "=" between two expressions.

$2x + 5 = 7$ is an equation. $2x + 5 = 9$ is an equation.

Term: A single number or a product of numbers and variables in an expression.

There are two terms in the expression $3x^2 + 6$, a polynomial or binomial.

There are three terms in the expression $3x^2 - 5x - 2$, a polynomial or trinomial.

There are four terms in the expression $xy + x + 2y - 3$, a polynomial.

Constant: A term without a variable. It is a number only. $x + 3$, 3 is a constant.

Coefficient: A number that is multiplied by a variable.

In the term $2y$, 2 is the coefficient. The coefficient of the term x is 1.

Monomial: An expression with only one term. $5x$ is a monomial. 12 is a monomial.

Polynomial: An expression formed by two or more terms.

Degree of an expression: The greatest of the degrees of its terms after it has been simplified.

The degree of $2x + 1$ is 1, a linear form. The degree of $3x^2 + x - 1$ is 2, a quadratic form.

The degree of $4x^3 + 5x$ is 3, a cubic form. The degree of $x^4 - 1$ is 4, a quartic form.

The degree of $4a^2b + 2a^3b^4 - 2$ is 7.

Evaluating an expression: The process of replacing variables with numbers in an algebraic expression and finding its value.

Examples

1. Evaluate $2x + 15$ when $x = 4$.

Solution:

$2x + 15 = 2(4) + 15 = 23$. Ans.

2. Evaluate $a^2 + b - 2$ when $a = 3$, $b = 1$.

Solution:

$a^2 + b - 2 = 3^2 + 1 - 2 = 9 + 1 - 2 = 8$. Ans.

3. Evaluate $\frac{1}{2}x + 3$ if $x = 3$.

Solution:

$\frac{1}{2}x + 3 = \frac{1}{2}(3) + 3 = 4\frac{1}{2}$. Ans.

4. Evaluate $\frac{1}{2}x - 3$ if $x = 3$.

Solution:

$\frac{1}{2}x - 3 = \frac{1}{2}(3) - 3 = -1\frac{1}{2}$. Ans.

EXERCISES

Evaluate each expression.

1. $2x+1$ if $x=1$.
2. $2x+1$ if $x=2$.
3. $2x+1$ if $x=3$.

4. $2x-1$ if $x=1$.
5. $2x-1$ if $x=2$.
6. $2x-1$ if $x=3$.

7. $2x+4$ if $x=4$.
8. $2x+4$ if $x=5$.
9. $2x+4$ if $x=6$.

10. $2x-6$ if $x=0$.
11. $2x-6$ if $x=3$.
12. $2x-6$ if $x=5$.

13. $4x-2$ if $x=-1$.
14. $4x-2$ if $x=-2$.
15. $4x-2$ if $x=-3$.

16. $4x+2$ if $x=-1$.
17. $4x+2$ if $x=-2$.
18. $4x+2$ if $x=-3$.

19. $-3x+5$ if $x=1$.
20. $-3x+5$ if $x=-1$.
21. $-3x+5$ if $x=3$.

22. $-2x-4$ if $x=3$.
23. $-2x-4$ if $x=-2$.
24. $-2x-4$ if $x=-3$.

25. $\frac{1}{2}x+5$ if $x=0$.
26. $\frac{1}{2}x+5$ if $x=3$.
27. $\frac{1}{2}x+5$ if $x=4$.

28. $\frac{1}{2}x-2$ if $x=4$.
29. $\frac{1}{2}x-2$ if $x=-3$.
30. $\frac{1}{2}x-2$ if $x=1$.

31. $\frac{3}{2}x+1$ if $x=2$.
32. $\frac{3}{2}x-3$ if $x=3$.
33. $\frac{3}{2}x-2$ if $x=-3$.

34. $2.5x+3$ if $x=4$.
35. $2.5x-3$ if $x=4$.
36. $2.5x-1.5$ if $x=-2$.

37. $2x^2-3x+5$ if $x=3$.
38. $2x^2-3x+5$ if $x=-2$.
39. x^3-x-3 if $x=-1$.

40. $2x^2+y-1$ if $x=1$, $y=2$.
41. a^2-2b+3 if $a=3$, $b=-2$.
42. $3a^2b$ if $a=2$, $b=3$.

43. The cost to buy a suit is given by the equation $c=p-0.20p$, where p is the regular price, and $0.20p$ is a 20% discount on p. Find the cost to buy the suit if the suit is regularly \$50.

44. The cost to rent a car is \$20 per day plus 20 cents per mile. Find the cost per day to rent the car if you drive 120 miles. (Hint: $c=20+0.20d$, where d is the mileage.)

45. The area of a square is given by the formula $A=s^2$, where s is the length of one side. Find the area of a square with one side 15 meters.

46. The distance in feet that an object falls in t seconds is given by the formula $d=16t^2$. Find the distance of an object falls in 8 seconds.

6-2 Adding and Subtracting Expressions

Like terms (Similar terms): Terms that contain the same form of variables in an algebraic expression.
$5x$ and $9x$ are like terms. $4a$ and $10a$ are like terms.
$3x^2y$ and $7x^2y$ are like terms.

Simplifying an expression: To simplify an expression, we combine the like terms.

To simplify an expression, we also follow the **order of operations**. (See Section 1-7)
To simplify an expression, we must use the distributive property to remove the parentheses. If there are like terms inside the parentheses, we combine them first.

Distributive Property: $a(b+c) = ab + ac$
$a(b-c) = ab - ac$

If there is only a negative sign in front of the parenthesis, we remove the parenthesis by writing the opposite of each term inside the parenthesis. Or, consider the negative sign as "–1".

Examples: $-(3x+4-2x) = -(x+4) = -1(x) + (-1)(4) = -x + (-4) = -x - 4$.
$-(x-4) = -1(x) + (-1)(-4) = -x + 4$.
$-(-x-4) = -1(-x) + (-1)(-4) = x + 4$.
$(-x-4)(-1) = -x(-1) + (-4)(-1) = x + 4$.

To add or subtract expressions, we remove the parentheses using the Distributive Property and then regroup like terms.
When we write an expression (polynomial), we always write it in **standard form**. The terms are ordered from the greatest exponent to the least.

Examples

1. $2x + 3x = 5x$. **2.** $2x - 3x = -x$. **3.** $-(2x - 3x) = -(-x) = x$.

4. $-(2x+5) = -2x - 5$. **5.** $-(2x-5) = -2x + 5$. **6.** $-(-2x-5) = 2x + 5$.

7. $(3x-5) + (5x-3) = 3x - 5 + 5x - 3 = (3x + 5x) + (-5 - 3) = 8x + (-8) = 8x - 8$.

8. $(3x-5) - (5x-3) = 3x - 5 - 5x + 3 = (3x - 5x) + (-5 + 3) = -2x + (-2) = -2x - 2$.

9. $(4x+3y) + (6x-2y) = 4x + 3y + 6x - 2y = (4x + 6x) + (3y - 2y) = 10x + y$.

10. $(4x+3y) - (6x-2y) = 4x + 3y - 6x + 2y = (4x - 6x) + (3y + 2y) = -2x + 5y$.

11. $3a^2 - 2a + 4 - (2a^2 + 3a) = 3a^2 - 2a + 4 - 2a^2 - 3a = (3a^2 - 2a^2) + (-2a - 3a) + 4$.
$= a^2 + (-5a) + 4 = a^2 - 5a + 4$.

12. $x^3 - 2x^2 + x - 4 + 2x^3 + 5x^2 + 7 = (x^3 + 2x^3) + (-2x^2 + 5x^2) + x + (-4 + 7)$
$= 3x^3 + 3x^2 + x + 3$.

EXERCISES

Simplify each expression.

1. $3x+6x$
2. $3x-6x$
3. $-3x-6x$
4. $-3x+6x$
5. $-(3x-5+6x)$
6. $-(3x+5-6x)$
7. $3x+(5+6x)$
8. $3x-(5-6x)$
9. $-3x+(5+6x)$
10. $3x-(-5-6x)$
11. $9-(4x-6)$
12. $7x-4y+2x+8y$
13. $7x+4y+(2x+8y)$
14. $(7x+4y)-(2x+8y)$
15. $(7x-4y)-(2x-8y)$
16. $(2a+9)-(6a-10)$
17. $(4a-6b)+(-3a-2b)$
18. $(3u-4w)+(6u-5w)$
19. $(5w-8u)-(2w-4u)$
20. $(5w+8u)+(4w-9u)$
21. $(5m+6)-(-4m-7)$
22. $6+x-(-x-7)$
23. $-(2-3y)-(2y+12)$
24. $n-(2n+7)$
25. $-(5-x)-(x-12)$
26. $\frac{1}{2}x+\frac{1}{3}y+\frac{1}{3}x+\frac{1}{3}y$
27. $\frac{1}{2}x-\frac{1}{3}y-\frac{1}{3}x+\frac{1}{2}y$
28. $\frac{2}{3}x+\frac{3}{4}y-\frac{1}{4}x-\frac{1}{5}y$
29. $(\frac{1}{5}x+\frac{2}{3}y)+(\frac{3}{5}x-\frac{1}{3}y)$
30. $(\frac{1}{5}x+\frac{2}{3}y)-(\frac{3}{5}x-\frac{1}{3}y)$
31. $(1.2x-3.5y)+(3.5x+1.2y)$
32. $(1.2x-3.5y)-(3.5x+1.2y)$
33. $4.5x-(6.8x-4)$
34. $2x^2+5+4x^2-6$
35. $4y^2-7-6y^2+9$
36. $3x^2-(-x^2+7)$
37. $(5x^2-x)+(4x^2-7x)$
38. $(-6x^2-4)-(2x^2+6)$
39. $-6n^2+(3n^2-4)$
40. $x-2x^2-(3x^2-4)$
41. $(-a^2-2)+(3a^2-7)$
42. $(-k^2+4)-(k^2-4)$
43. $(x+2y)+(2x-3y)-(3x+4y)$
44. $(3a-4b)-(4a-b)+(9-2b)$
45. $(m-2n+3)-(4+3n)+(2m-7)$
46. $(3p-4q)-(p-2q)-(3p+4)$
47. $(3x^2-2x+4)+(2x^2+4x-6)-7x^2$
48. $(3x^2-2x+4)-(2x^2+4x-6)+7x^2$
49. Write an expression for the perimeter.

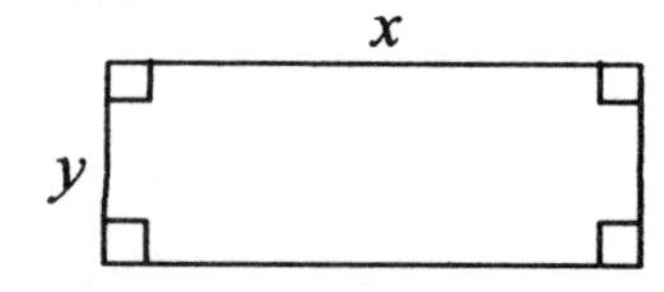

50. Write an expression for the perimeter.

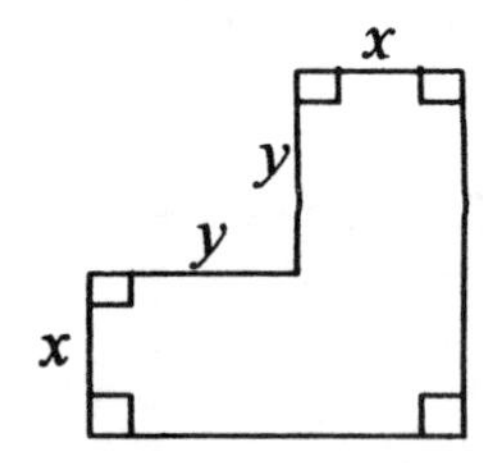

6-3 Multiplying and Dividing Expressions

To multiply any two expressions, we use the **distributive property** and **rule of exponents.**

Distributive Property: $a(b+c) = ab + ac$

$a(b-c) = ab - ac$

To divide an expression by a monomial (a divisor), each term in the expression is divided by the divisor.

1) $\dfrac{a+b}{c} = \dfrac{a}{c} + \dfrac{b}{c}$, $c \neq 0$ **2)** $\dfrac{a-b}{c} = \dfrac{a}{c} - \dfrac{b}{c}$, $c \neq 0$

Rules of Exponents (Powers): To multiply two powers having the same base, we add the exponents.
To divide two powers having the same base, we subtract the exponents.

1) $a^m \cdot a^n = a^{m+n}$ **2)** $\dfrac{a^m}{a^n} = a^{m-n}$

Examples:

1. $x^5 \cdot x^2 = x^{5+2} = x^7$.

2. $\dfrac{x^5}{x^2} = x^{5-2} = x^3$.

3. $4(3x-5) = 4(3x) + 4(-5) = 12x - 20$.

4. $5(-2a+1) = 5(-2a) + 5(1) = -10a + 5$.

5. $3x(2x^2 - 4x - 6) = 3x(2x^2) + 3x(-4x) + 3x(-6) = 6x^3 - 12x^2 - 18x$.

6. $\dfrac{4x+8}{2} = \dfrac{4x}{2} + \dfrac{8}{2} = 2x + 4$.

7. $\dfrac{12a-6}{3} = \dfrac{12a}{3} - \dfrac{6}{3} = 4a - 2$.

8. $\dfrac{4x^2 - 8x}{2x} = \dfrac{4x^2}{2x} - \dfrac{8x}{2x} = 2x - 4$.

9. $\dfrac{-28x^3 + 7x^2}{7x} = -\dfrac{28x^3}{7x} + \dfrac{7x^2}{7x} = -4x^2 + x$.

10. $\dfrac{20n^3 - 16n^2 - 12n}{4n} = \dfrac{20n^3}{4n} - \dfrac{16n^2}{4n} - \dfrac{12n}{4n} = 5n^2 - 4n - 3$.

11. Write an expression for the area.

$x+4$

x

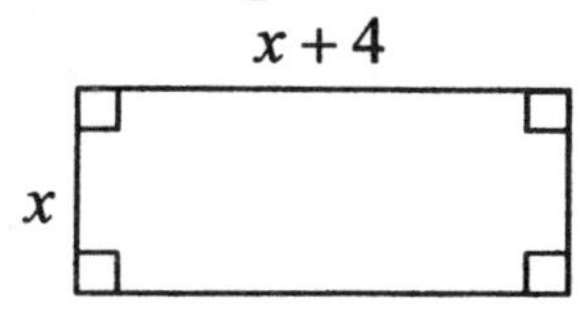

Solution:

Area $= x(x+4) = x^2 + 4x$. Ans.

12. Write an expression for the area.

$2n+5$

$2n$

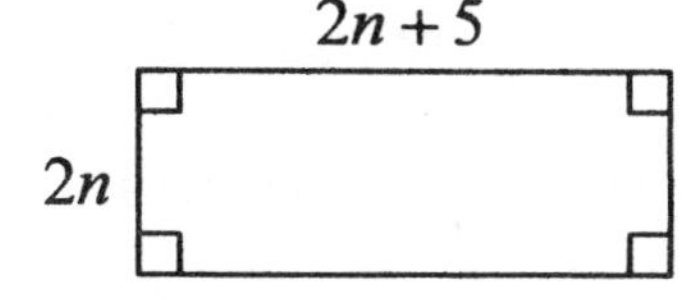

Solution:

Area $= 2n(2n+5) = 4n^2 + 10n$. Ans.

EXERCISES

Simplify each expression.

1. $x \cdot x$ **2.** $2x \cdot 3x$ **3.** $-4x \cdot 6x^3$ **4.** $-4x^5 \cdot x^7$ **5.** $2.4n^6 \cdot 3n^4$

6. $\frac{x}{x}$ **7.** $-\frac{2x}{3x}$ **8.** $\frac{6x^3}{4x}$ **9.** $\frac{x^7}{4x^5}$ **10.** $\frac{2.4n^6}{3n^4}$

11. $3x \cdot 2 + 4x \cdot 3$ **12.** $5a \cdot 3 - 2a \cdot 6$ **13.** $-4n \cdot 4 - 3n \cdot 3$

14. $-(4x-5)$ **15.** $-(-5+4x)$ **16.** $-(-2x^2-x)$

17. $-(3a^2-4a+6)$ **18.** $3x^2-(2x^2+6)$ **19.** $x^2-(4x^2-3)$

20. $2(3x+5)$ **21.** $-2(5-3x)$ **22.** $-4(3a+5)$

23. $5x(3x-6)$ **24.** $6a(3a-7)$ **25.** $4a(2a+5)$

26. $k(k-1)$ **27.** $k(2k-4)$ **28.** $3x(y-9z)$

29. $4a(2b-c)$ **30.** $5a(2b-7c)$ **31.** $2x(-y+x)$

32. $3(x^2-4x-6)$ **33.** $4(y^2+3y-2)$ **34.** $-5(2a^2+3a-7)$

35. $-2(3n^2-3n+5)$ **36.** $3y(2y^2-y+6)$ **37.** $2x(x^2-3x-9)$

38. $-4x(2x+y-z)$ **39.** $-3x(-5x-2y+7z)$ **40.** $-a(a-b+c)$

41. $5x^2-4(2-2x^2)$ **42.** $8x^2-4(x^2+2)$ **43.** $7a^2+3(2-4a^2)$

44. $\frac{12x-4}{4}$ **45.** $\frac{-12x+3}{3}$ **46.** $\frac{-12x^2+4x}{4}$ **47.** $\frac{-8x^2+6x}{-2}$

48. $\frac{-15a^2+6a}{-3}$ **49.** $\frac{a^3-a^2}{a}$ **50.** $\frac{4x^3-2x^2}{x}$ **51.** $\frac{6a^4-3a^2}{3a}$

52. $\frac{12x^3+15x^2+9x}{3x}$ **53.** $\frac{4a^3-8a^2-4a}{4a}$ **54.** $\frac{9a^2b-3ab^2}{3ab}$ **55.** $\frac{8x^2y^2-4xy}{xy}$

56. $\frac{4x+6}{3}$ **57.** $\frac{4x-3}{5}$ **58.** $\frac{12a-8b}{6}$ **59.** $\frac{4x-2x^2}{6x}$

60. Write an expression for the area.

a)

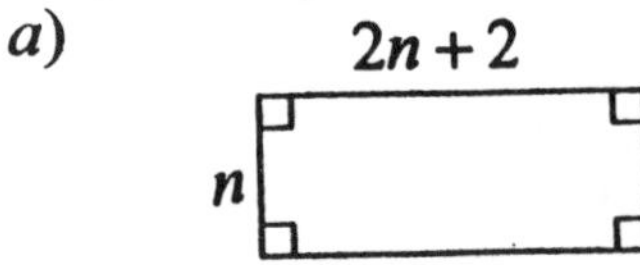

b)

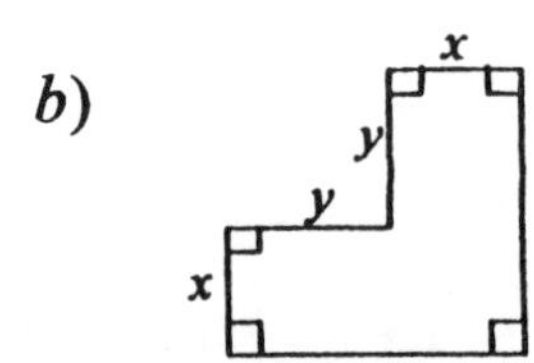

6-4 Solving One – Step Equations

An algebraic **equation** is a statement by placing an equal sign "=" between two expressions.

$x+5=9$ is an equation. $2y-3=7$ is an equation.

When a number which is substituted for (plugged into) the variable makes an equation a true statement, it is called a **solution** of the equation.

$x=4$ is a solution to the equation $x+5=9$.

$y=5$ is a solution to the equation $2y-3=7$.

To find the solution of an equation, we transfer the given equation into a simpler and equivalent equation that has the same solution. The last statement of the given equation after we transfer it should isolate the variable to one side of the equation in the form:

variable = a number, or **a number = variable**

To check a solution of an equation, we substitute the solution for the variable in the equation to see if we get a true statement.

The following four examples show the basic steps to solve simple **one-step equations**.

Examples

BY USING ADDITION

1. Solve $x-5=9$.

Solution:

$x-5=9$

Add 5 to each side:

$x-\cancel{5}+\cancel{5}=9+5$

$x=14$. Ans.

Check: $x-5=9$

$14-5=9$ ✓

BY USING SUBTRACTION

2. Solve $x+5=9$

Solution:

$x+5=9$

Subtract 5 from each side:

$x+\cancel{5}-\cancel{5}=9-5$

$x=4$. Ans.

Check: $x+5=9$

$4+5=9$ ✓

BY USING DIVISION

3. Solve $5x=40$.

Solution:

$5x=40$

Divide each side by 5:

$\frac{\cancel{5}x}{\cancel{5}}=\frac{40}{5}$

$x=8$. Ans.

Check: $5x=40$

$5(8)=40$ ✓

BY USING MULTIPLICATION

4. Solve $\frac{x}{5}=40$.

Solution:

$\frac{x}{5}=40$

Multiply each side by 5:

$(\cancel{5})\frac{x}{\cancel{5}}=40(5)$

$x=200$. Ans.

Check: $\frac{x}{5}=40$

$\frac{200}{5}=40$ ✓

-----Continued-----

Examples (Follow the same steps as shown in the examples 1 to 4.)

5. Solve $-12 = -3 + x$

Solution:

$$-12 = -3 + x$$
$$-12 + 3 = \cancel{-3} + x \cancel{+3}$$
$$-9 = x. \quad \text{Ans.}$$

6. Solve $15 = 3 - y$

Solution:

$$15 = 3 - y$$
$$15 - 3 = \cancel{3} - y \cancel{-3}$$
$$12 = -y \qquad \therefore -12 = y. \quad \text{Ans.}$$

7. Solve $24 = -8a$.

Solution:

$$24 = -8a$$
$$\frac{24}{-8} = \frac{-8a}{-8}$$
$$-3 = a. \quad \text{Ans.}$$

8. Solve $-24 = \frac{3}{5}n$.

Solution:

$$-24 = \tfrac{3}{5}n$$
$$\frac{5}{3} \cdot -24 = \frac{\cancel{3}}{\cancel{5}} n \cdot \frac{\cancel{5}}{\cancel{3}}$$
$$-40 = n. \quad \text{Ans.}$$

9. Solve $m + 2.5 = -3.8$.

Solution:

$$m + 2.5 = -3.8$$
$$m + \cancel{2.5} - \cancel{2.5} = -3.8 - 2.5$$
$$m = -6.3. \quad \text{Ans.}$$

10. Solve $y - 4.5 = -2.5$.

Solution:

$$y - 4.5 = -2.5$$
$$y - \cancel{4.5} + \cancel{4.5} = -2.5 + 4.5$$
$$y = 2. \quad \text{Ans.}$$

11. Solve $x - \frac{4}{5} = 2\frac{1}{5}$.

Solution:

$$x - \cancel{\frac{4}{5}} + \cancel{\frac{4}{5}} = 2\frac{1}{5} + \frac{4}{5}$$
$$x = \frac{11}{5} + \frac{4}{5} = \frac{15}{5} = 3$$
$$x = 3. \quad \text{Ans.}$$

12. Solve $x + 1\frac{3}{4} = \frac{2}{5}$.

Solution:

$$x + \cancel{1\frac{3}{4}} - \cancel{1\frac{3}{4}} = \frac{2}{5} - 1\frac{3}{4} = \frac{2}{5} - \frac{7}{4}$$
$$= \frac{8}{20} - \frac{35}{20} = -\frac{27}{20} = -1\frac{7}{20}$$
$$x = -1\frac{7}{20}. \quad \text{Ans.}$$

13. Solve $\frac{3}{7}x = \frac{5}{7}$.

Solution:

$$\frac{\cancel{7}}{\cancel{3}} \cdot \frac{\cancel{3}}{\cancel{7}} x = \frac{5}{\cancel{7}} \cdot \frac{\cancel{7}}{3}$$
$$x = \frac{5}{3} = 1\frac{2}{3}. \quad \text{Ans.}$$

14. Solve $-\frac{5y}{7} = \frac{1}{2}$

Solution:

$$-\frac{\cancel{7}}{\cancel{5}} \cdot -\frac{\cancel{5}y}{\cancel{7}} = \frac{1}{2} \cdot -\frac{7}{5}$$
$$y = -\frac{7}{10}. \quad \text{Ans.}$$

EXERCISES

Solve each equation. Check your solution.

1. $x-3=9$

2. $x+3=9$

3. $3x=9$

4. $\frac{x}{3}=9$

5. $n+6=12$

6. $n-6=12$

7. $\frac{n}{6}=12$

8. $6n=12$

9. $y-4=-2$

10.. $y+4=-2$

11. $\frac{y}{4}=-2$

12. $4y=-2$

13. $a-15=30$

14. $a+15=30$

15. $15a=30$

16. $\frac{a}{15}=30$

17. $-x=1$

18. $2=-x$

19. $-a=\frac{1}{2}$

20. $-4=-x$

21. $3-x=9$

22. $-3-x=9$

23. $-3x=9$

24. $-\frac{x}{3}=9$

25. $12=2+x$

26. $12=-2+x$

27. $12=-\frac{x}{2}$

28. $12=-2x$

29. $-12=2+x$

30. $-12=-2+x$

31. $-12=-\frac{x}{2}$

32. $-12=-2x$

33. $12=-2-x$

34. $-12=-2-x$

35. $-12=2x$

36. $2=-\frac{x}{12}$

37. $m+1.5=4.5$

38. $m-1.5=4.5$

39. $\frac{m}{1.5}=4.5$

40. $1.5m=4.5$

41. $2.5=5-n$

42. $2.5=-5+n$

-----Continued-----

43. $2.5 = -5n$

44. $2.5 = -\frac{n}{5}$

45. $y - 4.5 = -3.5$

46. $y + 4.5 = -3.5$

47. $\frac{y}{4.6} = -2.5$

48. $-2.5x = 4.6$

49. $x + \frac{1}{2} = 4$

50. $x - \frac{1}{2} = 4$

51. $x - \frac{1}{2} = -4$

52. $x + \frac{1}{2} = -4$

53. $y - \frac{2}{3} = \frac{2}{3}$

54. $y + \frac{2}{3} = 4$

55. $\frac{2}{3}y = \frac{1}{3}$

56. $\frac{3}{2}y = \frac{1}{3}$

57. $a + 1\frac{2}{5} = \frac{4}{5}$

58. $a - 1\frac{2}{5} = \frac{4}{5}$

59. $a - 1\frac{2}{5} = -\frac{4}{5}$

60. $a + 1\frac{2}{5} = -\frac{4}{5}$

61. $\frac{x}{-6} = 2.4$

62. $\frac{-x}{8} = 1.6$

63. $24t = 48$

64. $\frac{t}{24} = 48$

65. $\frac{k}{-4} = -2.5$

66. $-\frac{1}{24} = \frac{x}{2}$

67. $-\frac{k}{24} = 4.8$

68. $-24k = 4.8$

69. $-\frac{1}{3}x = \frac{1}{5}$

70. $-\frac{2x}{5} = \frac{3}{4}$

71. $\frac{2}{5} = -\frac{x}{4}$

72. $\frac{2x}{5} = \frac{1}{4}$

A kid said to his mother.
Kid: Did you wear contact lenses this morning ?
Mother: No. Why do you ask ?
Kid: I heard you yell to father this morning :
" After 20 years of marriage, I have finally seen what kind of man you are ! ".

6-5 Solving Two – Step Equations

To solve some equations, it is necessary to use more than one step.
To solve an equation containing like terms, we simplify the equation by combining like terms on each side of the equation, and then apply the basic steps as we did in Section 6-4.
If an equation has a variable on both sides, we use the addition and the subtraction to get the variable alone on one side.
If an equation has numbers on both sides, we use the addition and subtraction to get the numbers alone to one side.
An easier way to simplify an equation is to transfer (isolate) all variables to one side of the equation and transfer (isolate) all numbers to the other side. Reverse (change) the signs in the process. (See Example 11 and Example 12.)

The following examples show the basic steps to solve simple **two-step equations**.

Examples

1. Solve $2x + 3x = 15$.
Solution:

$$2x + 3x = 15$$
$$5x = 15$$
$$\frac{\cancel{5}x}{\cancel{5}} = \frac{15}{5}$$
$$\therefore x = 3. \quad \text{Ans.}$$

2. Solve $2x - 3x = 15$.
Solution:

$$2x - 3x = 15$$
$$-x = 15$$
$$\frac{-x}{-1} = \frac{15}{-1}$$
$$\therefore x = -15. \quad \text{Ans.}$$

3. Solve $4x + 6x = -5 + 8$.
Solution:

$$4x + 6x = -5 + 8$$
$$10x = 3$$
$$\frac{\cancel{10}x}{\cancel{10}} = \frac{3}{10}$$
$$\therefore x = \frac{3}{10}. \quad \text{Ans.}$$

4. Solve $-4x + 6x = -5 - 8$.
Solution:

$$-4x + 6x = -5 - 8$$
$$2x = -13$$
$$\frac{\cancel{2}x}{\cancel{2}} = \frac{-13}{2}$$
$$\therefore x = -6\frac{1}{2}. \quad \text{Ans.}$$

5. Solve $6y = 8 - 2y$.
Solution:

$$6y = 8 - 2y$$
$$6y + 2y = 8 - \cancel{2y} + \cancel{2y}$$
$$8y = 8$$
$$\frac{\cancel{8}y}{\cancel{8}} = \frac{8}{8}$$
$$\therefore y = 1. \quad \text{Ans.}$$

6. Solve. $2.4n = 1.8 - 1.26$
Solution:

$$2.4n = 1.8n - 1.26$$
$$2.4n - 1.8n = \cancel{1.8n} - 1.26 - \cancel{1.8n}$$
$$0.6n = -1.26$$
$$\frac{\cancel{0.6}n}{\cancel{0.6}} = \frac{-1.26}{0.6}$$
$$\therefore n = -2.1. \quad \text{Ans.}$$

-----Continued-----

Examples

7. Solve $8 = 7x - 4x$.

Solution:

$$8 = 7x - 4x$$
$$8 = 3x$$
$$\frac{8}{3} = \frac{\cancel{3}x}{\cancel{3}}$$
$$\frac{8}{3} = x \qquad \therefore x = \frac{8}{3} = 2\frac{2}{3}. \quad \text{Ans.}$$

8. Solve $8 = -7x - 4x$.

Solution:

$$8 = -7x - 4x$$
$$8 = -11x$$
$$\frac{8}{-11} = \frac{\cancel{-11}x}{\cancel{-11}}$$
$$\therefore -\frac{8}{11} = x. \quad \text{Ans.}$$

9. Solve $\frac{1}{3}a - 5 = 6$.

Solution:

$$\frac{1}{3}a - 5 = 6$$
$$\frac{1}{3}a - \cancel{5} + \cancel{5} = 6 + 5$$
$$\frac{1}{3}a = 11$$
$$(\cancel{3})\frac{1}{\cancel{3}}a = 11(3)$$
$$\therefore a = 33. \quad \text{Ans.}$$

10. Solve $\frac{y}{4} = -\frac{y}{2} - \frac{2}{5}$.

Solution:

$$\frac{y}{4} = -\frac{y}{2} - \frac{2}{5}$$
$$\frac{y}{4} + \frac{y}{2} = -\cancel{\frac{y}{2}} - \frac{2}{5} + \cancel{\frac{y}{2}}$$
$$\frac{y}{4} + \frac{2y}{4} = -\frac{2}{5} \rightarrow \frac{3y}{4} = -\frac{2}{5}$$
$$\frac{\cancel{4}}{\cancel{3}} \cdot \frac{\cancel{3}y}{\cancel{4}} = -\frac{2}{5} \cdot \frac{4}{3}$$
$$\therefore y = -\frac{8}{15}. \quad \text{Ans.}$$

11. Solve $3x - 12 = 18$.

Solution:

$$3x - 12 = 18$$
$$3x = 18 + 12$$
$$3x = 30$$
$$\therefore x = 10. \quad \text{Ans.}$$

12. Solve $5y = 7y - 18$

Solution:

$$5y = 7y - 18$$
$$5y - 7y = -18$$
$$-2y = -18$$
$$\therefore y = 9. \quad \text{Ans.}$$

EXERCISES

Solve each equation. Check your solution.

1. $2x+5x=28$	**2.** $5x-2x=27$	**3.** $-5x+2x=15$
4. $-5x-2x=21$	**5.** $2a=20-3a$	**6.** $3y=15-2y$
7. $4y-10=2y$	**8.** $3a-15=2a$	**9.** $5x-3x=2-8$
10. $5x-8x=8-2$	**11.** $3y-5y=2-8$	**12.** $8y-5y=-2-8$
13. $7a=9a+24$	**14.** $8a=7a-15$	**15.** $4m=3m+2$
16. $-4m=3m+14$	**17.** $5x-2=30$	**18.** $5x+2=30$
19. $-5x+2=30$	**20.** $-5x+2=-30$	**21.** $6y=4y-20$
22. $6y=-4y+20$	**23.** $-6n=4n-20$	**24.** $6n=-4n-20$
25. $5x=8+3x$	**26.** $5x=-3x-8$	**27.** $5y=19y+7$
28. $5y=19y-7$	**29.** $4=5p-7p$	**30.** $4=7p-5p$
31. $15=3c+7c$	**32.** $15=3c-7c$	**33.** $-8=7c-5c$
34. $-8=5c-7c$	**35.** $-9=6p-8p$	**36.** $-9=8p-6p$
37. $1.2x+3.4x=9.2$	**38.** $3.4x-1.2x=8.8$	**39.** $4.2y-3y=4.8$
40. $3y-4.2y=4.8$	**41.** $3.2=2y+3y$	**42.** $3.2=3y-2y$
43. $3.2=2k-3k$	**44.** $3.2=-k+5k$	**45.** $1.2=c+c$
46. $-2.4=2c+c$	**47.** $1.2=3n-n$	**48.** $2.4=n-3n$
49. $3x+4x=8.4$	**50.** $3x-4x=8.4$	**51.** $3x-4x=-8.4$
52. $4x-3x=-8.4$	**53.** $1.5x=0.5x-10$	**54.** $3.5x=-0.5x+10$

-----Continued-----

55. $3.5y = y - 12$

56. $3.5y = -y + 6.3$

57. $2.5n = 1.5n + 8$

58. $2.5n = -1.5n + 8$

59. $2m + 6.2 = 7.4$

60. $2m - 6.2 = 7.4$

61. $\frac{x}{2} + 2 = 4$

62. $\frac{x}{2} - 2 = 4$

63. $\frac{y}{3} + 4 = 7$

64. $\frac{y}{3} - 4 = 7$

65. $\frac{k}{5} - 6 = 9$

66. $\frac{k}{5} + 6 = 9$

67. $-\frac{b}{6} + 4 = -3$

68. $-\frac{b}{6} - 4 = 3$

69. $\frac{x}{8} + 1.2 = 3$

70. $\frac{x}{8} - 1.2 = 3.1$

71. $\frac{1}{2}x + \frac{1}{3}x = 10$

72. $\frac{1}{2}x - \frac{1}{3}x = 10$

73. $\frac{y}{2} = \frac{y}{3} - 4$

74. $\frac{y}{2} = -\frac{y}{3} - 4$

75. $\frac{y}{2} = \frac{y}{5} + 6$

76. $\frac{y}{2} = -6 - \frac{y}{5}$

77. $\frac{5}{3}x + \frac{2}{3} = \frac{4}{3}$

78. $\frac{3}{5}x - \frac{2}{15} = \frac{2}{3}$

79. $1\frac{1}{2}x - \frac{3}{4} = \frac{1}{6}$

80. $2\frac{2}{3}x + \frac{1}{12} = \frac{1}{2}$

81. $\frac{n}{3} = \frac{2n}{3} + \frac{1}{4}$

82. $\frac{n}{3} = -\frac{2n}{3} - \frac{1}{4}$

83. $\frac{3n}{4} = \frac{1}{5} - \frac{n}{4}$

84. $\frac{3n}{4} = -\frac{1}{5} + \frac{n}{4}$

85. $\frac{x+5}{4} = 12$

86. $\frac{y-3}{-2} = 10$

87. $\frac{d-7}{3} = -8$

A students said to a priest.
Student: I don't believe that heaven is as good as you said. You have never been there.
Priest: Have you ever heard of anyone who came back here because he didn't like it there.

6-6 Solving Multistep Equations

To solve an equation involving several steps, we simplify the equation by combining like terms on each side of the equation, and then apply the basic steps as we did in Section 6-4 and Section 6-5.

To solve an equation containing parentheses, we combine likes terms inside the parentheses and use the Distributive Property to remove the parentheses.

An easier way to simplify an equation is to transfer (isolate) all variables to one side of the equation and transfer (isolate) all numbers to the other side. Reverse (change) the signs in the process. (See Example 5 and Example 6.)

Examples

1. Solve $2x + 6 = 4x$.

Solution:

$$2x + 6 = 4x$$
$$2x + 6 - 4x = 4x - 4x$$
$$-2x + 6 = 0$$
$$-2x + 6 - 6 = -6$$
$$-2x = -6$$
$$\frac{-2x}{-2} = \frac{-6}{-2}$$
$$\therefore x = 3. \quad \text{Ans.}$$

2. Solve $2x - 5 = 7 - 6x$.

Solution:

$$2x - 5 = 7 - 6x$$
$$2x - 5 + 6x = 7 - 6x + 6x$$
$$8x - 5 = 7$$
$$8x - 5 + 5 = 7 + 5$$
$$8x = 12$$
$$\frac{8x}{8} = \frac{12}{8}$$
$$\therefore x = 1\tfrac{1}{2}. \quad \text{Ans.}$$

3. Solve $3(a - 5 + 2a) = 6$.

Solution:

$$3(a - 5 + 2a) = 6$$
$$3(3a - 5) = 6$$
$$9a - 15 = 6$$
$$9a - 15 + 15 = 6 + 15$$
$$9a = 21$$
$$\frac{9a}{9} = \frac{21}{9}$$
$$\therefore a = 2\tfrac{1}{3}. \quad \text{Ans.}$$

4. Solve $3(2y + 3) = 4y - 7$.

Solution:

$$3(2y + 3) = 4y - 7$$
$$6y + 9 = 4y - 7$$
$$6y + 9 - 4y = 4y - 7 - 4y$$
$$2y + 9 = -7$$
$$2y + 9 - 9 = -7 - 9$$
$$2y = -16, \quad \frac{2y}{2} = \frac{-16}{2}$$
$$\therefore y = -8. \quad \text{Ans.}$$

5. Solve $2x + 6 = 4x$.

Solution:

$$2x + 6 = 4x$$
$$2x - 4x = -6$$
$$-2x = -6$$
$$\therefore x = \frac{-6}{-2} = 3. \quad \text{Ans.}$$

6. Solve $2x - 5 = 7 - 6x$.

Solution:

$$2x - 5 = 7 - 6x$$
$$2x + 6x = 7 + 5$$
$$8x = 12$$
$$\therefore x = \frac{12}{8} = 1\tfrac{1}{2}. \quad \text{Ans.}$$

EXERCISES

Solve each equation. Check your solution.

1. $3x-16=x$ **2.** $2x+15=-3x$ **3.** $5x+12=3x$

4. $4x-9=x$ **5.** $-4y-7=3y$ **6.** $-6y+8=2y$

7. $2(x-5+3x)=14$ **8.** $5(6x+8-2x)=-10$ **9.** $4(-8+y+6)=4$

10. $2(5-2y-7)=8$ **11.** $3(m-5)=-30$ **12.** $4(2m+3)=28$

13. $-6(a+3)=9$ **14.** $-2(4-2x)=-16$ **15.** $-3(5+3x)=4$

16. $-4(2+5a)=32$ **17.** $-(-5+2x)=-13$ **18.** $5(2n-3)=-30$

19. $3x-6=4x-8$ **20.** $3x-6=4x+8$ **21.** $3x+6=-4x-8$

22. $y+4=5y-7$ **23.** $y-4=5y-7$ **24.** $y-4=-5y+8$

25. $6m-3=m+11$ **26.** $6m+3=m+11$ **27.** $-6m-3=m-11$

28. $-9+4n=10n-3$ **29.** $9-4n=10n-5$ **30.** $9-4n=-10n+15$

31. $-8-x=x+2$ **32.** $-8+y=-y-2$ **33.** $8+z=-z+2$

34. $20c+5=10c+25$ **35.** $20c-5=10c-25$ **36.** $20c-5=-10c+25$

37. $10+k=4k-20$ **38.** $10-k=4k-30$ **39.** $10-k=-4k-14$

40. $-12+b=3b+30$ **41.** $-12+b=-3b+32$ **42.** $12-b=3b-16$

43. $15+2p=39-p$ **44.** $15-2p=p-21$ **45.** $-15+2p=p+40$

46. $100x-200=50x+50$ **47.** $100x+200=50x-50$ **48.** $10x+20=50x-60$

49. $40y-40=-40y+40$ **50.** $40y+40=50y+50$ **51.** $40y+40=-50y-50$

52. $7(x+1)=5(x-3)$ **53.** $7(x-1)=5(x+3)$ **54.** $-7(x-1)=5(x+3)$

55. $10(y-2)=-2(2y+3)$ **56.** $10(y+2)=-2(2y-3)$ **57.** $-10(-y+2)=2(-2y)$

58. $2(x+3)-5=3x-6$ **59.** $2(x-3)-5=3x+6$ **60.** $2(x-3)+5=3x+6$

61. $5x-3(x+1)=x-10$ **62.** $5y-3(y-1)=4y+1$ **63.** $6a-4(2a+3)=7a$

64. $3x-2(5x-4)=4x-(2x-26)$ **65.** $7y-3(2y+6)=5y-(3y+10)$

6-7 Translating Words into Symbols

Before we learn how to solve word problems in algebra, we must learn how to translate words into symbols containing variables.

A. Translate word phrases into variable expressions.

5 more than x	$\to x+5$.	The sum of 5 and x	$\to 5+x$.
5 less than x	$\to x-5$.	The difference between 5 and x	$\to 5-x$.
x less than 5	$\to 5-x$.	The product of 5 and x	$\to 5x$.
x minus 5	$\to x-5$.	The quotient of 5 and x	$\to 5/x$.
x is less than 5	$\to x<5$.	x increased by 5	$\to x+5$.
x is greater than 5	$\to x>5$.	x decreased by 5	$\to x-5$.
x is less than or equal to 5	$\to x\leq 5$.	x times 5	$\to 5x$.
x is greater than or equal to 5	$\to x\geq 5$.	x divided by 5	$\to x/5$.

5 times the quantity x decreased by 3 $\to 5(x-3)$.
5 times the difference of x and 3 $\to 5(x-3)$.
5 times the sum of x and 3 $\to 5(x+3)$.
3 more than 5 times x $\to 5x+3$.
3 increased by 5 times x $\to 3+5x$.
The difference between 5 times x and 3 $\to 5x-3$.
3 decreased by 5 times x $\to 3-5x$.
3 less than 5 times x $\to 5x-3$.
3 less than half of x $\to \frac{1}{2}x-3$.
Half the difference between x and 3 $\to \frac{1}{2}(x-3)$
Twice the sum of x and 3 $\to 2(x+3)$.
Twice x, increased by 3 $\to 2x+3$.

Three consecutive integers (n is an integer) $\to n,\ n+1,\ n+2$.
Three consecutive even integers (n is an even integer) $\to n,\ n+2,\ n+4$.
Three consecutive odd integers (n is an odd integer) $\to n,\ n+2,\ n+4$.
Three consecutive positive multiples of 5 (n is a multiple of 5) $\to n,\ n+5,\ n+10$.

Examples: Translate each statement into a variable expression.

1. There are 15 fewer girls than boys in a class. How many girls are there if the number of boys is b ?
   Translation: $b-15$. Ans.

2. You paid $50 cash and made 10 equal monthly payments. What is the total cost of the product if the amount in dollars of each of the monthly payments is x ?
   Translation:
   $50+10x$. Ans.

-----Continued-----

B. Translate word sentences into equations.

Examples: Translate each sentence into an equation.

1. Eight less than a number n is 15.
Translation:
$n-8=15$. Ans.

2. Four more than a number n is 12.
Translation:
$n+4=12$. Ans.

3. Five more than a number n is equal to 9.
Translation:
$n+5=9$. Ans.

4. 5 times the quantity x decreased by 3 is 25.
Translation: $5(x-3)=25$. Ans.

5. Twice the sum of a number x and 3 is 22.
Translation:
$2(x+3)=22$. Ans.

6. A number n decreased by 12 is 30.
Translation:
$n-12=30$. Ans.

7. The result of a number x multiplied by 5 , decreased by 3 is 57.
Translation:
$5x-3=57$. Ans.

8. The sum of three consecutive numbers is 48.
Translation:
$n+(n+1)+(n+2)=48$. Ans.

9. Write an equation involving x if the perimeter of the figure is 24.

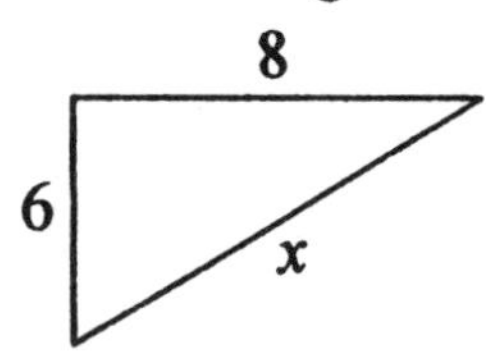

Solution: $x+8+6=24$
$x+14=24$. Ans.

10. Write an equation involving x if the perimeter of the figure is 32.

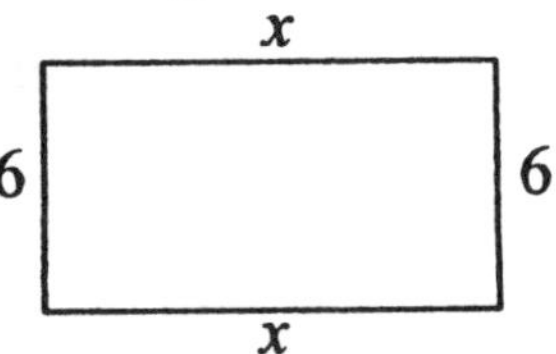

Solution: $x+x+6+6=32$
$2x+12=32$. Ans.

11. One-third of a pizza sold for \$3.50. Write an equation that represents the cost c of the whole pizza.

Solution: $\frac{1}{3}c=3.50$. Ans.

12. You paid \$50 cash and made 10 equal monthly payments. The total cost of the product was \$400. If the amount in dollars of each of the monthly payments is x, write an equation that represents the amount of each of the monthly payments.
Solution:
$400=50+10x$. Ans.

13. John picks up two-thirds of the books in the shelf. There are 12 books left in the shelf. Write an equation that represents the number n of the books in the shelf to begin with.

Solution: $n-\frac{2}{3}n=12$. Ans.

EXERCISES

Translate each phrase into a variable expression.

1. Five more than n
2. n more than five
3. Five less than n
4. n less than five
5. Five is greater than n
6. n is less than five
7. Five more than twice x
8. Five less than half of a number n
9. Five times n, decreased by ten
10. Five times n increased by ten
11. Five is less than or equal to m
12. Five is greater than or equal to m
13. Twice the sum of x and y
14. The sum of twice x and y
15. Five times the sum of n and two
16. The product of five and m
17. Twicc the sum of x and five
18. The sum of twice x and five
19. Five decreased by four times k
20. The difference between five and k
21. Five times the sum of x and two
22. The sum of five times x and two
23. Five times the difference of x and two
24. The difference of five times x and two
25. Two less than five times x
26. Two more than five times x
27. The sum of four times k and five
28. The product of $4k$ and five
29. The difference of four times k and five
30. The quotient of four times k and five
31. Five is greater than one-third of k
32. Five is less than or equal to half of k
33. Five plus the product of m and two
34. Five plus the quotient of m and two
35. Five more than three times n
36. Five less than three times n
37. Five increased by three times n
38. Five decreased by three times n
39. Five times the sum of n and three
40. Half the quotient of n and three

Translate each statement into a variable expression

41. There are 15 more boys than girls in a class. How many boys are there if the number of girl is g ?
42. Write four consecutive integers beginning with n.
43. Write four consecutive even integers beginning with n.
44. Write four consecutive odd integers beginning with n.
45. You paid $200 cash and made 20 equal monthly payments. Write a variable expression to represent the total cost of the product if the amount in dollars of each of the monthly payments is x.
46. John wants to rent a car for one day. The car rental costs $110 plus $0.25 a mile. Write a variable expression to represent the total rental costs if the mileage is x.
47. A rectangle has width 12 and length x. Write a variable expression to represent its perimeter.
48. A rectangle has width 12 and length x. Write a variable expression to represent its area.
49. The difference between my age and your age is 35. My age is n. Write a variable expression to represent your age.
50. The cost to buy an A-Plus Notes book is $19.50. How much will n books cost ?

-----**Continued**-----

Translate each sentence into an equation.

51. A number n increased by 21 is 28.
52. 8 less than a number n is 15.
53. A number n decreased by 12 is equal to 30.
54. A number n less than 8 is 15.
55. 5 more than a number x is 25.
56. 5 times a number n is 45.
57. 5 more than twice a number x is 13.
58. A number x less than 10 is 29.
59. Twice the sum of a number n and 5 is -35.
60. One-third of a number x is 12.
61. Six times the sum of n and two is fourteen.
62. 3 times x decreased by 4 is 8.
63. Half the quotient of x and three is ten.
64. 2 less than twice a number n is 6.
65. 12 more than 4 times a number n is twice the number n.
66. Nine times x is twice the difference between x and seven.
67. Five times the number which is two less than x is fifteen.
68. Four subtracted from one-third of the number x is three-eighths.

69. A rectangle has width 12 and length x. Its perimeter is 54. Write an equation involving x to represent its perimeter.
70. A rectangle has width 12 and length x. Its area is 180. Write an equation involving x to represent its area.
71. The sum of two consecutive integers is 31. The first integer is n. Write an equation involving n to represent the sum.
72. The sum of three consecutive even integers is 48. The first integer is n. Write an equation involving n to represent the sum.
73. The sum of three consecutive multiples of 5 is 60. The first integer is n. Write an equation involving n to represent the sum.
74. John rented a car. The car rental costs \$80 per day with no mileage fee. He paid \$240 when he returned the car. The number of days he traveled is x. Write an equation involving x to represent the total car rental charges.
75. John rented a car for one day. The car rental costs \$50 plus \$0.25 a mile. He paid \$120 when he returned the car. The mileage that he traveled is x. Write an equation involving x to represent the total rental charges.
76. In an isosceles triangle, the length of each leg exceeds the base by $4\,cm$. The base is $x\ cm$, and the perimeter is $92\,cm$. Write an equation involving x to represent the perimeter.
77. The length of a rectangle is twice its width. The perimeter is 30. The width is x. Write an equation involving x to represent the perimeter.
78. The width of a rectangle is x. Its length is $x+6$. The perimeter is 36. Write an equation involving x to represent the perimeter.
79. John has twice as much money as Carol. Together they have \$36. If x is the amount of money that Carol has, write an equation involving x to represent the total money that they have.
80. A rope is $27\,cm$ long. You cut it into two pieces. One piece is $8\,cm$ longer than the other piece. If the shorter piece is $x\ cm$, write an equation involving x to represent the length of the rope.

6-8 Solving Word Problems in One Variable

To solve a word problem using an equation in algebra, we use the following steps.

Steps for solving a word problem:

1. Read the problem carefully.
2. Choose a variable and assign it to represent the unknown (answer).
3. Write an equation based on the given fact in the problem.
4. Solve the equation and find the answer.
5. Check your answer with the problem (usually on a scratch paper).

Examples

1. Eight less than a number is 18. Find the number.

Solution:

Let $n =$ the number

The equation is

$$n - 8 = 18$$
$$n - \cancel{8} + \cancel{8} = 18 + 8$$
$$\therefore n = 26. \quad \text{Ans.}$$

Or:
$$n - 8 = 18$$
$$n = 18 + 8$$
$$n = 26. \quad \text{Ans.}$$

Check:
$$n - 8 = 18$$
$$26 - 8 = 18 \checkmark$$

2. A number increased by 15 is 32. Find the number.

Solution:

Let $n =$ the number

The equation is

$$n + 15 = 32$$
$$n + \cancel{15} - \cancel{15} = 32 - 15$$
$$\therefore n = 17. \quad \text{Ans.}$$

Or:
$$n + 15 = 32$$
$$n = 32 - 15$$
$$n = 17. \quad \text{Ans.}$$

Check:
$$n + 15 = 32$$
$$17 + 15 = 32 \checkmark$$

3. Find two consecutive integers whose sum is 37.

Solution:

Let $n = 1^{st}$ integer, $n + 1 = 2^{nd}$ integer

The equation is

$$n + (n + 1) = 37$$
$$2n + 1 = 37$$
$$2n = 37 - 1$$
$$2n = 36$$

$\therefore n = 18$ 1st integer.

$n + 1 = 19$ 2nd integer. Ans.

Check: $18 + 19 = 37 \checkmark$

-----**Continued**-----

Examples

4. Twice the sum of a number and 3 is 22. Find the number.

Solution:

Let $n =$ the number

The equation is

$$2(n+3) = 22$$
$$2n + 6 = 22$$
$$2n = 22 - 6$$
$$2n = 16$$
$$\therefore n = 8. \quad \text{Ans.}$$

Check: $2(n+3) = 22$

$2(8+3) = 22$ ✓

5. Find three consecutive even integers whose sum is 120.

Solution:

Let $n = 1^{st}$ even integer

$n+2 = 2^{nd}$ even integer

$n+4 = 3^{rd}$ even integer

The equation is

$$n + (n+2) + (n+4) = 120$$
$$3n + 6 = 120$$
$$3n = 114$$

$\therefore n = 38$ 1st integer

$n+2 = 40$ 2nd integer

$n+4 = 42$ 3rd integer

Ans: 38, 40, 42

Check: 38 + 40 + 42 = 120 ✓

6. John has twice as much money as Carol. Together they have $36. How much money does each have ?

Solution: Let $c =$ Carol's money

$2c =$ John's money

The equation is

$$c + 2c = 36$$
$$3c = 36$$

$\therefore c = 12 \rightarrow$ Carol's money.

$2c = 24 \rightarrow$ John's money. Ans.

Check: $12 + 24 = 36$ ✓

7. A number divided by 2 is equal to the number increased by 2. Find the number.

Solution: Let $n =$ the number

The equation is $\frac{n}{2} = n + 2$

Multiply each side by 2: $\not{2} \cdot \frac{n}{\not{2}} = 2(n+2)$, $n = 2n + 4$, $n - 2n = 4$, $-n = 4$

$\therefore n = -4$. Ans.

Check: $\frac{-4}{2} = -4 + 2$ ✓

-----Continued-----

Examples

8. John picked up two-thirds of the books in the shelf. There are 12 books left in the shelf. How many books were in the shelf to begin with ?

Solution:

Let $n =$ the number of books to begin with

There are $(1 - \frac{2}{3})\ n$ books left in the shelf after John picked up $\frac{2}{3}$ of the books.

The equation is

$$(1 - \tfrac{2}{3})n = 12$$

$$\tfrac{1}{3}n = 12$$

$$\therefore n = 36 \text{ books. Ans.}$$

Check: $\frac{2}{3}(36) = 24$

$36 - 24 = 12$ ✓

9. Roger picked up two-fifths of the books in the shelf. Maria picked up one-half of the remaining books. Jack picked up 6 books that were left. How many books were in the shelf to begin with ?

Solution:

Let $n =$ the number of books to begin with

There are $(1 - \frac{2}{5})\ n$ remaining books in the shelf after Roger picked up $\frac{2}{5}$ of the books.

Then, Maria picked up $\frac{1}{2}(1 - \frac{2}{5})n$ books after Roger picked up $\frac{2}{5}$ of the books.

The equation is

$$(1 - \tfrac{2}{5})n - \tfrac{1}{2}(1 - \tfrac{2}{5})n = 6$$

$$\tfrac{3}{5}n - \tfrac{1}{2} \cdot \tfrac{3}{5}n = 6$$

$$\tfrac{3}{5}n - \tfrac{3}{10}n = 6$$

$$\tfrac{6n-3n}{10} = 6$$

$$\tfrac{3n}{10} = 6$$

$$3n = 60$$

$$n = 20 \text{ books. Ans.}$$

Check: $\frac{2}{5}(20) = 8$

$20 - 8 = 12$

$\frac{1}{2}(12) = 6$

$12 - 6 = 6$ ✓

10. There are one-dollar bills and five-dollar bills in the piggy bank. It has 60 bills with a total value of \$228. How many ones and how many fives in the piggy bank ?

Solution:

Let $x =$ number of one-dollar bills

$6 - x =$ number of five-dollar bills

The equation is

$$1x + 5(60 - x) = 228$$

$$x + 300 - 5x = 228$$

$$-4x = -72$$

$$\therefore x = 18 \rightarrow \text{one-dollar bills}$$

$$60 - x = 42 \rightarrow \text{five-dollar bills}$$

Check:

$\$18 + 5(42) = \$228$ ✓

Examples

11. The length of a rectangle is 14 cm longer than the width. The perimeter is 76 cm. Find its length and width.

Solution:

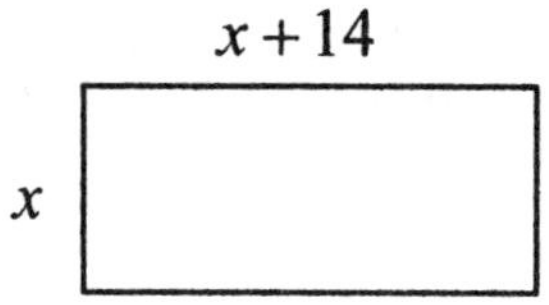

Let x = the width

$x+14$ = the length

The equation is

$$2x+2(x+14)=76$$
$$2x+2x+28=76$$
$$4x=76-28$$
$$4x=48$$
$$\therefore x=12\ cm \rightarrow \text{the width.}$$
$$x+14=26\ cm \rightarrow \text{the length.} \quad \text{Ans.}$$

Check: $(12+26)\times 2=76$ ✓

12. A suit is on sale for 20% discount. Roger paid \$8.82, including a 5% sales tax. Find the original price of the suit.

Solution:

Let x = the original price

The sales price = $80\% \cdot x = 0.80x$

The sales tax = $0.80x \cdot 5\% = 0.80x \cdot 0.05$

The equation is

$$0.80x+0.80x\cdot 0.05=8.82$$
$$0.80x+0.04x=8.82$$
$$0.84x=8.82$$
$$\therefore x=\tfrac{8.82}{0.84}=\$10.50. \quad \text{Ans.}$$

Check: $10.50 × 0.80 = $8.40
$8.40 × 0.05 = $0.42
$8.40 + $0.42 = $8.82 ✓

13. A 16-gallon salt-water solution contains 25% pure salt. How much water should be added to produce the solution to 20% salt ?

Solution:

$16\times 25\%=16\times 0.25=4$ gallons of salt in the original solution.

Let x = gallons of water added

The equation is

$$\frac{4}{16+x}=20\%,\quad \frac{4}{16+x}=0.20,$$
$$4=0.20(16+x)$$
$$4=3.2+0.20x$$
$$4-3.2=0.20x$$
$$0.8=0.20x \quad \therefore x=\tfrac{0.8}{0.20}=4 \text{ gallons of water added.} \quad \text{Ans.}$$

Check: $\frac{4}{16+4}=\frac{4}{20}=0.2=20\%$ ✓

EXERCISES

Solve each equation. Check your solution.

1. $x+6=15$

2. $x-6=-15$

3. $\frac{x}{5}=7$

4. $\frac{y}{-5}=7$

5. $7n=-42$

6. $-7n=42$

7. $15=x+3$

8. $15=-x-3$

9. $12=\frac{x}{3}$

10. $-12=\frac{x}{-3}$

11. $-12=3x$

12. $12=-3x$

13. $x+\frac{1}{2}=-3$

14. $\frac{k}{12}=1.5$

15. $-\frac{2}{9}x=\frac{4}{3}$

16. $5x=2x+6$

17. $5x=-2x-6$

18. $4x-5=19$

19. $-5y+7=-18$

20. $4x-6x=18$

21. $-12=3a-5a$

22. $2.5p=0.5p-10$

23. $2.5p=-0.5p+14$

24. $3.2x=1.8x-4.2$

25. $\frac{x}{2}=\frac{x}{4}-7$

26. $\frac{3}{4}x+\frac{1}{4}=2$

27. $\frac{x-4}{5}=-10$

28. $3(2x-4+x)=14$

29. $-2(3y+4-5y)=4$

30. $5(2n-5)=5$

31. $-5(5-3n)=5$

32. $3x+4=-5x-20$

33. $3a+5=a-4$

34. $7(x+2)=3(x-4)$

35. $5y-2(y+3)=4y$

36. A number increased by 21 is 28. Find the number.
37. 8 less than a number is 15. Find the number.
38. A number decreased by 12 is equal to 30. Find the number.
39. A number less than 8 is 15. Find the number.
40. 5 more than a number is 25. Find the number.
41. 5 times a number is 45. Find the number.
42. 5 more than twice a number is 13. Find the number.
43. Twice the sum of a number and 5 is –35. Find the number.
44. One-third of a number is 12. Find the number.
45. 3 times a number decreased by 4 is 18. Find the number.
46. Six times the sum of a number and two is fourteen. Find the number.
47. Half the quotient of a number and three is ten. Find the number.
48. 12 more than 4 times a number is twice the number. Find the number.
49. Nine times a number is twice the difference between the number and seven. Find the number.
50. Five times the number which is two less than x is fifteen. Find x.

-----Continued-----

51. Find two consecutive integers whose sum is 55.
52. Find two consecutive even integers whose sum is 114.
53. Find two consecutive odd integers whose sum is 116.
54. Find three consecutive integers whose sum is 69.
55. Find three consecutive even integers whose sum is 222.
56. Find four consecutive even integers whose sum is 172.
57. Find four consecutive odd integers whose sum is 152.
58. Find four consecutive multiples of 5, whose sum is 230.
59. Twice the sum of a number and 4 is 32. Find the number.
60. Roger has twice as much money as David. Together they have $63. How much does each have ?
61. A number divided by 3 is equal to the number decreased by 14. Find the number.
62. A number divided by 3 is equal to the number decreased by 3. Find the number.
63. John picked up one-third of the books in the shelf. There are 18 books left in the shelf. How many books were in the shelf to begin with ?
64. Roger picked up one-third of the books in the shelf. Maria picked up two-thirds of the remaining books. There are 8 books left in the shelf. How many books were in the shelf to begin with ?
65. The width of a rectangle is 6 inches shorter than the length. The perimeter is 48 inches. Find its length and width.
66. In an isosceles triangle, the length of each leg exceeds the base by 4 inches. The perimeter is 92 inches. Find the length of each leg.
67. A product is on sale for 30% discount. Maria paid $15.12, including a 8% sales tax. Find the original price of the product.
68. A 15-gallon salt-water solution contains 20 % pure salt. How much water should be added to produce the solution to 15% salt ?
69. A 200-gram chlorine-water solution contains 40% chlorine. How much water should be added to make a new solution that is only 25% chlorine ?
70. A rope $27\,cm$ long is cut into two pieces. One piece is $8\,cm$ longer than the other piece. Find the lengths of the pieces.
71. A board 39 inches long is cut into three pieces. The largest piece is 5 inches longer than the median piece. The shortest piece is 5 inches shorter than the median piece. Find the length of each piece.
72. Roger has twice as much money as Carol. Maria has $\frac{1}{2}$ as much money as Carol. Together they have $77. How much money does each have ?
73. There are one-dollar bills and five-dollar bills in the piggy bank. It has 54 bills with a total value of $174. How many ones and how many fives in the piggy bank ?
74. John gave three-fifths of his money to Joe and gave $40 to Janet. John had $8 left. How much money did John have to start with ?
75. Tom ate one-third of the candies in a box. Roger ate two-fifths of the remaining candies. There were 18 candies left. How many candies were in the box to start with ?

6-9 Literal Equations and Formulas

A literal equation is an equation that contains two or more variables.
In science and math, every formula is a literal equation.
In a literal equation, we can solve it for a given variable by **isolating** the variable to one side of the equation. We follow the same steps as we solve an equation.

The area of a rectangle is given by the formula: $A = \ell \cdot w$ (Area = length × width)

Solve it for ℓ, we divide each side by w: $\ell = \dfrac{A}{w}$.

Solve it for w, we divide each side by ℓ: $w = \dfrac{A}{\ell}$.

The perimeter of a rectangle is given by the formula: $p = 2(\ell + w)$

Solve it for w, we have : $p = 2(\ell + w)$

$$p = 2\ell + 2w$$
$$2w = p - 2\ell$$
$$\therefore w = \frac{p - 2\ell}{2} \text{ or } \frac{p}{2} - \ell. \text{ Ans.}$$

Examples

1. Solve $x = y + 5$ for y.

Solution:

$$x = y + 5$$
$$x - 5 = y$$
$$\therefore y = x - 5. \text{ Ans.}$$

2. Solve $x = 2y - 5$ for y.

Solution:

$$x = 2y - 5$$
$$x + 5 = 2y$$
$$\frac{x+5}{2} = y \quad \therefore y = \frac{x+5}{2}. \text{ Ans.}$$

3. The formula for the area of a triangle is $A = \frac{1}{2}bh$, where b is the base, and h is the height. Solve the formula for h.

Solution:

$$A = \tfrac{1}{2}bh$$

Multiply each side by 2, we have $2A = bh$ $\therefore h = \dfrac{2A}{b}$. Ans.

4. The formula to find the Celsius temperature for each Fahrenheit temperature is given by the formula $C = \dfrac{5}{9}(F - 32)$. Solve the formula for *F*.

Solution: $C = \dfrac{5}{9}(F - 32)$

Multiply each side by $\dfrac{9}{5}$, we have

$$\frac{9}{5}C = F - 32 \quad \therefore F = \frac{9}{5}C + 32. \text{ Ans.}$$

EXERCISES

Solve each equation for the indicated variable.

1. $x + y = 5$, for y
2. $x + y = 5$, for x
3. $x - y = 5$, for x

4. $x - y = 5$, for y
5. $2x + y = 10$, for y
6. $2x + y = 10$, for x

7. $2x + y = 10$, for y
8. $2x - y = 10$, for x
9. $2x - 4y = -5$, for y

10. $2a + b = c$, for b
11. $2a + b = c$, for a
12. $2a + 2b = c$, for a

13. $a - b = 3a + b$, for a
14. $ax = a - 2$, for x
15. $ax = a + 2$, for a

16. If $\frac{a}{b} = 1$, then $a - b = ?$

17. The formula for the circumference of a circle is $C = 2\pi r$, where $\pi \approx 3.14$, and r is the radius.

a. Solve the formula for r.
b. Find the radius of a circle with a circumference of 4 meters.

18. The formula for the area of a triangle is $A = \frac{1}{2}bh$, where b is the base, and h is the height.

a. Solve the formula for h.
b. Find the height of a triangle if $A = 20\,ft^2$, $b = 5\,ft$.

19. The formula for the perimeter of a rectangle is $p = 2(l + w)$, where l is the length, and w is the width.

a. Solve the formula for l.
b. Find the length of a rectangle if $p = 88\ cm$, $w = 16\ cm$.

20. The formula for the distance traveled in a single direction at a constant speed is $d = rt$, where r is the speed (rate), and t is the time traveled.

a. Solve the formula for r.
b. If a plane left an airport at a constant speed to a distance of 1,950 miles in three hours, what is its speed ?

21. The formula for the volume of a pyramid is $V = \frac{1}{3}bh$, where b is the area of the base, and h is the height. Solve the formula for h.

22. The formula for the total surface area of a right circular cylinder is $A = 2\pi r h + 2\pi r^2$, where r is the radius of the circular base and top, and h is the height of the cylinder. Solve the formula for h.

CHAPTER 6 EXERCISES

Evaluate each expression.

1. $3x-5$ if $x=4$
2. $3x-5$ if $x=-4$
3. $-3x+5$ if $x=-4$
4. $-3x-5$ if $x=4$
5. $\frac{1}{2}x-8$ if $x=6$
6. $\frac{1}{2}x-8$ if $x=-6$
7. $\frac{3}{4}y+9$ if $y=16$
8. $\frac{3}{4}y-9$ if $y=-16$
9. $\frac{3}{4}y-9$ if $y=0$
10. $4-2.5a$ if $a=3$
11. $4-2.5a$ if $a=1.4$
12. $4-2.5a$ if $a=-1.4$
13. $3x^2-2x+1$ if $x=5$
14. x^2+y-2 if $x=-5$, $y=2$
15. $3a^2b^3$ if $a=2$, $b=-3$

Simplify each expression.

16. $x+5x$
17. $5x-2x$
18. $-5x+2x$
19. $6x+(4x-2)$
20. $6x+(2-4x)$
21. $-6x+(4x+2)$
22. $7+(2x+5)$
23. $7-(2x+5)$
24. $7-(2x-5)$
25. $3x+7+(5x-8)$
26. $3x+7+(8-5x)$
27. $3x-7+(-5x+8)$
28. $3x+7-(5x-8)$
29. $3x+7-(8-5x)$
30. $3x-7-(-5x+8)$
31. $-(2+x)-(x-2)$
32. $x-2y-(3y-4x)$
33. $x-2y+(3y-4x)$
34. $\frac{1}{4}+\frac{2}{3}y-(1-\frac{1}{3}y)$
35. $\frac{2}{5}x+\frac{1}{3}y+(\frac{1}{5}x-\frac{2}{3}y)$
36. $\frac{2}{5}x+\frac{1}{3}y-(\frac{1}{5}x-\frac{2}{3}y)$
37. $\frac{2}{5}x+\frac{1}{5}y-(\frac{1}{3}x-\frac{2}{3}y)$
38. $1.2x+3.5y-(2y-4.5x)$
39. $5.6n-(4-2.5n)$
40. $2x^2-3-4x^2-6$
41. $2x^2-3-(4x^2-6)$
42. $3x^2-(-4x^2+6)$
43. $x+(2x-2)-(4y+3)$
44. $3p+(p-q)-(4p-3q)$
45. $2x^2-(4x^2-3x-5)$
46. $2x\cdot 4x$
47. $-2x\cdot 4x^2$
48. $-1.5a^3\cdot 3a^4$
49. $4x(2x-8)$
50. $-5(3x-4y+5)$
51. $-2a(a-2b+5)$
52. $4x-6(2x-3)$
53. $4x^2+5x-6(x^2+x-2)$
54. $4p^2-(2p^2-3p+1)$
55. $\dfrac{5x+10}{5}$
56. $\dfrac{5x^2-10x}{5x}$
57. $\dfrac{3a^3-2a^2}{a^2}$
58. $\dfrac{10x^3+5x^2+15x}{5x}$
59. $\dfrac{4a^3-12a^2-4a}{4a}$
60. $\dfrac{9x^3y-6xy^2}{3xy}$

-----Continued-----

Solve each equation. Check your solution.

61. $x-9=30$

62. $25=y+15$

63. $\frac{a}{15}=6$

64. $32=-5p$

65. $x+\frac{3}{4}=8$

66. $y-\frac{3}{4}=8$

67. $-\frac{4}{5}p=\frac{8}{25}$

68. $3x=6x+8$

69. $\frac{1}{3}x=\frac{3}{4}x-9$

70. $-5x+8=4x-10$

71. $3y-7=4y+7$

72. $4(2p-3)=6$

73. $2(x-1)=4(2x+5)$

74. $6y-3(3y-1)=5y$

75. $\frac{2x+15}{3}=-31$

76. $2x-y=6$, for y

77. $2x-y=6$, for x

78. $2x-4y=8$, for y

79. $4x+5y=-25$, for y

80. $2a+2b=3c$, for b

81. 25 less than a number is 40. Find the number.
82. A number less than 25 is 40. Find the number.
83. Five times a number is 95. Find the number.
84. The quotient of a number and 1.5 is 9. Find the number.
85. Eight times the difference between a number and 15 is 40. Find the number.
86. 12 less than 4 times a number is twice the number. Find the number.
87. A number divided by 6 is equal to the number increased by 6. Find the number.
88. Find two consecutive integers whose sum is 135.
89. Find two consecutive even integers whose sum is 174.
90. Find three consecutive multiples of 5, whose sum is 255.
91. A board is 75 inches. The carpenter cut it into two pieces. One piece is 11 inches less than the other piece. Find the lengths of pieces.
92. A board is 75 inches. The carpenter cut it into three pieces. The median piece is 5 inches longer than the shortest piece. The longest piece is 5 inches longer than the median piece. Find the lengths of the pieces.
93. The length of a rectangle is 7 feet longer than the width. The perimeter is 38 feet. Find its length and width.
94. A 20-gallon salt-water solution contains 15% pure salt. How much water should be added to make a new solution that is only 12% salt ?
95. In an isosceles triangle, the length of each leg exceeds the base by 14 cm. The perimeter is 73 cm. Find the length of the leg and the length of the base.
96. A suit is on sale for 25% discount. You paid \$93.15, including a 8% sales tax. Find the original price of the suit.
97. There are one-dollar bills and five-dollar bills in the piggy bank. It has 80 bills with a total value of \$184. How many ones and how many fives in the piggy bank ?
98. John's land is 500 ft^2 less than Roger's land. Use the symbol J for John's land, and R for Roger's land. Write an equation to represent John's land.

Additional Examples

1. A 16-gallon salt-water solution contains 4 gallons of pure salt. What is the percentage of salt in the solution ?
   Solution:

$$p = \frac{4}{16} = 0.25 = 25\%. \quad \text{Ans.}$$

2. A 16-gallon salt-water solution contains 25% pure salt. How much salt should be added to produce the solution to 32% salt ?
   Solution:

   Original solution: $16 \times 25\% = 16 \times 0.25 = 4$ gallons of salt.
   Let $x =$ gallons of salt added
   The equation is

$$\frac{4+x}{16+x} = 32\%, \quad \frac{4+x}{16+x} = 0.32$$

$$4 + x = 0.32(16 + x)$$
$$4 + x = 5.12 + 0.32x$$
$$x - 0.32x = 5.12 - 4$$
$$0.68x = 1.12 \quad \therefore x = \tfrac{1.12}{0.68} = 1.65 \text{ gallons of salt added. Ans.}$$

3. The sum of two numbers is 42. The larger number is 18 more than one-half of the smaller number. Find the two numbers.
   Solution:

   Let $n =$ the larger number
   $42 - n =$ the smaller number
   The equation is

$$n = 18 + \tfrac{1}{2}(42 - n)$$
$$n = 18 + 21 - \tfrac{1}{2}n$$
$$n + \tfrac{1}{2}n = 39$$
$$\tfrac{3}{2}n = 39$$

   Multiply each side by $\frac{2}{3}$

$$\therefore n = 39 \cdot \tfrac{2}{3} = 26 \rightarrow \text{the larger number.}$$
$$42 - 26 = 16 \rightarrow \text{the smaller number. Ans.}$$

-----Continued-----

Additional Examples

4. Find values for a, b, and c in the equation whose sides are identity (equivalent).

$$2ax - 3(x - by - 4) = 7x + 6y + 2c.$$

Solution:

$$2ax - 3(x - by - 4) = 7x + 6y + 2c$$

Simplify and combine like terms

$$2ax - 3x + 3by + 12 = 7x + 6y + 2c$$
$$(2a - 3)x + 3by + 12 = 7x + 6y + 2c$$

Compare with the coefficients on two sides, we have

$2a - 3 = 7$	$3b = 6$	$12 = 2c$
$2a = 7 + 3$	$b = 2$	$c = 6$
$2a = 10$		
$a = 5$		

Ans. $a = 5$, $b = 2$, $c = 6$.

5. If $x^3 + 2x^2 - x - 2x(4x^2 + 3x - 5) = ax^3 + bx^2 + cx$ for all values of x, what is the value of a, b, and c ?

Solution:

$$x^3 + 2x^2 - x - 2x(4x^2 + 3x - 5) = ax^3 + bx^2 + cx$$

Simplify and combine like terms

$$x^3 + 2x^2 - x - 8x^3 - 6x^2 + 10x = ax^3 + bx^2 + cx$$
$$(x^3 - 8x^3) + (2x^2 - 6x^2) + (-x + 10x) = ax^3 + bx^2 + cx$$
$$-7x^3 - 4x^2 + 9x = ax^3 + bx^2 + cx$$

Comparing with the coefficients on two sides, we have

$a = -7$, $b = -4$, $c = 9$. Ans.

6. The formula to find the Fahrenheit temperature for each Celsius temperature is $F = \frac{9}{5}C + 32$. Find the Fahrenheit temperature for the normal body temperature $37°C$ of human being.

Solution:

$$F = \tfrac{9}{5}C + 32 = \tfrac{9}{5}(37) + 32 = \tfrac{333}{5} + 32 = 66.6 + 32 = 98.6°$$

Ans. $98.6°F$

-----Continued-----

Additional Examples

7. A train left a town at 12:00 noon at a steady speed. A bus left the same town at 10 A.M and traveled the same route with an average speed 32 miles/h less than the train's speed. They arrived at the same destination at 4 P.M. What is the speed of the train ?
Solution:

Let $x =$ the speed of the train
$x - 32 =$ the speed of the bus
The equation is

$$4x = 6(x-32),$$
$$4x = 6x - 192, \quad 192 = 2x$$
$$\therefore x = 96 \text{ miles per hour.} \quad \text{Ans.}$$

8. John can complete a job in 6 hours. Erica can complete the same job in 9 hours. If they work together, how long will it take them to complete the job ?

Solution: John can do $\frac{1}{6}$ of the job per hour.

Erica can do $\frac{1}{9}$ of the job per hour.

Therefore $\frac{1}{6} + \frac{1}{9} = \frac{9}{54} + \frac{6}{54} = \frac{15}{54} = \frac{5}{18}$

If they work together, they can complete the $\frac{5}{18}$ of the job per hour.

$$1 \div \frac{5}{18} = \frac{18}{5} = 3\tfrac{3}{5} \text{ hours.}$$

They can complete the job in $3\frac{3}{5}$ hours if they work together. Ans.

9. John can complete a job in 6 hours. When John and Erica work together, they can complete the job in $3\frac{3}{5}$ hour. How many hours would it take Erica alone to complete the Job ?
Solution:

Let $x =$ The time in hours Erica needs to complete the job alone
When they work together, the job they can do per hour is

$$\frac{1}{6} + \frac{1}{x} = \frac{x}{6x} + \frac{6}{6x} = \frac{x+6}{6x}$$

The equation to complete the job if they work together is

$$\frac{6x}{x+6} = 3\frac{3}{5}, \quad \frac{6x}{x+6} = \frac{18}{5}, \quad 6x \cdot 5 = 18(x+6)$$
$$30x = 18x + 108, \quad 12x = 108, \quad \therefore x = 9 \text{ hours.}$$

It would take 9 hours for Erica to complete the job alone. Ans.

-----Continued-----

Additional Examples

Law of Lever (in science): If a lever is balanced, the **torques** (twist works) on one side of the lever is equal to the torques on the other side.

The **torque** due to a force about a pivot is equal to the force times its distance from the pivot. If m_1 and m_2 are the masses placed at distances d_1 and d_2 from the fulcrum, and the lever is balanced, then $m_1\,d_1 = m_2\,d_2$.

10. A 60-kg weight is placed on the end of a 5-meters board of negligible mass (as a lever) and the fulcrum (the support point) is placed as shown in the figure. Find the unknown weight on the other end if the lever is balanced ?

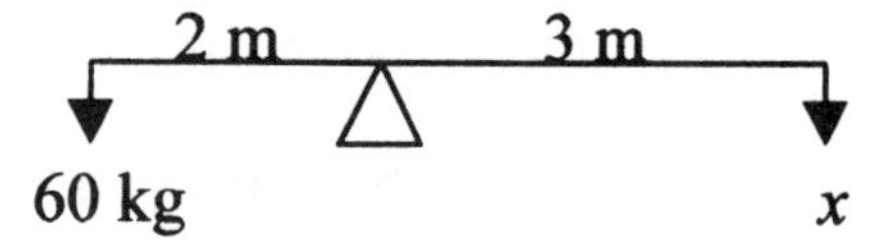

Solution:

Let $x =$ the unknown weight,

The equation is: $60 \cdot 2 = x \cdot 3$

$120 = 3x$

$\therefore x = 40$ kg. Ans.

11. A 20-pound weight is placed 4 feet from the fulcrum (the support point) on a lever. At what distance from the fulcrum must a 50-pound weight be placed to balance the lever ?

Solution:

4 x

20 50

Let $x =$ the distance of a 50-pound weight from the fulcrum

The equation is: $20 \cdot 4 = 50x$

$80 = 50x$ $\therefore x = 1.6$ feet. Ans.

12. A uniform board of 10-pound and 12 feet is used as a lever. Where must the support point be placed if a 50-pound weight at one end is to balance a 20-pound weight at the other end ?

Solution:

x 12 − x

6 x − 6

20 10 50

The equation is: $20x + 10(x-6) = 50(12-x)$

$20x + 10x - 60 = 600 - 50x$

$80x = 660$

$x = 8.25$ feet. Ans.

The support point should be placed 8.25 feet from the lighter-loaded end.

-----Continued-----

Additional Examples

13. A car and a bus left a town at the same time and traveled the same route to a city. The car traveled at 60 mph and arrived two hours late to the scheduled arriving time. The bus traveled at 90 mph and arrived two hours early. Find the distance between the town and the city.

Solution: Let $x =$ the time in hours needed to arrive on time.

The equation (the distance each traveled is the same) is:

$$60(x+2) = 90(x-2)$$
$$60x + 120 = 90x - 180$$
$$300 = 30x$$
$$x = 10 \text{ hours.}$$

The distance is $60(10+2) = 720$ miles. Ans.

14. David and Mario drove from home to office at the same time and traveled the same route. David drove at 42mph and arrived 5 minutes late. Mario drove 54mph and arrived 5 minutes early. Find the distance from home and office.

Solution:

Let $x =$ the time in minutes needed to arrive on time.

David's speed is $42 \div 60 = 0.7$ mile per minute.

Mario's speed is $54 \div 60 = 0.9$ miles per minutes.

The equation (the distance each traveled is the same) is:

$$0.7(x+5) = 0.9(x-5)$$
$$0.7x + 0.35 = 0.9x - 0.45$$
$$0.80 = 0.2x$$
$$\therefore x = 40 \text{ minutes.}$$

The distance is $0.7(40 + 5) = 31.5$ miles. Ans.

15. David rented a car. The car rental was \$50 plus \$0.25 per mile. If the total bill was \$120, how many miles did he drive by using this car ?

Solution:

Let $x =$ the mileage he drove.

The equation is:

$$50 + 0.25x = 120$$
$$0.25x = 70$$
$$x = \tfrac{70}{0.25} = 280 \text{ miles.}$$ Ans.

16. The sum of the digits of a three-digit number is 14. The hundreds digit is 4 times the tens digit and the tens digit is $\frac{1}{2}$ the unit digit. What is the number ?

Solution:

Let x = the unit digit.
$\frac{1}{2}x$ = the tens digit.
$4(\frac{1}{2}x)$ = the hundreds digit.

The equation is:

$$x + \frac{1}{2}x + 4(\frac{1}{2}x) = 14$$
$$x + \frac{1}{2}x + 2x = 14$$
$$3\frac{1}{2}x = 14$$
$$\frac{7}{2}x = 14$$
$$7x = 28$$
$$\therefore x = 4$$

Ans: 824.

OR: Let x = the tens digit.
$2x$ = the unit digit.
$4x$ = the hundreds digit.

The equation is:

$$x + 2x + 4x = 14$$
$$7x = 14$$
$$\therefore x = 2$$

Ans: 824.

In the gym, a man asks the coach.

Man: I want to be more attractive to young ladies.
What type of machines do I need to use ?
Coach: The ATM machine outside the gym.

Functions and Graphs

7-1 Number line and Coordinate Plane

In Section 1-1, we have learned the classification of real numbers. We can graph all of the real numbers on the **number line**. Each number is assigned a point on the line.

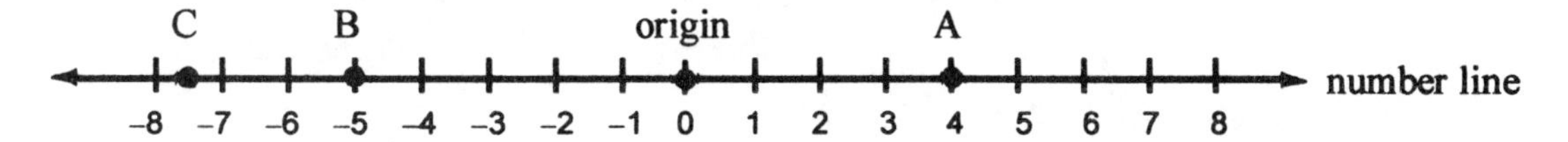

The arrowheads on the number line mean that the line continues in both directions. The number is called the **coordinate** of the point. The coordinate of the point tells the distance and direction from the origin of the line. The dot mark on the point is the graph of the point. The coordinate of point A is 4. The coordinate of point B is –5. The coordinate of point C is –7.5.

We can graph the solution of an equation on the number line.

Example 1: Solve $x-5=-8$.
Solution:

$$x-5=-8$$
$$x=-8+5$$
$$x=-3. \quad \text{Ans.}$$

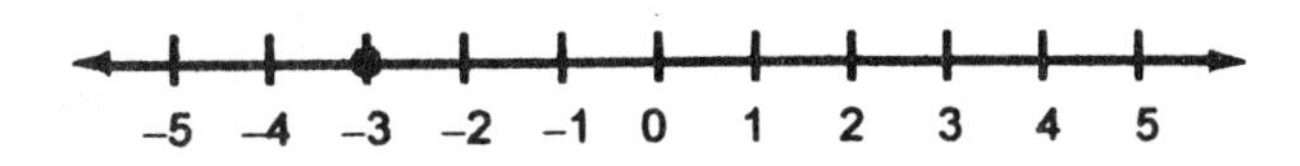

Example 2: Solve $2x+6=3$.
Solution:

$$2x+6=3$$
$$2x=3-6$$
$$2x=-3$$
$$x=-1\tfrac{1}{2} \text{ or } -1.5. \quad \text{Ans.}$$

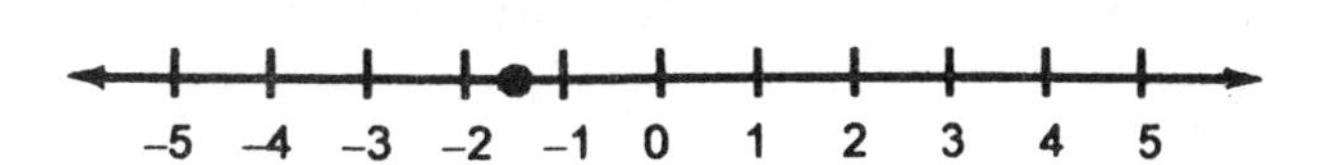

-----Continued-----

The graph of a point on a number line shows only the horizontal location of the point.
We can graph a point showed both the horizontal and vertical locations on a **coordinate plane**.

Cartesian Coordinate Plane:

It is a rectangular plane used to locate, or plot points $(x,\ y)$, lines, and graphs.
The horizontal line is called the $x-$axis.
The vertical line is called the $y-$axis.
Two number lines perpendicular to each other at the **origin** (O).
The coordinates of origin is (0, 0).
The axes separate the plane into four quadrants.
A point $(x,\ y)$, called **ordered pair**, in the plane, such as (4, 2), is the coordinates of the point.
x is the $x-$coordinate (or **abscissa**). y is the $y-$coordinate (or **ordinate**).

Ordered pair $(x,\ y)$: Ordered pair is used to indicate the location of a point on a plane.
x is the number of horizontal units moved from 0.
y is the number of vertical units moved from 0.
A **grid** is often used to find the ordered pair $(x,\ y)$ for each point.

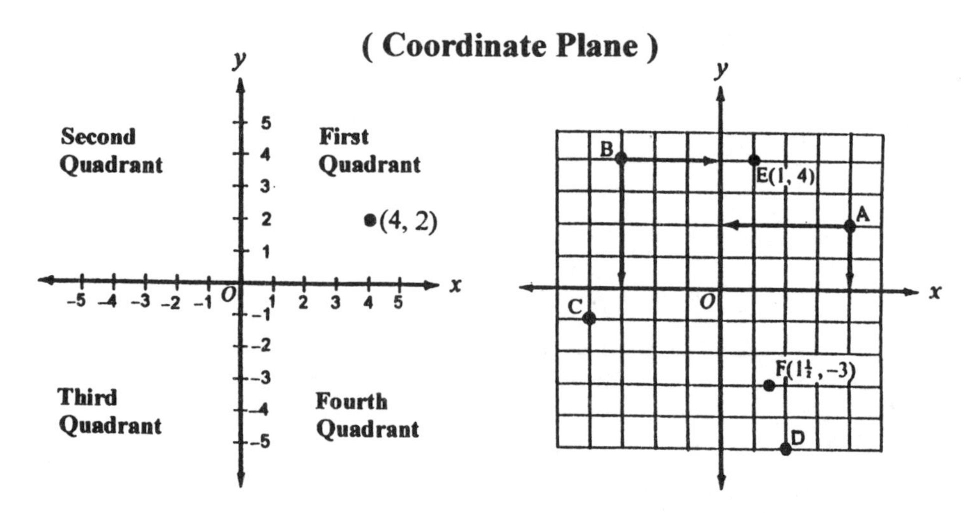

To find the coordinates of a point in the coordinate plane, first draw a vertical line from the point to the $x-$axis, next draw a horizontal line from the point to the $y-$axis. Together, we have the coordinates of the point.

Example 3: Give the coordinates of each of the points A, B, C, and D on the above grid.
Solution:

The coordinates of A is (4, 2). The coordinates of B is (–3, 4).
The coordinates of C is (–4, –1). The coordinates of D is (2, –5).

Example 4: Graph each of the ordered pairs on a coordinate plane: E(1, 4), $F(1\frac{1}{2}, -3)$.
Solution: The answers are shown on the above grid.

EXERCISES

Identify the coordinate of each point described on the number lines.

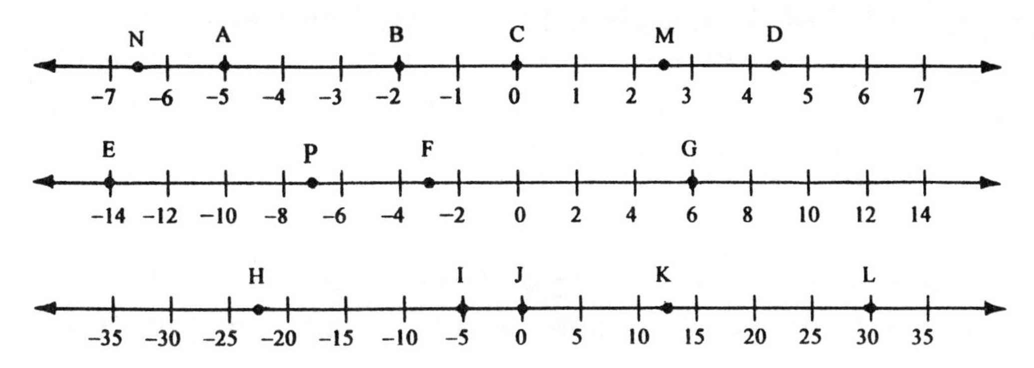

1. A **2.** B **3.** C **4.** D **5.** E **6.** F **7.** G **8.** H **9.** I **10.** J
11. K **12.** L **13.** M **14.** N **15.** P

16. The point whose distance is 4 points to the left of B.
17. The point half way from A to C.
18. The point whose distance is 25 points to the right of I.
19. The point whose distance is 14 points to the left of G.
20. The point whose distance is 15 points to the left of L.
21. The point to the right of B is twice as far as the distance from B to C.
22. The midpoint between E and G.
23. The point to the left of C is 2 times the distance from C to B.
24. The point to the right of I is 2 times the distance from I to J.
25. The point halfway between I and L.

26. In the diagram, D is the midpoint between C and E.
C is the midpoint between A and E.
D is 6 points to the left of E.
C is 5 points to the right of B.
The coordinate of A is –3. Find the coordinates of B, C, D, and E.

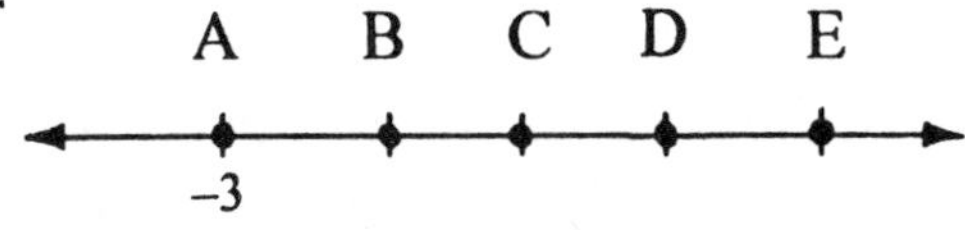

Solve each equation. Graph each solution on a number line.

27. $x - 4 = -2$ **28.** $6 = -y + 5$ **29.** $2.4a = -6$ **30.** $\frac{1}{3}x + 2 = 5$

Identify coordinates of each point described on the coordinate plane.

31. A **32.** B **33.** C **34.** D **35.** E
36. F **37.** G **38.** K **39.** P **40.** Q

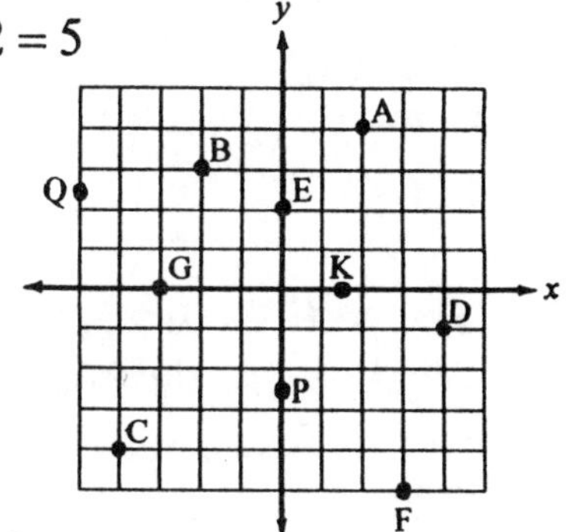

Graph and label each of the point on a coordinate plane.

41. $B(2, -2)$ **42.** $p(-3, 4)$ **43.** $n(-1, -5)$ **44.** $Q(3,\ 2\frac{1}{2})$ **45.** $A(-4, 0)$

-----Continued-----

Name the quadrant containing the point.

46. (–3, 5) **47.** (–4, –1) **48.** (7, 9) **49.** (5, –5) **50.** (–2, –2)

Graph each point and find the length of the segment between them.

51. A(–2, –3) and B(–2, 2) **52.** M(3, 0) and N(3, –4) **53.** P(–4, 5) and T(2, 5)
54. C(6, 3) and D(1, 3) **55.** E(4, 5) and F(4, –6) **56.** G(5, 4) and H(–6, 4)
57. J(–4, –3) and K(–4, –7) **58.** S(3, 0) and T(–5, 0) **59.** U(0, –2) and V(0, 7)
60. W(–8, 0) and Z(7, 0)

Graph the given points on a coordinate plane. Draw line segments to connect the points in the order given and to connect the first and last points. Name the closed figure.

61. (4, 3), (–4, 3), (–4, –3), (4, –3) **62.** (1, 3), (–1, –3), (1, –3)
63. (–2, 0), (–3, –2), (2, –2), (3, 0) **64.** (–3, 0), (1, 0), (1, 3), (–1, 3)
65. (2, 1), (4, 1), (4, –4), (2, –4)

Determine whether the line connecting the two points is horizontal, vertical, or neither.

66. (4, 1), (4, –5) **67.** (2, –3), (2, –4) **68.** (5, 0), (–4, 0) **69.** (2, 5), (5, 2)
70. (–2, –6), (–2, –3) **71.** (0, 4), (–5, 0) **72.** (3, –6), (–1, –6) **73.** (0, 5), (0, 0)
74. (5, 5), (–5, –5) **75.** (5, –5), (–5, –5)

Graph the triangle connecting the given points on a coordinate plane. Change the values of the coordinates as directed. Identify the change of the graph.

(1, 1), (5, 1), (3, 3)

76. Increase all $x-$coordinates by 2. **77.** Decrease all $x-$coordinates by 2.
78. Increase all $y-$coordinates by 2. **79.** Decrease all $y-$coordinates by 2.
80. Multiply all $x-$coordinates by -1. **81.** Multiply all $y-$coordinates by -1.

Graph the given points on a coordinate plane. State whether they lie on a straight line.

82. (1, 1), (2, 3), (3, 5), (0, –1) **83.** (1, –1), (–1, 1), (–3, 3)
84. (–2, –2), (0, 1), (2, 3) **85.** (–1, –5), (0, –3), (2, 1), (3, 3)

86. The vertices of a rectangle on a coordinate plane are (3, –1), (3, 3), (–2, 3), and (–2, –1). What are the dimensions of the rectangle ?

87. The vertices of a triangle on a coordinate plane are (1, 1), (1, –3.5), (–2, –3.5). What is the area of the triangle ?

7-2 Linear Equations

In Chapter 6, we have learned to solve equations that have only one variable.
In this chapter, we will learn to solve equations that have two variables of degree 1.
Here is an example in different forms:

$$2x - y = 3, \quad \text{or } 2x = y + 3, \quad \text{or } y = 2x - 3$$

The solutions to equations in one variable are numbers (see Chapter 6).
The solutions to equations in two variables are ordered pairs of numbers (x, y).

Example: Identify whether the ordered pair (0, –3) is a solution of $2x - y = 3$.
Solution:
We substitute $x = 0$ and $y = -3$ in the equation:
$2x - y = 2(0) - (-3) = 0 + 3 = 3$ ✓
Yes. It is a solution.

Therefore, $x = 0$ and $y = -3$ is a solution of $2x - y = 3$ and can be written as (0, –3).
The equation $2x - y = 3$ has infinitely many solutions. Here are some of its solutions:

(–2, –7), (–1, –5), (0, –3), (1, –1), (2, 1), …….

Each solution of the equation is a point (x, y) on the coordinate plane.

The graph of connecting all the solutions (points) of the equation $2x - y = 3$ forms a straight line. Therefore, we call the equation $2x - y = 3$ a **linear equation**.

We can write any linear equation into the form $ax + by = c$, where a, b, and c are real numbers with a and b not both zero.

The equation $ax + by = c$ is called the **standard form** of a linear equation. If a is negative, we change it to positive by multiplying each side of the equation by –1 (see example 3).

Examples

1. Write $2x = y + 3$ in standard form.
Solution:
$2x = y + 3$
Move y to the left side(change its sign):
$2x - y = 3$. Ans.

2. Write $-y = -2x + 3$ in standard form.
Solution:
$-y = -2x + 3$
Move $-2x$ to the left side(change its sign):
$2x - y = 3$. Ans.

3. Write $y = 2x - 3$ in standard form.
Solution:
$y = 2x - 3$
Move $2x$ to the left side(change its sign):
$-2x + y = -3$
Multiply each side by –1:
$2x - y = 3$. Ans.

4. Write $y + 3 = 2x$ in standard form.
Solution:
$y + 3 = 2x$
Move y to the right side(change its sign):
$3 = 2x - y$
$2x - y = 3$. Ans.

EXERCISES

Identify whether each ordered pair is a solution of $x+2y=6$.

1. (0, 3) **2.** (2, 1) **3.** (–2, 4) **4.** (6, 0) **5.** (–1, 3)

Identify whether each ordered pair of number is a solution of $2x-y=3$.

6. (2, 1) **7.** (3, 3) **8.** (4, 2) **9.** (–1, –5) **10.** (–2, 5)

Identify whether each ordered pair of number is a solution of the equation.

11. $x+y=3$, (1, 2) **12.** $x-y=3$, (3, 1) **13.** $2x-3y=2$, (4, 2)
14. $3x-2y=5$, (1, –1) **15.** $5a-2b=5$, (2, 2) **16.** $2m-4n=5$, $(\frac{1}{2},-1)$
17. $6s-t=2$, $(\frac{1}{3}, 1)$ **18.** $3a+4b=3$, (–2, 3) **19.** $x+3y=12$, (2, 3)
20. $8x-y=-15$, $(-\frac{3}{4}, 9)$ **21.** $2x+y=7$, (0, 5) **22.** $2x+y=7$, (0, 7)
23. $2x-y=7$, (0, –7) **24.** $-2x-3y=1$, (1, –1) **25.** $-2x+3y=-5$, (1, –1)
26. $y=2x-1$, (1, 1) **27.** $y=3x+2$, (2, 9) **28.** $y=\frac{1}{2}x-5$, (4, –3)
29. $y=\frac{2}{3}x$, (–6, –4) **30.** $y=-\frac{2}{3}x+1$, (–3, 3)

Write each of the linear equations in standard form.

31. $3x=y-5$ **32.** $y=-4x+7$ **33.** $3x=-4y+6$ **34.** $-4y=-5x-9$
35. $2x-8=-4y$ **36.** $-3y+20=5x$ **37.** $-4x=7y-1$ **38.** $-2x+4y=7$
39. $-2x=-5y-4$ **40.** $4y-6x=-5$

Find the values for y by substituting –2, 0, and 2 for x.

41. $y=2x$ **42.** $y=2x+1$ **43.** $y=2x-1$ **44.** $y=-2x+1$
45. $y=-2x-1$ **46.** $y=-x+5$ **47.** $y=-x-5$ **48.** $y=4-x$
49. $y=4-2x$ **50.** $y=-4+2x$

51. The cost to buy a card is \$1.50. Write an equation which expresses the relationship between the total cost (c) and the number (n) of cards bought.

52. A book regularly priced at p dollars is on sale at a 30% discount. Write an equation which expresses the relationship between the cost (c) of the book after the discount.

53. The price of a suit is regularly at p dollars. The sales-tax rate is 7.5%. What is the equation of the cost (c) including tax ?

54. The cost to rent a car is \$20 per day plus 20 cents per mile. What is the equation of the cost (c) to rent the car for one day with a mileage (m) ? What is the equation of the cost to rent the car for n days ?

55. A-Plus Company charges \$11.50 per copy of a book plus \$5 handling fee per order. Write an equation which expresses the relationship between the total charge (c) and the number (n) of books per order. What is the total charge if Barnes and Noble Bookstore places an order of 34 copies ?

7-3 Equations and Functions

The other way to describe equations in two variables is the application of **functions.**
If we rewrite a linear equation by solving for y in terms of x, we have a **linear function**.
Example: Solve the equation $3x+2y=6$ for y in terms of x.

Solution:

We write the equation with y on the left side and x on the other side:

$$3x+2y=6$$

$$2y=-3x+6$$

Divide each side by 2: $y=-\frac{3}{2}x+3$. Ans.

In the equation $y=-\frac{3}{2}x+3$, x is called the **independent variable**, and y is called the **dependent variable**. Each value of y depends on the value chosen for x.
To find the value of y in the equation, we could substitute the value chosen for x into the equation: $x=-1$, $y=-\frac{3}{2}(-1)+3=4\frac{1}{2}$. solution is $(-1,\ 4\frac{1}{2})$.
$x=2$, $y=-\frac{3}{2}(2)+3=0$. solution is (2, 0).
The equation $y=-\frac{3}{2}x+3$ has infinitely many solutions.
In the equation $y=-\frac{3}{2}x+3$, each value of x is assigned exactly one value of y. We say that "**y varies directly with x**", or "**y is a function of x**".
"y is a function of x" is written by **functional notation** $y=f(x)$.
Therefore, the equation $y=-\frac{3}{2}x+3$ can be written by **functional notation**:

$$f(x)=-\frac{3}{2}x+3$$

The other way to write a function is to use **arrow notation**:

$$f: x\rightarrow -\frac{3}{2}x+3$$

We may use any variable to define a function.

Examples:

1. Given $f(x)=2x-4$, find the value for each function.

a) $f(3)$ **b)** $f(-2)$ **c)** $f(0)$ **d)** $f(\frac{1}{2})$

Solution:

a) $f(3)=2(3)-4=2$. **b)** $f(-2)=2(-2)-4=-8$.
c) $f(0)=2(0)-4=-4$. **d)** $f(\frac{1}{2})=2(\frac{1}{2})-4=-3$. Ans.

2. Given $f: x\rightarrow 4-3x$, find the value for each function.

a) $f(-4)$ **b)** $f(1)$ **c)** $f(-\frac{1}{3})$ **d)** $f(\frac{1}{2})$

Solution:

Write the function: $f(x)=4-3x$
a) $f(-4)=4-3(-4)=4+12=16$. **b)** $f(1)=4-3(1)=1$.
c) $f(-\frac{1}{3})=4-3(-\frac{1}{3})=4+1=5$. **d)** $f(\frac{1}{2})=4-3(\frac{1}{2})=\frac{1}{2}$. Ans.

EXERCISES

Solve each linear equation for y in terms of x.

1. $x+y=5$ **2.** $x-y=5$ **3.** $-x+y=-5$ **4.** $-x-y=5$

5. $2x+y=6$ **6.** $2x-y=-6$ **7.** $-2x+2y=7$ **8.** $y-2x=-5$

9. $6x-3y=8$ **10.** $4x-5y=20$ **11.** $x-5y=10$ **12.** $3x-y=1$

13. $-x+\frac{1}{2}y=12$ **14.** $2x-\frac{2}{3}y=10$ **15.** $\frac{1}{2}x-\frac{3}{2}y=6$

Find the solutions for the given values of x.

16. $y=2x$ for $x=-2, 0, 2$ **17.** $y=2x+3$ for $x=-1, 0, 2$

18. $y=2x-3$ for $x=-2, 1, 3$ **19.** $y=\frac{1}{2}x+1$ for $x=-2, 0, 1$

20. $y=\frac{2}{3}x-1$ for $x=-3, 0, 1$

Write each of the equation in functional notation $f(x)$.

21. $y=2x-4$ **22.** $y=-2x+5$ **23.** $y=7-8x$ **24.** $y=-5+8x$

25. $x+y=4$ **26.** $x-y=4$ **27.** $2x+y=4$ **28.** $-2x-y=4$

29. $2x+2y=7$ **30.** $4x-2y=3$

Find the value for each function.

31. $f(x)=3x-6$ **a)** $f(-1)$ **b)** $f(0)$ **c)** $f(1)$ **d)** $f(2)$

32. $f(x)=2-5x$ **a)** $f(-2)$ **b)** $f(\frac{1}{2})$ **c)** $f(3)$ **d)** $f(5)$

33. $p(a)=a^2-2a$ **a)** $p(-\frac{1}{2})$ **b)** $p(0)$ **c)** $p(3)$ **d)** $p(6)$

34. $q(c)=2c^2+c-2$ **a)** $q(-2)$ **b)** $q(0)$ **c)** $q(1)$ **d)** $q(\frac{1}{2})$

35. $f: x \to 4x-5$ **a)** $f(-3)$ **b)** $f(-\frac{1}{2})$ **c)** $f(1)$ **d)** $f(1\frac{1}{4})$

36. Given $f(x)=4x-16$, solve $f(x)=0$.
37. Given $g(x)=3-4x$, solve $g(x)=0$.
38. Given $f(x)=2x-1$ and $g(x)=x^2$, find $f(1)+g(2)$.
39. Given $f(x)=3-2x$ and $g(x)=2x-1$, find $f(2)\cdot g(3)$.
40. Given $f(x)=x^2$ and $g(x)=3-2x$, find $f[g(2)]$ and $g[f(2)]$.
41. If $f(x)=x^2+18$ and n is a positive number such that $f(2n)=2f(n)$, find the value of n.

7-4 Relations and Functions

An equation in two variables has infinitely many solutions. Each solution can be represented by an ordered pair (x, y) of a point.
The set of all ordered pairs of solutions to an equation determines a **relation** between two sets of numbers (x – values and y – values).

We use braces $\{ \}$ to represent the relation of the set of ordered pairs of points (x, y).
For example, the equation $y = 2x - 3$ has some of its solution set of ordered pairs (x, y):

$$\{(-2, -7), (-1, -5), (0, -3), (1, -1), (2, 1), (3, 3), (4, 5)\}$$

The x – values in the set of ordered pairs are called the **domain** of the relation, and the y – values are called the **range** of the relation. For the above example:

the domain: $D = \{ -2, -1, 0, 1, 2, 3, 4 \}$.
the range: $R = \{ -7, -5, -3, -1, 1, 3, 5 \}$.

Definition of a relation: A relation is any set of ordered pairs (x, y).
Each value of x is paired with one or more values of y.
The solution set of any equation in two variables defines a relation.

Definition of a function: A function is a special kind of relation. It is a relation in which each value of x is paired with exactly one value of y.

The set of ordered pair $\{(1, 3), (3, 5), (4, 7)\}$ is a relation and a function.
The set of ordered pair $\{(2, 3), (2, 6), (3, 9)\}$ is a relation but not a function because $x = 2$ is assigned to two values of y, 3 and 6.

When a distance varies directly with time, we say that the distance (d) is a function of time (t). It is a function because each value of t can find exactly one value of d.

A car travels at 60 mph. The distance that the car travels in t hours is given by the function $d(t) = 60t$, where t represents the number of hours and d represents the total distance.
Notice that the domain (t) of the function is the set of positive real numbers (hours).
The range $d(t)$ denotes the value of distance (d) at t.

Examples

1. If the domain of $f(x) = 2x - 3$ is $D = \{-1, 0, 2, 3\}$, find the range of $f(x)$.
Solution:

$x = -1$, $f(-1) = 2(-1) - 3 = -5$
$x = 0$, $f(0) = 2(0) - 3 = -3$
$x = 2$, $f(2) = 2(2) - 3 = 1$
$x = 3$, $f(3) = 2(3) - 3 = 3$

The range of $f(x)$ is $R = \{-5, -3, 1, 3\}$. Ans.

-----Continued-----

Examples

2. The following points (x, y) represent a relation. Is the relation a function ? Explain.

$$\{(-3, 2), (-2, 3), (0, 1), (2, 3), (5, 2)\}$$

Solution:

Yes, the relation is a function because each value of x is assigned to exactly one value of y. Ans.

3. The following points represent the relation of the set of ordered pairs (x, y).

$$\{(-4, 1), (-1, 2), (1, 3), (2, 2), (1, 5)\}$$

Is the relation a function ? Explain.

Solution:

No, the relation is not a function because $x = 1$ is assigned to two values of y, 3 and 5. Ans.

4. Does the equation $y = 3x - 1$ determine y as a function of x ? Explain.

Solution:

Yes, y is a function of x because each value of x can find exactly one value of y. Ans.

5. Does the equation $y^2 = x$ determine y as a function of x ? Explain.

Solution:

No, y is not a function of x because each value of x can find two values of y. Ans. Hint: (1, 1), (1, –1), (4, 2), (4, –2), ...

The graph of a relation gives an easy way to identify whether or not a relation is a function.

A vertical line intersects the graph of a relation in one or more points.

A vertical line intersects the graph of a function in only one point.

The graph of each relation below shows that y is a function of x.

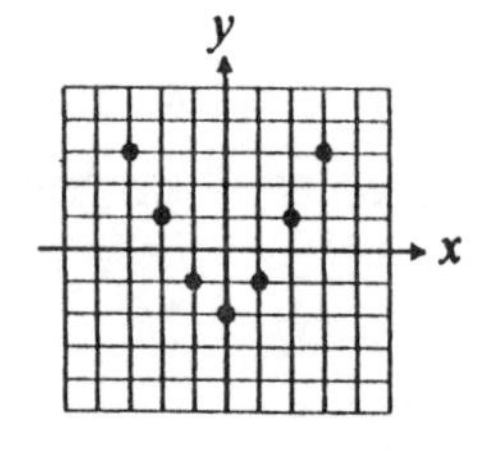

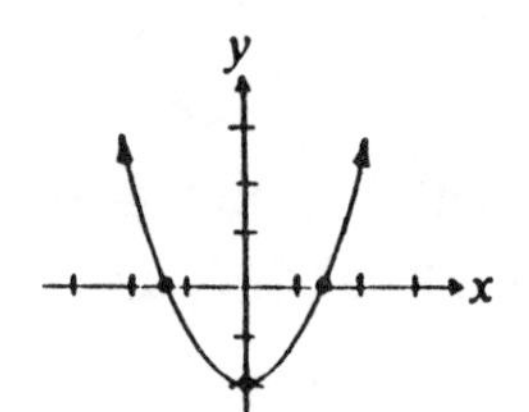

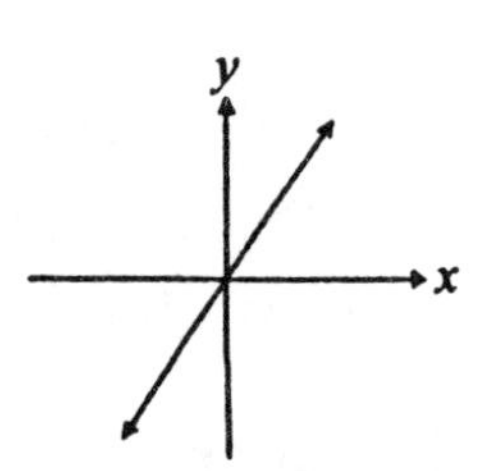

The graph of each relation below shows that y is not a function of x.

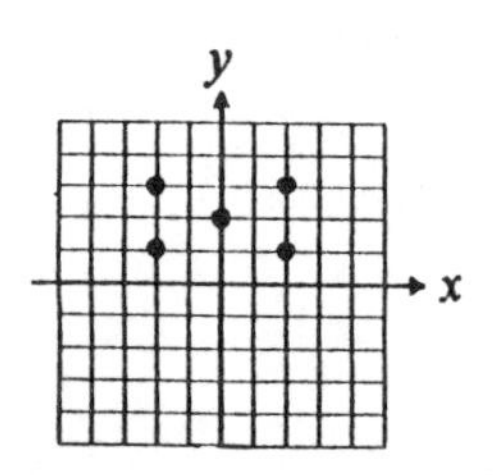

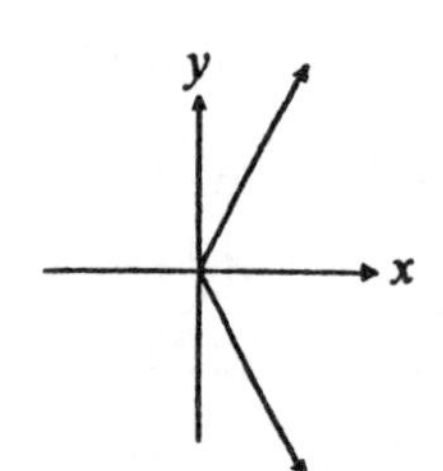

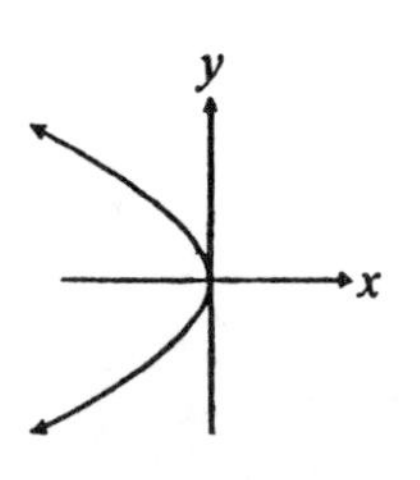

EXERCISES

Given the domain of each function, find the range.

1. $f(x) = x + 2$, D = {0, 1, 2, 3, 4}

2. $f(x) = x - 2$, D = {0, 1, 2, 3, 4}

3. $f(x) = 2x + 1$, D = {0, 1, 2, 3, 4}

4. $f(x) = 2x - 1$, D = {0, 1, 2, 3, 4}

5. $f(x) = x + 5$, D = {–2, –1, 0, 1, 2}

6. $f(x) = x - 5$, D = {–2, –1, 0, 1, 2}

7. $f(x) = 2x - 3$, D = {–3, –1, 0, 2, 4}

8. $f(x) = 3x - 1$, D = {–4, –2, 3, 5, 6}

9. $f(x) = x^2 + 2$, D = {–2, –1, 0, 2, 3}

10. $f(x) = |x| - 1$, D = {–2, –1, 0, 2, 3}

Find the domain and range of each relation (ordered pairs). Determine whether or not it is a function.

11. {(0, 1), (1, 3), (2, 4), (3, 6)}

12. {(–2, 3), {–1, 1), (2, 3), (3, 5)}

13. {(–3, 1), (–2, 1), (3, 3), (3, 4)}

14. {(–4, 2), (–1, 1), (0, 2), (4, 3)}

15. {(1, 3), (3, 4), (1, 5), (3, 7)}

16. {(5, 1), (3, 4), (4, 3), (5, 5)}

17. {(5, 7), (–6, 3), (4, 8), (–7, 2)}

18. {(2, 1), (2, –1), (3, –2), (3, 2)}

19. {(1, 0), (0, 1), (1, 3), (2, 1), (3, 4)}

20. {(0, 1), (1, 3), (2, 5), (4, 6), (3, 5)}

21. Find the domain and range of the equation $y = 2x - 1$.

22. Find the domain and range of the equation $|y| = x - 2$.

23. The following points represent the relation between x and y. Is the relation a function ? Explain. {(–3, 1), (–1, 2), (0, 3), (2, 5), (3, 7)}

24. The following points represent the relation between x and y. Is the relation a function ? Explain. {(–5, 2), (–2, 0), (–5, –1), (3, 1), (2, 0)}

25. Does the equation $y = 2x - 2$ determine y is a function of x ? Explain.

26. Does the equation $y^2 = 2x - 2$ determine y is a function of x ? Explain.

27. Does the equation $|y| = x + 1$ determine y is a function of x ? Explain.

28. Does the equation $y = |x| + 1$ determine y is a function of x ? Explain.

29. John drives at a speed of 60 miles per hour. Write a function for distance (d) as a function of the number of hours (h).

30. A-Plus Company charges \$19.50 per copy of book and \$2.50 for a handling fee per order. Write a function for the total cost (c) per order as a function of the number of books (n).

-----Continued-----

Given the graph of each relation between x and y. Is the relation a function ?

31.

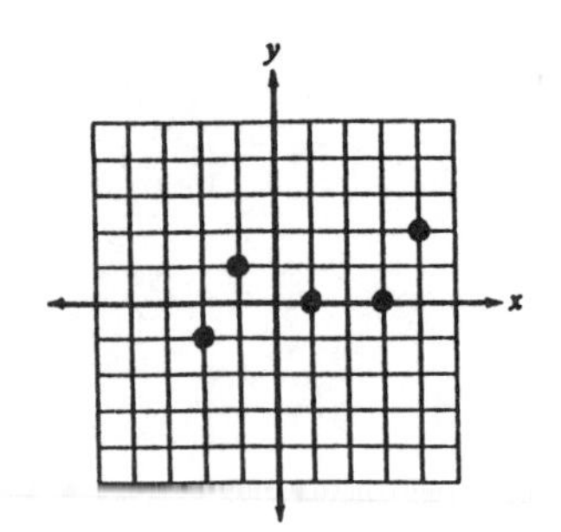

32.

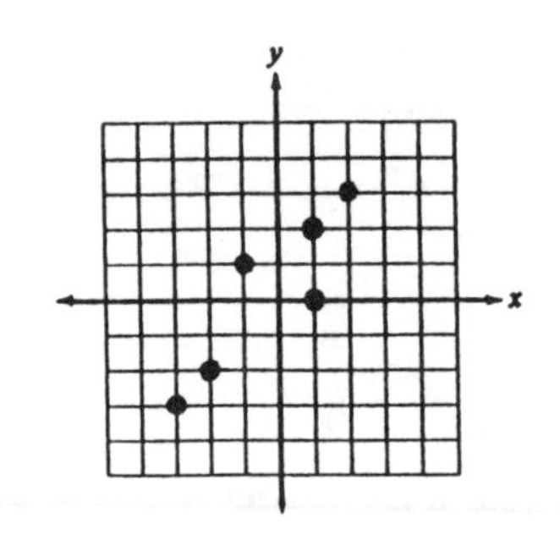

33.

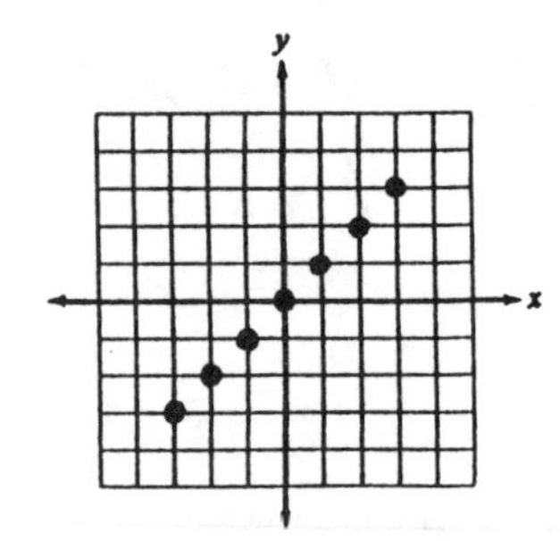

34.

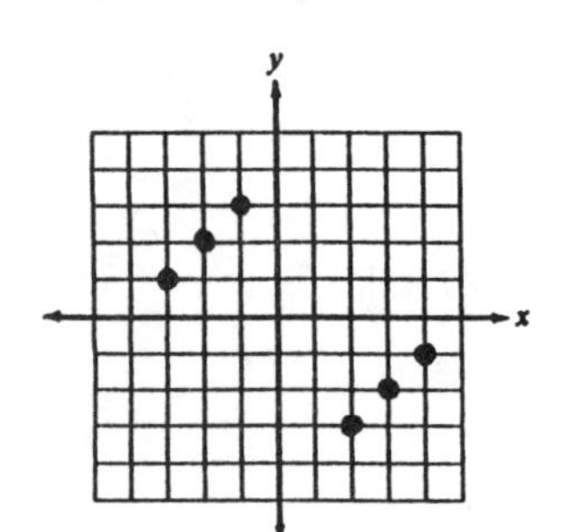

35.

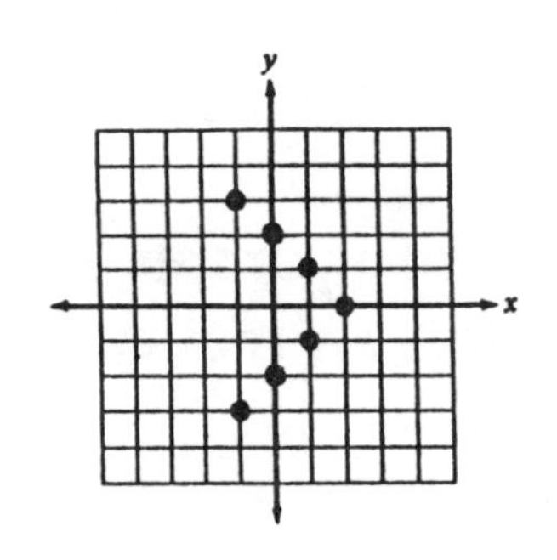

36.

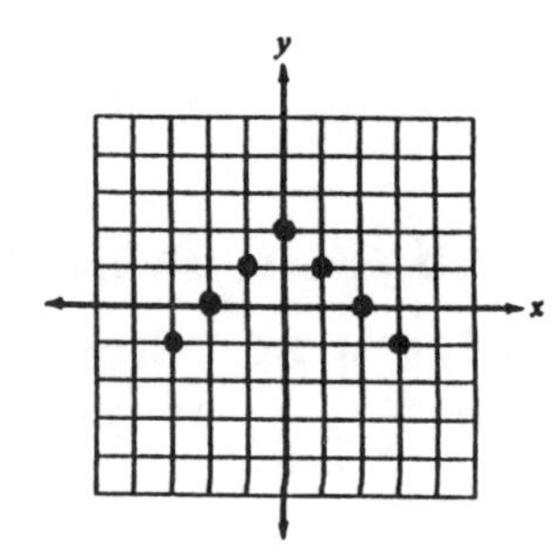

37.

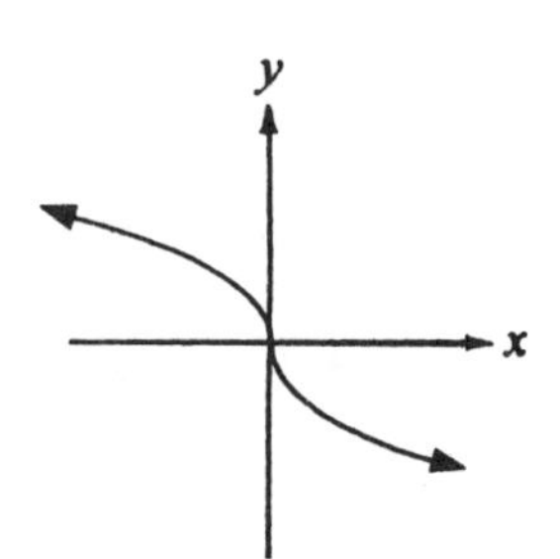

38.

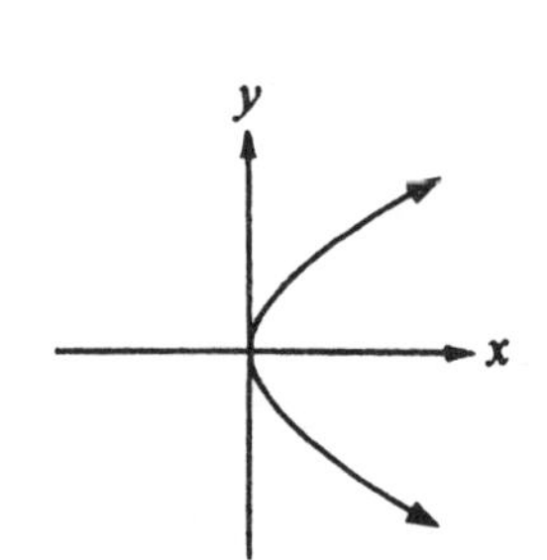

39.

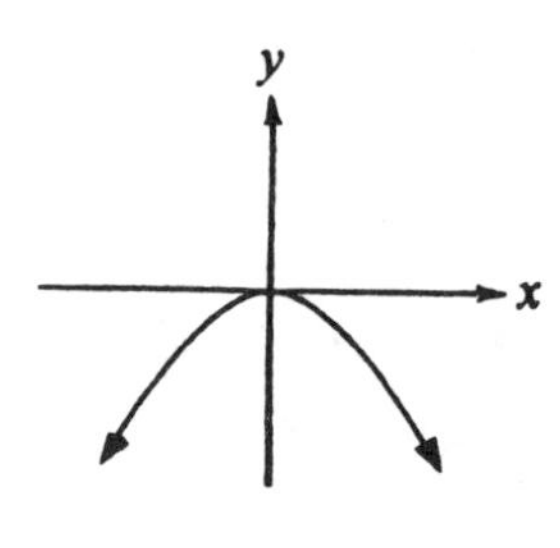

40.

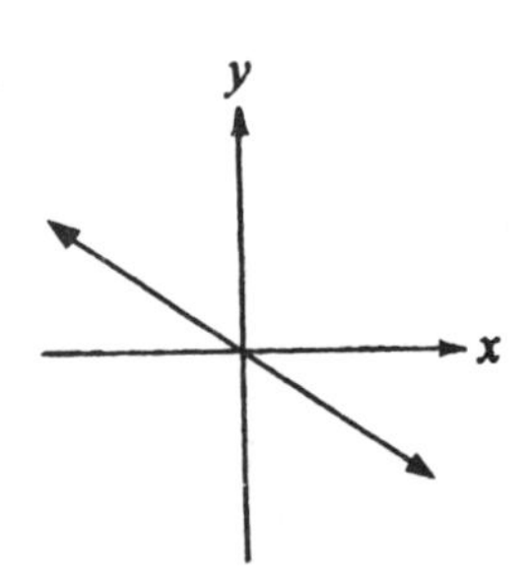

41.

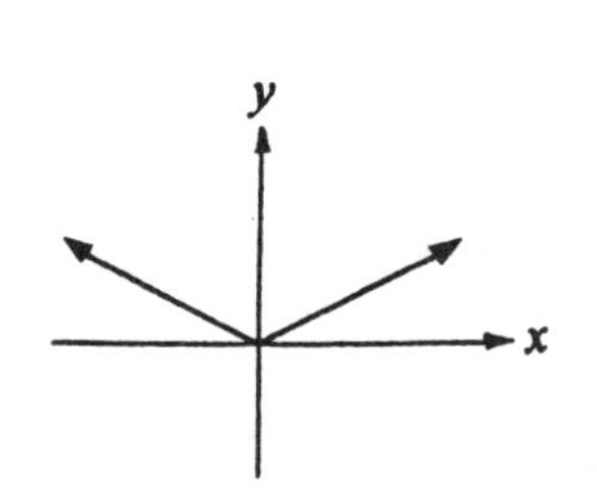

42.

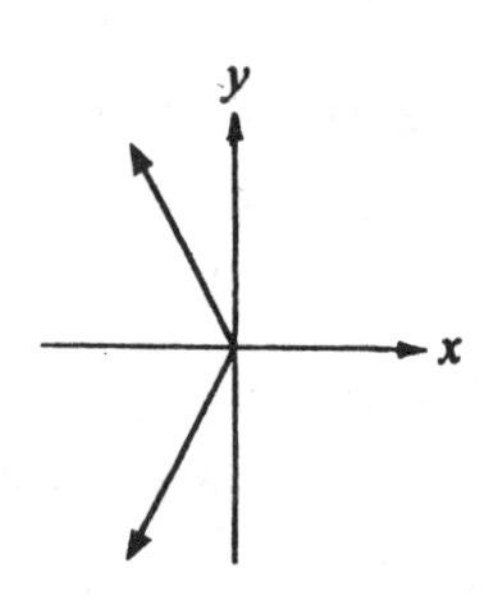

43.

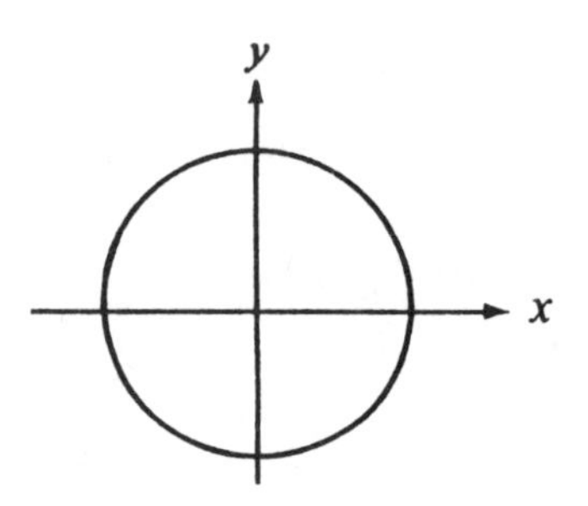

44.

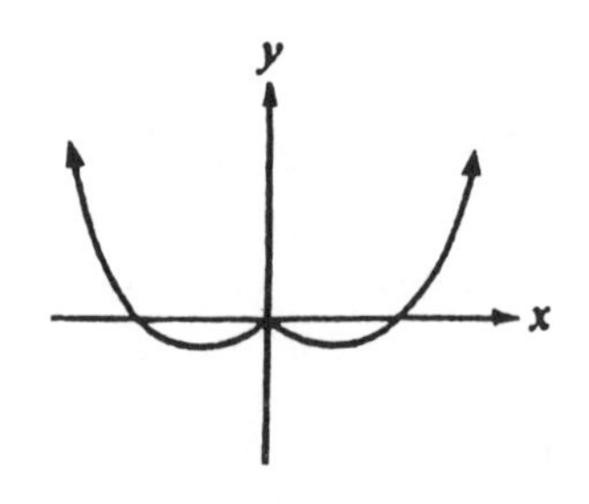

45.

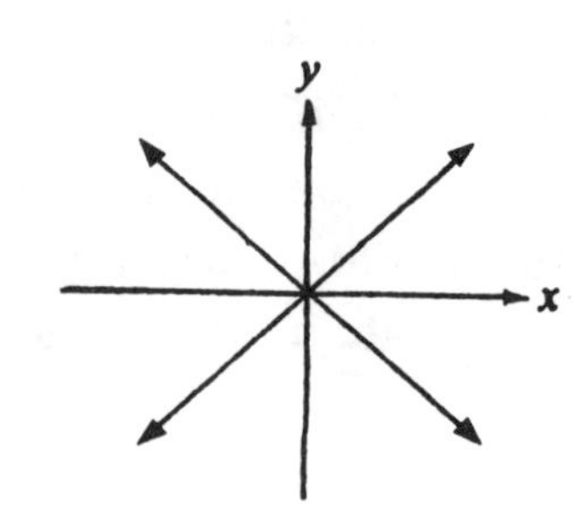

7-5 Graphing Linear Equations

An equation in two variables can be linear or not linear. In this section, we discuss how to graph a linear equation. We will discuss how to graph an equation which is not linear in the later chapters.

An equation in the form $ax + by = c$ is called a linear equation, because all its ordered pairs (points) lie on a straight line.
The graph of every linear equation in two variables is a straight line in a coordinate plane. Therefore, the equation $2x - y = 3$ is a linear equation.

If we solve the equation $2x - y = 3$ for y, we have a **linear function**:

$$y = 2x - 3 \quad \text{or} \quad f(x) = 2x - 3$$

A linear equation or a linear function can have infinitely many solutions. Each solution represents a point (x, y) on its graph. To graph a linear equation or a linear function , we select a few points and graph the points on the coordinate plane, draw a line that connects the points.

Examples

1. Graph $y = 2x$.
Solution:

x	$y = 2x$	(x, y)
0	$y = 2(0) = 0$	(0, 0)
1	$y = 2(1) = 2$	(1, 2)
2	$y = 2(2) = 4$	(2, 4)

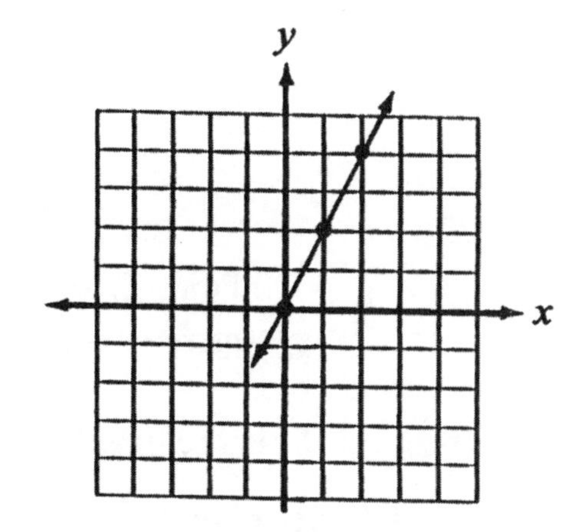

2. Graph $y = 2x - 3$.
Solution:

x	$y = 2x - 3$	(x, y)
0	$y = 2(0) - 3 = -3$	(0, −3)
1	$y = 2(1) - 3 = -1$	(1, −1)
2	$y = 2(2) - 3 = 1$	(2, 1)
3	$y = 2(3) - 3 = 3$	(3, 3)
4	$y = 2(4) - 3 = 5$	(4, 5)

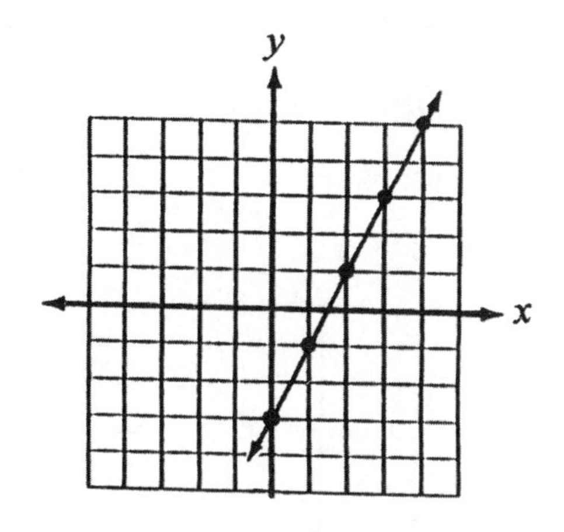

-----Continued-----

Examples

3. Graph $y = \frac{1}{2}x - 1$.

Solution:

y	$y = \frac{1}{2}x - 1$	$(x,\ y)$
0	$y = \frac{1}{2}(0) - 1 = -1$	$(0, -1)$
1	$y = \frac{1}{2}(1) - 1 = -\frac{1}{2}$	$(1, -\frac{1}{2})$
2	$y = \frac{1}{2}(2) - 1 = 0$	$(2, 0)$
3	$y = \frac{1}{2}(3) - 1 = \frac{1}{2}$	$(3, \frac{1}{2})$
4	$y = \frac{1}{2}(4) - 1 = 1$	$(4, 1)$

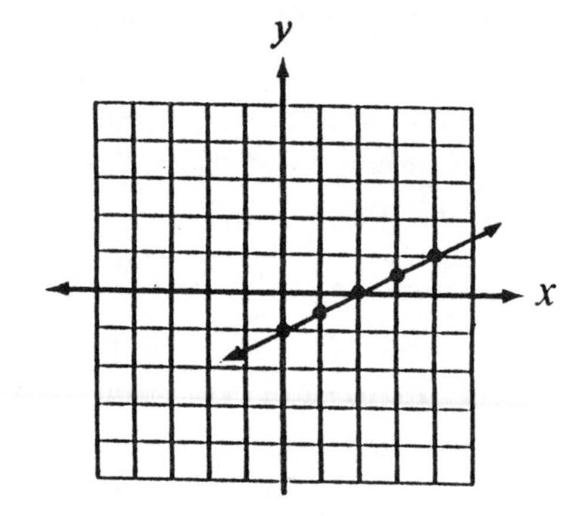

Since two points determine a straight line, we need to find only two points to graph a linear equation.
The points where the line crosses the $x-axis$ (let $y = 0$) and the $y-axis$ (let $x = 0$) are the easiest two points to find. We call them the $x-$intercept and the $y-$intercept.

4. Graph $2x - y = 4$.

Solution:

Let $y = 0$	Let $x = 0$
$2x - 0 = 4$	$2(0) - y = 4$
$2x = 4$	$-y = 4$
$x = 2$	$y = -4$
Point (2, 0)	Point (0, –4)
$x-$ intercept $= 2$	$y-$intercept $= -4$

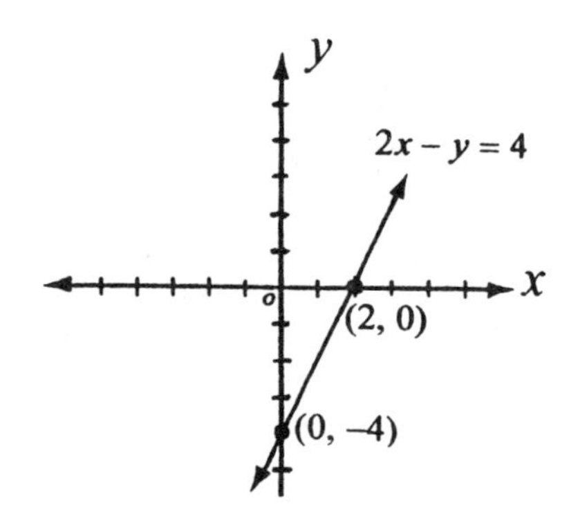

On the number line, the graph of an equation in one variable is a point (see Chapter 6).

On the coordinate plane, an equation in one variable is a linear equation. Its graph is a vertical line or a horizontal line.
A vertical line has no restriction on y. A horizontal line has no restriction on x.

5. a. Graph $x = 4$. **b.** Graph $y = 3$. **c.** Graph $x = -2$. **d.** Graph $y = -3$.

Solution:

a.

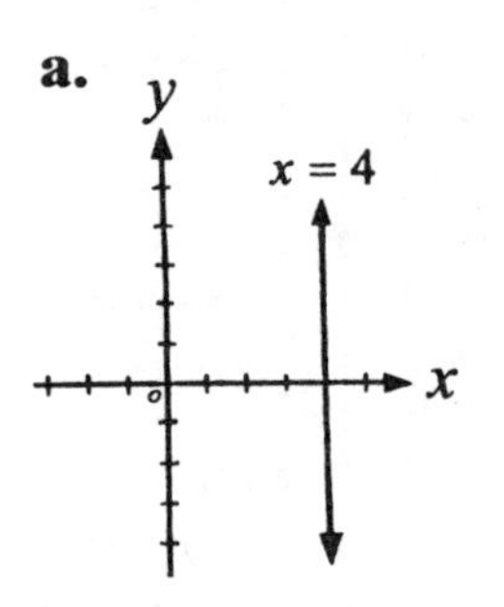

b. $y = 3$

c.

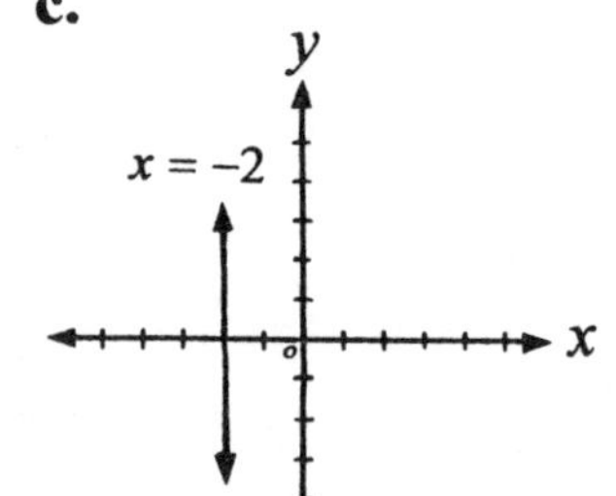

d. $y = -3$

EXERCISES

First complete the values in each table. Then graph the equation in a coordinate plane. (Hint: They are on a straight line.)

1. $y = 2x$

x	$y = 2x$	(x, y)
0	$y = 2(0) = 0$	(0, 0)
1		
2		
3		
4		

2. $y = -2x$

x	$y = -2x$	(x, y)
−1	$y = -2(-1) = 2$	(−1, 2)
0		
1		
2		
3		

3. $y = 2x - 1$

x	$y = 2x - 1$	(x, y)
−2	$y = 2(-2) - 1 = -5$	(−2, −5)
−1		
0		
1		
2		

4. $y = 3x + 2$

x	$y = 3x + 2$	(x, y)
−4		
−2		
0		
2		
3		

5. $y = \frac{1}{2}x - 1$

x	$y = \frac{1}{2}x - 1$	(x, y)
−2		
−1		
0		
2		
5		

Use the x - intercept **and** y - intercept **to graph each equation.**

6. $y = x + 1$

7. $y = x - 2$

8. $y = 2x - 4$

9. $y = 4 - x$

10. $y = 4 - 2x$

11. $y = 3x - 6$

12. $y = -3x + 6$

13. $y = -4x + 2$

14. $y = x + \frac{1}{2}$

15. $y = x - \frac{1}{2}$

16. $y = \frac{1}{2}x + 4$

17. $y = -\frac{1}{3}x - 3$

18. $y = 0.5x - 2$

19. $y = -0.5x + 2$

20. $x - y = 5$

21. $x + y = 5$

22. $2x + y = 4$

23. $2x - y = 4$

24. $x + 3y = 9$

25. $x - 3y = 9$

26. $4x - y = 2$

27. $-4x + y = 2$

28. $-2x + y = 1$

29. $-2x + 5y = 10$

30. $y = x$

31. $y = 3x$

32. $\frac{1}{2}x - \frac{1}{3}y = 3$

33. $\frac{1}{3}x - \frac{1}{2}y = 3$

34. $\frac{2}{3}x + \frac{1}{2}y = 1$

35. $\frac{1}{2}x + \frac{2}{3}y = 3$

-----Continued-----

Graph each equation in a coordinate plane.

36. $y = 3$ **37.** $x = 5$ **38.** $y = -1$ **39.** $x = -4$ **40.** $y = 6$

41. $y = -5$ **42.** $x = 7$ **43.** $x = -7$ **44.** $x = 0$ **45.** $y = 0$

46. Describe the graph of the equation $x = 0$ in a coordinate plane.
47. Describe the graph of the equation $y = 0$ in a coordinate plane.
48. For what value of k is the point (2, –3) in the graph of the equation $6x + ky = 3$?
49. For what value of k is the point (–2, 3) in the graph of the equation $kx - 5y = 1$?
50. For what value of k is the point (6, k) so that it is a solution of the equation $3x + 5y = 3$?
51. For what value of k is the point (k, – 4) so that it is a solution of the equation $2x + y = 1$?
52. Find an equation for the group of ordered pairs (1, 6), (2, 7), (3, 8).
53. Find an equation for the group of ordered pairs (1, $\frac{1}{2}$), (2, 1), (4, 2).
54. Find an equation for the group of ordered pairs (2, 3), (4, 7), (6, 11).
55. Write an equation for the data in the table below. Graph the equation.

x	0	1	2	3	4	5	6	7
y	4	6	8	10	12	14	16	18

56. The table shows the distance that a bus travels over time. Let the horizontal axis represent time (t) in hours. Let the vertical axis represent distance (d) in miles. Find and graph the equation for the data.

Time (hours)	0	1	2	3	4	5
Distance (miles)	0	60	120	180	240	300

57. The equation $d = 50\,t$ represents distance in miles and time in hours for a car traveling at a constant speed of 50 miles per hour. Graph the equation for the data.

t	0	2	4	6
d	0	100	200	300

In an English class, the teacher asked the kids to write a letter to his (her) mother regarding what he (she) did today in school. The teacher noticed that John was writing very slowly.
Teacher: John, why do you write so slowly ?
John: My mom could not read fast.

7-6 Slope of a Line

Slope is used to describe the **steepness** of a straight line. To find the slope of a line, we choose any two points on the line and form a right triangle which has the line as its hypotenuse. A line that rises more steeply has a greater slope. A line that runs more horizontally has a smaller slope. Therefore, the slope of a line is defined as the ratio of its vertical rise to its horizontal run. The letter m is commonly used to represent the slope.

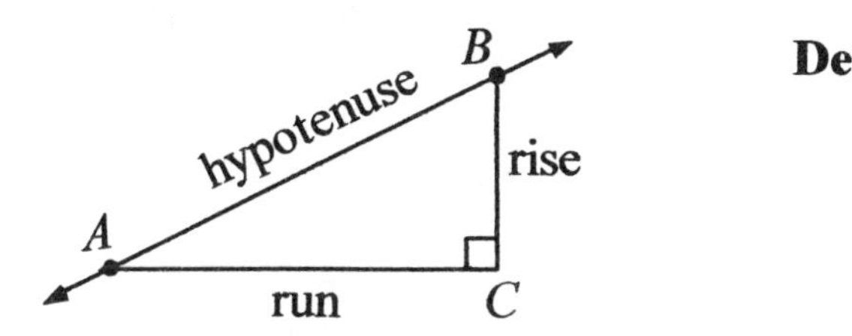

Definition of slope:

$$\text{Slope of } \overleftrightarrow{AB} = \frac{rise}{run} = \frac{\overline{BC}}{\overline{AC}}$$

In the coordinate plane, the slope of a line between two points (x_1, y_1) and (x_2, y_2) is given in the following slope formula:

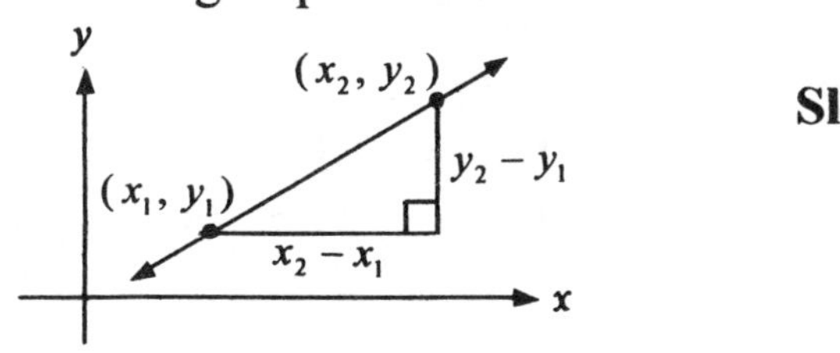

Slope Formula:

$$\text{Slope } (m) = \frac{y_2 - y_1}{x_2 - x_1}, \quad (x_1 \neq x_2)$$

When we use the slope formula to find the slope of a line, we can choose any point as the first point and the other as the second point. The result will be the same.

Examples

1. Find the slope of the line passing through the points (2, 3) and (6, 5).
   Solution:

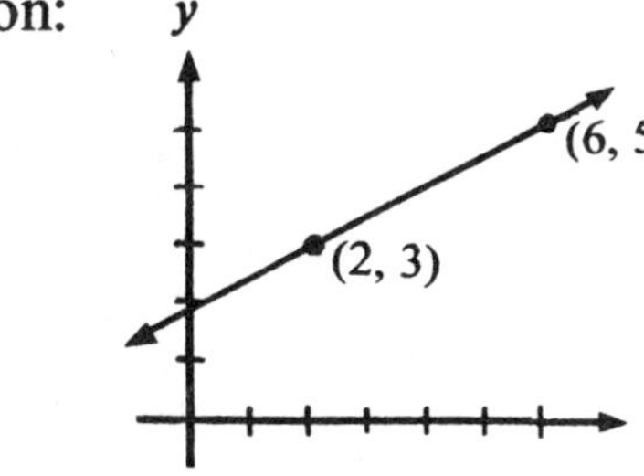

$$m = \frac{y_2 - y_1}{x_2 - x_1} = \frac{5-3}{6-2} = \frac{2}{4} = \frac{1}{2}. \quad \text{Ans.}$$

or: $$m = \frac{y_2 - y_1}{x_2 - x_1} = \frac{3-5}{2-6} = \frac{-2}{-4} = \frac{1}{2}. \quad \text{Ans.}$$

2. Find the slope of the line containing the points (–2, 4) and (5, 2).
   Solution:

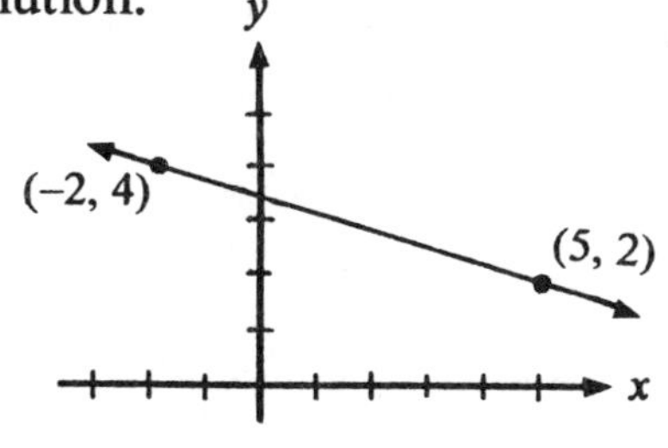

$$m = \frac{y_2 - y_1}{x_2 - x_1} = \frac{2-4}{5-(-2)} = \frac{-2}{7} = -\frac{2}{7}. \quad \text{Ans.}$$

or: $$m = \frac{y_2 - y_1}{x_2 - x_1} = \frac{4-2}{-2-5} = \frac{2}{-7} = -\frac{2}{7}. \quad \text{Ans.}$$

-----Continued-----

(See Examples 1 and 2): 1) If a line slants up from left to right, the slope is positive.
2) If a line slants down from left to right, the slope is negative.

Examples

3. Find the slope of the line passing the points (–7, 4) and (3, 4).
Solution:

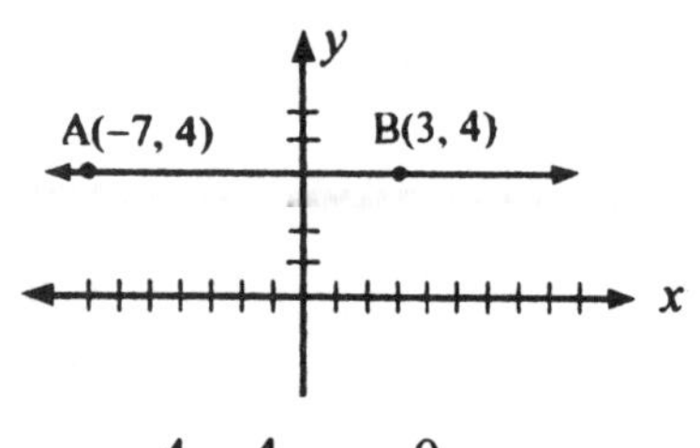

$$m = \frac{4-4}{3-(-7)} = \frac{0}{10} = 0.$$

Ans: The slope is 0.

4. Find the slope of the line passing through the points (7, 5) and (7, –3).
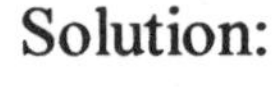
Solution:

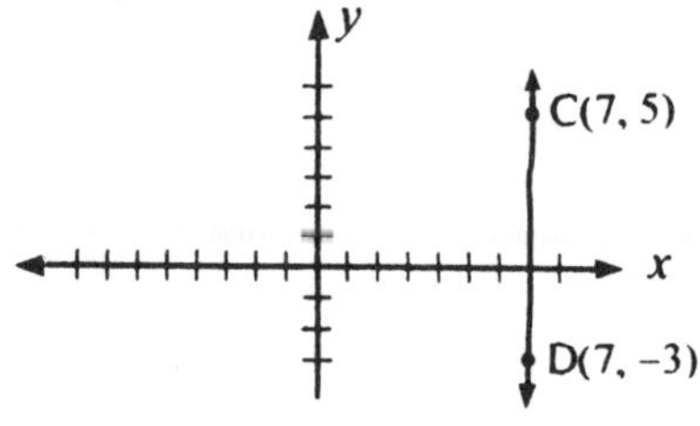

$$m = \frac{-3-5}{7-7} = \frac{-8}{0}.$$

Ans: The slope is undefined.

(See examples 3 and 4): 1) The slope of a horizontal line is 0.
2) The slope of a vertical line is undefined.
It means that the slope of a vertical line does not exist.

Slope-intercept form of a straight line

We can choose any two points on a straight line and find its slope by the slope formula (see examples 1 and 2).
However, an easier way to find the slope of a straight line is to rewrite the equation in the slope- intercept form:

$$y = mx + b$$

Then, the slope is m and $y-$ intercept is b.

5. Find the slope of the line $y + 2x = 3$.
Solution:

$$y + 2x = 3$$
$$y = -2x + 3$$
$$\therefore m = -2. \quad \text{Ans.}$$

6. Find the slope of the line $y - 2x = 3$
Solution:

$$y - 2x = 3$$
$$y = 2x + 3$$
$$\therefore m = 2. \quad \text{Ans.}$$

7. Find the slope and $y-$ intercept of the line whose equation is $2x + 3y - 6 = 0$.
Solution:

$$2x + 3y - 6 = 0$$
$$3y = -2x + 6$$
$$y = -\tfrac{2}{3}x + 2$$

Ans: slope $= -\frac{2}{3}$.
$y-$ intercept = 2.

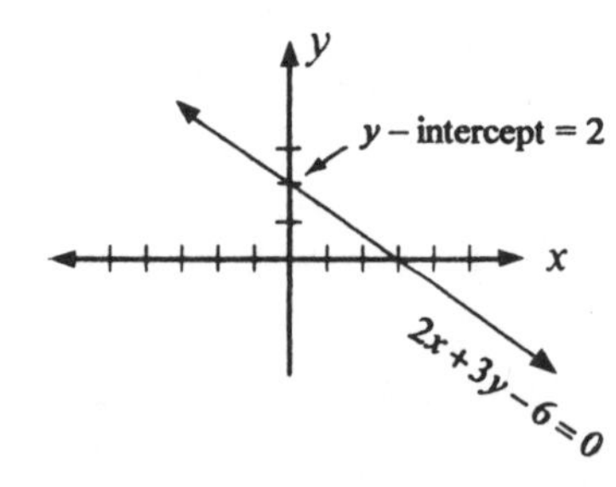

EXERCISES

Find the slope of each line using the given two points on the line.
(Hint: The rise of a line is negative if it falls from left to right.)

1.
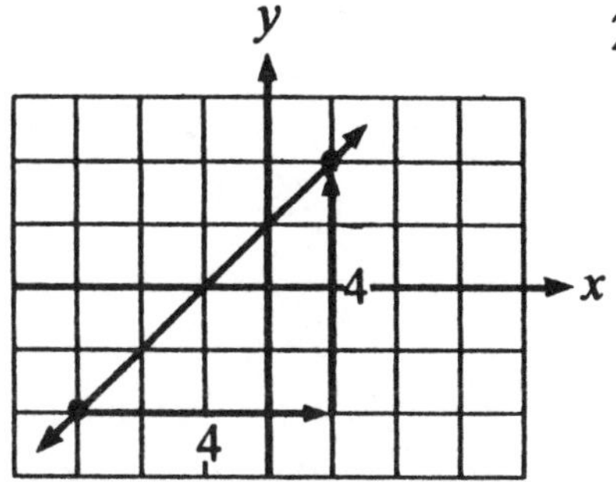

2.
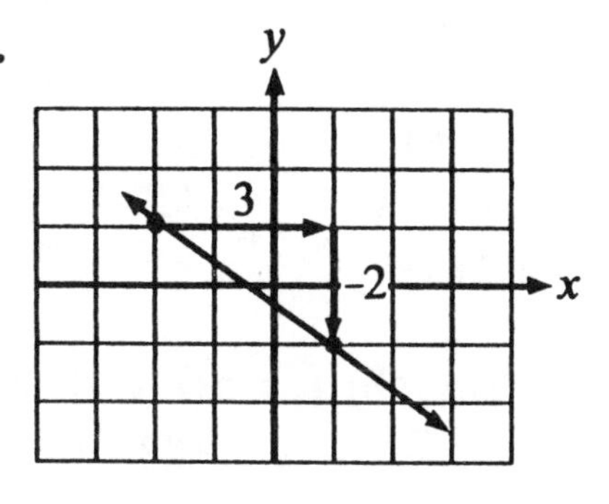

3.
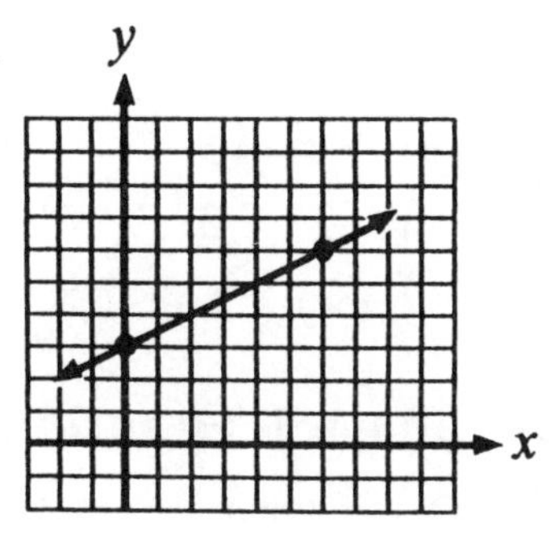

4.
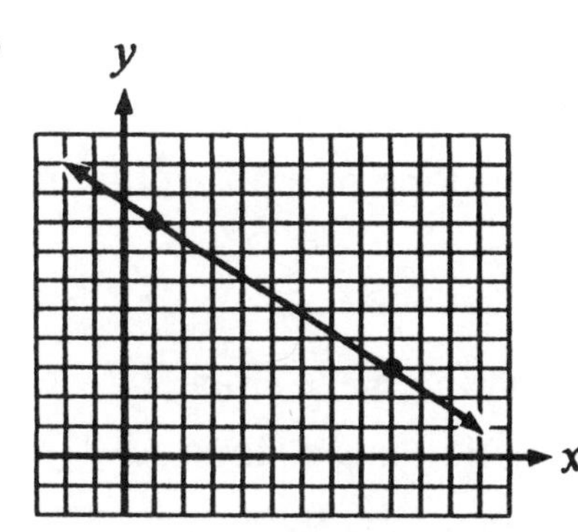

5.
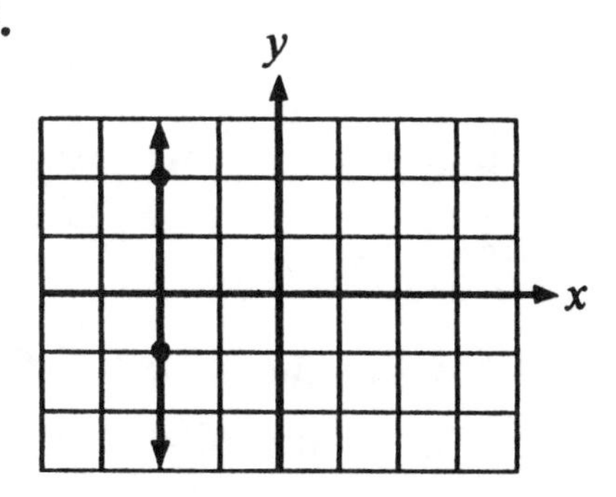

6.
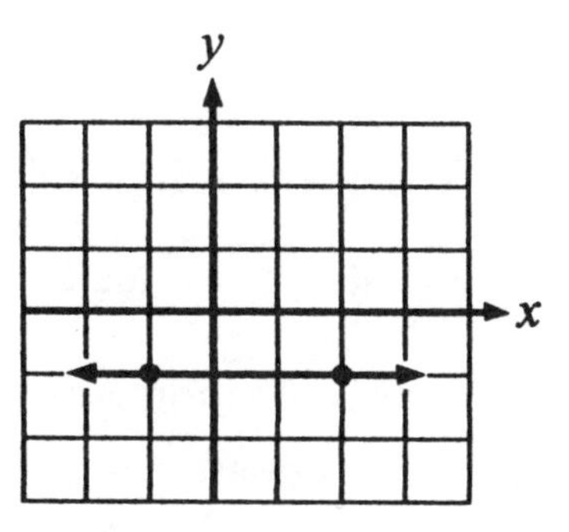

Find the slope of each line with its rise and run.

7. rise 3, run 1 **8.** rise 1, run 3 **9.** rise –4, run 2 **10.** rise 2, run 6
11. rise –2, run 6 **12.** rise 5, run 2 **13.** rise 2, run 5 **14.** rise 8, run 0
15. rise 0, run 8 **16.** rise $-\frac{1}{2}$, run 3 **17.** rise $\frac{1}{2}$, run $\frac{1}{3}$ **18.** rise $-1\frac{1}{2}$, run $\frac{3}{4}$
19. rise 1.5, run 3 **20.** rise 4.5, run 1.5

Find the slope of each line containing the given two points.

21. (4, 2) and (2, 4) **22.** (–4, 2) and (2, 4) **23.** (4, –2) and (2, 4)
24. (4, –2) and (6, –2) **25.** (–6, 0) and (0, 6) **26.** (0, –6) and (6, 0)
27. (–4, 0) and (–4, 8) **28.** (0, 0) and (–2, 4) **29.** (–3, –7) and (–7, –3)
30. (1.2, 5) and (0.6, 6.2) **31.** (–3.4, 1.4) and (0.4, 9) **32.** $(\frac{1}{2}, 1)$ and $(1, \frac{1}{2})$
33. $(\frac{1}{2}, \frac{1}{2})$ and $(\frac{1}{4}, \frac{1}{5})$ **34.** $(-\frac{1}{4}, 2)$ and $(2, \frac{1}{4})$ **35.** $(\frac{1}{4}, 4)$ and $(4, \frac{1}{2})$
36. (1, 4) and (–3, 0) **37.** (1, 5) and (–6, –2) **38.** (–2, –4) and (2, 4)
39. (–1, –3) and (4, –2) **40.** (–7, –8) and (–3, –4)

Draw a line passing through the given point with the given slope.

41. (1, 2), slope $\frac{1}{2}$ **42.** (–2, 4), slope $-\frac{1}{2}$ **43.** (–2, 3), slope 3 **44.** (–3, 2), slope $\frac{2}{3}$
45. (3, –1), slope 0 **46.** (3, –2), slope $-\frac{3}{5}$ **47.** (0, 3), slope $-\frac{2}{5}$ **48.** (–5, 2), slope 4
49. (–5, –2), slope $\frac{4}{5}$ **50.** (–3, 4), slope undefined

-----Continued-----

Find the slope and y–intercept **of each line.**

51. $y = 2x - 5$ **52.** $y = -2x + 5$ **53.** $y = -7x - 8$ **54.** $y = 7x - 8$

55. $y = \frac{2}{3}x + 4$ **56.** $y = -\frac{2}{3}x + 4$ **57.** $y = -\frac{1}{2}x - \frac{3}{4}$ **58.** $y = \frac{1}{2}x + \frac{3}{4}$

59. $y = -4x$ **60.** $y = -x + 5$ **61.** $y = x - 10$ **62.** $y = 7x$

63. $y = x$ **64.** $y = -x$ **65.** $y = 5$ **66.** $y = -2$

67. $y = 7 - 8x$ **68.** $y = 1 + 2x$ **69.** $y = -2 - \frac{1}{3}x$ **70.** $y = 8 - \frac{4}{5}x$

71. $y - 2x = 15$ **72.** $y + 2x = 15$ **73.** $2x + y = 6$ **74.** $2x - y = 6$

75. $2y - 2x = 3$ **76.** $2y + 4x = 1$ **77.** $4y + 2x = 5$ **78.** $4y - 2x = 3$

79. $3x - 2y = 6$ **80.** $3x + 2y = 7$

Graph each equation using the slope and y–intercept **.**

81. $y = 2x + 3$ **82.** $y = -2x + 1$ **83.** $y = 2x - 3$ **84.** $y = -2x - 1$

85. $y = 3x - 2$ **86.** $y = -3x + 1$ **87.** $y = -3x - 1$ **88.** $y = 3x - 4$

89. $2y - 4x = 5$ **90.** $2x - 4y = 5$

91. What is the slope of a vertical line ? **92.** What is the slope of a horizontal line ?

93. What is the value of $\frac{y_2 - y_1}{x_2 - x_1}$ if $x_1 = x_2$?

94. In the equation $ax + 4y = 6$, for what value of a that the equation has a slope 2 ?

95. In the equation $6x - ay = 3$, for what value of a that the equation has a slope –2 ?

96. Determine whether the given points (1, 3), (2, 5), and (4, 9) lie on the same line.

97. Determine whether the given points (2, –6), (0, –2), and (–2, 3) lie on the same line.

98. The points (2, -3), (4, 1), and (5, y) lie on the same line. Find the value of y.

99. The points (2, –3), (0, –2), and (x, 3) lie on the same line. Find the value of x.

100. Find the formula for the slope (m) and the formula for the y– intercept of the linear equation $ax + by = c$ in terms of a, b, and c.

101. A ladder hit the wall at a height of 39 feet with its base 6 feet from the wall. Find the slope of the ladder.

102. If the temperature in the City of Hawthorne drops from $96°F$ at 2 P.M. to $78°F$ at 8 P.M., what is the average rate of change in the temperature ?
Estimate the temperature at 12 P.M..
(Hint: The average rate of change is the slope connecting the points (2, 96) and (8, 78).

On the airplane, a three-year old boy asks his mother.
Boy: Where are we going ?
Mother: We are going to grandma's home.
Boy: Why are so many people going to grandma's home ?

7-7 Finding the Equation of a Line

We have learned how to find the points, slope, and intercepts of a given line.
Now we will learn how to find the equation of a line under certain given information.
Such as, find the equation of a line:

1) having its slope and $y-$ intercept.
2) having its slope and a point on the line.
3) having its two points on the line.

The following examples show how to find the equation of a line.
The equation of a line can be written in two different ways:

1) slope-intercept form $y = mx + b$. **2) standard form** $ax + by = c$.

Examples

1. Find the equation of the line having slope -5 and $y-$ intercept 2.

Solution:

$m = -5$, $b = 2$

$y = mx + b$

$\therefore y = -5x + 2$. Ans. (slope-intercept form)

or: $5x + y = 2$. Ans. (standard form)

2. Find the equation of the line having slope $\frac{3}{4}$ and $y-$ intercept $\frac{5}{2}$.

Solution:

$m = \frac{3}{4}$, $b = \frac{5}{2}$

$y = mx + b$

$y = \frac{3}{4}x + \frac{5}{2}$. Ans. (slope-intercept form)

or: Multiply each side by 4:

$4y = 3x + 10$

$-3x + 4y = 10$ $\quad\therefore 3x - 4y = -10$. Ans. (standard form)

3. Find the equation of the line having slope 4 and passing through the point (2, –3).

Solution:

$m = 4$

$y = mx + b$

$y = 4x + b$

To find b, substitute the point (2, –3) into the above equation.

$-3 = 4(2) + b$

$-11 = b$

$\therefore y = 4x - 11$. Ans. (slope-intercept form)

or: $4x - y = 11$. Ans. (standard form)

-----**Continued**-----

Examples

4. Find the equation of the line passing through the points (–2, 3) and (4, 6).
   Solution:

$$\text{Find the slope: } m = \frac{y_2 - y_1}{x_2 - x_1} = \frac{6-3}{4-(-2)} = \frac{3}{6} = \frac{1}{2}$$

$$y = mx + b$$

$$y = \tfrac{1}{2}x + b$$

To find b, substitute the point (–2, 3) into the above equation.

$$3 = \tfrac{1}{2}(-2) + b$$

$$3 = -1 + b$$

$$4 = b$$

$\therefore y = \frac{1}{2}x + 4$. Ans. (slope-intercept form)

or: $x - 2y = -8$. Ans. (standard form)

Point-Slope form of a line: If m is the slope of a line and (x_1, y_1) is one of its points, we can write the equation of the line in **point-slope form**.

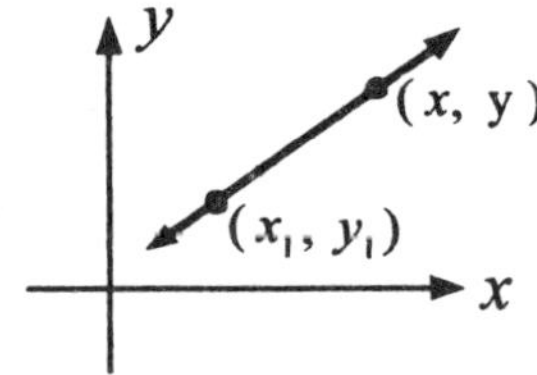

To write the equation, we choose any other point (x, y) on the line.

The slope is: $m = \dfrac{y - y_1}{x - x_1}$, $x \neq x_1$

Using cross multiplication, we have:

$y - y_1 = m(x - x_1)$ → **Point-Slope form of a line**

Examples :

5. Find the equation in point-slope form of a line having slope 4 and passing through the point (2, –3).
   Solution:

$$y - y_1 = m(x - x_1)$$

$$y - (-3) = 4(x - 2)$$

$\therefore y + 3 = 4(x - 2)$. Ans. (point-slope form)

6: Find the equation in point-slope form of a line passing through the points (–2, 3) and (4, 6). (Hint: There are two answers.)

Solution: Find the slope: $m = \dfrac{y_2 - y_1}{x_2 - x_1} = \dfrac{6-3}{4-(-2)} = \dfrac{3}{6} = \dfrac{1}{2}$

Substitute one of the points, say (–2, 3), and the slope into the equation.

$$y - y_1 = m(x - x_1)$$

$$y - 3 = \tfrac{1}{2}[x - (-2)]$$

$\therefore y - 3 = \frac{1}{2}(x + 2)$. Ans. (point-slope form)

If we choose (4, 6), the other answer is: $y - 6 = \frac{1}{2}(x - 4)$.

Both answers have the same standard form: $x - 2y = -8$.

EXERCISES

Find the equation of the line having the slope and $y-$intercept. Express each equation in standard form $ax+by=c$.

1. slope –3, $y-$intercept 5
2. slope 3, $y-$intercept –5
3. slope –3, $y-$intercept –5
4. slope 6, $y-$intercept –9
5. slope $\frac{1}{2}$, $y-$intercept 7
6. slope $-\frac{1}{2}$, $y-$intercept 7
7. slope $-\frac{2}{3}$, $y-$intercept 8
8. slope $\frac{3}{4}$, $y-$intercept 8
9. slope $\frac{2}{3}$, $y-$intercept –4
10. slope $-\frac{3}{4}$, $y-$intercept -4

Find the equation of the line having the slope and passing through the given point. Express each equation in slope-intercept form $y=mx+b$.

11. slope 2, (2, 5)
12. slope 2, (5, 2)
13. slope –2, (3, 4)
14. slope –2, (–3, 4)
15. slope –2, (3, –4)
16. slope 2, (–3, –4)
17. slope –2, (–3, –4)
18. slope 7, (1, 6)
19. slope 6, (–5, 8)
20. slope –1, (4, 9)
21. slope 1, (5, –1)
22. slope 0, (–7, 5)
23. slope 0, (3, 0)
24. slope undefined , (5, 0)
25. slope undefined, (0, 5)
26. slope $\frac{1}{2}$, (4, –2)
27. slope $-\frac{3}{5}$, (10, 3)
28. slope $-\frac{2}{3}$, (2, –1)
29. slope $\frac{4}{5}$, (0, 9)
30. slope $\frac{5}{4}$, (9, 0)

Find the equation of the line passing through the given two points. Express each equation in standard form $ax+by=c$.

31. (1, 2), (3, 4)
32. (2, 1), (4, 3)
33. (–2, 3), (0, 4)
34. (2, –3), (4, 0)
35. (0, 5), (2, 6)
36. (0, 0), (–2, 4)
37. (5, 0), (6, 3)
38. (–3, –4), (0, 7)
39. (–3, –6), (–4, 7)
40. (–6, –2), (7, 0)
41. (2.5, 3), (4, 7.5)
42. (3, 4.5), (2.5, 3.5)
43. $(1, \frac{1}{2}), (2, \frac{2}{3})$
44. $(2, 3), (\frac{1}{2}, \frac{1}{2})$
45. $(\frac{3}{2}, \frac{1}{2}), (-\frac{1}{2}, \frac{1}{3})$

Find the equation of each line described. Express each equation in point-slope form $y-y_1=m(x-x_1)$.

46. slope 3, passes through (2, 4)
47. slope –3, passes through (–2, 4)
48. slope 4, passes through (–1, –2)
49. slope –5, passes through (7, –5)
50. slope $\frac{3}{4}$, passes through (–1, 5)
51. slope $-\frac{1}{5}$, passes through (4, –7)
52. passes through (4, 5) and (8, –3)
53. passes through (5, –2) and (0, 2)
54. passes through (0, 5) and (5, 0)
55. passes through (0, 0) and (–3, –1)

-----Continued-----

56. Find the equation in standard form of a line passing through (4, 6) and parallel to the line $20x-2y=5$.
57. Find the equation in standard form of a line passing through (4, –6) and parallel to the line $20x-2y=5$
58. Find the equation in standard form of a line having slope 5 and x – intercept 4.
59. Find the equation in standard form of a line having slope –5 and x – intercept 4.
60. Find the equation in standard form of a line having x – intercept –6 and y – intercept 4.
61. Find the equation in standard form of a line having x – intercept 6 and y – intercept –4.
62. Find the equation of a horizontal line passing through (3, –2).
63. Find the equation of a vertical line passing through (3, –2).
64. Find the equation of a vertical line passing through (–2, 3).
65. Find the equation of a horizontal line passing through (–2, 3).
66. Find a if two points (3, 2) and (–2, a) are on a line having slope 2.
67. Find a if two points (3, 2) and (a, –2) are on a line having slope 2
68. Find a if two points (3, 2) and (–2, $a+1$) are on a line having slope –2.
69. Find a if three points (–1, 3), (–3, 2), and (–4, $a-5$) are on a line.
70. Find a

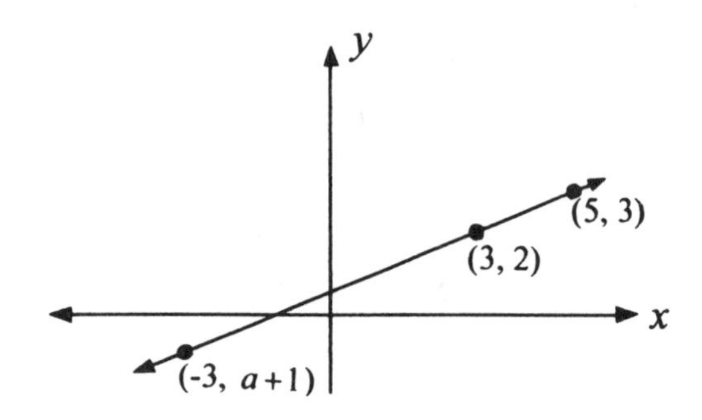

Roger was absent yesterday. Teacher: Roger, why were you absent ? Roger: My tooth was aching. I went to see my dentist. Teacher: Is your tooth still aching ? Roger: I don't know. My dentist has it.

7-8 Parallel and Perpendicular Lines

Now we will learn how to identify parallel lines and perpendicular lines by comparing their slopes.

Two different lines that have the same slope are **parallel.** Two different lines that intersect to form right angles(90°) are **perpendicular.**

Rules: 1. Two lines are parallel if and only if their slopes are equal. $m_1 = m_2$

2. Two lines are perpendicular if and only if the product of their slopes is -1.
(They are negative reciprocals of each other.)

$$m_1 \cdot m_2 = -1, \quad \text{or} \quad m_1 = -\frac{1}{m_2}, \quad m_2 = -\frac{1}{m_1}$$

Examples

1. Find the slope of a line that is perpendicular to the line $8x - 4y = 3$.

Solution:

$8x - 4y = 3$

Rewrite it in slope-intercept form: $-4y = -8x + 3$

Divide each side by -4, we have $y = 2x - \frac{3}{4}$. slope $m = 2$

The slope of the perpendicular line is the negative reciprocal of $2 = -\frac{1}{2}$. Ans.

2. Tell whether or not the following two lines are parallel, perpendicular, or neither.

$3x - 4y = -5$ and $6x - 8y = 7$

Solution:

To find their slopes, we rewrite each equation in slope-intercept form.

$3x - 4y = -5$, $-4y = -3x - 5$, $y = \frac{3}{4}x + \frac{5}{4}$ $\therefore m_1 = \frac{3}{4}$

$6x - 8y = 7$, $-8y = -6x + 7$, $y = \frac{3}{4}x - \frac{7}{8}$ $\therefore m_2 = \frac{3}{4}$

$m_1 = m_2$ Ans: They are parallel.

3. Tell whether or not the following two lines are parallel, perpendicular, or neither.

$x - 2y = 4$ and $2x + y = 5$

Solution:

$x - 2y = 4$, $-2y = -x + 4$, $y = \frac{1}{2}x - 2$ $\therefore m_1 = \frac{1}{2}$

$2x + y = 5$, $y = -2x + 5$ $\therefore m_2 = -2$

$m_1 \cdot m_2 = \frac{1}{2} \cdot (-2) = -1$ Ans: They are perpendicular.

4. Tell whether or not the following two lines are parallel, perpendicular, or neither.

$x + 3y = 4$ and $6y = 4x - 5$

Solution: $x + 3y = 4$, $3y = -x + 4$, $y = -\frac{1}{3}x + \frac{4}{3}$ $\therefore m_1 = -\frac{1}{3}$

$6y = 4x - 5$, $y = \frac{2}{3}x - \frac{5}{6}$ $\therefore m_2 = \frac{2}{3}$

$m_1 \neq m_2$ and $m_1 \cdot m_2 \neq -1$ Ans: Neither.

EXERCISES

Rewrite each equation in slope-intercept form $y = mx + b$.

1. $x + y = 5$ **2.** $x - y = 5$ **3.** $x + y = -5$ **4.** $-x + y = 5$

5. $2x + y = 6$ **6.** $2x - y = -6$ **7.** $-2x + 2y = 7$ **8.** $2y + 3x = -5$

Find the slope of a line that is perpendicular to the given line.

9. $6x - 3y = 8$ **10.** $4x - 5y = 20$ **11.** $x + 5y = 10$ **12.** $3x + y = 1$

13. $x + y = 5$ **14.** $x - y = 5$ **15.** $y - 2x = -5$ **16.** $2x + y = 9$

17. $y = 5$ **18.** $x = -3$ **19.** $2x - \frac{1}{3}y = 10$ **20.** $\frac{1}{2}x - \frac{3}{2}y = 6$

Tell whether or not each pair of equations are parallel, perpendicular, or neither.

21. $2x - 3y = 8$, $-2x + 3y = 14$ **22.** $2x - 3y = 8$, $2x + 3y = 14$ **23.** $2x - 3y = 8$, $3x + 2y = 14$

24. $y - 4x = -3$, $2y - 8x = 5$ **25.** $y - 4x = 8$, $2y + 8x = 3$ **26.** $5y - 3x = 25$, $3y + 5x = 15$

27. $x + y = 5$, $x - y = 5$ **28.** $x + y = 5$, $x + y = 7$ **29.** $x - y = 10$, $x - y = 9$

30. $5y - 4x = 10$, $5x + 4y = 1$ **31.** $4x + 8y = 7$, $2x + 4y = 9$ **32.** $4x - 2y = 13$, $2x - 4y = 11$

33. $4x - 2y = 13$, $2x + 4y = 11$ **34.** $\frac{2}{3}x + y = 15$, $y = \frac{3}{2}x - 10$ **35.** $2x - \frac{1}{2}y = 3$, $y = -\frac{1}{2}x - 3$

36. What is the slope of a line that is parallel to the line $5x - 4y = 7$?

37. What is the slope of a line that is perpendicular to the line $y = 2$?

38. What is the slope of a line that is parallel to the line $x = 2$?

39. What is the slope of a line that is perpendicular to a vertical line ?

40. What is the slope of a line that is parallel to a horizontal line ?

41. Write an equation in slope-intercept form for a line perpendicular to the line with slope 7 and y-intercept 15.

42. Find the equation in standard form of a line that passes the point (10, –2) and is parallel to the line $2x + 5y = 3$.

43. Find the equation in standard form of a line that passes the point (10, –2) and is perpendicular to the line $2x + 5y = 3$.

7-9 Direct Variations

A linear equation with the form $y = kx$ is called an equation having a **direct variation** with x and y. k is a nonzero constant. k is called the **constant of variation (constant of proportionality)**. In the equation $y = 3x$, 3 is a constant. We say that y varies directly as x. The graph of $y = 3x$ is a straight line with slope 3 and passes through the origin (0, 0). If (x_1, y_1) and (x_2, y_2) are two ordered pairs of an equation having a direct variation defined by $y = kx$ and that neither x_1 nor x_2 is zero, we have: $y_1 = kx_1$ and $y_2 = kx_2$,

$k = \frac{y_1}{x_1}$ and $k = \frac{y_2}{x_2}$. Therefore: $\frac{y_1}{x_1} = \frac{y_2}{x_2}$. **The ratios are equal.**

If $y = kx$, we say that y varies directly as x, or y is directly proportional to x.
In direct variation, the variables that vary may involve powers (exponents).

Note that the equation $y = 3x - 2$ does not show x and y as a direct variation because the ratios of its ordered pairs $\frac{y_1}{x_1}$ and $\frac{y_2}{x_2}$ are not equal (not proportional). However, the equation $y = 3x - 2$ shows x and $(y + 2)$ as a direct variation ($y + 2 = 3x$).

Joint Variation: An equation having one variable varies directly as the product of two or more other variables.
For $y = kxz$, we say that y varies jointly with x and z.

In science, there are many word problems which involve the concept of direct variations. Force (F) is directly proportional to acceleration (a) ($F = ma$, **Newton's Second Law**). The distance (d) that an object falls is directly proportional to the square of the time (t) of the fall ($d = 16t^2$, **Law of Free-Falling**). The square of the orbital period (T) of a planet varies directly with the cube of its mean distance (a) from the Sun ($T^2 = ka^3$, **Kepler's Third Law**). The stretched length (y) beyond its natural length of the wire spring is directly proportional to the mass (x) of the weight attached ($y = kx$, **Hook's Law**).

Examples

1. If y varies directly as x, and if $y = 6$ when $x = 2$, find the constant of variation.
Solution: Let $y = kx$, $6 = k \cdot 2$ $\therefore k = \frac{6}{2} = 3$. Ans.

2. If y varies directly as x, and if $y = 6$ when $x = 2$, find y when $x = 3$.
Solution: Let $y = kx$, $6 = k \cdot 2$ $\therefore k = \frac{6}{2} = 3$, $y = 3x$ is the equation.
When $x = 3$, $y = 3x = 3 \cdot 3 = 9$. Ans.

3. If (x_1, y_1) and (x_2, y_2) are ordered pairs of the same direct variation, find y_1.

$$x_1 = 3,\ y_1 = ?,\ x_2 = 12,\ y_2 = 8$$

Solution: The ratios are equal: $\frac{y_1}{x_1} = \frac{y_2}{x_2}$, $\frac{y_1}{3} = \frac{8}{12}$, $12y_1 = 24$ $\therefore y_1 = 2$. Ans.

Or: $y_2 = kx_2$, $8 = k \cdot 12$, $k = \frac{8}{12} = \frac{2}{3}$ $\therefore y_1 = kx_1 = \frac{2}{3}x_1 = \frac{2}{3} \cdot 3 = 2$. Ans.

EXERCISES

Find the constant of variation and write an equation of direct variation.
(Hint: "varies directly as" and "is directly proportional to" have the same meaning.

1. If y varies directly as x, and if $y = 10$ when $x = 2$.
2. If y is directly proportional to x, and if $y = 10$ when $x = 5$.
3. If y varies directly as x, and if $y = 5$ when $x = 2$.
4. If y varies directly as x, and if $y = 6.5$ when $x = 13$.
5. If n is directly proportional to m, and if $n = 200$ when $m = 20$.
6. If p varies directly as n, and if $p = 36$ when $n = 30$.
7. If s varies directly as t, and if $s = 15$ when $t = 30$.
8. If m varies directly as n, and if $m = 30$ when $n = 90$.
9. If y varies directly as $(x-4)$, and if $y = 15$ when $x = 9$.
10. If y is directly proportional to $(x+8)$, and if $y = 150$ when $x = 2$.

For each direct variation described, find each missing value.

11. If y varies directly as x, and if $y = 10$ when $x = 2$, find y when $x = 12$.
12. If y varies directly as x, and if $y = 10$ when $x = 2$, find x when $y = 12$.
13. If y is directly proportional to x, and if $y = 2$ when $x = 10$, find y when $x = 12$.
14. If n varies directly as m, and if $n = 150$ when $m = 6$, find n when $m = 20$.
15. If n varies directly as m, and if $n = 150$ when $m = 6$, find m when $n = 20$.

If (x_1, y_1) and (x_2, y_2) are ordered pairs of the same direct variation, find each missing value.

16. $x_1 = 1$, $y_1 = ?$, $x_2 = 3$, $y_2 = 12$
17. $x_1 = 3$, $y_1 = ?$, $x_2 = 4$, $y_2 = 80$
18. $x_1 = 6$, $y_1 = ?$, $x_2 = 8$, $y_2 = 9.6$
19. $x_1 = 6$, $y_1 = 4$, $x_2 = 9$, $y_2 = ?$
20. $x_1 = ?$, $y_1 = 1.5$, $x_2 = 15$, $y_2 = 9$
21. $x_1 = 1$, $y_1 = 7$, $x_2 = ?$, $y_2 = 17.5$
22. $x_1 = \frac{4}{5}$, $y_1 = ?$, $x_2 = \frac{2}{5}$, $y_2 = \frac{1}{10}$
23. $x_1 = \frac{1}{4}$, $y_1 = \frac{1}{5}$, $x_2 = \frac{5}{3}$, $y_2 = ?$
24. $x_1 = \frac{7}{8}$, $y_1 = ?$, $x_2 = \frac{1}{2}$, $y_2 = \frac{1}{7}$
25. $x_1 = \frac{1}{5}$, $y_1 = 5$, $x_2 = 5$, $y_2 = ?$
26. Does the equation $xy = 5$ define a direct variation ? Explain.
27. Are the ordered pairs (6, 4), (9, 6), (27, 18) in the same direct variation ? Explain.
28. The actual distance D is directly proportional to the length L on the map. If 0.5 cm represents 5 miles, write a formula for computing the actual distance.
29. The stretched length beyond its natural length of the wire spring is directly proportional to the mass of the weight attached. A 30 grams weight causes a wire spring stretched 7.5 cm beyond its natural length. What will be the stretched length of the wire spring beyond its natural length if a 58 grams weight is attached ? (It is called Hooke's Law.)
30. The simple interest you pay is directly proportional to the amount of money you borrow. If you paid \$245 interest on \$3,500, how much interest would you pay on \$6,300 in the same period of time ?

CHAPTER 7 EXERCISES

Find the length of the segment between two points.

1. (5, 7) and (5, –3) **2.** (–5, 4) and (3, 4) **3.** (–2, 7) and (–2, 1)
4. (–6, –1) and (–3, –1) **5.** (6, –8) and (6, –7) **6.** (1, 1) and (1, 9)

Identify whether each ordered pair of number is a solution of the equation.

7. $x+2y=3$, (1, 1) **8.** $x+2y=3$, (–1, 2) **9.** $x+2y=3$, (3, 1)
10. $2x-y=5$, (2, 1) **11.** $2x-y=5$, (2, –1) **12.** $2x-y=5$, $(\frac{1}{2}, -4)$
13. $2a-4b=1$, $(3, \frac{5}{4})$ **14.** $2a-4b=1$, $(-2, \frac{5}{3})$ **15.** $y=\frac{2}{3}x-4$, (3, –2)

Write each of the linear equations in standard form $ax+by=c$.

16. $2x=y-5$ **17.** $y=-2x+5$ **18.** $-y=2x-5$ **19.** $-y=-2x+5$
20. $4x=5y+7$ **21.** $3x-9=8y$ **22.** $6y+7=2x$ **23.** $\frac{2}{3}x=-2y+1$

Solve each linear equation for y in terms of x.

24. $2x+y=4$ **25.** $-2x+y=4$ **26.** $2x-y=4$ **27.** $y-4=2x$
28. $4x+2y=5$ **29.** $4x-2y=5$ **30.** $5x-3y=6$ **31.** $3x+5y=6$

Find the value for each function.

$f(x)=2x-7$ **32)** $f(1)$ **33)** $f(0)$ **34)** $f(-1)$ **35)** $f(2)$
$f(x)=7-2x$ **36)** $f(1)$ **37)** $f(0)$ **38)** $f(2)$ **39)** $f(-\frac{5}{4})$
$p(n)=n^2-n+1$ **40)** $p(2)$ **41)** $p(0)$ **42)** $p(5)$ **43)** $p(\frac{1}{2})$
$f: x \to 5x-4$ **44)** $f(2)$ **45)** $f(-2)$ **46)** $f(\frac{1}{5})$ **47)** $f(-1\frac{1}{5})$

Given the domain of each function, find the range.

48. $f(x)=3x-7$, D = {0, 1, 2, 3, 4} **49.** $f(x)=4x-2$, D = {–2, –1, 0, 1, 2}
50. $f(x)=x^2-2$, D = {–3, –1, 0, 1, 3} **51.** $f(x)=2x^2-2$, D = {–2, 0, 1, 2}
52. $f(x)=|x|-4$, D = {–5, –1, 0, 3} **53.** $f(x)=4-|x|$, D = {–5, –1, 0, 3}

Find the domain and range of each relation (ordered pairs).
Determine whether or not it is a function.

54. {(0, –2), (1, –1), (2, 1), (3, 4)} **55.** {(–2, 1), (–2, 3), (0, 2), (2, 1)}

56. {(–2, –2), (–1, 1), (1, 0), (1, 2)} **57.** {(–3, 1), (–1, 3), (2, –3), (3, –2)}

58. {(0, 2), (1, 1), (2, 0), (3, –1), (4, –2)} **59.** {(1, 2), (2, 3), (8, 9)}

60. {(1, 2), (–4, 7), (1, –2)}

-----Continued-----

Find the $x-$intercept and $y-$intercept. Use the $x-$intercept and $y-$intercept to graph each equation.

61. $y = x - 3$ **62.** $y = 3 - x$ **63.** $x + y = -3$ **64.** $y = 2x + 2$
65. $x + 2y = 6$ **66.** $x - 2y = 6$ **67.** $2x - 3y = 12$ **68.** $2x + 3y = 12$
69. $y = -2$ **70.** $x = 1$ **71.** $y = 2x$ **72.** $y = -2x$
73. $\frac{2}{3}x - y = 4$ **74.** $\frac{1}{2}x + \frac{1}{3}y = 4$ **75.** $\frac{1}{2}x - \frac{2}{3}y = 4$

Find the slope of each line with its rise and run.

76. rise 5, run 2 **77.** rise 2, run –5 **78.** rise 0, run 9 **79.** rise 9, run 0
80. rise 6, run $\frac{1}{2}$ **81.** rise $-\frac{2}{3}$, run $\frac{1}{6}$ **82.** rise $1\frac{1}{2}$, run $-\frac{5}{4}$ **83.** rise 2.5, run 0.5

Find the slope of each line containing the given two points.

84. (3, 5) and (5, 3) **85.** (–3, 5) and (3, –5) **86.** (–3, –5) and (3, 5)
87. (0, –4) and (0, 5) **88.** (–4, 0) and (5, 0) **89.** (–8, 7) and (–8, –7)
90. $(2, \frac{2}{3})$ and (4, 2) **91.** $(\frac{1}{2}, 2)$ and $(2, -4)$ **92.** $(\frac{2}{5}, -\frac{3}{5})$ and $(\frac{1}{5}, \frac{1}{5})$
93. (–1.2, –1.5) and (1.3, 3) **94.** (0, 0) and (4.5, 7.2) **95.** $(\frac{1}{2}, \frac{1}{2})$ and $(\frac{2}{3}, -\frac{1}{3})$

Find the slope (m) and $y-$intercept (b) of each line.

96. $y = -4x + 9$ **97.** $y = 9 + 4x$ **98.** $y = \frac{4}{5}x - 15$ **99.** $y = -\frac{4}{5}x + 15$
100. $8x + y = 10$ **101.** $8x - y = 10$ **102.** $y = 8x$ **103.** $y = -8$
104. $3x + 6y = 8$ **105.** $3x - 6y = -8$ **106.** $\frac{1}{2}y - 4x = 7$ **107.** $-5y + 10x = 7$

Find the equation in standard form $ax + by = c$ of each line described.

108. slope 5, $y-$intercept 7
109. slope -4, $y-$intercept -2
110. slope $\frac{3}{2}$, $y-$intercept 12
111. slope $-\frac{4}{5}$, $y-$intercept $-\frac{2}{3}$
112. slope 7, passes through (2, 3)
113. slope $-\frac{1}{2}$, passes through (4, –5)
114. slope $\frac{3}{4}$, passes through (–2, –1)
115. slope 0, passes through (–7, 9)
116. slope undefined, (5, 1)
117. passes through (2, –3) and (6, 5)
118. passes through (1, 7) and (–6, –7)
119. passes through $(\frac{2}{3}, 4)$ and (1, –5)
120. passes through (3, –1) and $(\frac{1}{3}, 3)$

Tell whether or not they are parallel, perpendicular, or neither.

121. $5x - 6y = 4$, $6x + 5y = 7$
122. $2x - 4y = 11$, $4x + 8y = 15$
123. $2x - 4y = 13$, $4x - 8y = 19$
124. $4x - 8y = 21$, $4x + 2y = 25$

125. If y varies directly as x, and $y = 48$ when $x = 4$. Find y when $x = \frac{2}{3}$.

126. If n varies directly as m, and $n = 210$ when $m = 21$. Find n when $m = 5\frac{1}{2}$.

127. If (x_1, y_1) and (x_2, y_2) are ordered pairs of the same variation, find y_2.

$x_1 = \frac{5}{8}$, $y_1 = \frac{1}{4}$, $x_2 = \frac{2}{3}$, $y_2 = ?$

Additional Examples

A graph can represent the relationship between two quantities, such as "distance" and "time". We can use the given information to write an equation and graph it on a coordinate plane. If the graph of the given information is a straight line, then we can extend the graph and get more information about the relationship from the graph or the equation.

1. The temperature at 2 A.M was $-3°C$. At 5 A.M, it was $1°C$. If the temperature rises at a constant rate: a) what will the temperature be at 8 A.M. ?
b) write the equation to represent the relationship between temperature and time.
c) what will the temperature be at 12 A.M. ?

Solution:

a)

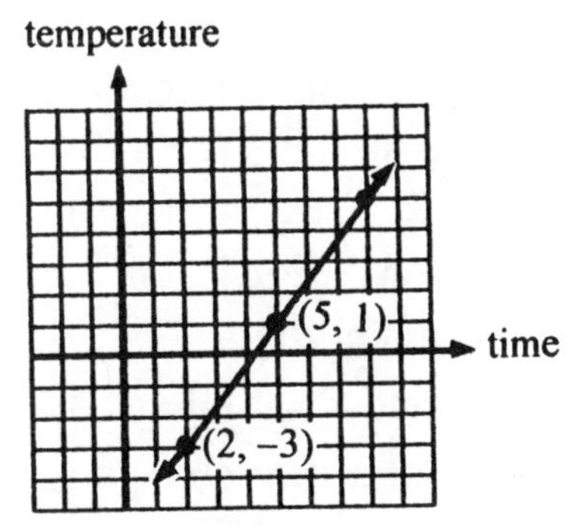

The temperature will be $5°C$ at 8 A.M. Ans.

b) Find the equation passing through the points (2, −3) and (5, 1).

The slope is: $m = \dfrac{1-(-3)}{5-2} = \dfrac{4}{3}$

The equation is: ($x =$ time, $y =$ temperature)

$$\frac{y-(-3)}{x-2} = \frac{4}{3}, \quad \frac{y+3}{x-2} = \frac{4}{3}, \quad 3(y+3) = 4(x-2)$$

$$3y + 9 = 4x - 8$$

$4x - 3y = 17$ is the equation. Ans.

c) $4x - 3y = 17$ and $x = 12$

$$3y = 4x - 17, \quad y = \frac{4x-17}{3} = \frac{4(12)-17}{3} = 10.33°C. \text{ Ans.}$$

2. Solve $3x + 2y - 14 = 0$ if x and y are whole numbers.

Solution: (Hint: Whole numbers are 0, 1, 2, 3, 4, 5, ...)

$$3x + 2y - 14 = 0, \; 2y = 14 - 3x$$

$$y = \frac{14-3x}{2} = 7 - \frac{3}{2}x$$

Only $x = 0$, 2, and 4 give values of y with whole numbers 7, 4, and 1.

Ans: (0, 7), (2, 4), (4, 1).

-----Continued-----

Additional Examples

3. Gary completed five math tests with an average score of 82. The average score for the first four tests was 78. What was the score in the fifth test ?
Solution:

Let $x =$ the score of the fifth test
We have the equation:

$$\frac{78(4)+x}{5}=82, \quad \frac{312+x}{5}=82$$

$$312+x=410$$

$$\therefore\ x=98. \quad \text{Ans.}$$

4. If the digits of a certain two-digit positive integer are interchanged, the result is 36 less than the original integer. Find all such integers.
Solution:

Let $x =$ the tens' digit
$y =$ the units' digit
The original two-digit integer is $10x+y$
The two-digit integer after interchange is $10y+x$
We have the equation:

$$10y+x+36=10x+y$$

$$9y-9x=-36$$

$$y-x=-4$$

$$\therefore x-y=4$$

Ans: 40, 51, 62, 73, 84, 95

5. If the digits of a certain two-digit positive integer are interchanged, the result increases 75% of the original integer. Find all such integers.
Solution:

Let $x =$ the tens' digit
$y =$ the units' digit
The original two-digit integer is $10x+y$
The two-digit integer after interchange is $10y+x$
We have the equation:

$$\frac{10y+x-(10x+y)}{10x+y}=75\%, \quad \frac{9y-9x}{10x+y}=\frac{3}{4}$$

$$4(9y-9x)=3(10x+y)$$

$$36y-36x=30x+3y$$

$$33y=66x$$

$$y=2x$$

Ans: 12, 24, 36, 48

Additional Examples

6. The rental cost of a car is \$20, plus \$15 per day. Graph the relationship between the cost and the number of days.

Solution:

The equation is: $c = 20 + 15d$

$d = 0$, $c = 20 + 15(0) = 20$

$d = 1$, $c = 20 + 15(1) = 35$

$d = 2$, $c = 20 + 15(2) = 50$

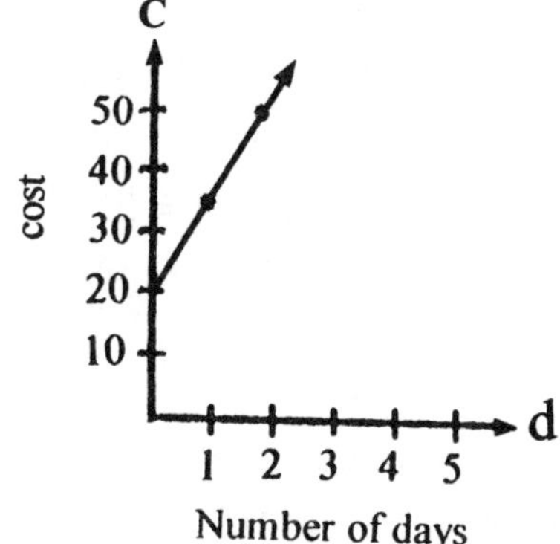

7. You bought a new car. The total cost was \$14,000, including tax and other charges. You paid \$6,000 down and made 60 monthly payments. Write an equation to represent the amount of dollars (m) for each monthly payment.

Solution:

Let m = each monthly payment

The equation is: $14{,}000 = 6{,}000 + 60\,m$. Ans.

8. What is the measure of each angle.

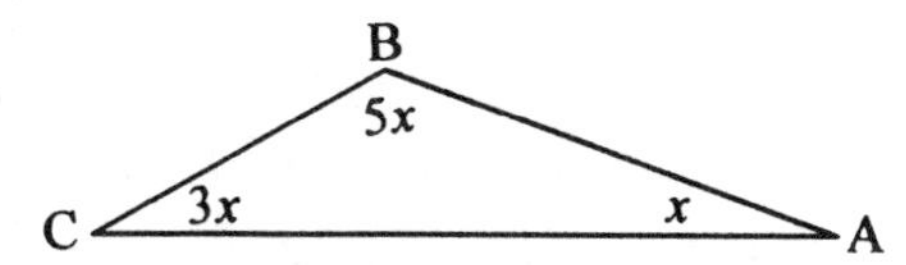

Solution:

The sum of the measures of the angles of a triangle is 180.

The equation is:

$$x + 3x + 5x = 180$$
$$9x = 180$$
$$x = 20^o \quad \text{(angle A). Ans.}$$
$$5x = 100^o \quad \text{(angle B). Ans.}$$
$$3x = 60^o \quad \text{(angle C). Ans.}$$

9. The hourly wage of a teacher is twice the hourly wage of his assistant. They were paid a total of \$266 for working on a project. The teacher worked 10 hours and his assistant worked 8 hours. Find the hourly wage in dollars of the assistant.

Solution:

Let x = the hourly wage of the assistant

$2x$ = the hourly wage of the teacher

The equation is:

$$10(2x) + 8x = 266$$
$$28x = 266 \quad \therefore x = \$9.50. \text{ Ans.}$$

10. The total length of the wire spring is directly proportional to the mass of the weight attached. A 25 grams weight causes a wire spring to a total length of 10 *cm*. What will be the total length of the wire spring if a 55 grams weight is attached ?

Solution:

Let x = the weight in grams, y = the total length of the spring in *cm*

We have: $y = kx$, $10 = k \cdot 25$, $k = \frac{10}{25} = 0.4$ $\therefore y = 0.4x$ is the formula.

When $x = 55$ grams, $y = 0.4x = 0.4(55) = 22$ *cm*. Ans.

Hint: It is called **Hook's Law**. k is called the coefficient of elasticity for a wire spring.

11. The stretched length beyond its natural length of the wire spring after stretched is directly proportional to the mass of the weight attached. A 4 kg weight causes a wire spring to a total length of 5 *cm*. A 8 kg weight causes this wire spring to a total length of 6 *cm*. Find a formula for the total length (L) of the wire spring as a function of the weight (w) attached.

Solution:

Stretched length $l = kw$, where k is the constant of variation.
(Hint: In physics, k is called the coefficient of elasticity for a wire spring.)

Natural length = total length after stretched – stretched length

Therefore

$$5 - 4k = 6 - 8k \quad \therefore k = \tfrac{1}{4}$$

Natural length = $5 - 4k = 5 - 4 \cdot \frac{1}{4} = 4$ *cm*

Total length (L) = stretched length + natural length

$$\therefore L = \tfrac{1}{4}w + 4. \quad \text{Ans.}$$

In the court room, three defendants are waiting for the judge to sentence them for the time they need to stay in jail for the crimes they have committed.

Judge: Do you have any good reason why you think I should sentence you lesser time in jail ?
Defendant A: I was your high school classmate.
Judge: 1 week.
Defendant B: I built your house.
Judge: 1 month.
Defendant C: I introduced a girl friend to you 20 years ago. She is your wife now.
Judge: 20 years.

Polynomials and Factoring

8-1 Laws of Exponents

In Section 1-6, we have learned the basic concept of exponents (powers).
Now we will learn more formulas and applications about exponents.
To simplify expressions involving exponents, we combine the like (similar) terms in the simplest form and use the following **Laws of Exponents (Rules of Powers)**:

Laws of Exponents (Rules of Powers)

If a and b are real numbers, m and n are positive integers, $a \neq 0$, $b \neq 0$, then:

1) $a^m \cdot a^n = a^{m+n}$ **2)** $\dfrac{a^m}{a^n} = a^{m-n}$ **3)** $(a^m)^n = a^{mn}$

4) $(ab)^m = a^m b^m$ **5)** $\left(\dfrac{a}{b}\right)^m = \dfrac{a^m}{b^m}$ **6)** $a^{-m} = \dfrac{1}{a^m}$

7) For any nonzero number a, $a^0 = 1$.

Examples

1. $a^5 \cdot a^3 = a^{5+3} = a^8$; $x^8 \cdot x^2 = x^{8+2} = x^{10}$; $n^2 \cdot n^3 \cdot n^7 = n^{2+3+7} = n^{12}$

2. $\dfrac{a^5}{a^3} = a^{5-3} = a^2$; $\dfrac{x^8}{x^2} = x^{8-2} = x^6$; $\dfrac{n^5 \cdot n^3}{n^7} = \dfrac{n^8}{n^7} = n$

3. $(a^5)^3 = a^{5 \cdot 3} = a^{15}$; $(x^8)^2 = x^{8 \cdot 2} = x^{16}$; $(2^3)^4 = 2^{3 \cdot 4} = 2^{12} = 4{,}096$

4. $(ab)^5 = a^5 b^5$; $(a^2b^3)^5 = a^{2 \cdot 5} b^{3 \cdot 5} = a^{10} b^{15}$; $(3x^2y)^4 = 3^4 (x^2)^4 (y)^4 = 81x^8y^4$

5. $\left(\dfrac{a}{b}\right)^5 = \dfrac{a^5}{b^5}$; $\left(\dfrac{a^5}{b^3}\right)^2 = \dfrac{(a^5)^2}{(b^3)^2} = \dfrac{a^{10}}{b^6}$; $\left(\dfrac{4}{5}\right)^3 = \dfrac{4^3}{5^3} = \dfrac{64}{125}$

6. $a^{-5} = \dfrac{1}{a^5}$; $x^5 \cdot x^{-8} = x^{5-8} = x^{-3} = \dfrac{1}{x^3}$; $\dfrac{3^{10}}{3^{14}} = 3^{10-14} = 3^{-4} = \dfrac{1}{3^4} = \dfrac{1}{81}$

7. $a^{-5} \cdot a^5 = a^{-5+5} = a^0 = 1$; $b^0 = 1, \cdots\cdots$; $1^0 = 1$, $2^0 = 1$, $3^0 = 1, \cdots\cdots, 99^0 = 1, \cdots\cdots$

8. $(3x)(4x^2) = 12x^3$; $(-3a^2b)(2ab^2) = -6a^3b^3$; $(-p^2q^2)(-5p^3q^4) = 5p^5q^6$

-----Continued-----

$(-1)^n = 1.$ if n is an even number. $(-1)^{30} = 1$

$(-1)^n = -1.$ if n is an odd number. $(-1)^{31} = -1$

$0^0 =$ undefined. It means that the power "0" does not apply to 0 itself.

$$\left(\frac{a}{b}\right)^{-m} = \left(\frac{b}{a}\right)^{m}. \quad \left(\frac{a}{b}\right)^{-5} = \left(\frac{b}{a}\right)^{5}, \quad \left(\frac{5}{3}\right)^{-2} = \left(\frac{3}{5}\right)^{2} = \frac{9}{25}$$

To change a negative exponent to a positive exponent, we use its reciprocal.

Examples

9. $(-x^2)^8 = (-1)^8 \cdot (x^2)^8 = 1 \cdot x^{16} = x^{16}$; $(-x^2)^9 = (-1)^9 \cdot (x^2)^9 = -1 \cdot x^{18} = -x^{18}$

10. $\dfrac{x^{12}}{x^{12}} = x^{12-12} = x^0 = 1$; $\dfrac{a^5b^6}{a^9b^6} = a^{5-9}b^{6-6} = a^{-4}b^0 = a^{-4} \cdot 1 = a^{-4} = \dfrac{1}{a^4}.$

11. $\dfrac{5^{20}}{5^{18}} = 5^{20-18} = 5^2 = 25$; $\dfrac{10^{14}}{10^{10}} = 10^{14-10} = 10^4 = 10{,}000$

12. $\dfrac{5^{18}}{5^{20}} = 5^{18-20} = 5^{-2} = \dfrac{1}{5^2} = \dfrac{1}{25}$; $\dfrac{10^{10}}{10^{14}} = 10^{10-14} = 10^{-4} = \dfrac{1}{10^4} = \dfrac{1}{1{,}000}$

13. $3^{-2} = \dfrac{1}{3^2} = \dfrac{1}{9}$; $\dfrac{1}{3^{-2}} = 3^2 = 9$; $\left(\dfrac{3}{4}\right)^{-2} = \left(\dfrac{4}{3}\right)^{2} = \dfrac{16}{9} = 1\dfrac{6}{7}$; $\dfrac{m^4}{n^{-3}} = m^4n^3$

14. $\dfrac{5a^6b^2}{15ab^7} = \dfrac{5}{15} \cdot a^{6-1} \cdot b^{2-7} = \dfrac{1}{3}a^5b^{-5} = \dfrac{a^5}{3b^5}$; $\dfrac{6x^9y3}{9x^9y} = \dfrac{6}{9} \cdot x^{9-9} \cdot y^{3-1} = \dfrac{2}{3}x^0y^2 = \dfrac{2y^2}{3}$

15. $\left(\dfrac{x^2}{y^3}\right)^4 = \dfrac{(x^2)^4}{(y^3)^4} = \dfrac{x^8}{y^{12}}$; $\left(\dfrac{2ab^3}{3c^2}\right)^{-5} = \left(\dfrac{3c^2}{2ab^3}\right)^5 = \dfrac{3^5c^{10}}{2^5a^5b^{15}} = \dfrac{243c^{10}}{32a^5b^{15}}$

16. Find the ratio of the area of a circle inscribed in the square to the area of the square.

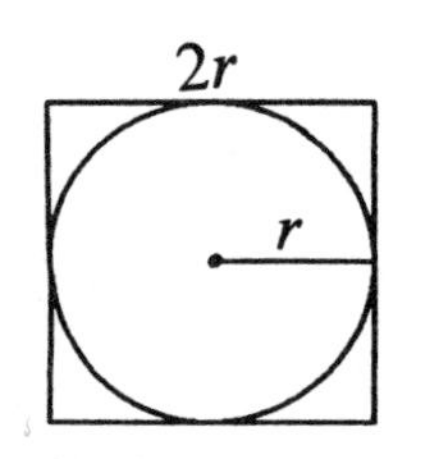

Solution: Let $A_1 =$ Area of the circle: $A_1 = \pi r^2$

$A_2 =$ Area of the square: $A_2 = (2r)^2 = 4r^2$

$$\frac{A_1}{A_2} = \frac{\pi r^2}{4r^2} = \frac{\pi}{4} \approx \frac{3.14}{4} \approx 0.79$$

Notice that the value of the ratio is a constant, regardless of the value of r.

EXERCISES

Evaluate each expression .

1. 2^3 **2.** 3^2 **3.** -2^3 **4.** $(-2)^3$ **5.** -2^4

6. $(-2)^4$ **7.** $(-5^n)^4$ **8.** 3^5 **9.** $(-5)^2$ **10.** -5^2

11. $(-3)^3$ **12.** $(-3)^4$ **13.** -3^4 **14.** 9 squared **15.** 92^0

16. 1^{15} **17.** $(-1)^{15}$ **18.** 8^0 **19.** 0^0 **20.** 9 cubed

Simplify each expression. Leave the product in exponent form.

21. $2^3 \cdot 2^4$ **22.** $(2^3)^4$ **23.** $a^3 \cdot a^4$ **24.** $(a^3)^4$ **25.** $m^{10} \cdot m^5$

26. $(m^{10})^5$ **27.** $a^n \cdot a^2$ **28.** $(a^n)^2$ **29.** $10^3 \cdot 10^5$ **30.** $(10^3)^5$

31. $x^6 \cdot x^7$ **32.** $-(a^3)^2$ **33.** $(-a^3)^2$ **34.** $a^8 \cdot a^{-3}$ **35.** $a^{-8} \cdot a^3$

36. $3^{10} \cdot 3^{-10}$ **37.** $a^{10} \cdot a^{-10}$ if $a \neq 0$ **38.** $x^a \cdot x^b$ **39.** $a^x \cdot a^y$ **40.** $(a^x)^y$

Divide. Assume that no denominator equals 0.

41. $\frac{2^8}{2^3}$ **42.** $\frac{2^3}{2^8}$ **43.** $\frac{a^8}{a^3}$ **44.** $\frac{a^3}{a^8}$ **45.** $\frac{10^8}{10^3}$

46. $\frac{2^{20}}{2^{17}}$ **47.** $\frac{5^{15}}{5^{12}}$ **48.** $\frac{2^{17}}{2^{20}}$ **49.** $\frac{5^{12}}{5^{15}}$ **50.** $\frac{x^5}{x^9}$

51. $\frac{x^9}{x^5}$ **52.** $\frac{x^2}{x^8}$ **53.** $\frac{a^6}{a^6}$ **54.** $\frac{x^5}{x^5}$ **55.** $\frac{x^{12}}{x^n}$

56. $\frac{x^n}{x^{12}}$ **57.** $\left(\frac{a}{b}\right)^4$ **58.** $\frac{x^2 \cdot x^n}{x^{2n}}$ **59.** $\left(\frac{a^2}{b^3}\right)^4$ **60.** $\left(\frac{x^n}{y^n}\right)^5$

Simplify each expression.

61. $3x + 4x$ **62.** $3x - 4x$ **63.** $3x \cdot 4x$

64. $(2x^2)(3x^5)$ **65.** $(3a)(6a^4)$ **66.** $(x^2y^3)(x^4y^5)$

67. $(3a^5b)(6a^3b^6)$ **68.** $(2x^2y^4)(-2y)$ **69.** $(-x^5y^5)(-2xy)$

70. $(pq)(2pq)(5p^2q^7)$ **71.** $(-xy)(-x^2y^3)(-y)$ **72.** $\left(\frac{3}{4}a^5\right)\left(\frac{8}{3}a^2\right)$

73. $(-4m^2n^3)(6n^6)$ **74.** $(-7m^6)(mp)$ **75.** $(2x^4)^3$

76. $(-\frac{1}{2}a^4)^5$ **77.** $(-2a^2)^4$ **78.** $3(5x^3)^3$

79. $-(3x^4y)^3$ **80.** $(p^2q^3)^4(pq)$ **81.** $(3xy^2)(2y^3)^4$

82. $(3a^4)^5(2a^2)^2$ **83.** $(2x^n)^4$ **84.** $(2a^n)^3(3a^n)^2$

85. $(\frac{2}{3}x^2)^2(\frac{3}{2}x^3)^4$

-----Continued-----

Simplify each expression. Assume that no denominator equals 0.

86. $\dfrac{5\times10^3}{15\times10^2}$ **87.** $\dfrac{15\times10^2}{5\times10^3}$ **88.** $\dfrac{4\times10^8}{2\times10^6}$ **89.** $\left(\dfrac{3}{5}\right)^2$ **90.** $\left(\dfrac{3}{5}\right)^{-2}$

91. $\dfrac{4x}{2}$ **92.** $\dfrac{4x}{2x}$ **93.** $\dfrac{4x^2}{2x}$ **94.** $\dfrac{2x}{4x}$ **95.** $\dfrac{2x}{4x^2}$

96. $\dfrac{3a^4}{15a}$ **97.** $\dfrac{8m^3}{2m^2}$ **98.** $\dfrac{4xy^3}{8y}$ **99.** $\dfrac{-6a^2b^4}{15ab}$ **100.** $\dfrac{-p^5q^3}{-p^2q^6}$

101. $\dfrac{x^4y^4}{x^4y}$ **102.** $\dfrac{3a^7b^2}{21a^4b^5}$ **103.** $\dfrac{mn^4}{m^5n^2}$ **104.** $\dfrac{12k^2}{-k^7}$ **105.** $\dfrac{-12x^3y^3}{4xy}$

106. $\dfrac{(2a^2)^3}{2a}$ **107.** $\dfrac{(3x)^4}{9x^2}$ **108.** $\dfrac{2x^3}{(4x^2)^3}$ **109.** $\dfrac{4m}{(2m)^3}$ **110.** $\dfrac{5p}{15p}$

111. $\dfrac{-2m^5}{10m}$ **112.** $\dfrac{(-2a^2)^3}{(-2a)^2}$ **113.** $\dfrac{-(2x)^2}{(-2x^2)^2}$ **114.** $\dfrac{-24r^7}{-32r^9}$ **115.** $\dfrac{16x^5}{-2x^4}$

116. $\dfrac{2x^2y^4}{(-xy)^2}$ **117.** $\dfrac{(4ab)^2}{4ab^2}$ **118.** $\dfrac{(2x^2)^4}{(2x^4)^2}$ **119.** $\dfrac{(-a)^{12}}{(-a)^9}$ **120.** $\dfrac{(-m^2)^5}{(-m^4)^3}$

121. $\left(\dfrac{2}{3}\right)^4$ **122.** $\left(\dfrac{2}{3}\right)^{-4}$ **123.** $\left(\dfrac{2a^2}{b^3}\right)^2$ **124.** $\left(\dfrac{xy}{x^3y^2}\right)^3$ **125.** $\left(\dfrac{xy}{x^3y^2}\right)^{-3}$

126. $\left(\dfrac{3ab^2}{2a^2b}\right)^3$ **127.** $\dfrac{a^5\cdot a^{-5}}{a^3}$ **128.** $\dfrac{a^5}{a^{-3}}$ **129.** $\dfrac{3x^{-4}}{6x^{-6}}$ **130.** $\dfrac{x^5y^{-3}}{x^{-2}y^2}$

Find the area of each shaded region.

131.

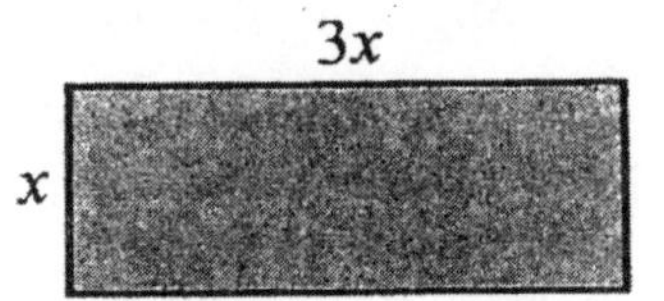

132.

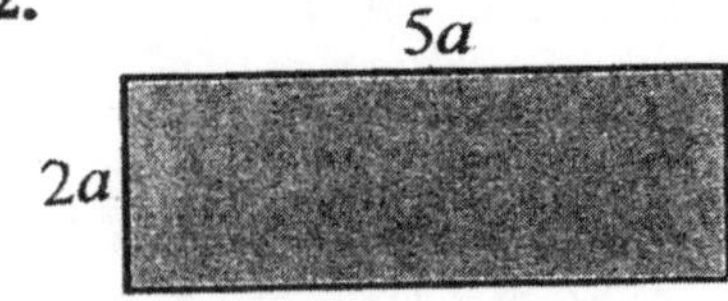

133.

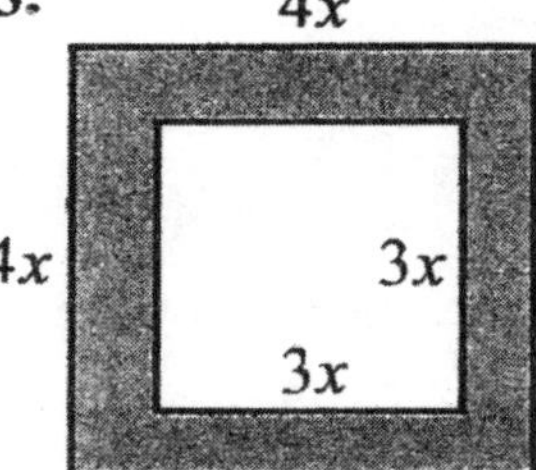

Find the total surface area and volume of each solid.

134.

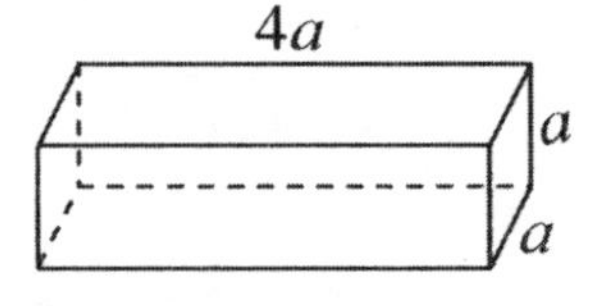

135.

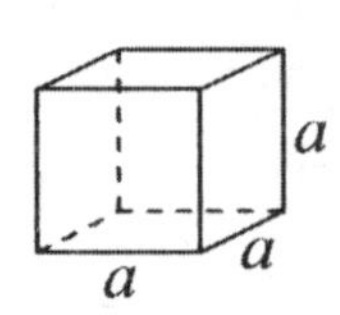

8-2 Multiplying Polynomials

In section 6-2, we have learned how to add and subtract expressions (polynomials).
In section 8-1, we have learned how to multiply and divide expressions (monomials).
Now we will learn how to multiply a polynomial by a monomial, and how to multiply two polynomials.
It is helpful to arrange the terms of the polynomials in either increasing degree or decreasing degree of a particular variable.

A) To multiply a polynomial by a monomial, we use the **distributive property** and the **laws of exponents.**

1. $4(x+3) = 4(x) + 4(3) = 4x + 12$

2. $-4x(x+3) = -4x(x) - 4x(3) = -4x^2 - 12x$

3. $(x+3)(5x) = x(5x) + 3(5x) = 5x^2 + 15x$

4. $-2x(3x^2 - 4x + 1) = -2x(3x^2) - 2x(-4x) - 2x(1) = -6x^3 + 8x^2 - 2x$

5. $4x^2y(2x^2 - 3xy + y^2) = 4x^2y(2x^2) + 4x^2y(-3xy) + 4x^2y(y^2)$
$= 8x^4y - 12x^3y^2 + 4x^2y^3$

B) To multiply two polynomials, we use the **distributive property** and the **laws of exponents**. Combine like terms.

6. $(x+4)(x+3) = (x+4)(x) + (x+4)(3)$
$= x^2 + 4x + 3x + 12$
$= x^2 + 7x + 12$

7. $(2x+4)(5x-3) = (2x+4)(5x) + (2x+4)(-3)$
$= 10x^2 + 20x - 6x - 12$
$= 10x^2 + 14x - 12$

8. $(2x-3)(3x^2 - 2x - 5) = (2x-3)(3x^2) + (2x-3)(-2x) + (2x-3)(-5)$
$= 6x^3 - 9x^2 - 4x^2 + 6x - 10x + 15$
$= 6x^3 - 13x^2 - 4x + 15$

Solving multistep equations

9. Solve $x(2x+4) + 8 = 2x^2$

Solution:

$x(2x+4) + 8 = 2x^2$
$\cancel{2x^2} + 4x + 8 = \cancel{2x^2}$
$4x + 8 = 0$
$4x = -8$
$\therefore x = -2$. Ans.

10. In an isosceles triangle, the length of each leg exceeds the base by 4 cm, and the perimeter is 23 cm. Find the length of the base.

Solution:

$x+4$ $x+4$ x

$2(x+4) + x = 23$
$2x + 8 + x = 23$
$3x = 15$
$\therefore x = 5\ cm$. Ans.

EXERCISES

Simplify each expression. (See Section 6-2, page 103)

1. $4x+6x$ **2.** $6x-4x$ **3.** $4x-6x$ **4.** $-(4+6x)$

5. $-(4-6x)$ **6.** $-(-4+6x)$ **7.** $-7x+(4+6x)$ **8.** $(7x-2)-(3x+6)$

9. $(7x+2)-(3x-6)$ **10.** $(a^2+2a-4)-(3a^2-4a+9)$

Multiply each expression. (See Section 8-1, page 175)

11. $x\cdot x$ **12.** $2x\cdot 3x$ **13.** $2x^2\cdot 3x^3$ **14.** $(-4a^3)(6a)$

15. $(2a)(-5a^5)$ **16.** $5x(x^3)$ **17.** $(-3x^4)(4x^5)$ **18.** $(2xy)(xy)$

19. $(3x^2y)(6xy^2)$ **20.** $-4a^2b(5a^3b)$

Multiply each expression.

21. $5(x-4)$ **22.** $5(x+4)$ **23.** $a(a+9)$ **24.** $-a(-a+9)$

25. $-a(a-9)$ **26.** $3x(2-4x)$ **27.** $x(5-6x)$ **28.** $-2m(m-4m^2)$

29. $xy(2x-3y)$ **30.** $2ab(3a-4b)$ **31.** $-2a^2b(ab+3ab^2)$ **32.** $3x^2y^3(xy+2x^2)$

33. $x(x^2-3x+5)$ **34.** $2x^2(x^2-4x-9)$ **35.** $2a(a^2-2a+6)$ **36.** $3a^2(a^3-2a^2+a)$

37. $\frac{1}{2}x^2(4x^2+2x)$ **38.** $\frac{1}{3}x(3x^2+6x-9)$ **39.** $\frac{2}{3}a(6a^2+3a-3)$ **40.** $\frac{4}{3}ab^2(12a^2b-9a)$

Simplify each expression.

41. $2a(a-8)+3a$ **42.** $-4a(2a-5)-3a$ **43.** $4x(2x-3)-6x$ **44.** $5x(6-4x)+7x$

45. $3x(x-2)+x(7-x)$ **46.** $5x(x-5)-2x(2x-1)$

47. $6p^2(2p-2)-3p(p^2-5p)$ **48.** $4a^2(2a+3)+a(-3a-4)$

49. $6n(n-1)-2n(4n-5)$ **50.** $2n(3-2n)-4n(n-5)$

Multiply by using the Distributive Property.

51. $(x+3)(x+2)$ **52.** $(x+3)(x-2)$ **53.** $(x-3)(x-2)$ **54.** $(2x+5)(x+4)$

55. $(2x+5)(x-4)$ **56.** $(2x-5)(x+4)$ **57.** $(3a+2)(3a-4)$ **58.** $(3a-2)(3a+4)$

59. $(2n-2)(3n^2-n-3)$ **60.** $(3a+1)(2a^2-4a+5)$

Solve each multistep equation.

61. $3(x-5)-4=2$ **62.** $2(3x+5)-4(x-5)=2$

63. $(x+4)(x+2)=(x-2)(x+9)$ **64.** $(x-3)(4x+5)=(2x-6)(2x-1)$

65. In an isosceles triangle, the length of each leg exceeds the base by 8 *in.*, and the perimeter is 34 *in.*. Find the length of the base and the length of each leg.

66. In a rectangle, the length is 2 *cm* longer than the width. If its length and width are both increased by 4 *cm*, its area is increased by 56 cm^2. Find the original length and width.

67. In a rectangle, the length is twice as long as the width. If its length and width are both increased by 6 *cm*, its area is increased by 162 cm^2. Find its original dimensions.

8-3 Multiplying Binomials (FOIL & Patterns)

In section 8-2, we have learned how to multiply two binomials by using the **distributive property**.

$$(2x-5)(3x-4) = (2x-5)(3x) + (2x-5)(-4)$$
$$= 6x^2 - 15x - 8x + 20$$
$$= 6x^2 - 23x + 20$$

Now we will learn a short way to multiply two binomials mentally. This short way is called the **FOIL** method (First Outside, Inside Last).

FOIL method:

1. Multiply the first terms: $6x^2$
2. Multiply the last terms: 20
3. Multiply the outside terms: $-8x$
   Multiply the inside terms: $-15x$
   Add these two products: $(-8x) + (-15x) = -23x$

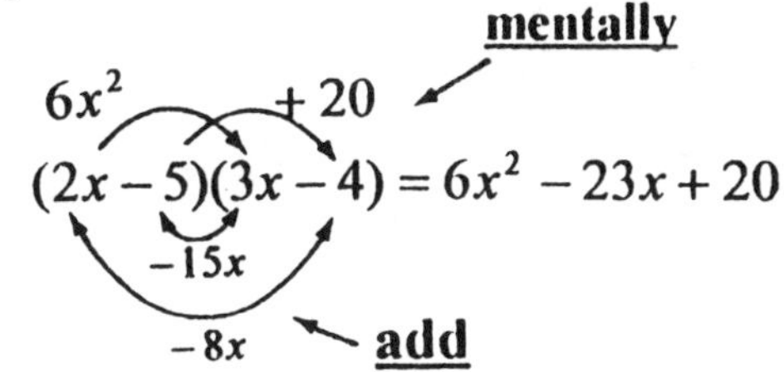

The following formulas are very useful to shorten the steps in multiplying two binomials which are in special multiplication patterns.

Formulas

1)	$(a+b)^2 = a^2 + 2ab + b^2$
2)	$(a-b)^2 = a^2 - 2ab + b^2$
3)	$(a+b)(a-b) = a^2 - b^2$

Examples

1. $(x+4)^2 = x^2 + 2 \cdot x \cdot 4 + 4^2 = x^2 + 8x + 16$
2. $(x-4)^2 = x^2 - 2 \cdot x \cdot 4 + 4^2 = x^2 - 8x + 16$
3. $(x+4)(x-4) = x^2 - 4^2 = x^2 - 16$

4. $(3x+2y)^2 = (3x)^2 + 2 \cdot 3x \cdot 2y + (2y)^2 = 9x^2 + 12xy + 4y^2$
5. $(3x-2y)^2 = (3x)^2 - 2 \cdot 3x \cdot 2y + (2y)^2 = 9x^2 - 12xy + 4y^2$
6. $(3x+2y)(3x-2y) = (3x)^2 - (2y)^2 = 9x^2 - 4y^2$

7. $(x^3+4)^2 = (x^3)^2 + 2 \cdot x^3 \cdot 4 + 4^2 = x^6 + 8x^3 + 16$
8. $(a^n - b^n)^2 = (a^n)^2 - 2 \cdot a^n \cdot b^n + (b^n)^2 = a^{2n} - 2a^n b^n + b^{2n}$
9. $(a^2 - b^2)(a^2 + b^2) = (a^2)^2 - (b^2)^2 = a^4 - b^4$
10. $(xy + 2z)(xy - 2z) = (xy)^2 - (2z)^2 = x^2y^2 - 4z^2$

We can use the distributive property or the FOIL method to multiply the above binomials. The result will be the same.

EXERCISES

Multiply by using the Distributive Property.

1. $(x+2)(x-4)$ **2.** $(x-6)(x+8)$ **3.** $(x-5)(x-7)$ **4.** $(x+9)(x+4)$
5. $(3x-4)(2x+5)$ **6.** $(4a+3)(a+8)$ **7.** $(5a-6)(3a-7)$ **8.** $(2a+5)(2a+5)$
9. $(2a-5)(2a-5)$ **10.** $(2a+5)(2a-5)$

Multiply by using the FOIL method.

11. $(x+3)(x-1)$ **12.** $(x+1)(x-3)$ **13.** $(a+5)(a+7)$
14. $(a-5)(a+7)$ **15.** $(2a-5)(a-7)$ **16.** $(2a+5)(a-7)$
17. $(2a+5)(2a+7)$ **18.** $(x-2)(x+2)$ **19.** $(x+2)(x+2)$
20. $(x-2)(x-2)$ **21.** $(2x-y)(3x+y)$ **22.** $(2x-y)(3x-y)$
23. $(a+2b)(a-b)$ **24.** $(a+2b)(a-2b)$ **25.** $(p+q)(p-q)$
26. $(p+q)(p-2q)$ **27.** $(2a+5)(2a+5)$ **28.** $(2a-5)(2a-5)$
29. $(2a+5)(2a-5)$ **30.** $(5-2a)(5-2a)$

Multiply by using the formulas for special patterns.

31. $(x+5)(x+5)$ **32.** $(x-5)(x-5)$ **33.** $(x+5)(x-5)$
34. $(a-6)(a+6)$ **35.** $(a-6)(a-6)$ **36.** $(a+6)(a+6)$
37. $(4-x)(4-x)$ **38.** $(4-x)(4+x)$ **39.** $(3+x)(3-x)$
40. $(-2x+3)(-2x+3)$ **41.** $(4+2x)(4-2x)$ **42.** $(3-x)(3-x)$
43. $(x+7)^2$ **44.** $(x-7)^2$ **45.** $(x+7)(x-7)$
46. $(2x+7)^2$ **47.** $(2x-7)^2$ **48.** $(2x+7)(2x-7)$
49. $(3x-y)^2$ **50.** $(3x+y)^2$ **51.** $(a-8)^2$
52. $(n+9)^2$ **53.** $(5y-6)^2$ **54.** $(x+4y)^2$
55. $(2x-5y)^2$ **56.** $(4a+1)^2$ **57.** $(4a+1)(4a-1)$
58. $(5x-2y)^2$ **59.** $(a+3b)^2$ **60.** $(2c-3d)^2$

Multiply by using any method.

61. $(2x+6y)(2x-6y)$ **62.** $(4x+8y)^2$ **63.** $(2a-5b)^2$
64. $(p-3q)^2$ **65.** $(4a-5b)(5a-4b)$ **66.** $(3x-4)(5x+6)$
67. $(k+5)(k-20)$ **68.** $(2a+1)(3a-12)$ **69.** $(c-2d)(c+2d)$
70. $(2c+4d)^2$ **71.** $(2x-8)^2$ **72.** $(2x-8y)^2$
73. $(7s-5)(2s+6)$ **74.** $(a^2+3)^2$ **75.** $(x^n-y^n)^2$
76. $(a-b^2)(a+b^2)$ **77.** $(a^3-b^3)(a^3+b^3)$ **78.** $(x^2-y)(x^2+y)$
79. $(ab-3c)(ab+3c)$ **80.** $(a^2-b)^2$

8-4 Factoring Polynomials

We have learned how to **multiply** a polynomial by a monomial, and how to multiply two polynomials by using the distributive property.

Multiply: $4x(x+5) = 4x \cdot x + 4x \cdot 5 = 4x^2 + 20x$

Working in the other direction, we can **factor** a polynomial by expressing it as a product of other **prime polynomials**. To check the factors of a polynomial, we can multiply the resulting factors to find the original expression.

Factor: $4x^2 + 20x = 4x(x+5)$

Hint: $4x$ is the greatest common factor of the two terms, $4x^2$ and $20x$.

Check: $4x(x+5) = 4x \cdot x + 4x \cdot 5 = 4x^2 + 20x$ ✓

To factor a polynomial, the first step is to find the **greatest common factor** of its terms, or **GCF** (see Section 3-4, page 49). Sometimes, we need to arrange factors that are opposite of each other (see example 13).

Examples:

1. $5x + 10 = 5(x+2)$
2. $8a - 6 = 2(4a-3)$
3. $5x^2 + 10x = 5x(x+2)$
4. $8a^2 - 6a = 2a(4a-3)$
5. $15x^3 - 20x^2 = 5x^2(3x-4)$
6. $12p^5 - 16p^3 = 4p^3(3p^2-4)$
7. $4xy + 8x = 4x(y+2)$
8. $4c^3 - 2c^2 + 6c = 2c(2c^2 - c + 3)$
9. $2x^5 + 4x^3 - 12x^2 = 2x^2(x^3 + 2x - 6)$
10. $8a^3b^2 + 20a^2b^3 = 4a^2b^2(2a+5b)$
11. $x(x-5) + 4(x-5) = (x-5)(x+4)$
12. $a(a+4) - 3(a+4) = (a+4)(a-3)$
13. $x(x-5) + 4(5-x) = x(x-5) - 4(x-5) = (x-5)(x-4)$

Factor by grouping

If a polynomial has four or more terms, we regroup the terms in pairs that have a common factor. We may need to try several groups before we get the right groups.

Examples:

14. $x^3 + x^2 + 4x + 4 = (x^3 + x^2) + (4x+4) = x^2(x+1) + 4(x+1) = (x+1)(x^2+4)$
15. $m^2 + 2m + mn + 2n = (m^2 + 2m) + (mn + 2n) = m(m+2) + n(m+2)$
$= (m+2)(m+n)$
16. $xy + 3y + x + 3 = (xy + 3y) + (x+3) = y(x+3) + (x+3) = (x+3)(y+1)$
17. $xy + 3y - x - 3 = (xy + 3y) - (x+3) = y(x+3) - (x+3) = (x+3)(y-1)$
18. $a^2 - bc - ab + ac = (a^2 - ab) + (ac - bc) = a(a-b) + c(a-b) = (a-b)(a+c)$
19. $2x - ay + ax - 2y = (2x + ax) - (ay + 2y) = x(2+a) - y(a+2) = (a+2)(x-y)$
20. $ax + by + ay + bx = (ax + bx) + (ay + by) = x(a+b) + y(a+b) = (a+b)(x+y)$

EXERCISES

Find the GCF for each set of numbers or algebraic expressions. (see Section 3-4)

1. 4, 8 **2.** 12, 4 **3.** 5, 9 **4.** 12, 20
5. 20, 35 **6.** 15, 24 **7.** $4,\ 2x$ **8.** $18x^2,\ 9x$
9. $12x,\ 18x$ **10.** $x^2y,\ xy^2$ **11.** $a^3b^2,\ ab^2$ **12.** $6ab,\ 15a$
13. $8xy^3,\ 12x^4y^2$ **14.** $5cd^3,\ 9cd$ **15.** $12y^3,\ 16y^2$

Factor each polynomial by using the GCF.

16. $2x+12$ **17.** $4x-12$ **18.** $4x^2-12$ **19.** $6x^2-15x$
20. $12a^3+18a^2$ **21.** $9a^2+21a$ **22.** $10m^2-2m$ **23.** $6x^2+10y^2$
24. $7p+14q$ **25.** $6x^2y+10y^2$ **26.** $2xy+4x^2y^2$ **27.** $6a^2b+9ab$
28. $4a^3b-6ab$ **29.** $5mn+20m^3n$ **30.** $16x^4-18x^2$ **31.** $4x^7-2x^3$
32. $2x^2+4x+6$ **33.** $2x^3+4x^2+6x$ **34.** $3p^2+6p^3+9p^4$ **35.** $4m^4-8m^2+4m$
36. $ab^3+3a^2b^2+a^3b$ **37.** $xy+x^2y^2+x^3y^3$ **38.** $4xy+6x^2y+8xy^2$ **39.** $xy-4xy^2+xy^3$
40. $x^2y^2z+xy^2z^2$ **41.** $6x^4-3x^3y+9x^2y^2$ **42.** $6a^2b^2c-14ab^2c^2$ **43.** $x^{n+2}+x^2$
44. $x^{n+2}+x^n$ **45.** $6x^{2n}+8x^{3n}$

Factor each polynomial.

46. $x(x+3)+2(x+3)$ **47.** $x(x+3)-2(x+3)$ **48.** $x(x-3)-2(x-3)$
49. $x(x-3)+2(3-x)$ **50.** $x(x-3)-2(3-x)$ **51.** $p(q-1)-5(q-1)$
52. $p(q-1)-5(1-q)$ **53.** $4a(a+5)-4(a+5)$ **54.** $9x(2x+3)+(2x+3)$
55. $4x(x-3y)-2(x-3y)$ **56.** $5a(2a-3b)+6(2a-3b)$ **57.** $2x(x-2)+3(2-x)$
58. $2x(x-2)-3(2-x)$ **59.** $2p(p-s)+s(s-p)$ **60.** $m(m-n)-2n(n-m)$

Factor by grouping.

61. $4x-4y+ax-ay$ **62.** $xy-x+2y-2$ **63.** $2xy-4x+2y-4$
64. $2x-2y-ax+ay$ **65.** $4a-12+a^3-3a^2$ **66.** $4ab-2b+4a-2$
67. $4ab-2b-4a+2$ **68.** $x^2+2ax+xy+2ay$ **69.** $s^2+s+ts+t$
70. $m^2+mn-2m-2n$ **71.** $4ab+16a-b-4$ **72.** $3ab-b+6a-2$
73. $x^2-2y+xy-2x$ **74.** $x^3-4x^2-3x+12$ **75.** p^3+3p^2-2p-6
76. $2x+by+bx+2y$ **77.** $x^2-4yz-2xy+2xz$ **78.** $ax+by-bx-ay$
79. $m^2+4n+mn+4m$ **80.** $2ax+by-2ay-bx$ **81.** $2cx+cy+8nx+4ny$
82. $x^3+2x^2-18x-36$

8-5 Factoring Quadratic Trinomials

We have learned how to multiply two binomials by using distributive property or the FOIL method. The result is a quadratic trinomial.

Multiply: $(x+2)(x+3) = x^2 + 5x + 6$

In this section we will learn how to factor a quadratic trinomial as a product of two binomials.

Factor: $x^2 + 5x + 6 = (x+2)(x+3)$

An expression in the form $ax^2 + bx + c$, $a \neq 0$ is called a **quadratic polynomial.** A quadratic trinomial can be factored as a product of the form $(px + r)(qx + s)$ by working backward with the FOIL method.

$$ax^2 + bx + c = (px + r)(qx + s)$$

We use the following steps to find the factors:

1) List all the pairs of factors of ax^2. $ax^2 = (px)(qx)$
2) List all the pairs of factors of c. $c = rs$
3) Test all possible factors of ax^2 and c to find out which factor produces the correct bx.

Examples

1. Factor $x^2 + 12x + 20$

Solution: The pair of factors of x^2 : x and x.
The pairs of factors of 20 : 1 and 20, 2 and 10, 4 and 5.
Test all factors to see which produces the correct term $12x$.

$$x^2 + 12x + 20 = (x+2)(x+10). \text{ Ans}$$

$+2x$, $+10x$ ← $2x + 10x = 12x$

2. Factor $14x^2 - 17x + 5$.

Solution: The pairs of factors of $14x^2$: $14x$ and x, $2x$ and $7x$, $-14x$ and $-x$.
The pairs of factors of 5: 1 and 5, -1 and -5.
Test all factors to see which produces the correct term $-17x$.

$$14x^2 - 17x + 5 = (2x-1)(7x-5). \text{ Ans.}$$

$-7x$, $-10x$ → $-7x - 10x = -17x$

3. Factor $x^2 - x - 20$.

Solution:

$$x^2 - x - 20 = (x+4)(x-5). \text{ Ans.}$$

$+4x$, $-5x$ → $4x - 5x = -x$

4. Factor $x^2 - 4xy - 12y^2$

Solution:

$$x^2 - 4xy - 12y^2 = (x+2y)(x-6y). \text{ Ans.}$$

$+2xy$, $-6xy$ $= -4xy$

EXERCISES

Factor. If the trinomial is not factorable, write *prime*.

1. x^2+3x+2 **2.** x^2-3x+2 **3.** x^2+6x+5
4. x^2-6x+5 **5.** x^2+8x+7 **6.** x^2-8x+7
7. x^2-7x-8 **8.** x^2+2x+3 **9.** x^2+2x-3
10. x^2-2x-3 **11.** x^2+3x-2 **12.** x^2+6x+9
13. $a^2+9a+14$ **14.** y^2+5y+6 **15.** $a^2-9a+14$
16. y^2-5y+6 **17.** $-y^2+5y+6$ **18.** $-y^2-5y+6$
19. y^2+5y+7 **20.** $c^2-9c+14$ **21.** $s^2-13s+40$
22. $n^2+11n+28$ **23.** $n^2-11n+28$ **24.** x^2-x-12
25. $p^2+29p-30$ **26.** $p^2+29p+30$ **27.** $a^2-13a-30$
28. a^2+3a+4 **29.** $n^2-10n-9$ **30.** x^2+x-56

Factor. If the trinomial is not factorable, write *prime*.

31. $x^2+3xy+2y^2$ **32.** $x^2-3xy+2y^2$ **33.** $x^2-2xy+y^2$
34. $a^2+5ab+6b^2$ **35.** $a^2-5ab-6b^2$ **36.** $x^2+7xy+12y^2$
37. $x^2-10xy+21y^2$ **38.** $m^2+23mn+42n^2$ **39.** $x^2-4xy+5y^2$
40. $a^2+6ab-5b^2$ **41.** $p^2-23pq-50q^2$ **42.** $r^2-14rs+49s^2$
43. $x^2+10xy+24y^2$ **44.** $x^2+10xy-24y^2$ **45.** $a^2-13ab+42b^2$

Factor. If the trinomial is not factorable, write *prime*.

46. $2x^2+3x+1$ **47.** $5x^2-3x-2$ **48.** $2x^2+5x-3$
49. $3x^2-7x-6$ **50.** $-6x^2-7x+3$ **51.** $3x^2-4x+5$
52. $-3n^2-5n+2$ **53.** $8x^2+2x-3$ **54.** $10a^2-13a-3$
55. $8c^2-10c-7$ **56.** $4x^2+5x+3$ **57.** $12x^2-11x-5$
58. $3x^2-7xy-6y^2$ **59.** $3x^2+2xy-8y^2$ **60.** $2a^2+5ab+2b^2$
61. $5a^2-ab-6b^2$ **62.** $2x^2+xy-3y^2$ **63.** $18c^2-9c-20$
64. $22y^2+17y-24$ **65.** $4x^4+4x^2-3$ **66.** $2x^4-3x^2-14$
67. $x^{2n}+5x^n+6$ **68.** $x^{4n}+2x^{2n}-15$ **69.** x^4+8x^2+15
70. $3x^4-11x^2-20$

Factor completely. Factor out the greatest monomial factor first.

71. x^3+3x^2+2x **72.** $2x^3-6x^2+4x$ **73.** $4x^3-12x^2-16x$
74. a^3-9a^2+14a **75.** $2x^3y+6x^2y^2+4xy^3$ **76.** $n^4-2n^3-3n^2$
77. $a^3b-5a^2b^2-6ab^3$ **78.** $3a^3b+15a^2b^2+18ab^3$ **79.** $4x^5+4x^3-3x$
80. $x^8-x^5-6x^2$

8-6 Factoring Special Quadratic Patterns

The special quadratic multiplication patterns in Section 8-3 are useful in factoring. Working backward with the formulas in Section 8-3, we have the following patterns:

Pattern 1: Factoring a perfect-square trinomial.

$$a^2 + 2ab + b^2 = (a+b)^2$$
$$a^2 - 2ab + b^2 = (a-b)^2$$

Pattern 2: Factoring a difference of two perfect-squares.

$$a^2 - b^2 = (a+b)(a-b)$$

Rules: A trinomial is a perfect square if :

1) The first term is a perfect square a^2.
2) The last term is a perfect square b^2.
3) The middle term is twice of $a \times b$.

If a binomial is a difference of two perfect-squares, we apply the rule in pattern 2.
Sometimes, we need more than one method to factor a polynomial.
Sometimes, we need to group its terms in a polynomial before factoring it.

Examples

1. Decide whether it is a perfect-square trinomial.

a. $x^2 - 14x + 49$. **b.** $9x^2 + 30x + 25$

Solution:

a. $x^2 - 14x + 49$
x^2 is a perfect square: x^2
49 is a perfect square: 7^2
The middle term $= 2 \cdot x \cdot 7 = 14x$ ✓

It is a perfect square. Ans
Hint: $x^2 - 14x + 49 = (x-7)^2$

b. $9x^2 + 30x + 25$
$9x^2$ is a perfect square: $(3x)^2$
25 is a perfect square: 5^2
The middle term $= 2 \cdot 3x \cdot 5 = 30x$ ✓

It is a perfect square. Ans
Hint: $9x^2 + 30x + 25 = (3x+5)^2$

Factor each polynomial.

2. $x^2 + 6x + 9 = (x+3)^2$

3. $x^2 - 6x + 9 = (x-3)^2$

4. $a^2 + 18a + 81 = (a+9)^2$

5. $x^2 - 18x + 81 = (x-9)^2$

6. $4x^2 + 12x + 9 = (2x+3)^2$

7. $4x^2 - 12x + 9 = (2x-3)^2$

8. $x^2 - 9 = x^2 - 3^2 = (x+3)(x-3)$

9. $4x^2 - 9 = (2x)^2 - 3^2 = (2x+3)(2x-3)$

10. $x^2 - 8x + 16 - 4y^2$
$= (x^2 - 8x + 16) - 4y^2$
$= (x-4)^2 - (2y)^2$
$= (x - 4 + 2y)(x - 4 - 2y)$
$= (x + 2y - 4)(x - 2y - 4)$

11. $x^2 + y^2 - 2xy - 9$
$= (x^2 - 2xy + y^2) - 9$
$= (x-y)^2 - 3^2$
$= (x - y + 3)(x - y - 3)$

12. $a^4 - 81$
$= (a^2)^2 - 9^2$
$= (a^2 + 9)(a^2 - 9)$
$= (a^2 + 9)(a+3)(a-3)$

EXERCISES

Multiply by using the formulas for special patterns. (see Section 8-3)

1. $(x+4)^2$ **2.** $(x-4)^2$ **3.** $(x+4)(x-4)$

4. $(3x+5)^2$ **5.** $(3x-5)^2$ **6.** $(3x+5)(3x-5)$

7. $(2x-3y)^2$ **8.** $(2x+3y)^2$ **9.** $(2x-3y)(2x+3y)$

10. $(2a+7b)^2$ **11.** $(2a-7b)^2$ **12.** $(2a-7b)(2a+7b)$

13. $(ab+3)^2$ **14.** $(2xy-z)^2$ **15.** $(-3x+5y)^2$

16. $(x^2+2y)(x^2-2y)$ **17.** $(ab+c^2)(ab-c^2)$ **18.** $(p^2+q^2)(p^2-q^2)$

19. $(2p^3+4)^2$ **20.** $(-2a^2+3b)^2$

State whether each polynomial is a perfect square. If it is, factor it by using the formulas. If it is not, factor it by the FOIL method.

21. x^2+4x+4 **22.** x^2-4x+4 **23.** x^2+3x+2

24. $a^2+8a+16$ **25.** $a^2-10a+16$ **26.** $a^2-8a+16$

27. $p^2+18p+81$ **28.** $y^2+22y+121$ **29.** $9x^2+24x+16$

30. $6x^2-5x-6$ **31.** $4x^2-12x+9$ **32.** $4x^2-2x-6$

33. $4a^2+12ab+9b^2$ **34.** $9x^2-30xy+25y^2$ **35.** $25x^2+40xy+16y^2$

State whether each polynomial is a difference of two squares. If it is, factor it by using the formula.

36. a^2-16 **37.** n^2+16 **38.** x^2-4

39. $2x^2-9$ **40.** $4x^2-9$ **41.** $4n^2-1$

42. $1-16x^2$ **43.** x^4-y^2 **44.** $9-c^2$

45. n^3-25 **46.** $16x^2-6y^2$ **47.** $49-4a^2$

48. $a^2b^2-c^4$ **49.** $4x^2y^2-9z^2$ **50.** $a^{2n}-b^{2n}$

Factor completely.

51. $x^2+16x+64$ **52.** $x^2-18x+81$ **53.** $4a^2-49$

54. $9n^2-1$ **55.** $1-9n^2$ **56.** $25n^2-60n+36$

57. $9x^2-12xy+4y^2$ **58.** $4p^2+20pq+25q^2$ **59.** $36x^2-49y^2$

60. $9a^2-4b^2$ **61.** $a^2b^2-4c^2$ **62.** x^3+10x^2+25x

63. $x^4-6x^3+9x^2$ **64.** $2x-8x^3$ **65.** $18c-2c^3$

66. $12a^3-36a^2+27a$ **67.** $x^2-6x+9-y^2$ **68.** $a^2+4a+4-9b^2$

69. x^2-y^2-6y-9 **70.** m^2-n^2+6n-9 **71.** x^6-8x^3+16

72. $x^2y^2-14xy+49$ **73.** x^4-y^4 **74.** a^4-16

75. $a^{2n}-16$ **76.** $x^{4n}-y^{4n}$ **77.** $4x^{2n}-9y^{4n}$

78. $x^2(5+x^2)-4(5+x^2)$ **79.** $x^2(x^2-9)+4(x^2-9)$ **80.** x^8-y^8

8-7 Solving Equations by Factoring

Many algebraic equations can be solved by **factoring** and the **zero-product theory.**
Zero-Product Theory: For all real numbers a and b:

$a \cdot b = 0$ **if and only if** $a = 0$ or $b = 0$.

The zero-product theory tells us when an equation is written as a product of factors, we can find its solutions by setting each of the factors equal to 0 and solve for its roots.

Double Roots: If an equation has two identical factors, the equation is said to have a double root (see example 4).

Examples

1. Find the zeros of $y = x^2 + x - 2$.

Solution:

Let $y = 0$, $x^2 + x - 2 = 0$

$(x+2)(x-1) = 0$

$x + 2 = 0 \;|\; x - 1 = 0$

$x = -2 \;|\; x = 1$

Ans: $x = -2$ and 1.

Check: $x = -2$: $(-2)^2 - 2 - 2 = 0$ ✔

$x = 1$: $1^2 + 1 - 2 = 0$ ✔

2. Solve $(2x-4)(x+5) = 0$.

Solution:

$(2x-4)(x+5) = 0$

$2x - 4 = 0 \;|\; x + 5 = 0$

$2x = 4 \;|\; x = -5$

$x = 2$

Ans: $x = 2$ and -5.

Check: $x = 2$: $(2 \cdot 2 - 4)(2+5) = 0$ ✔

$x = -5$: $(2 \cdot -5 - 4)(-5+5) = 0$ ✔

3. Solve $2x^2 + 3x = 9$.

Solution:

$2x^2 + 3x - 9 = 0$

$(2x-3)(x+3) = 0$

$2x - 3 = 0 \;|\; x + 3 = 0$

$2x = 3 \;|\; x = -3$

$x = 1\frac{1}{2}$

Ans: $x = 1\frac{1}{2}$ and -3.

4. Solve $x^2 - 14x + 49 = 0$.

Solution:

$x^2 - 14x + 49 = 0$

$(x-7)(x-7) = 0$

$x - 7 = 0 \;|\; x - 7 = 0$

$x = 7 \;|\; x = 7$

Ans: $x = 7$ **(a double root).**

5. Solve $2a^3 - 6a^2 - 20a = 0$.

Solution:

$2a^3 - 6a^2 - 20a = 0$

$2a(a^2 - 3a - 10) = 0$

$2a(a-5)(a+2) = 0$

$2a = 0 \;|\; a - 5 = 0 \;|\; a + 2 = 0$

$a = 0 \;|\; a = 5 \;|\; a = -2$

Ans: $a = 0$, 5, -2.

6. Solve $x^3 - 9x = 0$.

Solution:

$x^3 - 9x = 0$

$x(x^2 - 9) = 0$

$x(x+3)(x-3) = 0$

$x = 0 \;|\; x + 3 = 0 \;|\; x - 3 = 0$

$x = -3 \;|\; x = 3$

Ans: $x = 0$, -3, 3.

EXERCISES

Find the zeros of each function $y = f(x)$. (Hint: Let $y = 0$)

1. $y=(x+2)(x-1)$
2. $y=(x-10)(x+8)$
3. $y=(x-4)(x-4)$
4. $y=(x-12)(x-12)$
5. $y=(2x+4)(3x-3)$
6. $y=(2x-4)(3x+3)$
7. $y=(3x+3)(3x-3)$
8. $y=(4x-2)(3x-6)$
9. $y=2x(x-15)$
10. $y=x(2x-10)(3x+8)$

Solve by Factoring. Factor out the greatest monomial factor first.

11. $2x^2+3x=0$
12. $2x^2-3x=0$
13. $-4x^2+16x=0$
14. $-12x^2+8x=0$
15. $x^2-7x=0$
16. $35x^2-5x=0$
17. $5x^2-x^3=0$
18. $10a^2-20a^3=0$
19. $x^2=5x$
20. $x^3=5x^2$
21. $y^2=12y$
22. $4x^2-2x^3=0$
23. $p^2=p$
24. $n^3-n=0$
25. $10x^2-100x=0$

Solve.

26. $(a-5)(a+7)=0$
27. $(p+5)(p-5)=0$
28. $(x-9)(x-9)=0$
29. $(n+2)(n+12)=0$
30. $(2x+4)(x-6)=0$
31. $(3x+6)(4x-2)=0$
32. $x(x-2)(x+4)=0$
33. $3y(2y+1)(y-4)=0$
34. $2x(3x-6)(6x+3)=0$
35. $x^2+3x+2=0$
36. $x^2-3x+2=0$
37. $n^2-5n-6=0$
38. $x^2+16x+64=0$
39. $x^2-4x+4=0$
40. $a^2-10a+16=0$
41. $2x^2+5x+3=0$
42. $2x^2+5x-12=0$
43. $4x^2-8x+3=0$
44. $6x^2-13x-5=0$
45. $4x^2-17x-15=0$
46. $6x^2-13x+2=0$
47. $a^2-9=0$
48. $4a^2-9=0$
49. $2n^3-2n=0$
50. $x^2+16=8x$
51. $8a^2-1=7a$
52. $6y^2+y=2$
53. $x^3-x^2-6x=0$
54. $4x^3+10x^2-6x=0$
55. $y^4-8y^3+16y^2=0$
56. $p^5-13p^3+36p=0$
57. $(x+2)(x-3)=14$
58. $(x+6)(x-2)=-7$
59. $x(x-3)=2(x+7)$
60. $3x(x-1)=2(x+1)$

A high school boy is playing the violin.
His girlfriend sits next to him.
Boy: Do you like the song I played ?
Girl: Yes. It makes me miss my father.
Boy: Was your father a musician ?
Girl: No. He was a sawmill worker.

8-8 Solving Word Problems by Factoring

Many word problems can be written in polynomial equations and solved by factoring.
Steps for solving a word problem by factoring:

1. Read the problem.
2. Assign a variable to represent the unknown.
3. Write an equation based on the given facts.
4. Solve the equation by factoring to find the answers.
5. Check your answers (usually on a scratch paper).

Sometimes a solution may not satisfy the conditions of the problem. We need to check and reject the solutions that do not make sense (not permissible) for the problem.

Examples

1. If a number adds its square, the result is 42. Find the number.

Solution:

Let $n =$ the number.

The equation is:

$$n + n^2 = 42$$
$$n^2 + n - 42 = 0$$
$$(n+7)(n-6) = 0$$
$$n+7=0 \quad \Big| \quad n-6=0$$
$$n=-7 \quad \Big| \quad n=6$$

Ans: The number is -7 or 6.

Check: $-7+(-7)^2 = -7+49 = 42$ ✓

$6+6^2 = 6+36 = 42$ ✓

2. Find two consecutive positive integers whose product is 110.

Solution:

Let $n =$ the 1st positive integer.

$n+1 =$ the 2nd positive integer.

The equation is:

$$n(n+1) = 110$$
$$n^2 + n - 110 = 0$$
$$(n+11)(n-10) = 0$$

$n = -11$ not permissible | $n = 10$ (1st integer), $n+1 = 11$ (2nd integer)

Ans: The integers are 10 and 11.

3. A box has a square bottom, and the top is 5 *in.* high. Its total surface area is 192 $in.^2$. Find its volume.

Solution:

Let $x =$ one side of the equation.

The equation is:

$$2x^2 + 4(5x) = 192$$
$$2x^2 + 20x - 192 = 0$$
$$x^2 + 10x - 96 = 0$$
$$(x+16)(x-6) = 0$$
$$x+16=0 \quad \Big| \quad x-6=0$$
$$x=-16 \quad \Big| \quad x=6$$

not permissible | $V = 6 \cdot 6 \cdot 5 = 180\ in.^3$ .Ans

4. The height in feet of an object reaches in t seconds after it is thrown upward with an initial speed of v feet per second is given by the formula $h = vt - 16t^2$.
A rocket is fired upward with an initial speed of 208 feet per second. When will the rocket be at a height of 576 feet ?

Solution:

$$h = vt - 16t^2$$
$$576 = 208t - 16t^2$$
$$16t^2 - 208t + 576 = 0$$

Divide each side by 16:

$$t^2 - 13t + 36 = 0$$
$$(t-4)(t-9) = 0$$

$\therefore\ t = 4$ and 9 seconds. Ans.

EXERCISES

Solve by Factoring. If possible, factor out the greatest monomial factor first.

1. $6x^2+x-12=0$ **2.** $3x^2-13x-30=0$ **3.** $10x^2-x-3=0$

4. $10x^2-45x+35=0$ **5.** $24x^2+56x-48=0$ **6.** $80x^2-100x+30=0$

7. $20x^2-60x=80$ **8.** $15x^2=-15x+90$ **9.** $16x^2+48x+32=0$

10. $16t^2-176t+384=0$ **11.** $4.9t^2-39.2t+58.8=0$ **12.** $4.9t^2-49t+102.9=0$

13. The sum of a number and its square is 72. Find the number.
14. If a number is added to its square, the result is 156. Find the number.
15. If a number adds its square, the result is 132. Find the number.
16. If a number is subtracted from its square, the result is 110. Find the number.
17. Find two consecutive positive odd integers whose product is 143.
18. Find two consecutive positive even integers whose product is 168.
19. Find two consecutive negative integers whose product is 195.
20. The sum of the squares of two consecutive even integers is 164. Find the integers.
21. Find three consecutive integers whose sum is 36.
22. Find three consecutive even integers such that the sum of the product of the first two and the product of the last two is 392.
23. Find three consecutive positive integers such that the sum of the squares of the first two is 32 larger than the square of the third.
24. Find the dimensions of a rectangle having area $84\,m^2$ and whose length is $5\,m$ longer than its width.
25. A rectangular room has an area of $240\,ft^2$. Its width is shorter than its length by $8\,ft$. What is the length and the width ?
26. A box has a square bottom, and the top is $6\,cm$ high. Its total surface area is $378\,cm^2$. Find its dimensions and volume.
27. The height of a triangle is $12\,cm$ longer than the base. The area is $54\,cm^2$. Find the lengths of its base and height. The area of a triangle is $A=\frac{1}{2}bh$.
28. The height of a triangle is $10\,m$ longer than twice its base. Its area is $176\,m^2$. Find the lengths of its base and height. The area of a triangle is $A=\frac{1}{2}bh$.
29. The length of a rectangle is $3\,m$ longer than the width. If its length and width are both increased by $4\,m$, its area is increased by $76\,m^2$. Find its original dimensions.
30. A rectangle with its dimensions $5\,m$ by $12\,m$ is surrounded by a path of uniform width. How wide will this path be if the area of the rectangle is the same as the area of the path ?

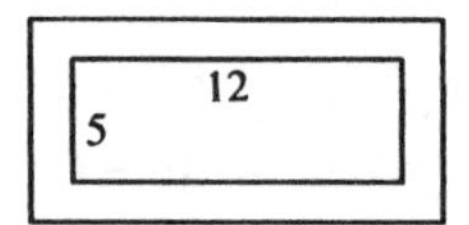

31. A ball is thrown upward with an initial speed of 160 feet per second. When will the ball be at a height of 336 feet ? The formula is $h=vt-16t^2$, where v is the initial speed and t is the time in seconds.
32. In question **31**, how many seconds will it take to reach the ground again ?

CHAPTER 8 EXERCISES

Simplify each expression.

1. 4^3
2. $(-4)^3$
3. -4^3
4. $(-4)^2$
5. $(4^2)^5$
6. $4^2 \cdot 4^5$
7. 7^0
8. 8^0
9. $(-1)^{15}$
10. $(-1)^{18}$
11. $(x^3)^4$
12. $x^3 \cdot x^4$
13. $(a^5)^4$
14. $a^5 \cdot a^4$
15. $a^9 \cdot a^{-6}$
16. $a^{-9} \cdot a^6$
17. $\frac{4^5}{4^2}$
18. $\frac{4^2}{4^5}$
19. $\frac{a^6}{a^9}$
20. $\frac{x^9}{x^6}$
21. $4x + 7x$
22. $4x - 7x$
23. $(4x)(7x)$
24. $(2a^2)(3a^4)$
25. $(2x)^3$
26. $(3x)^2$
27. $(3a^2b)(2ab)$
28. $(-xy)^2(xy^3)$
29. $(2x^2)^3(3x)^2$
30. $(\frac{2}{3}a^3)^2$
31. $(\frac{1}{3}n)^2(3n^2)^2$
32. $(-3p)^3(-2p)^2$
33. $\left(\frac{2}{5}\right)^2$
34. $\left(\frac{2}{5}\right)^{-2}$
35. $\frac{25\times10^5}{5\times10^2}$
36. $\frac{4x}{8}$
37. $\frac{4x}{8x^2}$
38. $\frac{5a^3b^4}{25ab^2}$
39. $\frac{12x^5y^2}{3x^3y^6}$
40. $\left(\frac{3n^2}{p^2}\right)^4$

Multiply by using the Distributive Property.

41. $4(x+6)$
42. $5(-x+3)$
43. $2x(x-3)$
44. $-a(3a-4)$
45. $2p(p-2p^2)$
46. $2(x+3)-4(x-4)$
47. $2x(x-4)+3x^2$
48. $(3a+4)(3a-4)$
49. $(x+4)(x-2)$
50. $(2a-5)(a+6)$

Multiply by using the FOIL method or the formulas for special patterns.

51. $(x+7)(x-3)$
52. $(a-5)(a+9)$
53. $(2a+3)^2$
54. $(2a-3)^2$
55. $(2a+3)(2a-3)$
56. $(2x-5)(4x-1)$
57. $(x-10)(x+10)$
58. $(c-3d)(c+3d)$
59. $(2x-y)^2$
60. $(3x+2y)^2$
61. $(3n+5)(4n-6)$
62. $(3a-4b)^2$
63. $(3p-3)(2p-5)$
64. $(3x+5)(3x+5)$
65. $(3x-5)(3x+5)$
66. $(a^2-b)(a^2+b)$
67. $(a^2+b)^2$
68. $(a-b^2)^2$
69. $(x^n+y^n)^2$
70. $(x^3+y^2)(x^3-y^2)$

Factoring each polynomial.

71. $2x-18$
72. $4x^2-12x$
73. $6x^2y+14xy^3$
74. $12x^5+10x^3$
75. $3x^3-6x^2-9x$
76. $2x^3y+6x^2y^2+4xy^3$
77. $2x(x-2)+3(x-2)$
78. $x(x-3)-2(3-x)$
79. $xy+3x-2y-6$
80. $x^2-6y+2xy-3x$

Factor. If it is not factorable, Write *prime*.

81. $x^2 - 7x + 10$
82. $x^2 + 3x - 10$
83. $x^2 + 3x + 10$
84. $a^2 + 4a - 45$
85. $p^2 - 5p + 6$
86. $p^2 - 5p - 6$
87. $y^2 - 13y + 40$
88. $x^2 + 39x - 40$
89. $-x^2 + 2x + 24$
90. $x^2 - xy - 2y^2$
91. $a^2 - 2ab - 3b^2$
92. $p^2 - 8pq - 20q^2$
93. $x^2 + 2xy - 48y^2$
94. $4x^2 + 4x - 15$
95. $12x^2 - 13x + 3$
96. $5x^2 + 3x - 4$
97. $2a^4 - 5a^2 - 3$
98. $4x^4 + 3x^2 - 10$
99. $4x^3 + 4x^2 - 3x$
100. $9y^3 + 39y^2 - 30y$

Factor by using special quadratic patterns.

101. $x^2 + 8x + 16$
102. $4x^2 - 12x + 9$
103. $a^2 - 18a + 81$
104. $4y^2 - 9$
105. $4a^2 - 4ab + b^2$
106. $16x^2 - 40xy + 25y^2$
107. $9x^2 - 4y^2$
108. $4a^2b^2 - c^2$
109. $25p^2 - 20pq + 4q^2$
110. $4x^2 - 20xy + 25y^2$
111. $x^4 - 8x^2 + 16$
112. $16n^2 - 40n + 25$
113. $x^2 + 6x + 9 - 4y^2$
114. $a^2 - b^2 + 4b - 4$
115. $a^{4n} - 9$
116. $a^6 - 10a^3 + 25$
117. $x^4 + 8x^2y + 16y^2$
118. $12x^3 - 27x$
119. $a^2(a+4) - 9(a+4)$
120. $2x(x^2 - 4) + 3(4 - x^2)$

Solve by factoring.

121. $(x+6)(x-12) = 0$
122. $(3x-6)(2x+1) = 0$
123. $2n(n-4)(2n-5) = 0$
124. $x^2 - 5x - 14 = 0$
125. $a^2 + 10a + 21 = 0$
126. $2x^2 - 5x - 3 = 0$
127. $5x^2 + 8x - 4 = 0$
128. $x^2 - 14x + 49 = 0$
129. $4x^2 + 12x + 9 = 0$
130. $9a^2 - 16 = 0$
131. $x^3 - 3x^2 - 18x = 0$
132. $2n^3 - 18n = 0$
133. $(x+4)(x-6) = -9$
134. $x(3x-1) = x + 5$
135. $2x(x+2) = 3(3x-1)$

136. Find an equation in standard form that has the given roots (solutions) 1 and -3.
137. Find an equation in standard form that has the given roots (solutions) -2 and $\frac{2}{3}$.
138. If a number is subtracted from its square, the result is 132. Find the number.
139. Find two consecutive positive integers whose product is 182.
140. Find three consecutive positive integers such that the sum of the squares of the first two is 45 larger than the square of the third.
141. In a rectangle, the length is twice as long as the width. If its length and width are both increased by $4\,cm$, its area is increased by $100\ cm^2$. Find the original dimensions.
142. Find the dimensions of a rectangle having an area $105\ cm^2$ and whose length is $8\ cm$ longer than its width ?
143. A ball is thrown upward with an initial speed of 80 feet per second. The height (h) of the ball is given by the formula $h = vt - 16t^2$, where v is the initial speed and t is the time in seconds. How many seconds will it take to reach the ground again ?
144. In question **143**, what will be the maximum height the rocket reaches ?

Newton's Law of Free-Falling Objects

Issac Newton (1642-1727, Creator of Calculus)

1. The distance (d) in feet that an object falls in t seconds is given by the formula $d = 16t^2$. Find the distance of an object falls in 5 seconds.
Solution:

$$d = 16t^2 = 16 \cdot 5^2 = 16 \cdot 25 = 400 \text{ feet. Ans.}$$

2. The height (h) in feet that an object falls from its initial height h_0 in t seconds is given by the formula $h = h_0 - 16t^2$.

a) If the initial height of a ball is 1,000 feet, find the height of the ball drops in 5 seconds.

b) How many seconds will it take the ball to reach the ground ?

Solution:

a) $h = h_0 - 16t^2 = 1000 - 16 \cdot 5^2 = 1000 - 16 \cdot 25 = 1000 - 400 = 600$ feet. Ans.

b)
$$h = 1000 - 16t^2$$
$$0 = 1000 - 16t^2$$
$$16t^2 = 1000$$
$$t^2 = \frac{1000}{16} \qquad \therefore t = \sqrt{\frac{1000}{16}} = \sqrt{62.5} \approx 7.91 \text{ seconds. Ans.}$$

3. A rocket is fired upward with an initial speed of 160 feet per second. The height (h) of the rocket is given by the formula $h = vt - 16t^2$, where v is the initial speed and t is the time in seconds.

a) Find the height of the rocket reaches in 3 seconds after being fired upward.

b) When will the rocket be at a height of 384 feet ?

c) How many seconds will it take to reach the ground again ?

d) What will be the maximum height the rocket reaches ?

Solution:

a) $h = vt - 16t^2 = 160(3) - 16(3)^2 = 480 - 144 = 336$ feet. Ans.

b)
$$h = vt - 16t^2$$
$$384 = 160t - 16t^2$$
$$16t^2 - 160t + 384 = 0$$
Divide each side by 16:
$$t^2 - 10t + 24 = 0$$
$$(t-4)(t-6) = 0$$
$\therefore t =$ 4 and 6 seconds. Ans.

c)
$$h = vt - 16t^2$$
$$h = 160t - 16t^2$$
Let $h = 0$
$$0 = 160t - 16t^2$$
$$16t^2 - 160t = 0$$
Factor out the GCF:
$16t(t-10) = 0 \quad \therefore t = 10$ seconds. Ans.

d) It is the time halfway between the starting time and the time to reach the ground again. Therefore, it reaches the maximum height when $t = 5$.
$$\therefore h = vt - 16t^2 = 160(5) - 16(5)^2 = 800 - 400 = 400 \text{ feet. Ans.}$$

-----Continued-----

4. A ball is thrown upward from the top of a tower 96 feet high with an initial upward speed of 80 feet per second. The height of the ball is given by the formula $h = 96 + vt - 16t^2$, where v is the initial speed and t is the time in seconds. How many seconds will it take the ball to reach the ground again ?

Solution:

$$h = 96 + vt - 16t^2$$
$$h = 96 + 80t - 16t^2$$
Let $h = 0$
$$0 = 96 + 80t - 16t^2$$
$$16t^2 - 80t - 96 = 0$$
$$t^2 - 5t - 6 = 0$$
$$(t+1)(t-6) = 0$$
$\therefore t = 6$ seconds. Ans.

5. An arrow is shot upward with an initial speed of 58.8 meter per second. The height of the arrow is given by the formula $h = vt - 4.9t^2$, where v is the initial speed and t is the time in seconds. How many seconds will it take the arrow to reach the ground again ? What will be the maximum height the arrow reaches ?

Solution:

$$h = vt - 4.9t^2 = 58.8t - 4.9t^2$$
Let $h = 0$, $58.8t - 4.9t^2 = 0$
$$4.9t(12 - t) = 0$$
$\therefore t = 12$ seconds. Ans.

$\therefore$ Max. height $h = 58.8(6) - 4.9(6)^2$
$= 352.8 - 176.4 = 176.4$ meters. Ans.

6. A rocket is fired upward from the top of a tower 20 feet. The height of the rocket after t seconds is given by the function $h(t) = c - (d - 2t)^2$, where c and d are positive numbers. The rocket reaches the maximum height of 120 feet at $t = 5$. What will be the height, in feet, of the rocket at $t = 1$?

Solution:

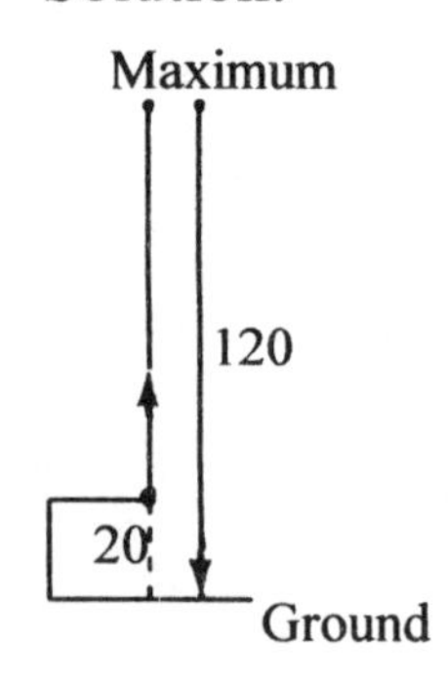

$$h(t) = c - (d - 2t)^2$$

First, we need to find the values of c and d from the given information.

At $t = 0$, $h = 20$, $\quad h(0) = c - d^2 = 20$

At $t = 5$, $h = 120$, $\quad h(5) = c - (d - 10)^2 = 120$

$$c - (d^2 - 20d + 100) = 120$$

We have: $c - d^2 + 20d - 100 = 120$ and $c - d^2 = 20$

$$20 + 20d - 100 = 120$$
$$20d = 200 \quad \therefore d = 10$$
$$c - d^2 = 20, \quad c - 100 = 20 \quad \therefore c = 120$$

We have the function:

$$h(t) = 120 - (10 - 2t)^2$$

At $t = 1$, $h(1) = 120 - (10 - 2)^2 = 56$ feet. Ans.

Additional Examples

1. A rectangular field has an area of 1925 square meters. Its perimeter is 180 meters. What is the length of the rectangular field ?

Solution:

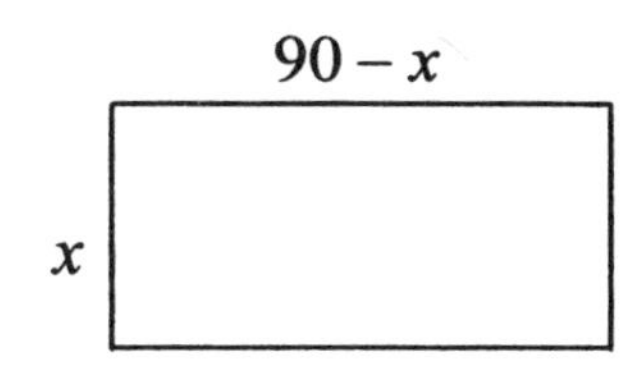

The equation is: $x(90-x)=1925$

$$90x - x^2 = 1925$$
$$x^2 - 90x + 1925 = 0$$
$$(x-35)(x-55)=0$$
$$x = 35 \text{ meters. (width)}$$
$$\therefore 90 - x = 90 - 35 = 55 \text{ meters. (length)} \quad \text{Ans.}$$

2. The dimensions of a rectangular field are 10 meters by 13 meters. If the length and width are both increased by x meters.

a) Find the expression to represent its area after the dimensions are increased.

b) Find the expression to represent how much the area is increased.

Solution:

a) $(10+x)(13+x) = x^2 + 23x + 130$. Ans.

b) $(10+x)(13+x) - (10\times 13) = x^2 + 23x + 130 - 130 = x^2 + 23x$. Ans.

3. Factor $x^4 + x^2y^2 + y^4$.

Solution:

$$x^4 + x^2y^2 + y^4 = x^4 + 2x^2y^2 + y^4 - x^2y^2$$
$$= (x^2+y^2)^2 - x^2y^2$$
$$= (x^2+y^2+xy)(x^2+y^2-xy). \quad \text{Ans.}$$

4. $(999998)^2 - 2(999994)^2 + (999990)^2$

$= (x+4)^2 - 2x^2 + (x-4)^2$

$= x^2 + 8x + 16 - 2x^2 + x^2 - 8x + 16$

$= 32$. Ans.

10. $(1234567)^2 - 2(1234565)^2 + (1234563)^2$

$= (x+2)^2 - 2x^2 + (x-2)^2$

$= x^2 + 4x + 4 - 2x^2 + x^2 - 4x + 4$

$= 8$. Ans.

5. Find the value of k that will make the following statement a true statement.

$(kx-2)^2 = k^2x^2 - 32x + 4$

Solution:

$$(kx-2)^2 = k^2x^2 - 32x + 4$$
$$\cancel{k^2x^2} - 4kx + \cancel{4} = \cancel{k^2x^2} - 32x + \cancel{4}$$
$$-4kx = -32x$$

Comparing the coefficient:

$$4k = 32 \quad \therefore k = 8. \quad \text{Ans.}$$

Space for Taking Notes

CHAPTER NINE

Inequalities and Absolute Values

9-1 Inequalities on the Number Line

An **inequality** is formed by placing an inequality sign or symbol ($>$, $<$, $\leq$, $\geq$) between numerical or variable expressions.

$5 > 3$ is read " 5 is greater than 3."

$3 < 5$ is read " 3 is less than 5."

$x > -4$ is read " x is greater than -4."

An inequality containing a variable is called **an open sentence.**

The inequality $x > -4$ indicates that all numbers greater than -4 are solutions of x.

$x \geq -3$ is read "x is greater than or equal to -3."

$x \leq 2$ is read "x is less than or equal to 2."

$-3 < x < 3$ is read "x is greater than -3 and less than 3."

$-3 \leq x < 2$ is read "x is greater than or equal to -3 and less than 2."

We can graph an inequality on a number line.

On a number line, the number on the right is greater than the number on the left.

To graph an inequality on a number line, we use a **close circle** " • " to show "included", and use an **open circle** " o " to show "not included".

$x > -4$

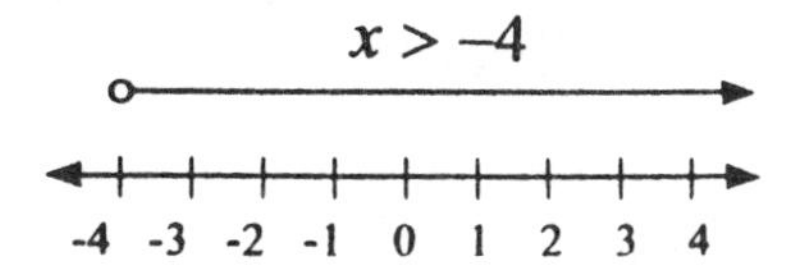

$-3 \leq x < 2$

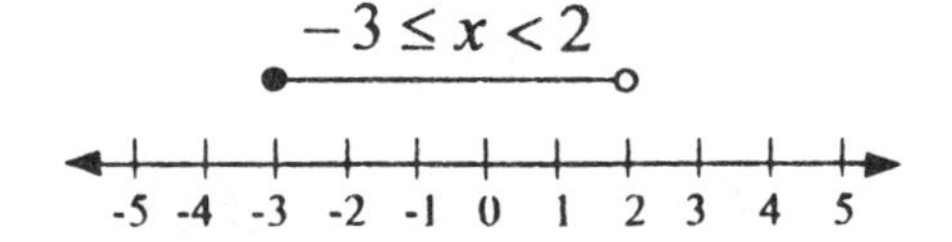

Properties for inequalities:

1. Adding or subtracting the same number to or from each side of an inequality does not change the direction (order) of its inequality sign.
2. Multiplying or dividing each side of an inequality by the same positive number does not change the direction (order) of its inequality sign.
3. Multiplying or dividing each side of an inequality by the same negative number reverses (changes) the direction (order) of its inequality sign.

Examples

1. Add 6 to each side of $5 > 2$.
Then write a true inequality.
Solution:

$$5 > 2$$
$$5 + 6 > 2 + 6$$
$$11 > 8 \text{ (true).}$$

2. Multiply each side of $5 > 2$ by -6.
Then write a true inequality.
Solution:

$$5 > 2$$
$$5(-6) < 2(-6)$$
$$-30 < -12 \text{ (true).}$$

EXERCISES

Compare each pair of numbers, using the symbols <, >, =.

1. 6 and 4 **2.** 3 and 8 **3.** –3 and 5 **4.** –3 and –5
5. –7 and –4 **6.** 4.5 and 6 **7.** –12 and –12 **8.** –4.2 and –4.5
9. –18 and –24 **10.** –32 and –15

Classify each statement as true or false.

11. $7 < 5$ **12.** $9 > 5$ **13.** $-5 > -4$ **14.** $-3 > -8$
15. $-5 \geq -5$ **16.** $-2 < 3 < 9$ **17.** $-3 < -6 < 8$ **18.** $|-0.36| < 0.24$
19. $|5-8| > |8-6|$ **20.** $|5-8| \leq |8-5|$

Perform the indicated operation on each side of the inequality. Then write a true inequality.

21. $6 < 9$, add 4 **22.** $6 < 9$, subtract 4 **23.** $6 < 9$, multiply by 4
24. $6 < 9$, divide by 4 **25.** $6 < 9$, multiply by –4 **26.** $6 < 9$, divide by –4
27. $-5 < 6$, multiply by –2 **28.** $2 > -3$, multiply by –5 **29.** $-4 < 6$, divide by 2
30. $-6 > -9$, divide by –3 **31.** $-6 > -9$, multiply by –3 **32.** $-6 > -9$, divide by 3
33. $2x > 4$, divide by 2 **34.** $-2x > 4$, divide by –2 **35.** $\frac{1}{2}x \leq 4$, multiply by 2
36. $-\frac{1}{2}x \leq 4$, multiply by –2 **37.** $\frac{2}{3}x > 6$, multiply by $\frac{3}{2}$ **38.** $-\frac{2}{3}x \leq 6$, multiply by $-\frac{3}{2}$
39. $-4x \geq 6$, divide by –4 **40.** $-4x \leq -6$, divide by –4

Translate each statement into symbols.

41. 8 is less than 15.
42. –8 is greater than –15.
43. –7 is less than –2.
44. –7 is between –9 and 2.
45. –2 is greater than –7 and less than 3.
46. –7 is greater than –10 and less than –2.
47. x is greater than 15.
48. x is less than or equal to 15.
49. n is greater than –8 and less than 2.
50. n is greater than 4 and less than or equal to 9.
51. The absolute value of n is less than 5.
52. The absolute value of x is greater than 7.
53. The absolute value of x is greater than or equal to 7.
54. The absolute value of n is less than or equal to 5.

Graph each inequality on a number line.

55. $x < -1$ **56.** $x \geq 4$ **57.** $x \geq -3$ **58.** $-5 < n < 3$
59. $-4 < n \leq 2$ **60.** $x < -3$ or $x \geq 3$ **61.** $-3 < x < 3$ **62.** $-2 \leq x \leq 5$

9-2 Solving Inequalities in One Variable

To solve an inequality, we use similar methods which are used to solve equations. Simply transfer all variables to one side of the inequality (usually to the left side), and transfer the others to the right side. Reverses (changes) the signs (+, –) in the process. Reverses the direction (order) of the inequality sign if we **multiply** (or **divide**) each side by **the same negative number**.

$$-\tfrac{1}{2}x > 4$$

$$(-2)\cdot -\tfrac{1}{2}x < 4\cdot(-2) \text{ (That is, } x < -8\text{)}.$$

If the final statement is a "true statement", the inequality has solutions for all real numbers. If the final statement is a "false statement", the inequality has no solution. (See Ex. 5 & 6)

Examples

1. Solve $4x-1<11$.

Solution:

$4x-1<11$

$4x \quad <11+1$

$4x \quad <12$

$\therefore x<3$. Ans.

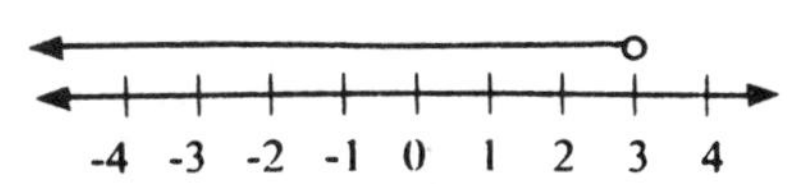

2. Solve $4x-1\geq 11$.

Solution:

$4x-1\geq 11$

$4x \quad \geq 11+1$

$4x \quad \geq 12$

$\therefore x\geq 3$. Ans.

3. Solve $-4x-1<11$.

Solution:

$-4x-1<11$

$-4x \quad <12$

Divide each side by –4:

$\therefore x>-3$. Ans.

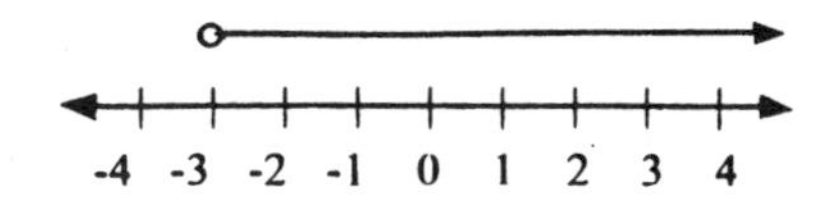

4. Solve $-\frac{1}{4}x-1\geq 11$.

Solution:

$-\frac{1}{4}x-1\geq 11$

$-\frac{1}{4}x \quad \geq 12$

Multiply each side by –4:

$\therefore x\leq -48$. Ans.

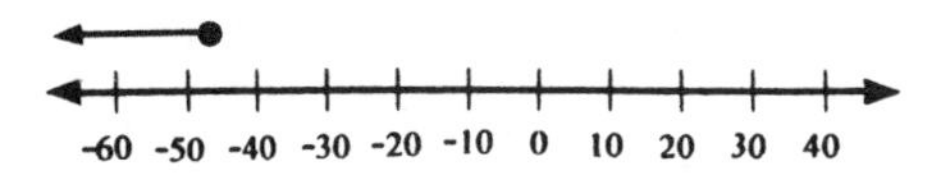

5. Solve $3x<3(x+4)$.

Solution:

$3x<3(x+4)$

$3x<3x+12$

$3x-3x<12$

$0<12$ (true)

The inequality is true for all real numbers. Ans.

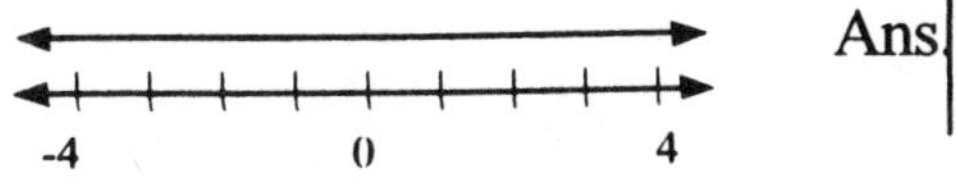

6. Solve $3x>3(x+4)$.

Solution:

$3x>3(x+4)$

$3x>3x+12$

$3x-3x>12$

$0>12$ (false)

The inequality has no solution. Ans.
The solution set is ϕ (empty).

EXERCISES

Solve each inequality. Graph each solution on a number line.

1. $x-3<8$ **2.** $x+3>8$ **3.** $x+4\geq -5$
4. $x-5\leq -2$ **5.** $a-4>7$ **6.** $n-3\geq -9$
7. $y+7<-3$ **8.** $y-7\leq -3$ **9.** $p-5\leq -8$
10. $k-4>0$ **11.** $x+5\geq 0$ **12.** $v-4<-1$
13. $2x<8$ **14.** $-2x<8$ **15.** $-2x\geq 8$
16. $2x>-8$ **17.** $-2x\leq -8$ **18.** $\frac{1}{2}x\geq 8$
19. $-\frac{1}{2}x\geq 8$ **20.** $-\frac{1}{3}n\leq -1$ **21.** $-\frac{1}{5}n>-2$
22. $8p<-24$ **23.** $7a>-21$ **24.** $-8a<-24$
25. $-21a>7$ **26.** $-24p\leq 8$ **27.** $-15y\geq 15$
28. $\frac{1}{5}x<-2$ **29.** $\frac{1}{4}x>5$ **30.** $\frac{1}{6}x\leq -2$
31. $-\frac{2}{3}x\geq 4$ **32.** $-\frac{1}{4}x>5$ **33.** $-\frac{2}{3}x\leq 6$
34. $-\frac{4}{5}p<16$ **35.** $\frac{5}{6}p>15$ **36.** $\frac{3}{10}a<-12$
37. $-\frac{3}{7}k\geq -9$ **38.** $-\frac{2}{5}k\leq -1$ **39.** $\frac{3}{4}x>\frac{3}{2}$
40. $-\frac{3}{8}x<\frac{3}{2}$ **41.** $-\frac{3}{5}x\geq \frac{1}{5}$ **42.** $-\frac{5}{6}x\leq \frac{5}{9}$
43. $2x-5>3$ **44.** $2x+5>3$ **45.** $-2x+5>3$
46. $4x-6<10$ **47.** $-4x-6\leq 10$ **48.** $-4x+6\geq -10$
49. $\frac{1}{3}x+2>-4$ **50.** $\frac{1}{4}x+5<-3$ **51.** $\frac{1}{5}x-8\leq 7$
52. $4x<4x+5$ **53.** $4x>4x-5$ **54.** $4x<4(x-3)$
55. $6x>6(x+1)$ **56.** $-5x+8>14$ **57.** $-6x-7<8$
58. $-\frac{1}{2}n+12<10$ **59.** $-\frac{2}{3}n-6\geq 8$ **60.** $-\frac{3}{4}p+4\leq 4$
61. $4x-2<3x$ **62.** $5x+4>-3x$ **63.** $2x-3\leq -3x+1$
64. $x+8\geq 4x-1$ **65.** $3x-4<6x+2$ **66.** $\frac{1}{2}x-8>\frac{3}{4}x+4$
67. $\frac{4}{3}y-9>-1$ **68.** $3(x-2)\leq 5x+6$ **69.** $4(2x-3)\geq 6x+3$
70. $3x-2\leq -2(x+1)+3$ **71.** $2(2x+3)<3(x-1)-1$ **72.** $6x-3(3x+5)>-1$
73. $2-(6-2n)>4n-(n-1)$ **74.** $4p+3(p-2)\geq 4(p+1)$ **75.** $2.5y+2<3y-5$

76. If $a>b$, identify the values of c for which the statement is true: $a+c>b+c$.
77. If $a>b$, identity the values of c for which the statement is true: $ac>bc$.
78. If $a>b$, identity the values of c for which the statement is true: $ac<bc$.
79. If $a>b$, identify whether the statement is true or false: $a^2>b^2$.
80. If $a>b$, identify whether the statement is true or false: $a^3>b^3$.
81. If $a>b>0$, identify whether the statement is true or false: $a^2>b^2$.
82. Identify the values of y for which the statement is true: $3^y>2^y$.

9-3 Solving Combined Inequalities

A **combined inequality** (or **compound inequality**) is an inequality formed by joining two inequalities with "**and**" or "**or**".
There are two forms of combined inequalities, **conjunction** and **disjunction**.

1. **Conjunction of Inequality**

   It is an inequality formed by two inequalities with the word "**and**".
   A conjunction is true when **both inequalities are true**.
   Here is an example:

   $-5 < n$ and $n < 3$
   (or $-5 < n < 3$)

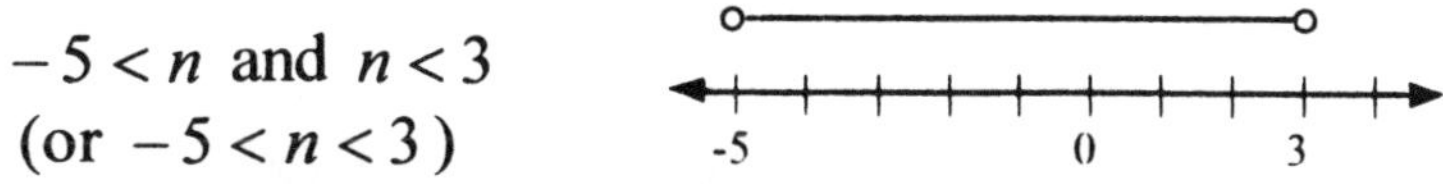

   To solve a conjunction of inequality, we isolate the variable between the two inequality signs. Or, solve each part separately.

2. **Disjunction of Inequality**

   It is an inequality forms by two inequalities with the word "**or**".
   A disjunction is true when **at least one of the inequalities is true**.
   Here is an example:

   $n < -1$ or $n \geq 3$

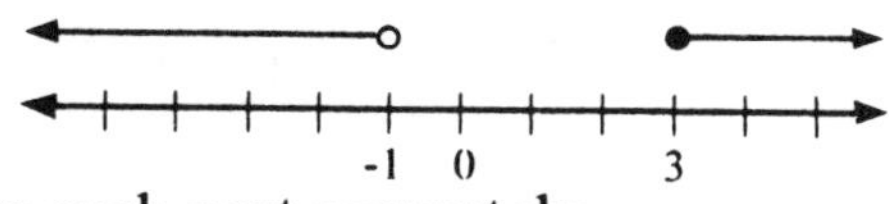

   To solve a disjunction of inequality, we solve each part separately.

Examples

1. Solve $2 < \frac{1}{3}x + 4 \leq 6$.

   Solution:

   Method 1:
   (Solve each part separately.)

   $2 < \frac{1}{3}x + 4$ and $\frac{1}{3}x + 4 \leq 6$

   $-2 < \frac{1}{3}x$ | $\frac{1}{3}x \leq 2$

   $-6 < x$ | $x \leq 6$

   $\therefore -6 < x \leq 6$. Ans.

   Method 2:
   (Solve between the inequality signs.)

   $2 < \frac{1}{3}x + 4 \leq 6$

   Subtract 4 to each expression:

   $-2 < \frac{1}{3}x \leq 2$

   Multiply each expression by 3:

   $\therefore -6 < x \leq 6$. Ans.

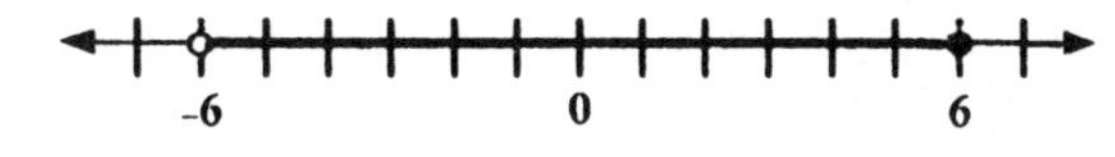

2. Solve $1 - 2x < -3$ or $3x + 14 \leq 2 - x$.

   Solution:

   Solve each part separately:

   $1 - 2x < -3$ or $3x + 14 \leq 2 - x$

   $-2x < -4$ | $4x \leq -12$

   $x > 2$ | $x \leq -3$

   Ans: $x \leq -3$ or $x > 2$.

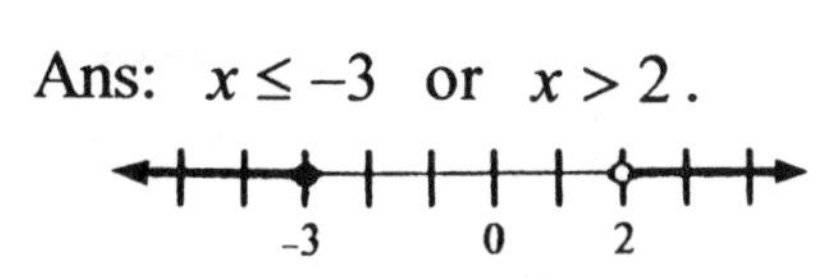

-----continued----

Examples

3. Solve $n-2 \le 2n \le 3n-1$.

Solution:

$n-2 \le 2n$ and $2n \le 3n-1$

$-2 \le n$ $\qquad$ $1 \le n$

It is a conjunction "and" inequality.
Both must be true.

Ans: $n \ge 1$.

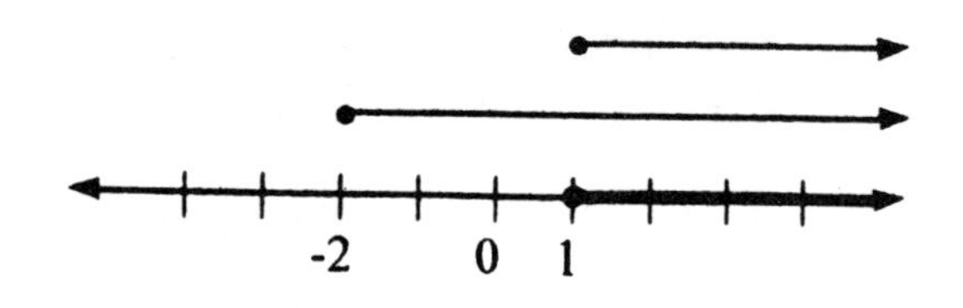

4. Solve $3x-1 \le 2x$ or $3x+10 \ge -2x$.

Solution:

$3x-1 \le 2x$ or $3x+10 \ge -2x$

$x \le 1$ $\qquad$ $5x \ge -10$

$x \ge -2$

It is a disjunction "or" inequality.
At least one is true.

Ans: All real numbers of x.

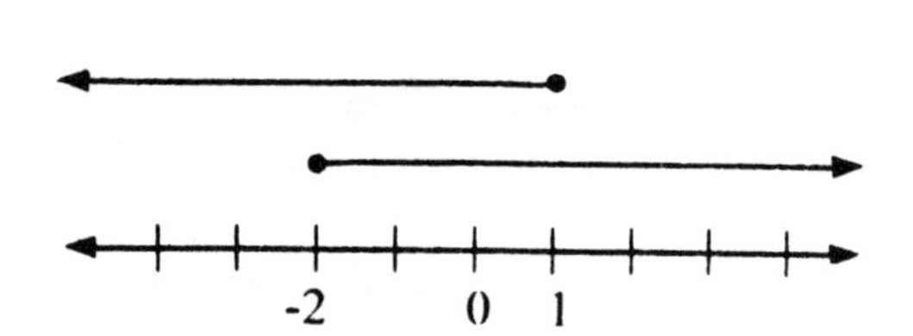

5. Solve $3a+5 < a+7$ and $3-a < 1$.

Solution:

$3a+5 < a+7$ and $3-a < 1$

$2a < 2$ $\qquad$ $-a < -2$

$a < 1$ $\qquad$ $a > 2$

It is a conjunction "and" inequality.
Both must be true.

Ans: No solution.

ϕ (The solution set is empty.)

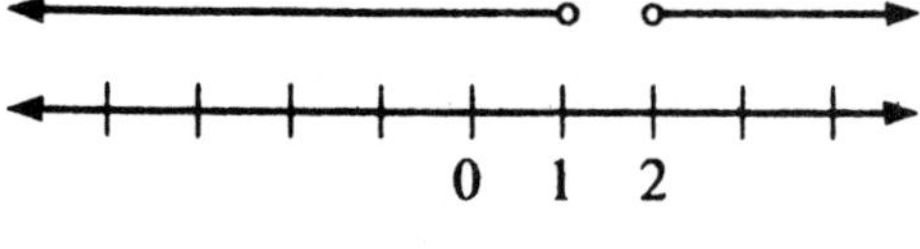

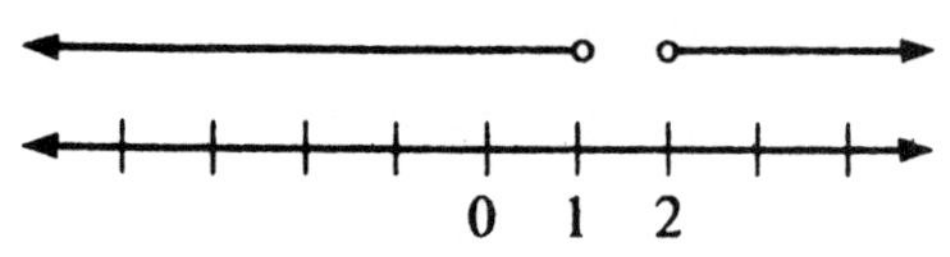

6. Solve $x-1 \le 0$ and $x+2 \ge 0$.

Solution:

$x-1 \le 0$ and $x+2 \ge 0$

$x \le 1$ $\qquad$ $x \ge -2$

It is a conjunction "and" inequality.
Both must be true.

Ans: $-2 \le x \le 1$.

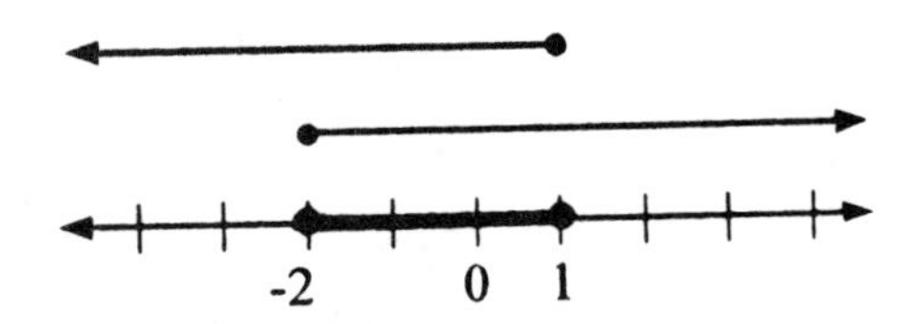

7. Solve $x-1 \le 0$ or $x+2 \ge 0$

Solution:

$x-1 \le 0$ or $x+2 \ge 0$

$x \le 1$ $\qquad$ $x \ge -2$

It is a disjunction "or" inequality.
At least one is true.

Ans: All real numbers of x.

-2 0 1

EXERCISES

Match each inequality with its graph.

1. $x \geq -1$ **2.** $x < 1$ **3.** $x \leq 1$

4. $-1 < x \leq 5$ **5.** $x < -2$ or $x \geq 1$ **6.** $x > -3$ or $x < 4$

7. $-3 < x < 3$ **8.** $-4 \leq x < 2$ **9.** $x \geq -4$ and $x < 2$

10. $x \geq -4$ or $x < 2$ **11.** $x \geq -1$ or $x \geq 1$ **12.** $x < -2$ and $x \geq 1$

A.

−6 −4 −2 0 2 4 6

B.

C.

−6 −4 −2 0 2 4 6

D.

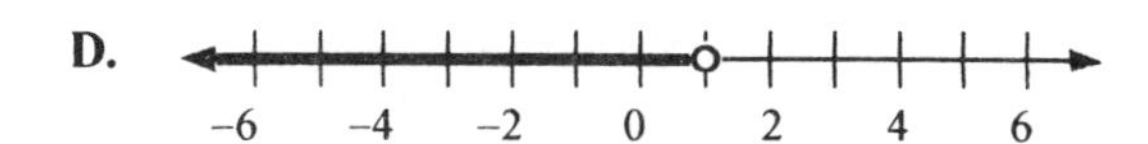

E.

−6 −4 −2 0 2 4 6

F.

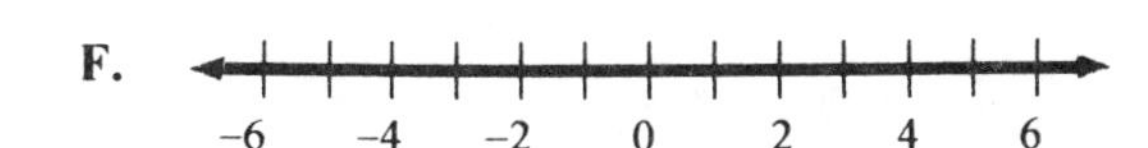

G.

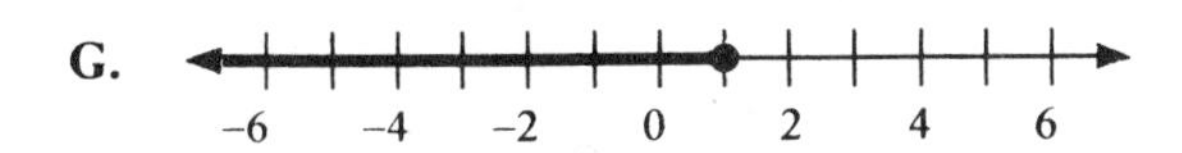

H.

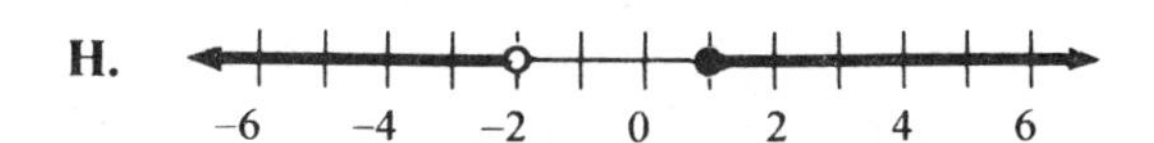

Write each inequality with an equivalent inequality without "and" or "or".

13. $x < 5$ and $x > -2$ **14.** $x = 8$ or $x > 8$ **15.** $x > 1$ and $x \leq 4$

16. $x > -3$ and $x \leq 7$ **17.** $x \geq -5$ and $x \leq -1$ **18.** $x < 9$ or $x = 9$

19. $n \leq 4$ and $n \geq -6$ **20.** $a \leq -4$ and $a > -12$

Solve each inequality and graph its solution set.

21. $x + 3 < -2$ **22.** $x - 3 < -2$ **23.** $3x + 6 \leq 12$

24. $3x - 6 \geq 12$ **25.** $-3x + 6 > -12$ **26.** $-3x - 6 > 12$

27. $3x + 5 > -\frac{3}{2}$ **28.** $-\frac{2}{3}x \leq x - 4$ **29.** $-4 \geq 5 + \frac{x}{2}$

30. $4(2 - a) > -(a - 5)$

-----Continued-----

Graph each inequality on a number line.

31. $-1 \le x \le 5$ **32.** $x < -2$ or $x \ge 1$ **33.** $x \le -2$ or $x > 1$

34. $-4 < x < 2$ **35.** $0 \le a \le 5$ **36.** $-1 < a \le 6$

37. $n \le 0$ or $n \ge 5$ **38.** $n < -3$ or $n \ge 4$ **39.** $x > 1$ or $x = 1$

40. $x > 1$ and $x = 1$

Solve each combined (compound) inequality. Graph its solution set.

41. $2 < x - 5 < 9$ **42.** $-2 \le x + 5 < 9$

43. $0 < a + 4 \le 7$ **44.** $-1 < a - 4 \le 2$

45. $-5 < -2 + n \le 3$ **46.** $1 \le 4 + n < 6$

47. $-4 < -x < 8$ **48.** $4 < -x < -8$

49. $-4 < -x + 5 < 7$ **50.** $-2 \le 2y + 6 \le 8$

51. $-4 \le 3y - 1 < 5$ **52.** $5 < 2b - 1 < 9$

53. $4 < 4p - 4 < 12$ **54.** $-6 < -2x + 2 \le 8$

55. $-7 \le -3x - 1 < 5$ **56.** $2x - 1 > -5$ and $3x < 9$

57. $4x + 2 > -10$ and $3x - 2 < 4$ **58.** $4x + 2 > 10$ and $2x < -10$

59. $4x + 2 > -10$ and $-3x - 2 < 4$ **60.** $4x + 2 > 10$ or $2x < -10$

61. $4x + 2 > -10$ or $3x - 2 < 4$ **62.** $4x + 2 > 10$ or $-2x < 10$

63. $4x + 2 > -10$ or $-3x - 2 < 4$ **64.** $4x + 2 \le -10$ or $2x \ge 10$

65. $4x + 2 \le -10$ and $-3x + 2 < -4$ **66.** $-4x + 2 \le -10$ and $2x \le 10$

67. $n - 4 \le 3n \le 2n + 4$ **68.** $n - 4 < 3n < 4n + 4$

69. $3p + 5 \ge -1$ and $p - 4 < -3p + 4$ **70.** $3p + 5 > -1$ and $-2p + 4 < -p + 1$

71. $3p + 5 \ge -1$ or $p - 4 < -3p + 4$ **72.** $3p + 5 \ge -1$ or $-2p + 4 < -p + 1$

73. $1.5 < 3.2x + 7.9 \le 12.7$ **74.** $-8 \ge \frac{5}{3}x - 3$ or $3x \ge 5$

75. $0.2x + 2.4 < 1.5$ or $0.5x > 1.7$ **76.** $-11 \le \frac{2}{3}x + 1 \le 9$

9-4 Solving Word Problems involving Inequalities

Steps for Solving a word problem involving inequality:

1. Read the problem.
2. Assign a variable to represent the unknown.
3. Determine which inequality sign ($<, >, \le, \ge$) needed.
4. Write an inequality based on the given facts.
5. Solve the inequality to find the answer.
6. Check your answer (usually on a scratch paper).

Examples

1. You want to rent a car for a rental cost of \$50 plus \$30 per day, or a van for \$80 plus \$25 per day. For how many days at most is it cheaper to rent a car ?
Soluion:

Let $d =$ days needed
The inequality is:
$50 + 30d \le 80 + 25d$
$5d \le 30$
$\therefore d \le 6$
Ans: At most 6 days (or less).

Check: $50 + 30d = 50 + 30(6) = \$230$ ✓
$80 + 25d = 80 + 25(6) = \$230$ ✓
The rental costs are the same for 6 days. It is cheaper to rent the car for less than 6 days. It is cheaper to rent the van for more than 6 days.

2. Your scores on your 4 math tests were 71, 74, 75 and 86. What is the lowest score you need in your next test in order to have an average score of at least 80 ?
Solution:

Let $x =$ score for next test
The inequality is:
$\frac{71 + 74 + 75 + 86 + x}{5} \ge 80$, $\frac{306 + x}{5} \ge 80$, $306 + x \ge 400$ $\therefore x \ge 94$

Ans: 94 (or more).

3. John wants to rent a car and pay at most \$250. The car rental costs \$150 plus \$0.25 a mile. How far can he travel ?
Solution:

Let $x =$ mileage he can travel
The inequality is:
$150 + 0.25x \le 250$
$0.25x \le 100$ $\therefore x \le 400$ Ans: 400 miles or less.

4. The sum of two consecutive integers is less than 79. Find the integers with the greatest sum.
Solution: Let $n = 1^{st}$ integer. $n + 1 = 2^{nd}$ integer.
The inequality is:
$n + (n + 1) < 79$, $2n < 78$, $\therefore n < 39$
The greatest integer in n is 38. Ans. 38 and 39.

EXERCISES

Solve each inequality. Graph each solution on a number line.

1. $x-4<9$ **2.** $x+4<9$ **3.** $2x+5\geq 97$

4. $2n-8\leq 78$ **5.** $40+10n\leq 90+5n$ **6.** $20+15n\geq 80+10n$

7. $45+0.25m>150$ **8.** $115+0.15m<220$ **9.** $2w+2(w-5)>70$

10. $2w+2(2w+10)>41$ **11.** $\frac{235+x}{4}\geq 80$ **12.** $\frac{335+x}{5}\geq 85$

13. You want to rent a car and pay at most \$125. The car rental costs \$80 plus \$0.15 a mile. How far can you travel ?

14. You want to rent a car for a rental cost of \$40 plus \$25 per day, or a truck for \$75 plus \$20 per day. For how many days at most is it cheaper to rent a car ?

15. John wants to rent a car for one day. Company A charges \$65 plus \$0.25 a mile. Company B charges \$110 with no mileage fee. How many miles John would need to travel to rent the car from Company A with the less expensive charges ?

16. Your scores on your 2 tests were 75 and 74. What is the lowest score you need in your next test in order to have an average score of at least 80 ?

17. Your scores on your 4 tests were 92, 86, 87 and 90. What is the lowest score you need in next test in order to have an average score of at least 90 ?

18. The sum of two consecutive integers is less than 97. Find the integers with the greatest sum.

19. The sum of two consecutive integers is larger than 97. Find the integers with the smallest sum.

20. The sum of two consecutive integers is no more than 97. Find the integers with the greatest sum. (Hint: "no more than" or "at most" is "$\leq$".)

21. The sum of three consecutive integers is no less than 129. Find the integers with the smallest sum. (Hint: "no less than" or "at least" is "$\geq$".)

22. Find all the sets of three consecutive, positive and odd integers whose sum is less than 23.

23. The width of a rectangle is 12 *cm* shorter than the length. The perimeter is at least 52 *cm*. What is the smallest possible dimensions of the rectangle ?

24. The length of a rectangle is 6 *cm* longer than the width. The perimeter is at least 48 *cm*. What is the smallest possible dimensions of the rectangle ?

25. The length of a rectangle is 6 *cm* longer than the width. The perimeter in no more than 48 *cm*. What is the greatest possible dimensions of the rectangle ?

26. The width of a rectangle is 15 in. shorter than twice the length. The perimeter is at least 45 in.. What is the smallest possible dimensions of the rectangle if each dimension is an integer ?

27. There are quarters and dimes in a box. Their total value is at most \$5.40. Quarters are five times as many as dimes. How many quarters and how many dimes are at most in the box ?

28. At most how many grams of pure salt must be added with 54 grams of water to produce an salt-water solution that is no more than 20% pure salt ?

9-5 Operations with Absolute Values

Absolute Value: The positive number of any real number a. It is written by $|a|$.

If $a > 0$, $|a| = a$. Example: $|5| = 5$.

If $a = 0$, $|a| = |0| = 0$

If $a < 0$, $|a| = -a$. Example: $|-5| = -(-5) = 5$.

The absolute value of a negative number is the opposite of the number.

Absolute value is commonly used as the distance between two points on a number line. The distance between two points is always a positive number. It can never be a negative number. Therefore, we use the absolute value to represent the distance.

A C B

-4 0 3 Number line

$\overline{AB} = |3-(-4)| = |3+4| = |7| = 7$. Or: $\overline{AB} = |-4-3| = |-7| = 7$.

The absolute value of a number is its distance from zero on a number line.

$\overline{BC} = |3-0| = |3| = 3$. $\overline{AC} = |-4-0| = |-4| = 4$.

The domain and range of the absolute-value function

$y = |x|$ is the most basic absolute-value function. Its domain is all real numbers of x.

Its range is all real numbers of $y \geq 0$, or all non-negative real numbers of y.

The graph of the absolute-value function

To graph an absolute-value function, we can plot its ordered pairs.
Or, we can plot its separated equations without absolute-value sign.

Examples

1. Find the domain and range of $y = |x-2|$.

   Solution: D = all real numbers. R = all non-negative real numbers.

2. Find the domain and range of $y = |x| - 2$.

   Solution: D = all real numbers. R = all real numbers of $y \geq -2$.

3. Find the domain and range of $y = -|x-2|$.

   Solution: D = all real numbers. R = all non-positive real numbers.

4. Graph $y = |x|$.

   Solution: **Method 1:**

x	-3	-2	-1	0	1	2	3
y	3	2	1	0	1	2	3

Method 2:

For $x > 0$, $y = x$

For $x < 0$, $y = -x$

For $x = 0$, $y = 0$

$y = -x$ y $y = x$

x

EXERCISES

Evaluate each absolute-value.

1. $|15|$ **2.** $|-15|$ **3.** $|-7.2|$ **4.** $|-\frac{4}{5}|$

5. $|4-5|$ **6.** $|5-4|$ **7.** $|-4-4|$ **8.** $|6-6|$

9. $|12-15|$ **10.** $|15-12|$ **11.** $|1-10|$ **12.** $|0-(-8)|$

13. $|-9+9|$ **14.** $|-9-9|$ **15.** $|-12-(-4)|$

Identify each statement as true or false.

16. $|3|>2$ **17.** $|-3|>2$ **18.** $|-3|<-2$

19. $|-5|<4$ **20.** $|-5|>-4$ **21.** $|0.24|<0.21$

22. $|-0.24|>0.21$ **23.** $|-0.27|<0.19$ **24.** $|-0.35|>0.25$

25. $|-0.29|<0.19$ **26.** $|5-2|\le|2-5|$ **27.** $|5-2|<|2-5|$

28. $|1.2-3|>|2-1.5|$ **29.** $|3.5-5.8|\ge|6.5-5.5|$ **30.** $|\frac{1}{2}-\frac{2}{3}|<|\frac{3}{4}-\frac{1}{4}|$

Find the domain and range of each function.

31. $y=|x+8|$ **32.** $y=|x|+8$ **33.** $y=|x-8|$

34. $y=|x|-8$ **35.** $y=-|x+8|$ **36.** $y=-|x-8|$

37. $y=-|x|+8$ **38.** $y=-|x|-8$ **39.** $y=8|x|$

40. $y=-8|x|$ **41.** $y=-\frac{1}{8}|x|$ **42.** $y=8|x-1|$

43. $y=8|x|-1$ **44.** $y=|x+8|+1$ **45.** $y=-|x+8|+1$

Graph each absolute-value function.

46. $|x|=3$ **47.** $y=-|x|$ **48.** $y=|x-3|$

49. $y=|x+3|$ **50.** $y=-|x-3|$ **51.** $y=|x|+3$

52. $y=|x|-3$ **53.** $y=-|x|+3$ **54.** $y=|x|+x$

55. $y=|x|+3x$

56. Identity the statement as true or false: $|a-b|=|b-a|$.

57. Identify the statement as true or false: $|x+5|=|x|+5$.

58. Identify the value of x for which the statement is true: $|x+5|=|x|+5$.

59. Identify the values of a and b for which the statement is true: $|a|+|b|=|a+b|$.

60. Identify the following statement as true or false:

"The absolute value of every real number is a positive number"

9-6 Absolute-Value Equations and Inequalities

To solve equations and inequalities involving absolute values, we start with the following most basic patterns. Following these patterns, we can solve all inequalities easily.

Rules: On the number line:

$|x| = 5$ means " the distance from x to 0 is 5 ".

$|x - c| = 5$ means " the distance from x to c is 5 ".

$|x + c| = 5$ means " the distance from x to $-c$ is 5 ".

$|2x - c| = 5$ means " the distance from $(2x)$ to c is 5 ". (See Example 4)

These rules are also valid for most absolute-value inequalities ($<, \leq, >, \geq$).

1. Solve $|x| = 5$.

Solution: $|x| = 5$

Ans: $x = 5$ or -5. (or $x = \pm 5$)

-5 0 5

2. Solve $|x| > 5$.

Solution: $|x| > 5$

Ans: $x < -5$ or $x > 5$.

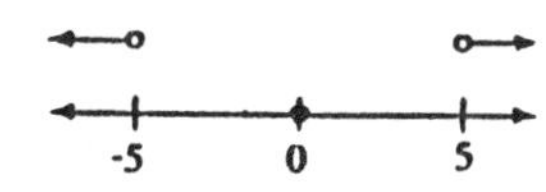

3. Solve $|x| < 5$.

Solution: $|x| < 5$

Ans: $-5 < x < 5$.

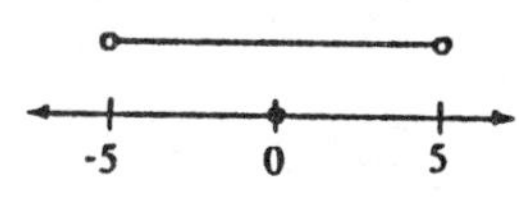

Examples

1. Solve $|x - 2| = 8$.

Solution: $x - 2 = \pm 8$

$x = \pm 8 + 2$

Ans: $x = 10$ or -6.

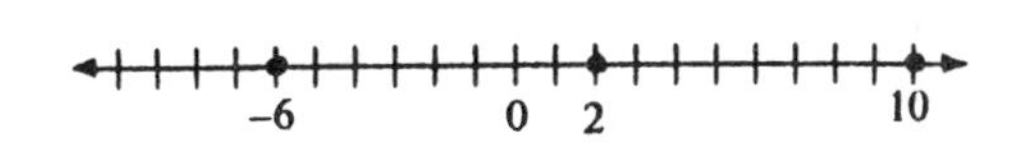

2. Solve $|x - 2| > 8$.

Solution: $x - 2 < -8$ or $x - 2 > 8$

$x < -6$ | $x > 10$

Ans: $x < -6$ or $x > 10$.

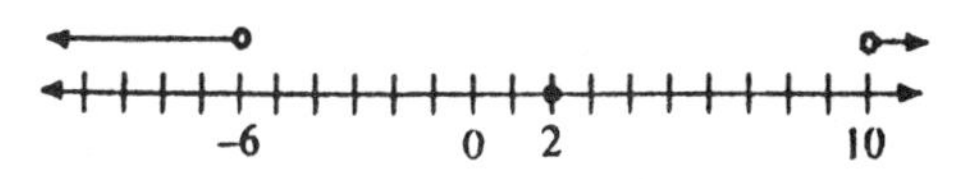

3. Solve $|x + 2| < 8$.

Solution: $-8 < x + 2 < 8$

$-10 < x < 6$. Ans.

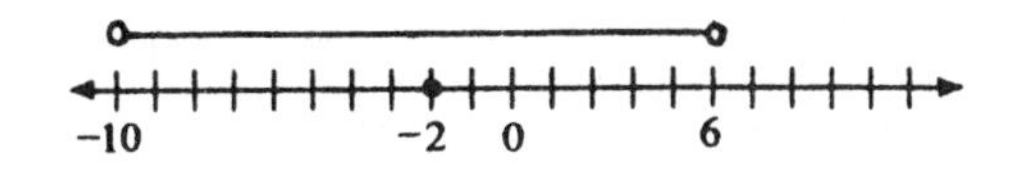

4. Solve $|2x - 7| = 5$.

Solution: $2x - 7 = \pm 5$, $2x = \pm 5 + 7$

$2x = 12$ or 2

Ans. $x = 6$ or 1

2x 2x
0 2 7 12

5. Solve $|3n + 5| - 3 \geq 7$.

Solution: $|3n + 5| \geq 10$

$3n + 5 \leq -10$ or $3n + 5 \geq 10$

$3n \leq -15$ | $3n \geq 5$

$\therefore n \leq -5$ or $n \geq \frac{5}{3}$. Ans.

6. Solve $|20 - 5y| \leq 25$.

Solution: $-25 \leq 20 - 5y \leq 25$

$-45 \leq -5y \leq 5$

$9 \geq y \geq -1$

Ans: $-1 \leq y \leq 9$.

EXERCISES

Solve each equation or inequality involving absolute value.

1. $|x| = 10$

2. $|x| > 10$

3. $|x| < 10$

4. $|x| \ge 10$

5. $|x| \le 10$

6. $|x - 4| = 10$

7. $|x + 4| = 10$

8. $|x - 4| \ge 10$

9. $|x + 4| \le 10$

10. $|5x - 7| = 13$

11. $|5x + 7| < 13$

12. $|6x + 5| < 7$

13. $|6x - 5| > 7$

14. $|4y + 12| + 5 > 9$

15. $|4y + 12| - 5 \le 7$

16. $2|x| = 12$

17. $3|x + 2| = 15$

18. $4|x - 2| > 16$

19. $5|3a - 4| + 1 < 6$

20. $5|2a + 6| - 3 = 7$

21. $|2n - \frac{1}{3}| = \frac{2}{3}$

22. $|p - \frac{2}{3}| \ge \frac{5}{3}$

23. $|\frac{1}{2}x - 2| \ge 2$

24. $|\frac{1}{4}y - 5| \le 7$

25. $|0.5x + 0.25| > 0.75$

26. $|0.5x - 0.25| < 0.75$

27. $|0.25 - 0.5x| \le 0.75$

28. $|-3x| > 6$

29. $|-4x| < 12$

30. $|4 - 2x| < 16$

31. $|5 - 4x| > 7$

32. $|6 - 3x| \ge 9$

33. $|2 - x| \ge 7$

34. $|4 - x| \le 12$

35. $|x| > 0$

36. $|x| < 0$

37. $|x| \ge 0$

38. $|x| \le 0$

39. $|x - 4| > 0$

40. $|7x + 14| > 0$

41. $|5n - 20| \le 0$

42. $|5n - 20| < 0$

43. $1 \le |x| \le 5$

44. $2 < |x + 1| \le 3$

45. $0 < |2 - x| < 4$

46. In a random survey in a city election that 46% of the voters will choose candidate A for mayor. The result of the survey is considered to be accurate within a range of $\pm 5\%$. Write an absolute-value inequality that represents the upper and lower boundaries of the percent (p) of voters who will choose candidate A.

47. In question **46**, find the boundaries (the upper and the lower) of the percents.

48. The normal human body temperature is $98.6^o\ F$ within a range of $\pm 1^o\ F$. Write an inequality that represents the range of normal body temperatures (t).

49. Solve $|x - a| \le b$ for x. Assume a and b are positive numbers.

50. Solve $|x + a| \ge b$ for x. Assume a and b are positive numbers.

CHAPTER 9 EXERCISES

Classify each statement as true or false.

1. $-5 > -3$

2. $-4 < -2$

3. $-2 < 0 < 1$

4. $-7 < -5 > -4$

5. $-1 \leq -1$

6. $-1 \geq -1$

7. $|-4-7| = |-4| + |-7|$

8. $|-4+7| = |-4| + |7|$

9. $|3-9| < |9-4|$

10. $|-17+5| \geq |17-5|$

Translate each statement into symbols.

11. 9 is greater than 1.

12. -9 is less than -1.

13. 2 is between -3 and 7.

14. -4 is greater than -8 and less than 1.

15. x is less than -1.

16. n is greater than or equal to 15.

17. x is greater than 0 and less than 12.

18. n is less than -3 or greater than 3.

19. The absolute value of x is greater than or equal to 25.

20. The absolute value of $x-a$ is less than or equal to b.

Solve each inequality.

21. $x+5 \geq 26$

22. $x-5 \leq 26$

23. $-x+5 \geq 26$

24. $-x-5 \leq 26$

25. $3x-6 > 9$

26. $3x+6 < -9$

27. $5-4x > 17$

28. $8-5x < 3$

29. $\frac{4}{5}x \geq 24$

30. $-\frac{4}{5}x \geq 24$

31. $0.2a > 10$

32. $-0.2a < 20$

33. $3x-6 \leq x+4$

34. $4x+8 \leq 7x-1$

35. $6x+8 > 6x+2$

36. $6(x+2) < 6x$

37. $\frac{4}{5}x+10 < 35$

38. $-\frac{4}{5}x-10 > 35$

39. $5x+4 \geq 2(x-3)-2$

40. $3(x+4) \leq 6(2x-1)-9$

41. $4-(n-3) > 7(n+2)$

42. $3 < a+4 < 10$

43. $-3 \leq a-7 < 8$

44. $0 < 2x-6 < 10$

45. $-6 < -2x < 8$

46. $-1 < -2n-3 \leq 7$

47. $7x+6 > -8$ and $4x-3 < 9$

48. $6x-5 > 13$ or $5x < -25$

49. $7x+6 > -8$ or $4x-3 < 9$

50. $6x-5 > 13$ and $5x < -25$

-----Continued-----

Graph each absolute-value function.

51. $|x| = 2$
52. $|y| = 2$
53. $y = |x+2|$
54. $y = |x-2|$
55. $y = -|x-2|$
56. $y = -|x+2|$
57. $y = |x| - 5$
58. $y = -|x| + 5$
59. $y = |x| - 2x$
60. $y = |2x| - x$

Solve each equation or inequality involving absolute value.

61. $|x| = -5$
62. $|x| = 15$
63. $|x+7| = 12$
64. $|x+7| < 12$
65. $|x+7| > 12$
66. $|2x-8| = 16$
67. $|2x-8| < 16$
68. $|2x-8| > 16$
69. $6|3x+6| = 18$
70. $|7x-6| \geq 1$
71. $|7x-6| \leq 1$
72. $4|5x-2| \leq 32$
73. $|k - \frac{4}{5}| \geq \frac{1}{5}$
74. $|a+0.5| > 5.5$
75. $|1.5x-3| < 2.5$
76. $|-5x| > 15$
77. $|-8x| \leq 16$
78. $|3-5x| < 8$
79. $|4-x| \geq 20$
80. $1 < |x-1| < 5$

81. If $a < b$, identify the values of c for which the statement is true: $ac > bc$.
82. If $a < b < 0$, identify whether the state is true or false: $a^2 < b^2$.
83. Find the domain and range of $y = |x+5|$.
84. Find the domain and range of $y = |x| + 5$.
85. Find the domain and range of $y = -|x+5|$.
86. Solve $|x-a| = b$ for x. Assume a and b are positive numbers.
87. Solve $|x-a| < b$ for x. Assume a and b are positive numbers.
88. Your scores on your 4 test were 74, 82, 77 and 78. What is the lowest score you need in your next test in order to have an average score of at least 80 ?
89. You want to rent vehicle for one day. The rental cost for a car is $45 plus $0.25 per mile. The rental cost for a truck is $60 plus $0.15 per mile. How many miles you would need to travel to rent the car with least expensive cost ?
90. The sum of three consecutive even integers is no less than 218. Find the integers with the smallest sum.
91. The length of a rectangle field is 7 feet longer than the width. The perimeter is no more than 350 feet. What is the greatest possible dimensions of the field ?
92. At most how many grams of pure salt must be added with 84 grams of water to produce an salt-water solution that is no more than 5% pure salt ?
93. A alcohol-water solution is produced with 5% pure alcohol. The acceptable error is $\pm 0.2\%$. Write an absolute-value inequality that represents the acceptable values of percent (p).

Additional Examples

1. Graph $y = |x| + 2$.

   Solution:

   If $x > 0$, then $y = x + 2$
   If $x < 0$, then $y = -x + 2$

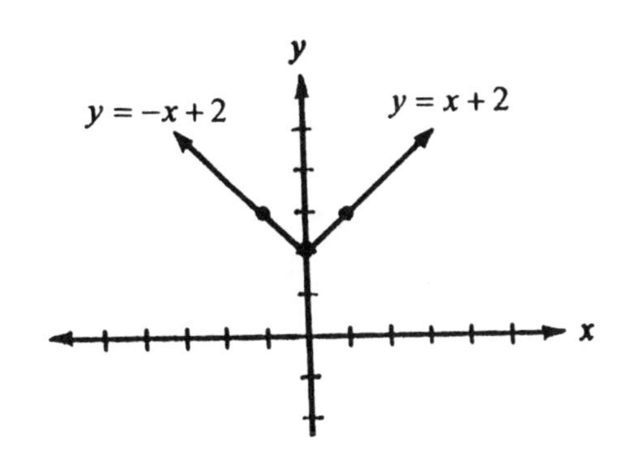

2. Graph $y = |x + 2|$.

   Solution:

   If $(x+2) > 0$, then $y = x + 2$
   We have: For $x > -2$, $y = x + 2$
   If $(x+2) < 0$, then $y = -(x+2)$
   We have: For $x < -2$, $y = -x - 2$

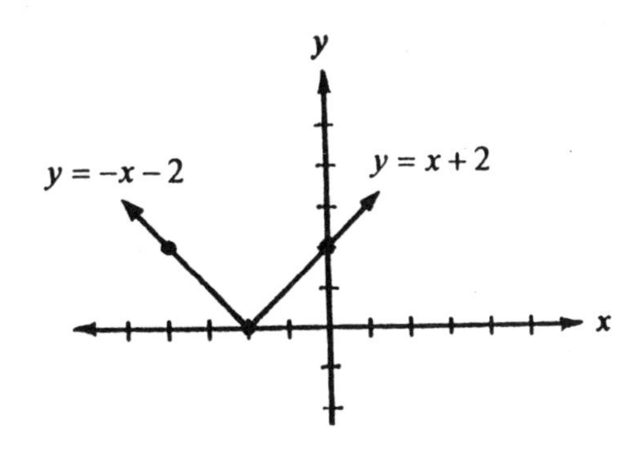

3. Graph $y = |x| - 2x$.

   Solution:

   If $x > 0$, then $y = x - 2x$, $y = -x$
   If $x < 0$, then $y = -x - 2x$, $y = -3x$

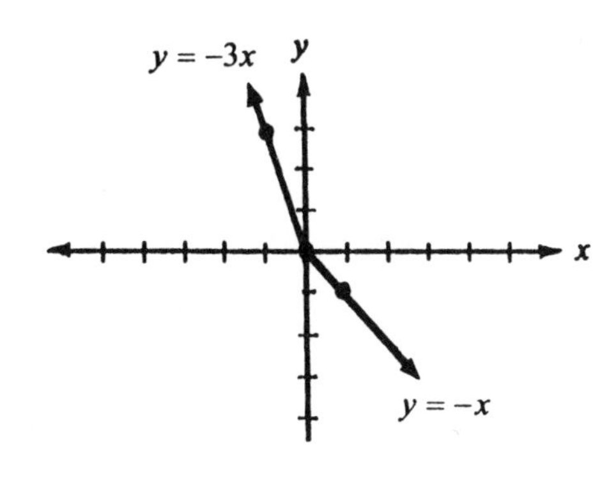

A middleaged single lady told her friend.
Lady: I prefer to marry an archaeologist rather than a doctor or a lawyer.
Friend: Why do you want to marry an archaeologist ?
Lady: If my husband is an archaeologist, the more I get older, the more he is interested in me.

Space for Taking Notes

In a senior house, a couple is watching two pigeons playing at the ground.

Wife: You see. They are kissing each other everyday. They love to be closed to each other.
Husband: Yes, but I notice that the female pigeon is a different one everyday.

Systems of Equations and Inequalities

10-1 Solving Systems by Graphing

The graph of a linear equation $ax + by = c$ is a **straight line.**

The graphs of two linear equations (two straight lines) in the same coordinate plane must show one of the following:

1) Intersect in one point (they are consistent and independent).

2) Intersect in no point (parallel, they are inconsistent).

3) Intersect in unlimited number of points (coincide, they are consistent and dependent).

System of equations: Two or more equations with the same variables.

System of inequalities: Two or more inequalities with the same variables.

A solution to a system of equations in two variables is the intersection point (x, y) that satisfies both equations. A system is also called a system of simultaneous equations. To find the solution of a system, we can graph both equations on the same coordinate plane. To graph a linear equation, we find the x – intercept and the y – intercept. See how to graph a linear equation on Section 7 ~ 5.

Examples

1. Solve by graphing. $\begin{cases} 2x - y = 4 \\ 2x + 5y = 16 \end{cases}$

Solution:

$2x - y = 4$	$2x + 5y = 16$
Let $y = 0$, $2x = 4$, $x = 2$	Let $y = 0$, $2x = 16$, $x = 8$
x – intercept = 2	x – intercept = 8
Let $x = 0$, $-y = 4$, $y = -4$	Let $x = 0$, $5y = 16$, $y = 3\frac{1}{5}$
y – intercept = –4	y – intercept = $3\frac{1}{5}$

Graph the two equations on the same coordinate plane.

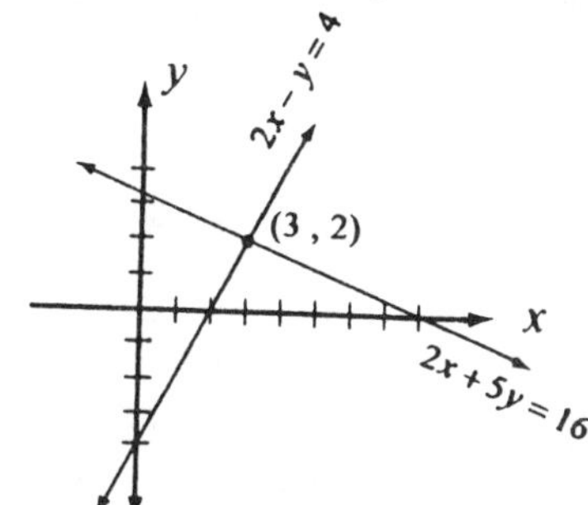

From the graph, we have the solution:

$x = 3$, $y = 2$. Ans.

or (3, 2). Ans.

-----Continued-----

Examples

2. Solve by graphing. $\begin{cases} 4x - 2y = 7 \\ 10x - 5y = 6 \end{cases}$

Solution:

Find the x – intercept and y – intercept of each equation. Graph the two equations. From the graph, they are parallel (no intersection).

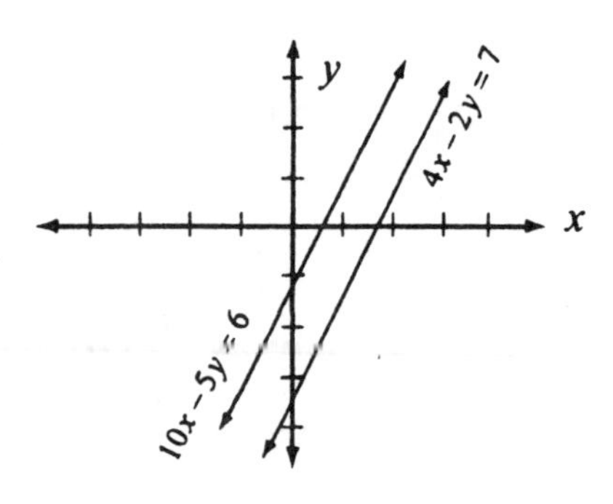

Ans: The system has no solution.

3. Solve by graphing $\begin{cases} 4x + 6y = 9 \\ 12x + 18y = 27 \end{cases}$

Solution:

Find the x – intercept and y – intercept of each equation. Graph the two equations. From the graph, they coincide.

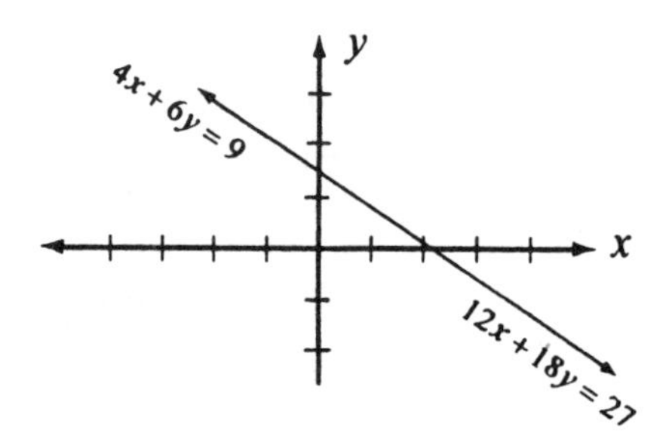

Ans: The system has unlimited number of solutions.

Graphing Linear inequalities

In a coordinate plane, the graph of a linear inequality in two variables is an **open** or **close half-plane**.

For example, the graph of the linear equation $x + 2y = 4$ (a straight line) separates the coordinate plane into two **open half-planes**.

The upper open half-plane is the region of the inequality $x + 2y > 4$.

The lower open half-plane is the region of the inequality $x + 2y < 4$.

$x + 2y \geq 4$ and $x + 2y \leq 4$ are called the **close half-planes**.

$x + 2y = 4$ is the **associated equation** of each linear inequality.

To find out which half-plane region (upper or lower) is the region of each linear inequality, we use the origin (0, 0) to test it. Simply substitute (0, 0) in the inequality and see if (0, 0) is included in the region of the inequality.

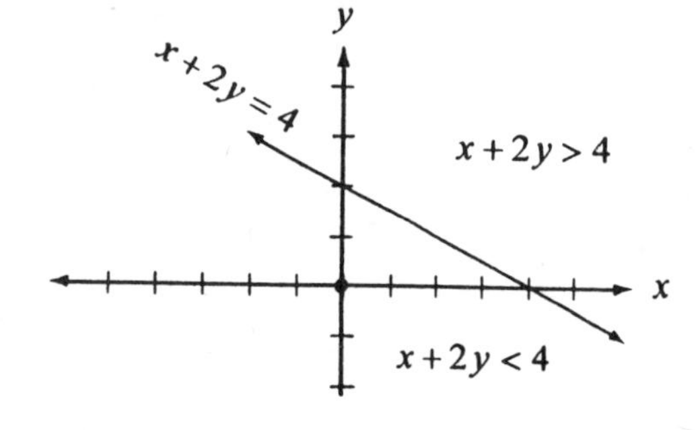

Substitute (0, 0) in $x + 2y < 4$

$0 + 2(0) < 4$

$0 < 4$ (True)

Therefore, (0, 0) is included in the region of $x + 2y < 4$.

-----**Continued**-----

Examples (Hint: Graph the boundary line first.)

4. Graph $4x+5y<20$.
Solution:

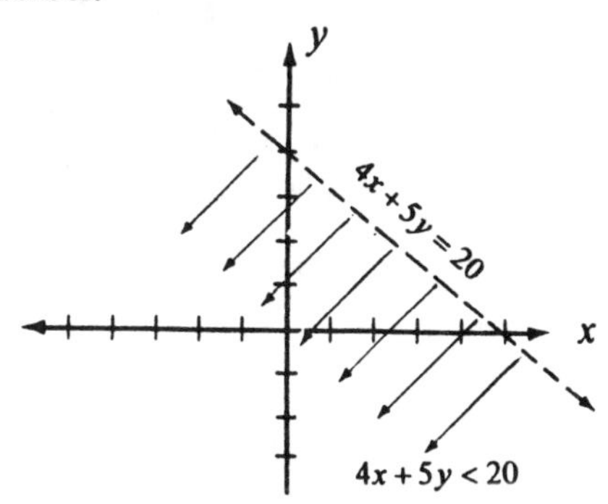

Its graph covers the lower region of the half-plane below the line. The boundary line $4x+5y=20$ is not included. The line is dashed.

5. Graph $4x+5y \geq 20$.
Solution:

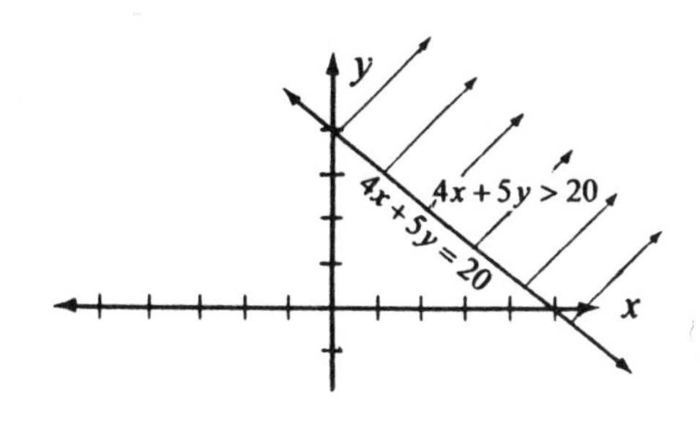

Its graph covers the upper region of the half-plane and the region on the line. The boundary line $4x+5y=20$ is included. The line is solid.

Solving a system of two linear inequalities

To solve a system of two linear inequalities, we graph the two boundary lines. Use a solid boundary line to represent $\leq$ or $\geq$ and a dashed boundary line to represent $<$ or $>$. The solution of the system is the intersection of two overlap (combined) regions of the combined graph. Every point in the intersection region satisfies both inequalities.

6. Solve by graphing. $\begin{cases} 4x+5y \geq 20 \\ 4x-5y<10 \end{cases}$

Solution:

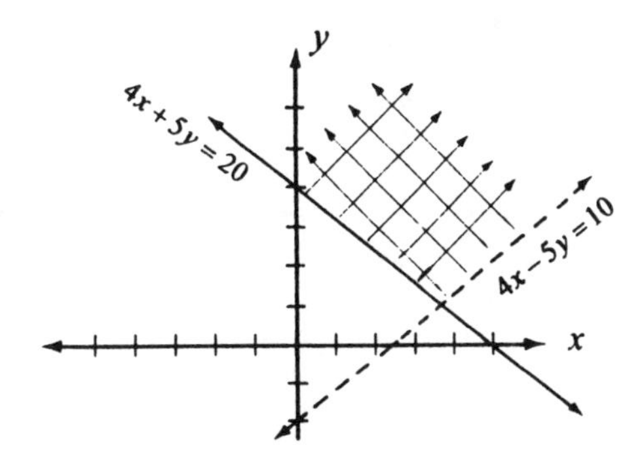

Its graph covers the overlap region above and on the line $4x+5y=20$ and above the line $4x-5y=10$.

EXERCISES

Find the x – intercept **and** y – intercept. **Graph each equation.**

1. $x - y = 3$ **2.** $x + 2y = 4$ **3.** $x - 2y = 4$

4. $2x - y = 6$ **5.** $2x + 5y = 16$ **6.** $4x + 5y = 20$

7. $4x - 5y = 10$ **8.** $4x - 2y = 7$ **9.** $4x + 6y = 9$

10. $y = 4x + 2$ **11.** $-2x + y = 1$ **12.** $y = \frac{1}{2}x + 4$

Graph each inequality.

13. $x - y > 3$ **14.** $x - y \le 3$ **15.** $x + 2y < 4$

16. $x + 2y \ge 4$ **17.** $-2x + y < 4$ **18.** $-2x + y > 4$

19. $3x - 5y > 15$ **20.** $3x - 5y \le 15$ **21.** $-4x - y > 2$

22. $4x - y > -2$ **23.** $-2x + 5y \ge 12$ **24.** $-2x + 5y \le 12$

Solve each system of equations by graphing.

25. $\begin{cases} y = x - 1 \\ y = -x + 3 \end{cases}$ **26.** $\begin{cases} y = x + 2 \\ y = -x + 4 \end{cases}$ **27.** $\begin{cases} x + y = 5 \\ x - y = 1 \end{cases}$

28. $\begin{cases} x - y = 6 \\ 3x + y = 10 \end{cases}$ **29.** $\begin{cases} 3a - b = -7 \\ 2a + b = -3 \end{cases}$ **30.** $\begin{cases} 2x + y = 7 \\ -x - 2y = 1 \end{cases}$

31. $\begin{cases} x + 2y = 4 \\ 2x + 4y = 8 \end{cases}$ **32.** $\begin{cases} m + 2n = 4 \\ 2m + 4n = 10 \end{cases}$ **33.** $\begin{cases} 2x - 5y = 10 \\ 4x - 10y = 4 \end{cases}$

34. $\begin{cases} 2x - 5y = 2 \\ 4x - 10y = 4 \end{cases}$ **35.** $\begin{cases} 6s - 2t = -5 \\ 4s - \frac{1}{2}t = 0 \end{cases}$ **36.** $\begin{cases} y = -\frac{1}{3}x + \frac{1}{6} \\ 4x - 8y = 12 \end{cases}$

Solve each system of inequalities by graphing. Indicate the solution set with crosshatching lines or shading.

37. $\begin{cases} x + y \ge 2 \\ x - y \le 1 \end{cases}$ **38.** $\begin{cases} 2x - y \le 4 \\ x \le 3 \end{cases}$ **39.** $\begin{cases} 2x - y \ge 4 \\ 3x + 4y < 12 \end{cases}$

40. $\begin{cases} 5x - 4y \ge -20 \\ 5x + 4y \ge 20 \end{cases}$ **41.** $\begin{cases} 2x + 3y > 0 \\ 2x - 3y \le -6 \end{cases}$ **42.** $\begin{cases} 4x + 5y \le 20 \\ 4x + 5y \ge 10 \end{cases}$

43. $\begin{cases} y < -1 \\ x > 3 \end{cases}$ **44.** $\begin{cases} 5x + y > 5 \\ 2x - y > -4 \end{cases}$ **45.** $\begin{cases} y < 2x + 4 \\ y > 2x - 5 \end{cases}$

10-2 Solving Systems Algebraically

In Section 10-1, we have learned how to solve systems of equations by graphing. However, it is sometimes difficult to find an exact solution from a hand-drawn graph. Graphing method estimates an approximate solution of a problem of systems having decimals as its solution.
We can find an exact solution to a system of equations by using algebraic methods. There are two algebraic methods for solving systems of equations.

1) **The substitution method**
2) **The elimination method** (or **The combination method)**

We may choose any method of the above two methods to find the solution of a system. In general, it is good to choose the **substitution method** if the coefficient of a variable is 1 or −1.
The main idea of the **elimination method** is to eliminate one variable by **adding or subtracting** the two equations to obtain a new equation with only one variable.

Examples

1. Solve $\begin{cases} 2x - y = 4 \cdots\cdots\cdots ① \\ 2x + 5y = 16 \cdots\cdots ② \end{cases}$

Solution:

Method 1: (Substitution Method)

From equation ①, we have $y = 2x - 4$.
Substitute $y = 2x - 4$ in equation ②, we have:

$$2x + 5y = 16$$
$$2x + 5(2x - 4) = 16$$
$$2x + 10x - 20 = 16$$
$$12x = 36$$
$$\therefore x = 3$$

Substitute $x = 3$ in equation ①:

$$2x - y = 4$$
$$2(3) - y = 4$$
$$6 - y = 4$$
$$\therefore y = 2$$

Ans: $x = 3$, $y = 2$ or (3, 2).

Method 2: (Combination Method)

equation ② – equation ①
We have: (x will be eliminated.)

$$\cancel{2x} + 5y = 16$$
$$-)\ \cancel{2x} - y = 4$$
$$6y = 12$$
$$\therefore y = 2$$

Substitute $y = 2$ in equation ①

$$2x - y = 4$$
$$2x - 2 = 4$$
$$2x = 6$$
$$\therefore x = 3$$

Ans: $x = 3$, $y = 2$ or (3, 2).

(The graph is showed in example 1, Section 10-1.)

-----Continued-----

Sometimes it is necessary to multiply one or both of the equations by a number to obtain the same coefficients(or differ in sign) for one of the variables. We can eliminate this variable.
If the final statement is a "**true statement**", the system of equations has unlimited number of solutions. The system is **consistent** and **dependent**.
If the final statement is a "**false statement**", the system of equations has no solution. The system is **inconsistent**.

Examples

2. Solve $\begin{cases} 3x-2y=10 \text{ .... ①} \\ 5x+4y=13 \text{ .... ②} \end{cases}$

Solution:

equation ① × 2: $\quad 6x - \cancel{4y} = 20$
equation ②: $\quad +)\ 5x + \cancel{4y} = 13$
$11x = 33$
$\therefore x = 3$

Substitute $x=3$ in equation ①:

$3x - 2y = 10$
$3(3) - 2y = 10$
$9 - 2y = 10$
$-2y = 1$
$\therefore y = -\frac{1}{2}$

Ans: $x = 3$, $y = -\frac{1}{2}$ or $(3, -\frac{1}{2})$.

3. Solve $\begin{cases} 3x-5y=12 \text{ .... ①} \\ 4x-2y=\frac{20}{3} \text{ .... ②} \end{cases}$

Solution:

① × 4: $\quad \cancel{12x} - 20y = 48$
② × 3: $\quad -)\ \cancel{12x} - 6y = 20$
$-14y = 28$
$\therefore y = -2$

Substitute $y=-2$ in ①:

$3x - 5y = 12$
$3x - 5(-2) = 12$
$3x + 10 = 12$
$3x = 2$
$\therefore x = \frac{2}{3}$

Ans: $x = \frac{2}{3}$, $y = -2$ or $(\frac{2}{3}, -2)$.

4. Solve $\begin{cases} 4x-2y=7 \text{ .... ①} \\ 10x-5y=6 \text{ .... ②} \end{cases}$

Solution:

① × 5 $\quad \cancel{20x} - \cancel{10y} = 35$
② × 2 $\quad -)\ \cancel{20x} - \cancel{10y} = 12$
$0 = 23$ (false)

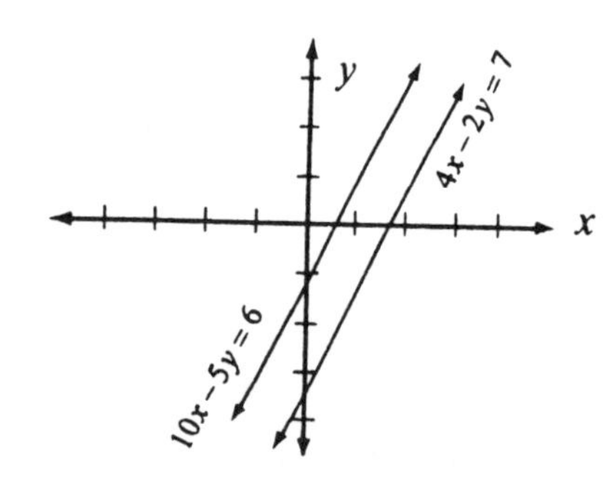

Ans: No solution (they are parallel).

5. Solve $\begin{cases} 4x+6y=9 \text{ .... ①} \\ 12x+18y=27 \text{ .... ②} \end{cases}$

Solution:

① × 3 $\quad \cancel{12x} + \cancel{18y} = \cancel{27}$
② $\quad -)\ \cancel{12x} + \cancel{18y} = \cancel{27}$
$0 = 0$ (true)

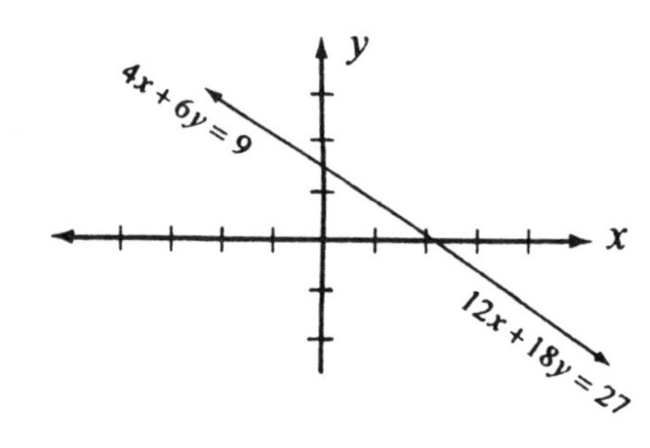

Ans: Unlimited number of solutions (they coincide).

EXERCISES

Solve each system by using the substitution method. Check your answers.

1. $\begin{cases} y = x \\ x + y = 4 \end{cases}$

2. $\begin{cases} y = 2x \\ x + 2y = 10 \end{cases}$

3. $\begin{cases} y = -3x \\ x - y = 4 \end{cases}$

4. $\begin{cases} y = -x \\ x - 2y = 2 \end{cases}$

5. $\begin{cases} x = 2y \\ 2x - y = 9 \end{cases}$

6. $\begin{cases} y = -2x \\ x - y = 6 \end{cases}$

7. $\begin{cases} x = 7 \\ 3x - y = 9 \end{cases}$

8. $\begin{cases} y = -8 \\ 2x + 4y = -22 \end{cases}$

9. $\begin{cases} x = 2y - 1 \\ -4x + 6y = 2 \end{cases}$

10. $\begin{cases} y = 4 - 2x \\ 2x - 5y = 16 \end{cases}$

11. $\begin{cases} y = x - 1 \\ x + y = 3 \end{cases}$

12. $\begin{cases} y = x + 2 \\ x + y = 4 \end{cases}$

13. $\begin{cases} x + y = 5 \\ x - y = 1 \end{cases}$

14. $\begin{cases} x - y = 6 \\ 3x + y = 10 \end{cases}$

15. $\begin{cases} 3a - b = -7 \\ 2a + b = -3 \end{cases}$

16. $\begin{cases} 2x + y = 7 \\ x + 2y = -1 \end{cases}$

17. $\begin{cases} x + 2y = 4 \\ 2x + 4y = 8 \end{cases}$

18. $\begin{cases} m + 2n = 4 \\ 2m + 4n = 10 \end{cases}$

19. $\begin{cases} x - y = 5 \\ 2x - 2y = 10 \end{cases}$

20. $\begin{cases} 2x + y = 7 \\ 4x + 2y = 5 \end{cases}$

21. $\begin{cases} 8x - y = 0 \\ 6x - 2y = -5 \end{cases}$

22. $\begin{cases} 2x + 6y = 1 \\ x - 2y = 3 \end{cases}$

23. $\begin{cases} x + y = 1 \\ 3x + 6y = 4 \end{cases}$

24. $\begin{cases} x - 2y = -2 \\ y = -\frac{3}{2}x \end{cases}$

Solve each system by using the combination method. Check your answers.

25. $\begin{cases} x + y = 8 \\ x - y = 10 \end{cases}$

26. $\begin{cases} -x + y = 7 \\ x + y = 11 \end{cases}$

27. $\begin{cases} a - b = 2 \\ a + b = 12 \end{cases}$

28. $\begin{cases} m - 2n = 12 \\ 3m + 2n = 4 \end{cases}$

29. $\begin{cases} 4p - 6q = 9 \\ 4p - 3q = 3 \end{cases}$

30. $\begin{cases} 12x + 3y = 10 \\ 2x + 3y = 0 \end{cases}$

31. $\begin{cases} -3x + 5y = 7 \\ 3x - y = 5 \end{cases}$

32. $\begin{cases} 2c - 4d = 15 \\ 3c + 4d = -5 \end{cases}$

33. $\begin{cases} 10a - b = -46 \\ -10a - b = 54 \end{cases}$

34. $\begin{cases} 5x - 7y = 35 \\ 5x - 7y = 15 \end{cases}$

35. $\begin{cases} 5x - 7y = 35 \\ 5x + 7y = 15 \end{cases}$

36. $\begin{cases} 4s - 7t = 1 \\ 2s - 7t = 0 \end{cases}$

37. $\begin{cases} 2x - 3y = 12 \\ 4x + 2y = 8 \end{cases}$

38. $\begin{cases} 4x + 5y = 16 \\ 3x - 10y = -37 \end{cases}$

39. $\begin{cases} 5x - 6y = -16 \\ 2x - 2y = -4 \end{cases}$

40. $\begin{cases} 3x - 7y = 2 \\ 6x - 14y = 4 \end{cases}$

41. $\begin{cases} 2x + 8y = 1 \\ 4x + 16y = 3 \end{cases}$

42. $\begin{cases} 3x + 4y = 4 \\ 12x - 8y = -2 \end{cases}$

43. $\begin{cases} 3x - 7y = 13 \\ 6x - 14y = 26 \end{cases}$

44. $\begin{cases} 3x + 2y = -4 \\ 4x + 5y = 11 \end{cases}$

45. $\begin{cases} 2x - 5y = 46 \\ 5x + 4y = -17 \end{cases}$

46. $\begin{cases} 6a + 12b = 7 \\ 4a - 9b = -1 \end{cases}$

47. $\begin{cases} 9a + 2b = -2 \\ 6a - 5b = 24 \end{cases}$

48. $\begin{cases} 4p - 3q = -40 \\ 6p - 2q = 40 \end{cases}$

49. $\begin{cases} 0.5x + 1.5y = 6.25 \\ 2.5x + 3y = 7.5 \end{cases}$

50. $\begin{cases} \frac{1}{3}x + \frac{1}{2}y = 6 \\ \frac{2}{5}x - \frac{3}{4}y = -\frac{18}{5} \end{cases}$

-----Continued-----

EXERCISES

Solve each system by using any algebraic method. Check your answers.

51. $\begin{cases} x+3y=-2 \\ x-y=6 \end{cases}$

52. $\begin{cases} 2x-y=5 \\ x-2y=1 \end{cases}$

53. $\begin{cases} 2x+4y=18 \\ 2x-y=-17 \end{cases}$

54. $\begin{cases} x-5y=17 \\ 2x+5y=-26 \end{cases}$

55. $\begin{cases} 3x-2y=5 \\ x-2y=-9 \end{cases}$

56. $\begin{cases} -4x+y=-26 \\ -4x+3y=-6 \end{cases}$

57. $\begin{cases} 2x-3y=20 \\ 3x-4y=27 \end{cases}$

58. $\begin{cases} 12a-13b=-38 \\ -10a-13b=-16 \end{cases}$

59. $\begin{cases} 8a+5b=7 \\ -8a-10b=18 \end{cases}$

60. $\begin{cases} 13c+15d=70 \\ -14c-15d=-80 \end{cases}$

61. $\begin{cases} -11m+5n=5 \\ -11m+4n=-7 \end{cases}$

62. $\begin{cases} -11m+3n=15 \\ 11m-7n=9 \end{cases}$

63. $\begin{cases} 3x-6y=39 \\ 5x+7y=-37 \end{cases}$

64. $\begin{cases} 6x-2y=-14 \\ 9x-4y=-29 \end{cases}$

65. $\begin{cases} 10x+8y=-4 \\ 15x-4y=6 \end{cases}$

66. $\begin{cases} 0.3x+0.8y=7 \\ 0.3x+0.6y=6 \end{cases}$

67. $\begin{cases} 0.04x-0.08y=4.8 \\ 0.02x+0.08y=4.4 \end{cases}$

68. $\begin{cases} 0.5x+0.6y=0.45 \\ 0.8x-0.4y=0.04 \end{cases}$

69. $\begin{cases} x=\frac{1}{3}y \\ 9x+5y=16 \end{cases}$

70. $\begin{cases} y=-5x \\ 20x-3y=21 \end{cases}$

71. $\begin{cases} a=4b \\ \frac{1}{3}a+\frac{4}{3}b=-2 \end{cases}$

72. $\begin{cases} \frac{1}{2}x+\frac{1}{3}y=2 \\ \frac{1}{2}x-\frac{4}{3}=-3 \end{cases}$

73. $\begin{cases} \frac{3}{5}x-\frac{1}{4}y=4 \\ \frac{4}{5}x-\frac{1}{4}y=5 \end{cases}$

74. $\begin{cases} \frac{2}{3}x+\frac{3}{4}y=2 \\ \frac{4}{3}x-\frac{5}{4}y=-18 \end{cases}$

A patient said to his dentist.
Patient: Why are you charging me $300 ?
Last time, you charged me only $100.
Dentist: Two patients in the waiting room ran away when they heard you screaming.

10-3 Solving Word Problems by using Systems

In Section 6-8, we have learned to solve a word problem using one equation in one variable. Now, we can use a system of two equations to solve a word problem in two variables.

The steps for solving a word problem in two variables are similar to the steps for solving a word problem in one variable (see section 6-8).

We assign two variables to represent the unknowns.

Examples

1. John has twice as much money as Carol. Together they have \$36. How much money does each have ? (See example 6, Section 6-8)

Solution:

(Assign two variables)

Let $x =$ John's money

$y =$ Carol's money

The system of equations is:

$$\begin{cases} x = 2y \cdots\cdots ① \\ x + y = 36 \cdots\cdots ② \end{cases}$$

Substitute $x = 2y$ in ②:

$2y + y = 36$

$3y = 36$

$\therefore y = 12$

Substitute $y = 12$ in ①:

$x = 2(12)$

$\therefore x = 24$

Ans: John's money \$24.
Carol's money \$12.

2. There are one-dollar bills and five-dollar bills in the piggy bank. It has 60 bills with a total value of \$228. How many ones and how many fives in the piggy bank ? (See example 10, Section 6-8)

Solution:

(Assign two variables)

Let $x =$ number of one-dollar bills

$y =$ number of five-dollar bills

The system of equations is:

$$\begin{cases} x + y = 60 \cdots\cdots ① \\ x + 5y = 228 \cdots\cdots ② \end{cases}$$

②–①: $4y = 168$

$\therefore y = 42$

Substitute $y = 42$ in ①:

$x + 42 = 60$

$\therefore x = 18$

Ans: 18 one-dollar bills.
42 five-dollar bills.

3. Flying with the tail wind, an airplane flies 600 km per hour. Flying with the same wind, it flies 500 km per hour to make the return flight against the wind. Find the speed of the wind and the speed of the airplane.

Solution:

Let $x =$ Airplane speed

$y =$ Wind speed

We have:

$$\begin{cases} x + y = 600 \cdots\cdots ① \\ x - y = 500 \cdots\cdots ② \end{cases}$$

① + ②: $2x = 1100$ $\therefore x = 550$

Substitute $x = 550$ in ①:

$550 + y = 600$ $\therefore y = 50$

Ans: Airplane speed = 550 km/hr.
wind speed = 50 km/hr.

-----Continued-----

Examples

4. Two hamburgers and one bagel cost \$4.82. At the same price, one hamburger and two bagels cost \$3.70. How much does each hamburger and each bagel cost ?

Solution:

Let $x =$ cost of each hamburger

$y =$ cost of each bagel

We have:

$$\begin{cases} 2x + y = 4.82 \;\cdots\cdots① \\ x + 2y = 3.70 \;\cdots\cdots② \end{cases}$$

From ①: $y = 4.82 - 2x$

Substitute $y = 4.82 - 2x$ in ②:

$$x + 2(4.82 - 2x) = 3.70$$
$$x + 9.64 - 4x = 3.70$$
$$-3x = -5.94$$
$$\therefore x = 1.98$$

Substitute $x = 1.98$ in ①:

$2(1.98) + y = 4.82 \quad \therefore y = 0.86$

Ans: Each hamburger \$1.98.
Each bagel \$0.86.

5. A teacher distributes a box of pencils to the students. If each student receives 5 pencils, there are 15 pencil left. If each students receives 6 pencils, there are 22 pencils short. How many students are there ? How many pencils are there ?

Solution:

Let $x =$ number of students

$y =$ number of pencils

We have:

$$\begin{cases} y - 5x = 15 \;\cdots\cdots① \\ 6x - y = 22 \;\cdots\cdots② \end{cases}$$

① + ②: $x = 37$

Substitute $x = 37$ in ①

$$y - 5(37) = 15$$
$$y - 185 = 15$$
$$\therefore y = 200$$

Ans: 37 students, 200 pencils.

6. A student wants to mix a salt-water solution containing 10% pure salt with another salt-water solution containing 15% pure salt. The total mixture must at least 3 gallons and must contain at most 0.6 gallons of pure salt. Graph the possible solution region to find the amount of each solution needed.

Solution:

Let $x =$ the amount of 10% solution

$y =$ the amount of 15% solution

We have:

$$x + y \geq 3$$
$$0.10x + 0.15y \leq 0.6$$

Rewrite the above system:

$$\begin{cases} x + y \geq 3 \\ 2x + 3y \leq 12 \end{cases}$$

Solve the system by graphing:

$2x + 3y = 12$

$x + y = 3$

Since x and y are nonnegative, the solution region is in the first quadrant. Any point in the region will be the amounts of each solution to satisfy the required mixture.

-----Continued-----

Linear Programming

Linear programming techniques are used to solve many business problems involving decision-making on seeking maximum profit and minimum cost.
A linear programming problem has an objective function that is to be maximized or minimized under certain conditions (or constraints). The constraints can be represented by a system of inequalities. The possible solutions is the points in the feasible region bounded by the graph. To find the point that maximizes (or minimizes) the objective function, we need only to test the corner points in the feasible region.

Examples

7. A factory makes two products, I and II. Each requires the use of two machines, A and B.
Each Product I requires 2 hours on Machine A and 3 hours on Machine B.
Each Product II requires 4 hours on Machine A and 2 hours on Machine B.
The profit of each Product I is \$60. The profit of each Product II is \$50.
Machine A has only 80 hours available. Machine B has only 60 hours available.
What is the amount of each product to maximize the profit ?
Solution:

Let $x =$ the amount of Product I
$y =$ the amount of Product II

We have:

Maximize the objective function:
Max. $p = 60x + 50y$

Subject to the conditions(constraints):

$$\begin{cases} 2x + 4y \le 80 \\ 3x + 2y \le 60 \end{cases}$$

$x \ge 0,\ y \ge 0$

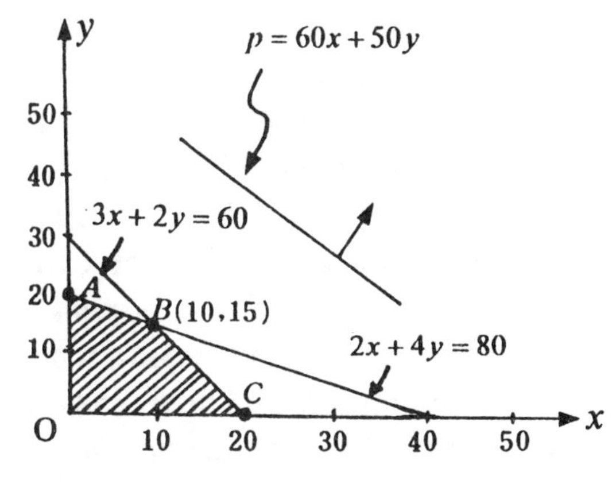

Graph the above system and find the corner points in the feasible region.
Evaluate the profit at each corner point.

$p = 60x + 50y$

A(0, 20), $p = 60\ (0) + 50\ (20) = \$1{,}000$
B(10, 15), $p = 60\ (10) + 50\ (15) = \$1{,}350$ ✓
C(20, 0), $p = 60\ (20) + 50\ (0) = \$1{,}200$

Ans: Max. P = \$1,350
Product I = 10 units
Product II = 15 units

In advanced math, we can solve linear programming problems involving many variables by using a method called **Simplex Method**. This method is introduced in a college book "Operations Research: Linear and integer programming" written by Rong Yang, 1986.

EXERCISES

Hint: We can find the solutions of most of the following word problems either by using one equation with one variable or by using a system of two equations.
Here you should practice to find the solutions by using a system of two equations.
Assign two variables to represent the two unknowns.

1. John has twice as much money as David. Together they have $156. How much money does each have ?
2. There are one-dollar bills and five-dollar bills in the piggy bank. It has 64 bills with a total value of $244. How many ones and how many fives in the piggy bank ?
3. There are quarters and dimes in a box. Their total value is $5.40. Quarters are five times as many as dimes. How many quarters and how many dimes are in the box ?
4. Flying with the tail wind, an airplane flies 860 km per hour. Flying with the same wind, it flies 740 km per hour to make the return flight against the wind. Find the speed of the wind and the speed of the airplane.
5. Twice a number is 20 more than another number. One fourth of the first number is 29 less than the second number. Find the two numbers.
6. You want to rent a vehicle for one day. The rental cost for a car is $65 plus $0.25 per mile. The rental cost for a truck is $110 plus $0.10 per mile. How many miles that you travel will cost the same for both options ?
7. The width of a rectangle is $12\,cm$ shorter than the length. The perimeter is $52\,cm$. What is the dimensions of the rectangle ?
8. The width of a rectangle is $15\,in.$ shorter than twice the length. The perimeter is $45\,in.$. What is the dimensions of the rectangle ?
9. In a school fair, Roger bought hamburgers at a cost of $9 per dozen and bagels at a cost of $3 per dozen. He bought a total of 45 dozens and paid a total of $255. How many dozens of hamburgers and how many dozens of bagels did he buy ?
10. Two hamburgers and one bagel cost $4.70. At the same prices, one hamburger and two bagels cost $3.70. How much does each hamburger and each bagel cost ?
11. In a school fair, John bought 30 boxes of hamburgers and bagels. Each box contains 20 hambergers. Each box contains 25 bagels. A total of 690 hamburgers and bagels was bought. How many boxes of hamburgers did he buy ? How many boxes of bagels did he buy ?
12. A balloon at 900 feet high is descending at a rate of 4 feet per second. Another balloon is rising from the ground at a rate of 2 feet per second. In how many seconds will the two balloons be at the same altitude ?
13. A teacher distributes a box of pencils to the students. If each student receives 4 pencils, there are 8 pencils left. If each student receives 5 pencils, there are 20 pencils short. How many students are there ? How many pencils are in the box ?
14. You have a $20 bill. If you buy 24 pens and 12 pencils, you will get $1.40 change. If you buy 12 pens and 24 pencils, you will get $4.40 change. How much does each pen and each pencil cost ?
15. There are 3,192 students in Hawthorne High School. There are 168 more boys than girls. How many boys and how many girls are there ?

CHAPTER 10 EXERCISES

Find the x – intercept and y – intercept. Graph each equation and inequality.

1. $x+y=4$
2. $x+y>4$
3. $x+y\le 4$
4. $2x-y=4$
5. $2x-y\ge 4$
6. $2x-y<4$
7. $4x-6y=3$
8. $4x-6y>3$
9. $4x-6y<3$
10. $-3x+5y=12$
11. $-3x+5y\ge 12$
12. $-3x+5y\le -12$

Solve each system of equations by graphing.

13. $\begin{cases} y=-2x+7 \\ y=x-5 \end{cases}$
14. $\begin{cases} y=\frac{1}{2}x-4 \\ y=-2 \end{cases}$
15. $\begin{cases} y=3x-6 \\ y=3x-1 \end{cases}$
16. $\begin{cases} x+\frac{1}{2}y=3 \\ 2x+y=6 \end{cases}$
17. $\begin{cases} 3a-2b=4 \\ 2a-3b=1 \end{cases}$
18. $\begin{cases} x+y=-5 \\ x-y=6 \end{cases}$
19. $\begin{cases} x+y=0 \\ 2x-y=3 \end{cases}$
20. $\begin{cases} 3x+2y=5 \\ 2x-y=0 \end{cases}$
21. $\begin{cases} 4m+2n=-4 \\ 2m+3n=6 \end{cases}$
22. $\begin{cases} x-4y=-6 \\ x+4y=2 \end{cases}$
23. $\begin{cases} 6x+2y=-5 \\ 8x-2y=12 \end{cases}$
24. $\begin{cases} 4x+5y=4 \\ 2x+4y=5 \end{cases}$

Graph each system of inequalities. Indicate the region of solution set with crosshatching lines or shading.

25. $\begin{cases} 2x+y\ge 4 \\ x-y\le 2 \end{cases}$
26. $\begin{cases} 2x+y\le 4 \\ x-y\le 2 \end{cases}$
27. $\begin{cases} 2x+y\ge 4 \\ x-y\ge 2 \end{cases}$
28. $\begin{cases} 2x+y\le 4 \\ x-y\ge 2 \end{cases}$
29. $\begin{cases} 2x+y<4 \\ x-y\ge 2 \end{cases}$
30. $\begin{cases} 2x+y\ge 4 \\ x-y<2 \end{cases}$
31. $\begin{cases} 2x-4y>-6 \\ x+2y>4 \end{cases}$
32. $\begin{cases} 2x-4y<-6 \\ x+2y>4 \end{cases}$
33. $\begin{cases} x+2y\le 4 \\ 2x-4y>-6 \end{cases}$
34. $\begin{cases} 4x+5y>4 \\ 2x+4y<5 \end{cases}$
35. $\begin{cases} 4x+5y\ge 4 \\ 2x+4y\le 5 \end{cases}$
36. $\begin{cases} 4x+5y\le 4 \\ 2x+4y>5 \end{cases}$
37. $\begin{cases} 2x+y\ge 4 \\ x-y\le 2 \\ x\ge 2 \end{cases}$
38. $\begin{cases} x\ge 2 \\ y\ge 3 \\ x+y>4 \end{cases}$
39. $\begin{cases} 2x-4y>-6 \\ x+2y\ge 4 \\ x\le 3 \end{cases}$

-----Continued-----

Solve each systems by using any algebraic method. Check your answers.

40. $3x - 5y = 37$
$2x + y = 16$

41. $x - 3y = -29$
$4x + 10y = 38$

42. $6x - y = -3$
$3x + 5y = 37$

43. $2x - 6y = -54$
$4x - 5y = -17$

44. $5x - 8y = -37$
$-3x + 4y = 23$

45. $6a + 3b = 4$
$8a - 3b = 3$

46. $0.5x - 0.5y = -0.1$
$0.1x + 0.4y = 0.38$

47. $0.4x + 0.6y = 0.22$
$1.6x - 1.8y = -1.22$

48. $4x + 5y = 4$
$5x + 4y = 4.1$

49. $a = \frac{1}{2}b$
$8a - 3b = 10$

50. $\frac{1}{2}x - \frac{1}{2}y = 1$
$\frac{1}{4}x + y = 3$

51. $\frac{2}{3}x - \frac{1}{2}y = -3$
$\frac{1}{3}x - \frac{3}{2}y = -4$

52. The sum of two numbers is 24. The difference of them is 10. Find the numbers

53. The sum of two numbers is 119. One number is 14 more than half another number. Find the numbers.

54. John has one-dollar bills and five-dollar bills. He has 43 bills worth $115. How many one-dollar bills and how many five-dollar bills does he have ?

55. Roger has nickels and quarters in the piggy bank. There are 73 coins with a total value of $7.45. How many nickels and how many quarters does he have ?

56. The width of a rectangle is 3 cm more than half the length. The perimeter is 96 cm. What is the dimensions of the rectangle ?

57. David bought hamburgers at a cost of $2.15 each and bagels at a cost of $0.60. He bought a total of 75 hamburgers and bagels and paid a total of $107. How many hamburgers and how many bagels did he buy ?

58. A teacher distributes a box of pencils to the students. If each student receives 7 pencils, there are 9 pencils left. If each students receives 8 pencils, there are 24 short. How many students are there ? How many pencils in the box ?

59. Lita had $7,000. She opened a bank account paying 4% interest per year and invested the rest in stocks paying 6% interest per year. She earned a total of $330 interest within one year. How much money did she put in each investment ?

60. An airplane flies 3,300 miles with the wind in 3 hours 18 minutes. It takes 4 hours and 24 minutes to make the return flight against the same wind. Find the speed of the airplane and the speed of the wind.

61. A balloon at 1,200 feet high is descending at a rate of 6 feet per second. Another balloon is rising from the ground at a rate of 4 feet per second. In how many seconds will the two balloons be at the same altitude ?

62. A soccer team needs at least 14 points to get into the playoffs. A win is worth 3 points and a tie is worth 1 point. There are 10 games to play. Write and graph a system of inequalities that shows how many wins (w) and ties(t) will get the team into the playoffs.

63. The sum of two positive integers x and y is greater than 30 and less than 55. Write and graph a system of inequalities that shows the desired integers to this system.

64. Solve the linear programming problem.
Objective function: Min. $C = 3x + 8y$
Constraints: $\begin{cases} x + y = 200 \\ 0 \le x \le 80 \\ y \ge 60 \end{cases}$

Additional Examples

1. Graph the system. $\begin{cases} 3x+5y \le 15 \\ x \ge 0 \\ y \ge 0 \end{cases}$

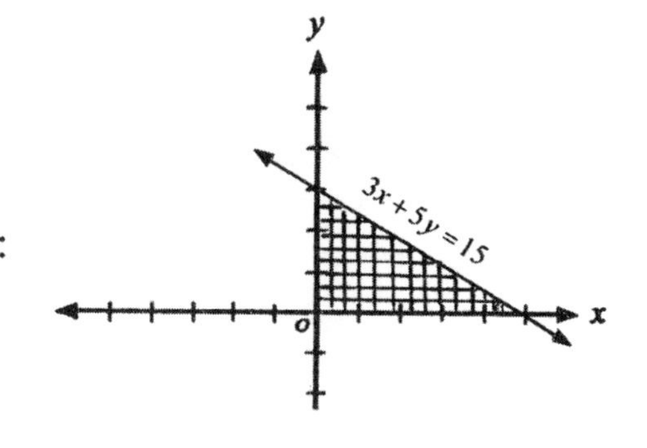

Solution:

The graph is the combined (overlap) region:
on and below the line $3x+5y=15$.
on and to the right of the $y-$axis ($x=0$).
on and above the $x-$axis ($y=0$).

2. Graph the system. $\begin{cases} x+y \ge 1 \\ x \ge -1 \\ y > -2 \end{cases}$

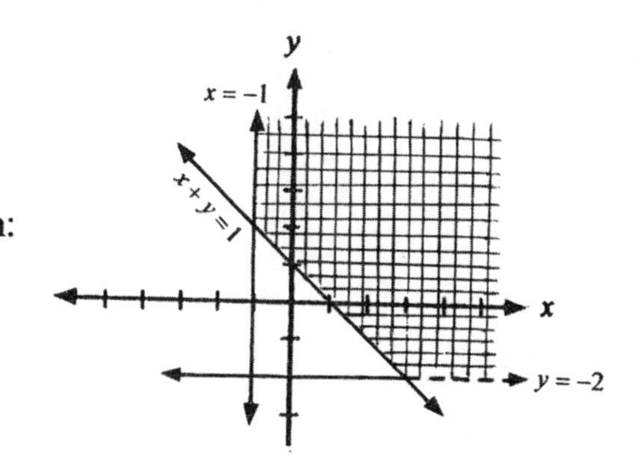

Solution:

The graph is the combined (overlap) region:
on and above the line $x+y=1$.
on and to the right of the line $x=-1$.
above the line $y=-2$.

A lady said to her friend.
Lady: I always keep quiet whenever my husband argues with me.
Friend: How do you control your anger ?
Lady: I clean the toilet.
Friend: How does that help ?
Lady: I use his toothbrush.

Space for Taking Notes

CHAPTER ELEVEN

Radicals

11-1 Square Roots and Radicals

Because $5^2 = 25$, we say that the square root of 25 is 5. We write $\sqrt{25} = 5$.
Because $5^3 = 125$, we say that the cubed root of 125 is 5. We write $\sqrt[3]{125} = 5$.
Because $2^4 = 16$, we say that the 4th root of 16 is 2. We write $\sqrt[4]{16} = 2$.

The symbol $\sqrt[n]{b}$ is called a **radical**. We read $\sqrt[n]{b}$ as "**the n th root of b**".
b is called the **radicand**, n is called the **index**.
Index "2" of the square root is usually omitted from the radical. $\sqrt{b}$ means $\sqrt[2]{b}$.

To find the square root of a given number, we find a number that, when squared, equals the given number. However, $\sqrt[n]{b}$ means only the **principal root** of b.

$\sqrt{4} = \sqrt{2 \cdot 2} = \sqrt{2^2} = 2$. $\quad$ $\sqrt{9} = \sqrt{3 \cdot 3} = \sqrt{3^2} = 3$.

$\sqrt{4} = \sqrt{(-2)^2} = -2$ is incorrect. $\quad$ $\sqrt{9} = \sqrt{(-3)^2} = -3$ is incorrect.

$\sqrt{(-2)^2} = \sqrt{4} = 2$. $\quad$ $\sqrt{(-3)^2} = \sqrt{9} = 3$.

$\sqrt{0.01} = \sqrt{(0.1)(0.1)} = \sqrt{(0.1)^2} = 0.1$. $\quad$ $\sqrt{1.44} = \sqrt{(1.2)(1.2)} = \sqrt{(1.2)^2} = 1.2$

To find the n th root of a given number, we follow the same idea as the above examples, and so on.

$\sqrt[3]{8} = \sqrt[3]{2 \cdot 2 \cdot 2} = \sqrt[3]{2^3} = 2$. $\quad$ $\sqrt[4]{16} = \sqrt[4]{2 \cdot 2 \cdot 2 \cdot 2} = \sqrt[4]{2^4} = 2$.

$\sqrt[3]{-8} = \sqrt[3]{-2 \cdot -2 \cdot -2} = \sqrt[3]{(-2)^3} = -2$. $\quad$ $\sqrt[4]{(-2)^4} = \sqrt[4]{16} = 2$, **not** -2.

The square root of a given number that is not a perfect square is an **irrational number** (a nonrepeating infinite decimal). We use a calculator to find its decimal approximation, depending on what decimal place we round to.

$\sqrt{3} \approx 1.732$. $\quad$ $\sqrt{5} \approx 2.236$. $\quad$ $\sqrt{7} \approx 2.646$. $\quad$ $\sqrt{8} \approx 2.828$.

$\sqrt{0.15} \approx 0.387$. $\quad$ $\sqrt{17} \approx 4.123$. $\quad$ $\sqrt{140} \approx 11.832$.

To find the square root of a fraction, we write the square root of the numerator over the square root of the denominator.

$$\sqrt{\frac{4}{9}} = \frac{\sqrt{4}}{\sqrt{9}} = \frac{2}{3}. \quad \text{Or: } \sqrt{\frac{4}{9}} = \sqrt{\frac{2}{3} \cdot \frac{2}{3}} = \frac{2}{3}.$$

$$\sqrt{\frac{12}{35}} = \frac{\sqrt{12}}{\sqrt{35}} \approx \frac{3.464}{5.916} \approx 0.586. \quad \text{Or: } \sqrt{\frac{12}{35}} \approx \sqrt{0.343} \approx 0.586.$$

EXERCISES

Evaluate each square root. Check you answers by squaring.

(The symbol "±" means "plus or minus", or "positive or negative".)

Hint: If you cannot find the square root of the given number, you may factor the given number by using a perfect-square factor, starting with 4, 9, 16, 25, 36, 49, 64, 81, ….

$\sqrt{1936} = \sqrt{4 \cdot 484} = \sqrt{4 \cdot 4 \cdot 121} = 4\sqrt{121} = 4\sqrt{11 \cdot 11} = 4 \cdot 11 = 44$

Or: $\sqrt{1936} - \sqrt{16 \cdot 121} = \sqrt{16} \cdot \sqrt{121} = 4 \cdot 11 = 44$

1. $\sqrt{16}$
2. $\sqrt{25}$
3. $-\sqrt{36}$
4. $\sqrt{49}$
5. $\sqrt{64}$
6. $-\sqrt{81}$
7. $\pm\sqrt{144}$
8. $\sqrt{10{,}000}$
9. $-\sqrt{196}$
10. $\pm\sqrt{400}$
11. $-\sqrt{900}$
12. $\sqrt{625}$
13. $\sqrt{784}$
14. $\sqrt{1{,}296}$
15. $\pm\sqrt{2{,}304}$
16. $\sqrt{11{,}025}$
17. $\sqrt{0.09}$
18. $-\sqrt{0.16}$
19. $-\sqrt{0.25}$
20. $\pm\sqrt{0.81}$
21. $\sqrt{1.44}$
22. $\sqrt{0.0016}$
23. $\sqrt{0.0025}$
24. $\sqrt{0.0121}$
25. $\sqrt{0.0144}$
26. $\pm\sqrt{0.0169}$
27. $\sqrt{0.0004}$
28. $\sqrt{0.000036}$
29. $\sqrt{\frac{9}{4}}$
30. $-\sqrt{\frac{25}{49}}$
31. $\pm\sqrt{\frac{1}{36}}$
32. $\sqrt{\frac{81}{400}}$
33. $\sqrt{\frac{100}{121}}$
34. $\pm\sqrt{\frac{16}{64}}$
35. $-\sqrt{\frac{24}{54}}$
36. $\sqrt{\frac{5}{125}}$

Evaluate each radical. If the root is an irrational number, round the decimal to the nearest thousandth.

37. $\sqrt[3]{27}$
38. $\sqrt[3]{64}$
39. $-\sqrt[4]{16}$
40. $\sqrt[4]{81}$
41. $\sqrt[5]{32}$
42. $\pm\sqrt{10}$
43. $\sqrt{12}$
44. $-\sqrt{40}$
45. $\sqrt{75}$
46. $\sqrt{27}$
47. $-\sqrt{32}$
48. $\sqrt{125}$
49. $\pm\sqrt[3]{0.001}$
50. $\sqrt[3]{0.008}$
51. $\sqrt{0.12}$
52. $\pm\sqrt{325}$
53. $2\sqrt[6]{64}$
54. $-\sqrt[4]{256}$
55. $5\sqrt{70}$
56. $3\sqrt{108}$
57. $\sqrt{0.9}$
58. $\sqrt{3.5}$
59. $\pm 8\sqrt{48}$
60. $12\sqrt{4.5}$
61. $\sqrt{\frac{15}{25}}$
62. $\pm\sqrt{\frac{11}{6}}$
63. $-\sqrt{\frac{54}{8}}$
64. $\sqrt{\frac{140}{18}}$

11-2 Simplifying Radicals

Rules of Radicals

For all nonnegative real numbers a and b:

1) $\sqrt{a^2} = a$ **2)** $\sqrt{ab} = \sqrt{a} \cdot \sqrt{b}$ **3)** $\sqrt{\frac{b}{a}} = \frac{\sqrt{b}}{\sqrt{a}}$, where $a \neq 0$

4) $\sqrt[n]{a} = a^{\frac{1}{n}}$ **5)** $\sqrt[n]{a^m} = a^{\frac{m}{n}}$

For all real numbers a, b, and x:

6) $\sqrt[n]{a^n} = |a|$ **if** n is even. **7)** $\sqrt{x^2} = |x|$

$\sqrt[n]{a^n} = a$ **if** n is odd. **8)** $\sqrt{x^3} = x\sqrt{x}$

The formula $\sqrt{x^2} = |x|$ shows that when we are finding **square root** of a variable expression, we must use absolute value sign to **ensure** the answer is positive.

$\sqrt{(-3)^2} = \sqrt{9} = 3$, not -3. $\sqrt[3]{-27} = \sqrt[3]{(-3)^3} = -3$. (Cubed root could be negative.)

$\sqrt{4x^2} = \sqrt{4} \cdot \sqrt{x^2} = 2|x|$. $\sqrt{4x^6} = \sqrt{4} \cdot \sqrt{(x^3)^2} = 2|x^3|$. $\sqrt[3]{8x^3} = \sqrt[3]{8} \cdot \sqrt[3]{x^3} = 2x$.

$\sqrt{4x^4} = \sqrt{4} \cdot \sqrt{(x^2)^2} = 2|x^2| = 2x^2$. ($x^2$ is always nonnegative. Omit the symbol $|\ |$.)

$\sqrt{4x^8} = \sqrt{4} \cdot \sqrt{(x^4)^2} = 2|x^4| = 2x^4$. ($x^4$ is always nonnegative. Omit the symbol $|\ |$.)

In the formula $\sqrt{x^3} = x\sqrt{x}$, we must assume that x is nonnegative. The radical $\sqrt{x^3}$ has no meaning (not a real number) if $x < 0$. Therefore, we omit the symbol $|\ |$.

$\sqrt{x^3} = \sqrt{x^2 \cdot x} = \sqrt{x^2} \cdot \sqrt{x} = |x| \cdot \sqrt{x} = x\sqrt{x}$ Similarly: $\sqrt[4]{x^5} = x\sqrt[4]{x}$

To simplify a square root in simplest radical form, we factor the radicand by using a perfect-square factor and leave any factor that is not a perfect-square in the radical form. To simplify a radical, we must **rationalize the denominator** by eliminating the radical from the denominator. No radical is allowed in denominator (see example **14**).

Examples. (Simplify each radical in simplest radical form)

1. $\sqrt{75} = \sqrt{25 \cdot 3} = 5\sqrt{3}$ **2.** $\sqrt{32} = \sqrt{16 \cdot 2} = 4\sqrt{2}$ **3.** $\sqrt[3]{16} = \sqrt[3]{2^3 \cdot 2} = 2\sqrt[3]{2}$

4. $\sqrt{25x^2} = 5|x|$ **5.** $\sqrt{81y^2} = 9|y|$ **6.** $\sqrt{24a^2} = \sqrt{4 \cdot 6 \cdot a^2} = 2\sqrt{6}|a|$

7. $\sqrt{16a^3} = 4\sqrt{a^2 \cdot a} = 4a\sqrt{a}$ **8.** $\sqrt{a^2 - 6a + 9} = \sqrt{(a-3)^2} = |a-3|$

9. $\sqrt[4]{x^4 y^8} = \sqrt[4]{x^4} \cdot \sqrt[4]{(y^2)^4} = |x|y^2$ **10.** $\sqrt[3]{-8x^6} = \sqrt[3]{(-2x^2)^3} = -2x^2$

11. $\sqrt[5]{-x^{10}} = \sqrt[5]{(-x^2)^5} = -x^2$ **12.** $\sqrt{18a^4b^5} = \sqrt{9 \cdot 2 \cdot (a^2)^2 \cdot (b^2)^2 \cdot b} = 3a^2b^2\sqrt{2b}$

13. $\sqrt{\frac{18x^2y^3}{z^2}} = \frac{\sqrt{9 \cdot 2 \cdot x^2 \cdot y^2 \cdot y}}{\sqrt{z^2}} = \frac{3|x|y\sqrt{2y}}{|z|}$, $z \neq 0$ (Hint: y is nonnegative.)

14. $\sqrt{\frac{2}{5}} = \frac{\sqrt{2}}{\sqrt{5}} \cdot \frac{\sqrt{5}}{\sqrt{5}} = \frac{\sqrt{10}}{\sqrt{25}} = \frac{\sqrt{10}}{5}$ **15.** $\sqrt[4]{9} = \sqrt[4]{3^2} = 3^{\frac{2}{4}} = 3^{\frac{1}{2}} = \sqrt{3}$

EXERCISES

Simplify each radical in simplest radical form

1. $\sqrt{12}$
2. $\sqrt{18}$
3. $-\sqrt{24}$
4. $\pm\sqrt{27}$
5. $\sqrt{32}$
6. $\sqrt{28}$
7. $\sqrt{80}$
8. $\sqrt{48}$
9. $\sqrt{150}$
10. $\pm\sqrt{108}$
11. $\sqrt{120}$
12. $-\sqrt{162}$
13. $\sqrt{396}$
14. $\sqrt{1800}$
15. $\sqrt{3200}$
16. $\sqrt{1875}$
17. $5\sqrt{72}$
18. $-4\sqrt{44}$
19. $\pm 7\sqrt{125}$
20. $10\sqrt{363}$
21. $3\sqrt{264}$
22. $9\sqrt{648}$
23. $8\sqrt{1200}$
24. $12\sqrt{1875}$
25. $-\sqrt[3]{54}$
26. $\sqrt[3]{-54}$
27. $\pm\sqrt[3]{-16}$
28. $\sqrt[3]{81}$
29. $\sqrt[3]{-81}$
30. $\sqrt[4]{81}$
31. $\sqrt[4]{(-3)^4}$
32. $\sqrt[5]{(-3)^5}$
33. $\sqrt{\frac{16}{4}}$
34. $\pm\sqrt{\frac{8}{49}}$
35. $\frac{3}{\sqrt{6}}$
36. $-6\sqrt{\frac{5}{6}}$
37. $\sqrt{\frac{18}{24}}$
38. $\sqrt{\frac{121}{8}}$
39. $3\sqrt{\frac{24}{9}}$
40. $\sqrt{3\frac{2}{5}}$

Simplify each radical in simplest radical form. Assume that all variables are nonnegative real numbers and that all denominators are nonzero.

41. $\sqrt{4a^2}$
42. $\sqrt{100x^4}$
43. $\sqrt{28a^4}$
44. $\sqrt{x^2-8x+16}$
45. $\sqrt{a^2+2ab+b^2}$
46. $-\sqrt{32b^2}$
47. $\sqrt{18a^3}$
48. $\sqrt{24x^5}$
49. $\sqrt{24a^2b^4}$
50. $\sqrt{\frac{x^4y^5}{4z^2}}$
51. $\pm\sqrt{\frac{8a^3b^2}{c^4}}$
52. $\sqrt{80m^7n^5}$
53. $\sqrt[3]{a^6b^5}$
54. $\sqrt[4]{a^{12}b^8}$
55. $\sqrt[3]{-27x^3}$
56. $\sqrt[5]{-32x^5}$

Simplify each radical in simplest radical form.

57. $\sqrt{4a^2}$
58. $\sqrt{100x^4}$
59. $\pm\sqrt{28a^4}$
60. $\sqrt{x^2-8x+16}$
61. $\sqrt{a^2+2ab+b^2}$
62. $-\sqrt{32b^2}$
63. $\sqrt{18a^3}$
64. $\sqrt{24x^5}$
65. $\sqrt{24a^2b^4}$
66. $\sqrt{\frac{x^4y^5}{4z^2}}$
67. $\pm\sqrt{\frac{8a^3b^2}{c^4}}$
68. $\sqrt{80m^7n^5}$
69. $\sqrt[3]{a^6b^5}$
70. $\sqrt[4]{a^{12}b^8}$
71. $\sqrt[3]{-27x^3}$
72. $\sqrt[5]{-32x^5}$

11-3 Simplifying Radical Expressions

To simplify radical expressions, we use the following steps:

1. Simplify each radical in simplest form.
2. Combine (add or subtract) with like radicands.
3. Use the **Rules of Radicals** to multiply or divide two radicals:

$$\sqrt[n]{a}\cdot\sqrt[n]{b}=\sqrt[n]{ab}\,,\qquad \frac{\sqrt[n]{b}}{\sqrt[n]{a}}=\sqrt[n]{\frac{b}{a}}$$

4. Multiply binomials by using the distributive property, FOIL, or formulas.
5. Rationalize the denominator. (No radicals are in the denominator.)

Examples (Assume that all variables are nonnegative real numbers.)

1. $4\sqrt{2}-\sqrt{2}=3\sqrt{2}$ **2.** $6\sqrt{5}-4\sqrt{5}=2\sqrt{5}$ **3.** $3\sqrt{7}-2\sqrt{11}+5\sqrt{7}=8\sqrt{7}-2\sqrt{11}$

4. $5\sqrt{3}+\sqrt{12}=5\sqrt{3}+2\sqrt{3}=7\sqrt{3}$ **5.** $\sqrt{24}-\sqrt{54}=2\sqrt{6}-3\sqrt{6}=-\sqrt{6}$

6. $3\sqrt{18}+2\sqrt{8}-6\sqrt{50}=3\cdot3\sqrt{2}+2\cdot2\sqrt{2}-6\cdot5\sqrt{2}=9\sqrt{2}+4\sqrt{2}-30\sqrt{2}=-17\sqrt{2}$

7. $5\sqrt{7}+6\sqrt{3}-\sqrt{7}+2\sqrt{3}=(5\sqrt{7}-\sqrt{7})+(6\sqrt{3}+2\sqrt{3})=4\sqrt{7}+8\sqrt{3}$

8. $\sqrt{2}\cdot\sqrt{16}=\sqrt{2\cdot16}=\sqrt{32}=4\sqrt{2}$ **9.** $\sqrt[3]{2}\cdot\sqrt[3]{16}=\sqrt[3]{2\cdot16}=\sqrt[3]{32}=2\sqrt[3]{4}$

10. $2\sqrt{5}\cdot6\sqrt{12}=2\cdot6\cdot\sqrt{5\cdot12}=12\sqrt{60}=12\cdot2\sqrt{15}=24\sqrt{15}$

11. $\sqrt{3}(\sqrt{4}+\sqrt{6})=\sqrt{3}\cdot\sqrt{4}+\sqrt{3}\cdot\sqrt{6}=\sqrt{12}+\sqrt{18}=2\sqrt{3}+3\sqrt{2}$

12. $(4+\sqrt{2})(5-\sqrt{2})=20-4\sqrt{2}+5\sqrt{2}-2=18+\sqrt{2}$

13. $(\sqrt{2}+\sqrt{3})^2=(\sqrt{2})^2+2\cdot\sqrt{2}\cdot\sqrt{3}+(\sqrt{3})^2=2+2\sqrt{6}+3=5+2\sqrt{6}$

14. $(\sqrt{2}-\sqrt{3})^2=(\sqrt{2})^2-2\cdot\sqrt{2}\cdot\sqrt{3}+(\sqrt{3})^2=2-2\sqrt{6}+3=5-2\sqrt{6}$

15. $(\sqrt{2}+\sqrt{3})(\sqrt{2}-\sqrt{3})=(\sqrt{2})^2-(\sqrt{3})^2=2-3=-1$

16. $\sqrt{\frac{5}{3}}\cdot\sqrt{\frac{2}{5}}=\sqrt{\frac{\cancel{5}}{3}\cdot\frac{2}{\cancel{5}}}=\frac{\sqrt{2}}{\sqrt{3}}\cdot\frac{\sqrt{3}}{\sqrt{3}}=\frac{\sqrt{6}}{3}$ **17.** $\frac{\sqrt{15}}{\sqrt{6}}=\sqrt{\frac{15}{6}}=\sqrt{\frac{5}{2}}=\frac{\sqrt{5}}{\sqrt{2}}\cdot\frac{\sqrt{2}}{\sqrt{2}}=\frac{\sqrt{10}}{2}$

18. $\sqrt[3]{\frac{5}{3}}\cdot\sqrt[3]{\frac{2}{5}}=\sqrt[3]{\frac{\cancel{5}}{3}\cdot\frac{2}{\cancel{5}}}=\frac{\sqrt[3]{2}}{\sqrt[3]{3}}\cdot\frac{\sqrt[3]{9}}{\sqrt[3]{9}}=\frac{\sqrt[3]{18}}{\sqrt[3]{27}}=\frac{\sqrt[3]{18}}{3}$

19. $4\sqrt{x}+5\sqrt{x}=9\sqrt{x}$ **20.** $4x\sqrt{3x}+7x\sqrt{3x}=(4x+7x)\sqrt{3x}=11x\sqrt{3x}$

21. $\sqrt{63x}-\sqrt{7x}=3\sqrt{7x}-\sqrt{7x}=2\sqrt{7x}$ **22.** $\sqrt{2x}\cdot\sqrt{9x}=\sqrt{18x^2}=3\sqrt{2}\,x$

23. $\sqrt{ab^2}\cdot\sqrt{4a}=\sqrt{4a^2b^2}=2ab$ **24.** $\sqrt{x}(2-\sqrt{x})=2\sqrt{x}-\sqrt{x^2}=2\sqrt{x}-x$

25. $\sqrt{\frac{b}{a}}\cdot\sqrt{\frac{8b}{a}}=\sqrt{\frac{b}{a}\cdot\frac{8b}{a}}=\sqrt{\frac{8b^2}{a^2}}=\frac{\sqrt{8b^2}}{\sqrt{a^2}}=\frac{2\sqrt{2}b}{a}$, $a\neq0$

EXERCISES

Simplify each radical expression in simplest radical form.

1. $5\sqrt{3}+2\sqrt{3}$ **2.** $5\sqrt{3}-2\sqrt{3}$ **3.** $2\sqrt{3}-5\sqrt{3}$

4. $2\sqrt{6}-\sqrt{6}+7\sqrt{6}$ **5.** $9\sqrt{5}+12\sqrt{5}-4\sqrt{5}$ **6.** $-\sqrt{18}-\sqrt{8}$

7. $\sqrt{20}+\sqrt{45}$ **8.** $\sqrt[3]{4}+\sqrt[3]{32}$ **9.** $5\sqrt{32}-7\sqrt{50}$

10. $\sqrt{150}+2\sqrt{96}$ **11.** $3\sqrt{28}-2\sqrt{7}+\sqrt{63}$ **12.** $\sqrt{11}+\sqrt{11}-\sqrt{13}$

13. $\sqrt{5}\cdot\sqrt{10}$ **14.** $\sqrt[3]{9}\cdot\sqrt[3]{6}$ **15.** $2\sqrt{5}\cdot 4\sqrt{8}$

16. $6\sqrt{2}\cdot 3\sqrt{18}$ **17.** $\sqrt{2}\cdot\sqrt{3}\cdot\sqrt{12}$ **18.** $\sqrt{6}\cdot\sqrt{6}\cdot\sqrt{24}$

19. $(4\sqrt{5})^2$ **20.** $(4\sqrt[3]{5})^3$ **21.** $\sqrt{5}(\sqrt{5}-\sqrt{3})$

22. $3\sqrt{2}(2\sqrt{3}+\sqrt{2})$ **23.** $(3+\sqrt{2})(5-\sqrt{2})$ **24.** $(6-\sqrt{5})(7-2\sqrt{5})$

25. $(\sqrt{5}+\sqrt{4})^2$ **26.** $(\sqrt{5}-\sqrt{4})^2$ **27.** $(\sqrt{5}+\sqrt{4})(\sqrt{5}-\sqrt{4})$

28. $\sqrt{\frac{3}{2}}-\sqrt{\frac{2}{3}}$ **29.** $\sqrt{\frac{7}{11}}+\sqrt{\frac{11}{7}}$ **30.** $\sqrt{\frac{4}{5}}-\sqrt{\frac{2}{10}}$

31. $\sqrt{\frac{2}{5}}+\sqrt{10}$ **32.** $3\sqrt{6}-\sqrt{\frac{2}{3}}$ **33.** $\sqrt{\frac{8}{3}}\cdot\sqrt{\frac{5}{4}}$

34. $\sqrt{\frac{5}{4}}\cdot\sqrt{\frac{4}{5}}$ **35.** $\frac{\sqrt{15}}{\sqrt{10}}$ **36.** $\frac{\sqrt[3]{15}}{\sqrt[3]{10}}$

Simplify each radical expression in simplest radical form. Assume that all variables are nonnegative real numbers and that all denominators are nonzero.

37. $\sqrt{x}+\sqrt{x}$ **38.** $\sqrt{3x}+\sqrt{3x}$ **39.** $3\sqrt{x}+4\sqrt{x}$

40. $5\sqrt{2x}-2\sqrt{2x}$ **41.** $7\sqrt{5x}-11\sqrt{5x}$ **42.** $\sqrt{3x}-\sqrt{75x}$

43. $3\sqrt{20x}+4\sqrt{45x}$ **44.** $7x\sqrt{x}-5x\sqrt{x}$ **45.** $\sqrt{2x^2}+3\sqrt{8x^2}$

46. $2\sqrt{18x^3}-\sqrt{8x^3}$ **47.** $\sqrt{\frac{x^2}{4}+\frac{x^2}{16}}$ **48.** $\sqrt{\frac{a}{x}}-\sqrt{\frac{x}{a}}$

49. $\sqrt{3m}\cdot\sqrt{8m}$ **50.** $\sqrt{xy^3}\cdot\sqrt{xy^5}$ **51.** $(10\sqrt{ab^2})(-2\sqrt{a^3})$

52. $\sqrt{2a}(\sqrt{8a}-\sqrt{18a^3})$ **53.** $\sqrt{6x}(\sqrt{4x}+2\sqrt{2x^2})$ **54.** $(3\sqrt{4x})^2$

55. $(4\sqrt{3x})^3$ **56.** $\sqrt[3]{56x}+\sqrt[3]{7x}$ **57.** $x\sqrt[4]{x^{11}}-\sqrt[4]{x^{15}}$

58. $\frac{\sqrt{x^2y}}{\sqrt{2xy}}$ **59.** $\frac{\sqrt{2ab}}{\sqrt{6b}}$ **60.** $\frac{\sqrt[3]{2ab}}{\sqrt[3]{6b}}$

11-4 Solving Equations involving Radicals

To find the solution of the equation $x^2 = 25$, we know that $x = 5$ and -5 are the solutions because $5^2 = 25$ and $(-5)^2 = 25$. The numbers, 5 and -5, are the solutions of $x^2 = 25$. We can write the solutions as $x = \pm 5$, which is read as "plus or minus 5", or "positive or negative 5".

Therefore, we solve the equation $x^2 = 25$ by the following steps:

Example 1: Solve $x^2 = 25$.

Solution: $x = \pm\sqrt{25} = \pm 5$. Ans.

Consider next example to find the solution of the equation $x^2 = -9$. Since x^2 is always nonnegative, there is no real-number value of x to be the square root of -9.

Example 2: Solve $x^2 = -9$.

Solution: $x = \pm\sqrt{-9}$. Ans: No real-number solution.

Rules: If $x^2 = a$, where $a \geq 0$ (positive or 0), it has the solutions: $x = \pm\sqrt{a}$.

If $x^2 = a$, where $a < 0$ (negative), it has no real-number solution.

To solve an equation having a variable in its radicand, we isolate the radical on one side (usually on the left side) of the equation and then remove the radical through raising its power equal to the index of the radical. To solve a radical equation, we must check each solution in the original equation and eliminate the solutions which are not permissible.

Example 3: Solve $\sqrt{x} - 3 = 0$.

Solution:

	Check:
$\sqrt{x} = 3$	$\sqrt{x} - 3$
$(\sqrt{x})^2 = 3^2$	$= \sqrt{9} - 3$
$\therefore x = 9$	$= 3 - 3 = 0$ ✓

Ans: $x = 9$.

Example 4: Solve $\sqrt{x} + 3 = 0$

Solution:

	Check:
$\sqrt{x} = -3$	$\sqrt{x} + 3$
$(\sqrt{x})^2 = (-3)^2$	$= \sqrt{9} + 3$
$\therefore x = 9$	$= 3 + 3 \neq 0$

Ans: No real-number solution.

Example 5: Solve $\sqrt{x^2 + 12} - 2x = 0$.

Solution:

$$\sqrt{x^2 + 12} - 2x = 0$$

$$\sqrt{x^2 + 12} = 2x$$

Square both sides: $x^2 + 12 = 4x^2$

$$3x^2 = 12$$

$$x^2 = 4$$

$$\therefore x = \pm\sqrt{4} = \pm 2.$$

Check: $x = 2$

$$\sqrt{2^2 + 12} - 2(2) = 4 - 4 = 0 \checkmark$$

$x = -2$

$$\sqrt{(-2)^2 + 12} - 2(-2) = 4 + 4 \neq 0$$

Ans: $x = 2$.

EXERCISES

Solve each equation.

1. $x^2 = 49$ **2.** $x^2 = 81$ **3.** $x^2 = 121$

4. $x^2 = 8$ **5.** $a^2 = 24$ **6.** $x^2 = 48$

7. $x^2 = -49$ **8.** $x^2 = -8$ **9.** $x^2 = -121$

10. $x^2 - 64 = 0$ **11.** $x^2 - 16 = 0$ **12.** $x^2 + 16 = 0$

13. $x^2 - 144 = 0$ **14.** $m^2 - 25 = 0$ **15.** $x^2 + 4 = 0$

16. $a^2 - 100 = 0$ **17.** $x^2 - 32 = 0$ **18.** $y^2 - 72 = 0$

19. $x^2 - 28 = 0$ **20.** $x^2 - 150 = 0$ **21.** $2x^2 - 50 = 0$

22. $3x^2 - 48 = 0$ **23.** $4x^2 - 9 = 0$ **24.** $9x^2 - 4 = 0$

25. $(x-2)^2 = 16$ **26.** $(x+4)^2 = 25$ **27.** $x^2 + 4x + 4 = 9$

28. $x^2 - 4x + 4 = 36$ **29.** $x^3 = 8$ **30.** $x^3 = -8$

31. $64x^3 - 27 = 0$ **32.** $x^4 = 16$ **33.** $81x^4 - 16 = 0$

Solve each radical equation.

34. $\sqrt{x} = 6$ **35.** $\sqrt{x} = 8$ **36.** $\sqrt{x} = -7$

37. $\sqrt{2x} = 4$ **38.** $\sqrt{3x} = 6$ **39.** $3\sqrt{x} = 12$

40. $5\sqrt{4x} = 20$ **41.** $\sqrt{a} + 7 = 9$ **42.** $\sqrt{x} - 9 = 0$

43. $\sqrt{x+4} - 9 = 0$ **44.** $\sqrt{x+1} = 3$ **45.** $\sqrt{2x+3} = 5$

46. $\sqrt{10-x} = 1$ **47.** $\sqrt{4x-3} = 5$ **48.** $\sqrt{6-p} = p$

49. $\sqrt{x^2+72} = 3x$ **50.** $\sqrt{x^2+24} = 2x$ **51.** $\sqrt{x+11} = x-1$

52. $\sqrt{x^2+5x-2} = x$ **53.** $\sqrt{x^2-8x} = 3$ **54.** $\sqrt{x^2+6x} = 4$

55. $\sqrt[3]{a} - 4 = 0$ **56.** $\sqrt[3]{a-4} = 4$ **57.** $\sqrt[4]{2a-4} = 2$

58. The square root of two times a number is 6. Find the number.

59. When a number is subtracted from the square root of 2 subtracted from three times the number, the result is 0. Find the number.

60. Find the length of one side of a square field whose area is 50 m^2.

61. In a rectangle, the length is twice as long as the width. The area is 242 cm^2. Find its length and width.

62. The distance d in feet that an object falls in t seconds is given by the formula $d = 16t^2$. How long does it take an object falling from rest to travel 1,024 feet ?

11-5 The Pythagorean Theorem

Right angle: An angle with measure 90°. It is indicated by a small square.
Right triangle: A triangle having one right (or 90°) angle.
The hypotenuse: The side opposite the right angle in a triangle.
The other two sides are the **legs**.

The Pythagorean Theorem:
In any right triangle, the square of the length of the hypotenuse equals the sum of the squares of the lengths of the two legs.

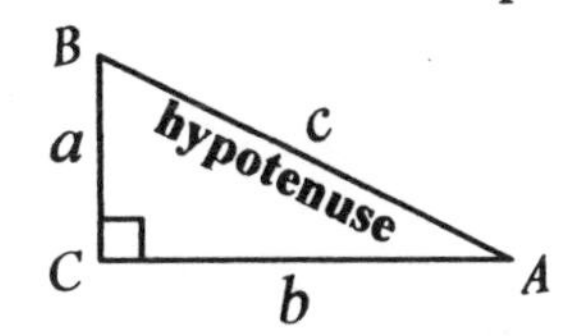

Formula: $a^2 + b^2 = c^2$

We can use the **Pythagorean Theorem** to find the length of one side of a right triangle when the lengths of the other two sides are known. To apply the Pythagorean Theorem, we need the knowledge of squares and square roots (see Section 11 – 1).

To determine whether or not the triangle with the given lengths of its three sides is a right triangle, we apply the Pythagorean Theorem. If the sum of the squares of the lengths of the two shorter sides is equal to the square of the length of the longest side, then it is a right triangle.

Examples

1. Find the length of the unknown side.

a)
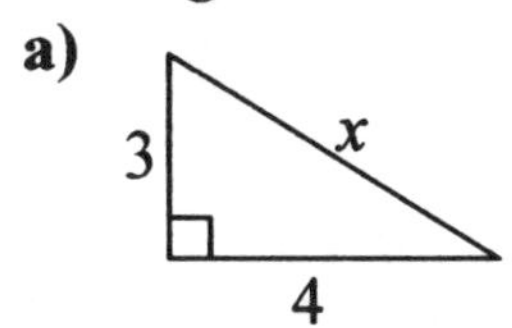

b)
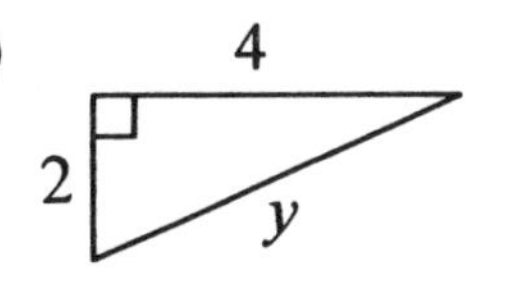

c)
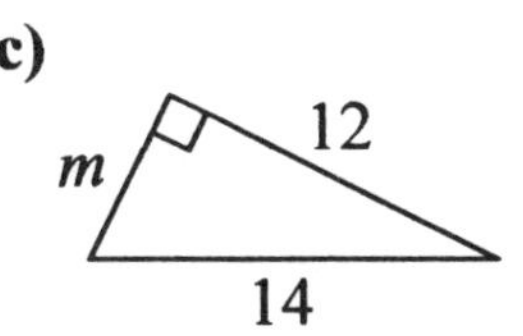

Solution:

a) $x^2 = 3^2 + 4^2$
$x^2 = 9 + 16$
$x^2 = 25$
$\therefore x = \sqrt{25} = 5$.

b) $y^2 = 2^2 + 4^2$
$y^2 = 4 + 16$
$y^2 = 20$
$\therefore y = \sqrt{20} = 4.47$.

c) $m^2 + 12^2 = 14^2$
$m^2 + 144 = 196$
$m^2 = 52$
$\therefore m = \sqrt{52} = 7.21$.

2. Determine whether or not the triangle with the given lengths of its three sides is a right triangle.

a) 6, 8, 10 **b)** 4, 5, $5\sqrt{2}$ **c)** 6, $6\sqrt{2}$, 6

Solution:

a) $6^2 + 8^2 = 100$
$10^2 = 100$
$6^2 + 8^2 = 10^2$ **Yes**

b) $4^2 + 5^2 = 41$
$(5\sqrt{2})^2 = 50$
$4^2 + 5^2 \neq (5\sqrt{2})^2$ **No**

c) $6^2 + 6^2 = 72$
$(6\sqrt{2})^2 = 72$
$6^2 + 6^2 = (6\sqrt{2})^2$ **Yes**

EXERCISES

Evaluate. Round to the nearest hundredth.

1. $\sqrt{49}$ **2.** $\sqrt{64}$ **3.** $\sqrt{121}$ **4.** $\sqrt{18}$ **5.** $\sqrt{24}$

6. $\sqrt{144}$ **7.** $\sqrt{98}$ **8.** $\sqrt{75}$ **9.** $(2\sqrt{3})^2$ **10.** $(3\sqrt{2})^2$

11. $(4\sqrt{5})^2$ **12.** $(5\sqrt{2})^2$ **13.** $\sqrt{8^2+6^2}$ **14.** $\sqrt{10^2-6^2}$ **15.** $\sqrt{(6\sqrt{2})^2-6^2}$

Find the length of the unknown side. Round to the nearest hundredth.

16.

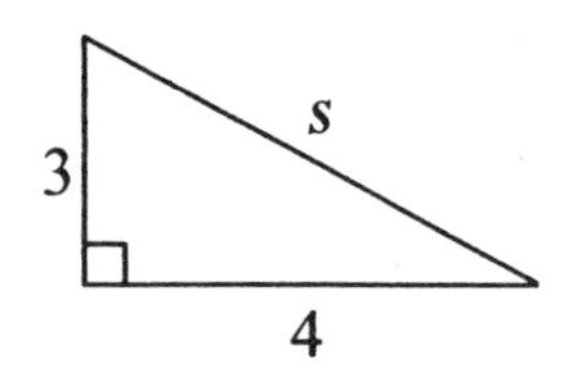

17.

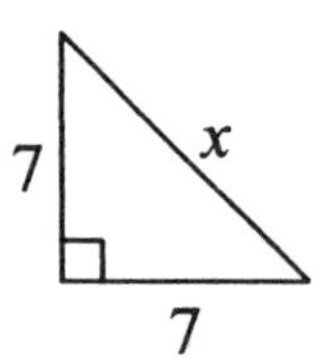

18.

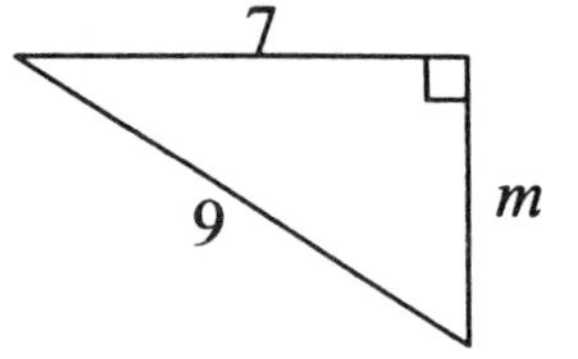

Find the length of the unknown side in the right triangle ABC. Round to the nearest hundredth.

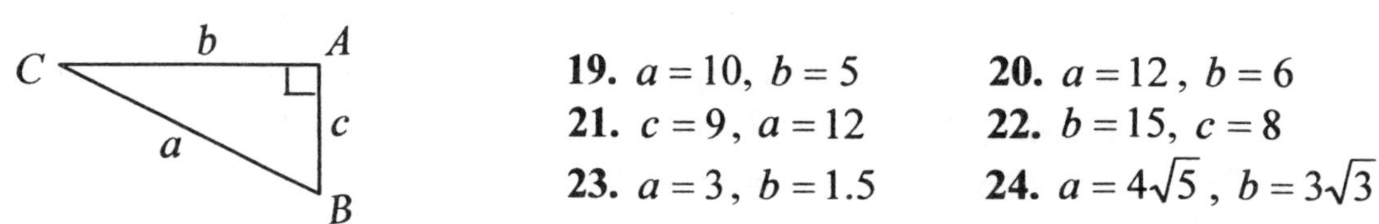

19. $a=10,\ b=5$ **20.** $a=12,\ b=6$
21. $c=9,\ a=12$ **22.** $b=15,\ c=8$
23. $a=3,\ b=1.5$ **24.** $a=4\sqrt{5},\ b=3\sqrt{3}$

Determine whether or not the triangle with the given lengths of its three sides is a right triangle.

25. 1, 1, 2 **26.** 1, 1, $\sqrt{2}$ **27.** 1, 2, $\sqrt{3}$ **28.** 2, 4, $\sqrt{6}$ **29.** 5, 5, $5\sqrt{2}$
30. 15, 20, 25 **31.** 12, 16, 20 **32.** 20, 7, 25 **33.** $\sqrt{3}$, $\sqrt{4}$, $\sqrt{7}$ **34.** $4\sqrt{3}$, 8, 4

35. A baseball field is shaped like a square diamond. Each side is 100 feet long. Find the distance from home plate to second base.

36. The length and the width of a rectangle field are 40 by 60 meters. Find the length of its diagonal.

37. The length of each leg of an isosceles right triangle is $8\,cm$. Find the length of the hypotenuse.

38. The length of each leg of an isosceles right triangle is s units. Find the length of the hypotenuse in terms of s. Write the answer in simplest radical form.

39. The sides of an equilateral triangle are each 6 inches long. Find the area.

40. The sides of an equilateral triangle are each s units long.
Find the altitude (h) in terms of s.
Write the answer in simplest radical form.

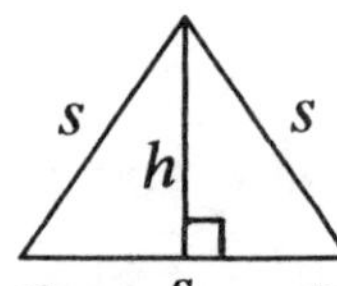

11-6 Distance and Midpoint Formulas

The distance between two points on a horizontal line or on a vertical line is the absolute value of the difference between them. The distance between two points, A and B, is indicated by $\overline{AB}$, which is the segment between A and B.

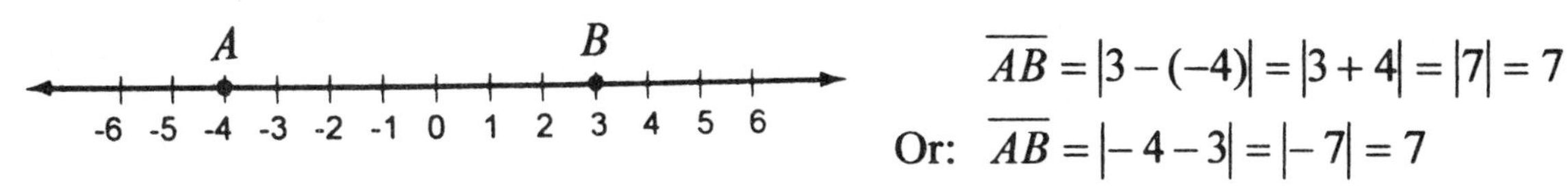

$\overline{AB} = |3-(-4)| = |3+4| = |7| = 7$

Or: $\overline{AB} = |-4-3| = |-7| = 7$

To find the distance between two points not on a horizontal line or a vertical line, we apply the **Pythagorean Theorem**:

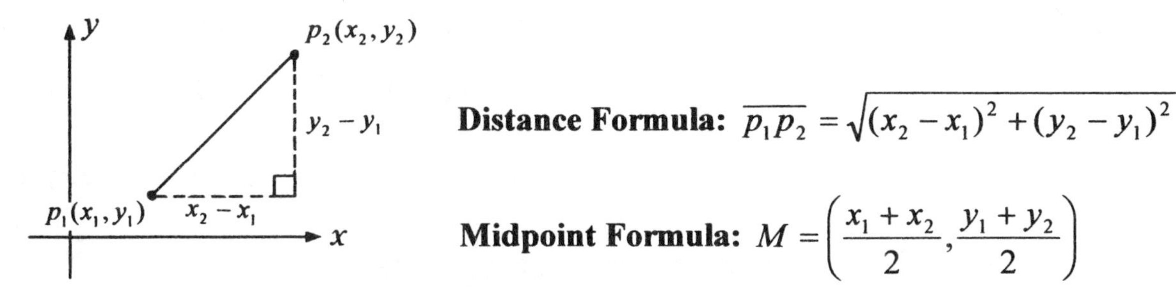

Distance Formula: $\overline{p_1p_2} = \sqrt{(x_2-x_1)^2+(y_2-y_1)^2}$

Midpoint Formula: $M = \left(\dfrac{x_1+x_2}{2}, \dfrac{y_1+y_2}{2}\right)$

Examples

1. Find the distance and midpoint between $A(-2, 1)$ and $B(5, 4)$.

Solution:

$$\overline{AB} = \sqrt{[5-(-2)]^2+(4-1)^2} = \sqrt{49+9} = \sqrt{58} \approx 7.62.$$

$$M = \left(\frac{-2+5}{2}, \frac{1+4}{2}\right) = (1.5,\ 2.5).$$

2. Find the distance and midpoint between $(4, 3\sqrt{3})$ and $(2, -\sqrt{3})$.

Solution:

$$d = \sqrt{(2-4)^2+(-\sqrt{3}-3\sqrt{3})^2} = \sqrt{(-2)^2+(-4\sqrt{3})^2} = \sqrt{4+48} = \sqrt{52} \approx 7.21.$$

$$M = \left(\frac{4+2}{2}, \frac{3\sqrt{3}+(-\sqrt{3})}{2}\right) = (3, \sqrt{3}) = (3,\ 1.73).$$

3. If $M\,(1, 0)$ is the midpoint of the segment $\overline{AB}$ and $B\,(-1, -2)$ is the coordinates of point B. Find the coordinates of point A.

Solution: $A(x, y)$, $M(1, 0)$, $B(-1, -2)$

$$\frac{x+(-1)}{2} = 1, \quad x-1 = 2, \quad \therefore x = 3$$

$$\frac{y+(-2)}{2} = 0, \quad y-2 = 0, \quad \therefore y = 2$$

Ans: The coordinates of point A is $A(3, 2)$.

EXERCISES

Find the distance between the given two points to the nearest hundredth.

1. A(8, 0), B(2, 0) **2.** C(−8, 0), D(2, 0) **3.** M(−8, 0), N(−2, 0)

4. A(0, 6), C(0, −4) **5.** E(0, −6), F(0, 4) **6.** G(0, −6), H(0, −4)

7. P(8, 0), Q(0, 2) **8.** S(−8, 0), T(0, −2) **9.** P(0, −6), R(−3, 0)

10. A(3, 2), B(7, 8) **11.** A(2, −3), C(8, 7) **12.** P(4, 1), Q(1, 4)

13. (10, 5), (−7, 0) **14.** (8, −2), (4, −6) **15.** (−2, 1), (−5, −4)

16. (−7, −9), (2, −9) **17.** (−7, 10), (−7, 6) **18.** (4, −12), (−8, 1)

19. (−1, −1), (−9, −7) **20.** (−4, −2), (−8, −5) **21.** $(2, \frac{2}{3}), (4, 1)$

22. $(1, \frac{2}{3}), (4, -\frac{1}{2})$ **23.** $(2, \sqrt{3}), (6, -2\sqrt{3})$ **24.** $(2, -\sqrt{3}), (6, \sqrt{3})$

Find the midpoint between the given two points.

25. A(8, 0), B(2, 0) **26.** C(−8, 0), D(2, 0) **27.** E(0, −6), F(0, 4)

28. A(3, 2), B(7, 8) **29.** P(−4, 1), Q(−1, 4) **30.** A(2, −3), C(8, 7)

31. (10, 5), (−7, 0) **32.** (−7, 10), (−7, 6) **33.** (4, −12), (−8, 1)

34. (−4, −2), (−8, −5) **35.** $(2, \frac{2}{3}), (4, 1)$ **36.** $(2, \sqrt{3}), (6, -2\sqrt{3})$

Given the coordinates of the midpoint (M) and one endpoint (P) of $\overline{PQ}$, find the coordinates of the other endpoint (Q).

37. $M(5, 0)$, $P(2, 0)$ **38.** $M(-3, 0)$, $P(2, 0)$ **39.** $M(5, 2)$, $P(2, -3)$

40. $M(0, -1)$, $P(0, -6)$ **41.** $M(5, 5)$, $P(3, 2)$ **42.** $M(-7, 8)$, $P(-7, 6)$

43. $M(\frac{3}{2}, \frac{5}{2})$, $P(10, 5)$ **44.** $M(3, \frac{5}{6})$, $P(2, \frac{2}{3})$ **45.** $M(4, -\frac{\sqrt{3}}{2})$, $P(2, \sqrt{3})$

Given the three vertices of a triangle, determine whether or not it is a right triangle.

46. A(4, 2), B(2, 2), C(4, 0) **47.** P(3, 4), Q(3, 1), R(−1, 4)
48. A(−3, 2), B(0, 4), C(0, 2) **49.** P(2, 3), Q(3, 1), R(4, 5)
50. A(1, 5), B(−1, 3), C(5, 2) **51.** P(1, 1), Q(−2, −2), R(4, −1)
52. A(0, 5), B(−3, 2), C(2, 3) **53.** P(0. 5), Q(−2, 3), R(2, 2)

CHAPTER 11 EXERCISES

Evaluate each radical. If the root is an irrational number, round the decimal to the nearest hundredth.

1. $\sqrt{9}$ **2.** $\sqrt{1}$ **3.** $\pm\sqrt{121}$ **4.** $\sqrt{-9}$

5. $\pm\sqrt{-121}$ **6.** $\sqrt{1600}$ **7.** $\sqrt{3.61}$ **8.** $\sqrt{1.21}$

9. $\pm\sqrt{0.0009}$ **10.** $\sqrt{0.0625}$ **11.** $\sqrt{15}$ **12.** $-\sqrt{20}$

13. $\sqrt{\frac{16}{9}}$ **14.** $\sqrt{\frac{1}{25}}$ **15.** $\sqrt{\frac{25}{81}}$ **16.** $\sqrt{\frac{20}{45}}$

17. $5\sqrt{25}$ **18.** $\pm 7\sqrt{12}$ **19.** $-12\sqrt{4.8}$ **20.** $8\sqrt{7.2}$

21. $\sqrt{\frac{3}{5}}$ **22.** $\sqrt{\frac{5}{3}}$ **23.** $\sqrt{\frac{8}{9}}$ **24.** $\sqrt{\frac{9}{8}}$

25. $\sqrt[3]{125}$ **26.** $\sqrt[4]{625}$ **27.** $\sqrt[3]{-125}$ **28.** $\sqrt[4]{-625}$

Simplify each radical in simplest radical form. Assume that all variables are nonnegative real numbers.

29. $\sqrt{20}$ **30.** $\sqrt{200}$ **31.** $\sqrt{75}$ **32.** $7\sqrt{72}$

33. $\sqrt{\frac{3}{5}}$ **34.** $\sqrt{\frac{5}{3}}$ **35.** $\sqrt{\frac{8}{9}}$ **36.** $\sqrt{\frac{9}{8}}$

37. $\sqrt[3]{250}$ **38.** $\sqrt[4]{405}$ **39.** $\sqrt[5]{64}$ **40.** $\sqrt[4]{64}$

41. $\sqrt{9x^2}$ **42.** $\sqrt{16x^4}$ **43.** $\sqrt{64x^3}$ **44.** $\sqrt{18x^3}$

45. $\sqrt{a^2-10a+25}$ **46.** $\sqrt{4a^2+4a+1}$ **47.** $\sqrt[3]{24x^5}$ **48.** $\sqrt[3]{-16n^3}$

49. $\sqrt[4]{32m^4n^5}$ **50.** $\sqrt[5]{-32m^4n^5}$ **51.** $\sqrt{\frac{x^2y^6}{9z^4}}$ **52.** $\sqrt{\frac{18a^4b}{4c^6}}$

Simplify each radical in simplest radical form.

53. $\sqrt{9x^2}$ **54.** $\sqrt{16x^4}$ **55.** $\sqrt{64x^3}$ **56.** $\sqrt{18x^3}$

57. $\sqrt{a^2-10a+25}$ **58.** $\sqrt{4a^2+4a+1}$ **59.** $\sqrt[3]{24x^5}$ **60.** $\sqrt[3]{-16n^3}$

61. $\sqrt[4]{32m^4n^5}$ **62.** $\sqrt[5]{-32m^4n^5}$ **63.** $\sqrt{\frac{x^2y^6}{9z^4}}$ **64.** $\sqrt{\frac{18a^4b}{4c^6}}$

-----Continued-----

Simplify each radical expression in simplest radical form. Assume that all variables are nonnegative real numbers.

65. $\sqrt{5}+4\sqrt{5}$ **66.** $4\sqrt{6}-7\sqrt{6}+5\sqrt{6}$ **67.** $\sqrt{12}+\sqrt{48}-\sqrt{75}$

68. $\sqrt[3]{16}-\sqrt[3]{54}$ **69.** $\sqrt{5}\cdot\sqrt{6}\cdot\sqrt{3}$ **70.** $5\sqrt{10}\cdot 2\sqrt{15}$

71. $(3\sqrt{7})^2$ **72.** $(3\sqrt[3]{7})^3$ **73.** $3\sqrt{2}(\sqrt{8}-8\sqrt{6})$

74. $(2-\sqrt{5})(3+\sqrt{5})$ **75.** $(2\sqrt{3}-\sqrt{2})^2$ **76.** $(2\sqrt{3}+\sqrt{2})(2\sqrt{3}-\sqrt{2})$

77. $\sqrt{\frac{3}{5}}+\sqrt{\frac{5}{3}}$ **78.** $\sqrt{\frac{5}{8}}\cdot\sqrt{\frac{3}{5}}$ **79.** $\frac{\sqrt[3]{15}}{\sqrt[3]{12}}$

80. $4\sqrt{5x}-9\sqrt{5x}$ **81.** $\sqrt{80x}+\sqrt{45x}$ **82.** $\sqrt{6ab}\cdot\sqrt{12a^5b}$

83. $\frac{\sqrt{2ab^2}}{\sqrt{8ab}}$ **84.** $\frac{\sqrt{5xy}}{\sqrt{2x}}$ **85.** $\frac{\sqrt[3]{5xy}}{\sqrt[3]{2x}}$

Solve each equation. Write the answer in simplest radical form.

86. $x^2=5$ **87.** $\sqrt{x}=5$ **88.** $x^2=-5$

89. $\sqrt{x}=-5$ **90.** $x^3=27$ **91.** $x^3=-27$

92. $\sqrt{3x}-1=5$ **93.** $9x^2-4=0$ **94.** $27x^3+8=0$

95. $\sqrt{5x-4}=4$ **96.** $(x+4)^2=9$ **97.** $\sqrt{x+2}=x-4$

98. $\sqrt{4x^2+7x-3}=2x$ **99.** $\sqrt[3]{x}=3$ **100.** $\sqrt[4]{3x}=3$

101. The lengths of two sides of a right triangle are $12\,cm$ and $16\,cm$. Find the length of the hypotenuse.

102. The length of the hypotenuse of a right triangle is 18 feet. The length of one side is 12 feet. Find the length of the unknown side in simplest radical form.

103. A baseball field is shaped like a square diamond. Each side is 90 feet long. Find the length from home plate to second base.

104. Find the distance and midpoint between points $A(4,-3)$ and $B(1,7)$.

105. Find the distance and midpoint between points $P(6,\ 4\sqrt{2})$ and $Q(9,\ 2\sqrt{2})$. Write the answer in simplest radical form.

106. If $M(3,\ -1)$ is the midpoint of the segment $\overline{AB}$ and $A(8,\ 12)$ is the coordinates of A. Find the coordinates of point B.

107. Determine whether or not the points $A(5,\ 5)$, $B(-1,\ -1)$, and $C(2,\ -4)$ are the vertices of a right triangle. Explain.

Rationalize the denominator of each radical expression:

108. $\frac{1}{\sqrt{x}-2}$ **109.** $\frac{\sqrt{3}-1}{\sqrt{3}+1}$ **110.** $\frac{\sqrt{3}-\sqrt{2}}{\sqrt{3}+\sqrt{2}}$

Proofs of the Pythagorean Theorem

The Pythagorean Theorem is probable the most famous theorem in the history of mathematics and has been used for thousands of years. It can help us solve many real-life problems in sports, sciences, and architecture. The Pythagorean Theorem was first proved by Pythagoras, a famous Greek mathematician who lived about 550 B.C.
There are numerous ways to prove this theorem. Five major proofs are introduced here. Read the following examples and find a new proof on your own. Send your proof to us. Your proof will be included in this book in the next printing.

Example 1: A proof of the Pythagorean Theorem: $a^2 + b^2 = c^2$

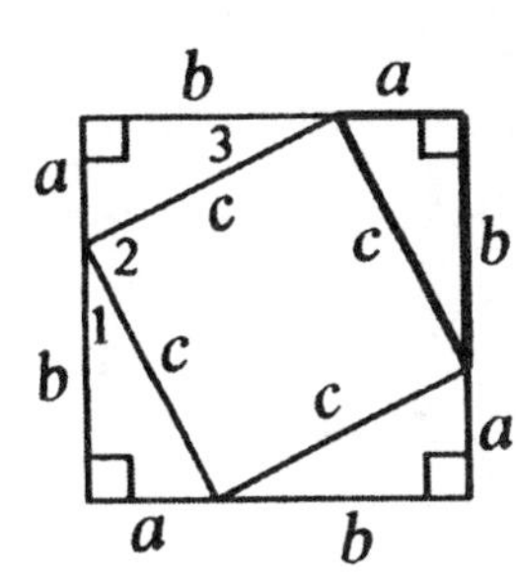

Draw a right triangle with legs a and b, and the hypotenuse c.
Construct a diagram shown at the left.
Since $\angle 1 + \angle 2 = \angle 3 + 90°$ and $\angle 1 = \angle 3$, we have $\angle 2 = 90°$.
The diagram shows a square with side $(a+b)$ units and a smaller square with side c units.
The diagram shows four equal right triangles. The area of each right triangle is $\frac{1}{2}ab$.

The area of the square with side $(a+b)$ units equals the sum of the area of the square with side c units plus the areas of the four right triangles.

$$(a+b)^2 = c^2 + 4(\tfrac{1}{2}ab)$$

$$a^2 + \cancel{2ab} + b^2 = c^2 + \cancel{2ab}$$

$\therefore a^2 + b^2 = c^2$. The proof is complete.

Example 2: A proof of the Pythagorean Theorem: $a^2 + b^2 = c^2$

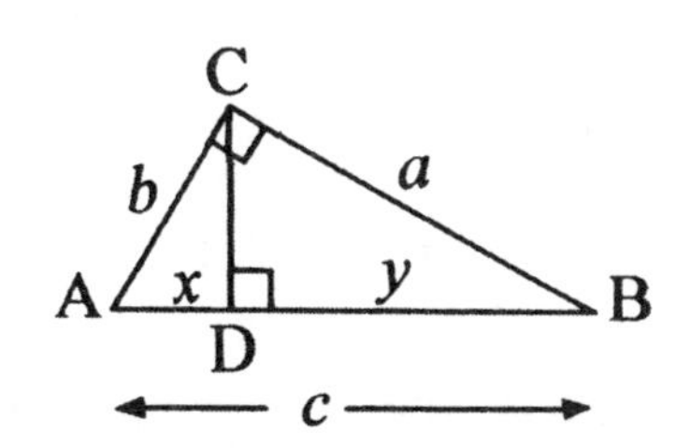

ΔABC is a right triangle with legs a and b, and the hypotenuse c. CD is the altitude.

If two triangles are similar, the corresponding sides are proportional.

ΔABC is similar to ΔBCD, we have: $\frac{c}{a} = \frac{a}{y}$ $\quad \therefore a^2 = cy$

ΔABC is similar to ΔACD, we have: $\frac{c}{b} = \frac{b}{x}$ $\quad \therefore b^2 = cx$

Therefore: $a^2 + b^2 = cy + cx$

$$a^2 + b^2 = c(y + x)$$

$$a^2 + b^2 = c \cdot c$$

$\therefore a^2 + b^2 = c^2$. The proof is complete.

-----Continued-----

Example 3: A proof of the Pythagorean Theorem: $a^2 + b^2 = c^2$

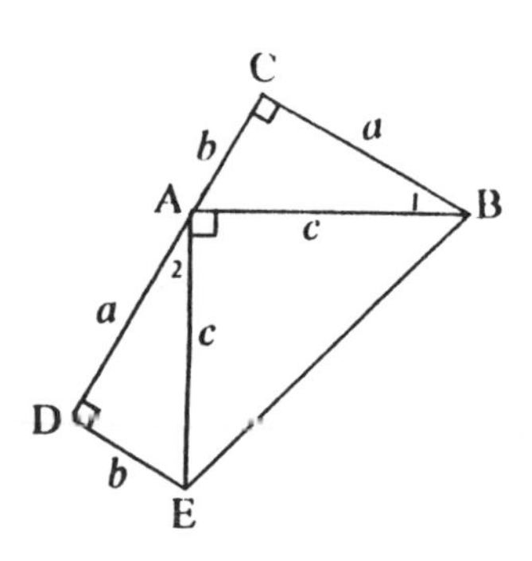

Draw a right triangle ABC with legs a and b, and the hypotenuse c.
Extend AC to point D so that AD = BC = a.
Draw segment DE so that DE ⊥ CD and DE = AC = b.
Draw segment AE and BE.
Construct a diagram shown at the left.

Based on the **SAS Postulate**, we have $\triangle ABC \cong \triangle ADE$.
Therefore, we have AE = AB = c and $\angle 1 = \angle 2$.
Since $\angle 2 + \angle BAE = \angle 1 + 90^\circ$, we have $\angle BAE = 90^\circ$.
The area of the trapezoid BCDE equals the sum of the areas of of the three right triangles. $\frac{1}{2}(a+b)(a+b) = \frac{1}{2}ab + \frac{1}{2}ab + \frac{1}{2}c^2$

$$\tfrac{1}{2}(a^2 + 2ab + b^2) = ab + \tfrac{1}{2}c^2$$

$a^2 + \cancel{2ab} + b^2 = \cancel{2ab} + c^2 \quad \therefore a^2 + b^2 = c^2$. The proof is complete.

Example 4: A proof of the Pythagorean Theorem: $a^2 + b^2 = c^2$

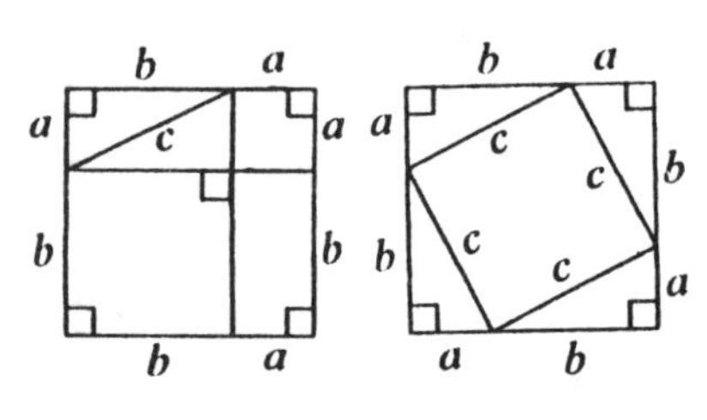

Draw a right triangle with legs a and b, and the hypotenuse c.
Construct two diagrams shown at the left.
Each diagram shows a square with ($a+b$) units on one side.
The area of the diagram with two squares and two rectangles equals the area of the diagram with one square and four equal right triangles. $a^2 + b^2 + 2ab = c^2 + 4(\frac{1}{2}ab)$

$a^2 + b^2 + \cancel{2ab} = c^2 + \cancel{2ab} \quad \therefore a^2 + b^2 = c^2$. The proof is complete.

Example 5: A proof of the Pythagorean Theorem: $a^2 + b^2 = c^2$

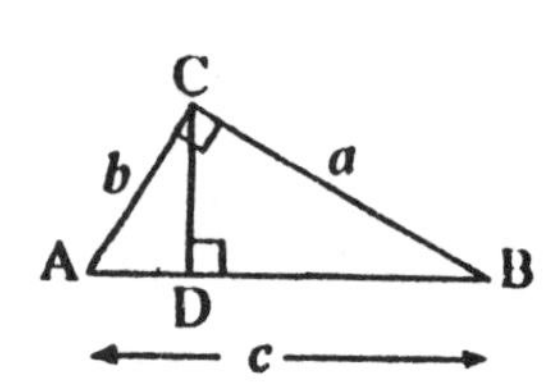

$\triangle ABC$ is a right triangle with legs a and b, and the hypotenuse c.
CD is the altitude.
If two triangles are similar, the ratio of their areas is the square of the ratio of any two corresponding sides. Therefore:

$\triangle ABC$ is similar to $\triangle BCD$, we have: $\dfrac{Area(\triangle BCD)}{Area(\triangle ABC)} = \left(\dfrac{a}{c}\right)^2$

$\triangle ABC$ is similar to $\triangle ACD$, we have: $\dfrac{Area(\triangle ACD)}{Area(\triangle ABC)} = \left(\dfrac{b}{c}\right)^2$

$$\frac{Area(\triangle BCD)}{Area(\triangle ABC)} + \frac{Area(\triangle ACD)}{Area(\triangle ABC)} = \frac{a^2}{c^2} + \frac{b^2}{c^2}, \quad \frac{Area(\triangle BCD) + Area(\triangle ACD)}{Area(\triangle ABC)} = \frac{a^2 + b^2}{c^2},$$

$1 = \dfrac{a^2 + b^2}{c^2} \quad \therefore a^2 + b^2 = c^2$. The proof is complete.

CHAPTER TWELVE

Quadratic Equations and Functions

12-1 Quadratic Equations involving Perfect Squares

Quadratic Equation: An equation of the form $ax^2 + bx + c = 0$ $(a \neq 0)$.
Quadratic Function: A function of the form $f(x) = ax^2 + bx + c$ $(a \neq 0)$.
In Section 8-7 , we have learned to solve quadratic equations by factoring.
In Section 11-4, we have learned to solve basic quadratic equations of the form $x^2 = a$.
In this section, we will learn more to solve any quadratic equation involving a perfect-square expression, such as $(x-3)^2$, $(4x-2)^2$, or $2(x-3)^2$.

We use the following rules to find the solutions (roots) of $x^2 = a$:

1. **If $a \geq 0$ (positive or 0), it has the solutions: $x = \pm\sqrt{a}$.**
2. **If $a < 0$ (negative), it has no real-number solution.**

The formulas of the special patterns in Section 8-3 are useful to solve the quadratic equations involving perfect squares (see example **3**).

1. $(a+b)^2 = a^2 + 2ab + b^2$ **2.** $(a-b)^2 = a^2 - 2ab + b^2$

Examples

1. Solve $x^2 = 36$.

Solution: $x^2 = 36$

$x = \pm\sqrt{36}$

$x = \pm 6$

$\therefore x = 6$ or -6. Ans.

2. Solve $(x-3)^2 = 36$.

Solution:

$(x-3)^2 = 36$

$x - 3 = \pm\sqrt{36}$

$x - 3 = \pm 6$

$x = 3 \pm 6$

$\therefore x = 9$ or -3. Ans.

3. Solve $x^2 - 6x + 9 = 36$.

Solution: $x^2 - 6x + 9 = 36$

$(x-3)^2 = 36$

$x - 3 = \pm 6$

$x = 3 \pm 6 = 9$ or -3. Ans.

4. Solve $(4x-2)^2 = 36$.

Solution: $(4x-2)^2 = 36$

$4x - 2 = \pm 6$

$4x = 2 \pm 6$, $4x = 8$ or -4

$\therefore x = 2$ or -1. Ans.

5. Solve $2(x-3)^2 = 36$.

Solution: $2(x-3)^2 = 36$

$(x-3)^2 = 18$

$x - 3 = \pm\sqrt{18}$

$x = 3 \pm \sqrt{18}$

$x = 3 \pm 3\sqrt{2}$

$\therefore x = 3 + 3\sqrt{2}$ or $3 - 3\sqrt{2}$. Ans.

6. Solve $2(x-3)^2 + 46 = 36$.

Solution: $2(x-3)^2 + 46 = 36$

$2(x-3)^2 = -10$

$(x-3)^2 = -5$

Ans: No real-number solution.

EXERCISES

Write each perfect-square expression as a trinomial.
(Formulas: $(a+b)^2 = a^2 + 2ab + b^2$ and $(a-b)^2 = a^2 - 2ab + b^2$**)**

1. $(x-4)^2$ **2.** $(x+4)^2$ **3.** $(x+12)^2$ **4.** $(x-12)^2$

5. $(2y-3)^2$ **6.** $(2y+3)^2$ **7.** $(x+\frac{1}{2})^2$ **8.** $(x-\frac{1}{2})^2$

9. $(2x-\frac{3}{4})^2$ **10.** $(\frac{1}{2}a-\frac{1}{5})^2$

Write each expression as a perfect-square of a binomial.

11. x^2-2x+1 **12.** x^2+2x+1 **13.** x^2+4x+4 **14.** x^2-4x+4

15. $x^2+8x+16$ **16.** $x^2-8x+16$ **17.** $y^2-20y+100$ **18.** $y^2+20y+100$

19. $a^2+\frac{1}{4}a+\frac{1}{64}$ **20.** $4x^2-2x+\frac{1}{4}$

Solve each equation. Write answers in the simplest radical forms if necessary.

21. $x^2=16$ **22.** $x^2=-16$ **23.** $x^2=144$ **24.** $x^2=54$

25. $x^2=200$ **26.** $a^2=\frac{16}{81}$ **27.** $a^2+5=25$ **28.** $y^2-5=19$

29. $y^2=\frac{16}{100}$ **30.** $7x^2=56$ **31.** $2x^2-6=40$ **32.** $4x^2+19=3$

33. $(x+2)^2=16$ **34.** $(x-2)^2=54$ **35.** $(a+5)^2=28$ **36.** $(a-5)^2=28$

37. $(y+2)^2=-16$ **38.** $(y-6)^2-8=8$ **39.** $(x+7)^2-18=0$ **40.** $4(x+3)^2+3=19$

41. $5(m-9)^2=30$ **42.** $7(n+1)^2=56$ **43.** $x^2+6x+9=16$ **44.** $x^2+16x+64=4$

45. $x^2-14x+49=1$ **46.** $a^2+2a+1=121$ **47.** $\frac{1}{2}x^2-\frac{9}{50}=0$ **48.** $\frac{1}{5}x^2-4=\frac{4}{5}$

49. $4(x-6)^2=25$ **50.** $8(x+5)^2=\frac{7}{8}$ **51.** $m^3-49m=0$ **52.** $3a^3-54a=0$

53. Find the length of one side of a square field whose area is 80 m^2.

54. The area of a circle is given by the formula $A=\pi r^2$ (where $\pi \approx 3.14$ and r is the radius). Find the radius if the area is 50.24 cm^2.

55. The distance d in feet that an object falls in t seconds is given by the formula $d=16t^2$. How long does it take an object falling from rest to travel 1,296 feet ?

56. An object falls from a height 1,296 feet above the ground. The height h in feet after t seconds is given by the formula $h=1296-16t^2$. How long does it take for an object to reach a height of 272 feet ? How long does it take for an object to reach the ground ?

12-2 Completing the Square

In Section 12-1, we found that it is easy to solve a quadratic equation if the equation has the form of perfect-square: (The left side is a perfect-square.)

$$x^2 - 6x + 9 = 36$$

$$(x-3)^2 = 36, \quad x - 3 = \pm 6 \quad \therefore x = 3 \pm 6 = 9 \text{ or } -3.$$

In this section, we will learn to solve a quadratic equation that does not have the form of perfect-square. We use a method called **completing the square**. The main step of this method is to transform the equation $ax^2 + bx + c = 0$ so that the left side is a perfect-square.

Rules of Completing the Square for $ax^2 + bx + c = 0$ **:**

1. If $a = 1$, we add $(\frac{b}{2})^2$ on both sides to complete the square on the left side.
   (Add the square of half of the coefficient of the x – term **to both sides.)**
2. If $a \neq 1$, we divide both side by a so that the coefficient of x^2 will be 1. Then, follow the Rule **1** to complete the square on the left side (see example **5**).

Examples

1. Write the value that makes a perfect-square when added to the form $x^2 + 8x$.
   Solution:
   $(\frac{b}{2})^2 = (\frac{8}{2})^2 = 16$. Ans. Check: $x^2 + 8x + 16 = (x+4)^2$ ✔

2. Write the value that makes a perfect-square when added to the form $x^2 - \frac{1}{2}x$.
   Solution:
   $(\frac{b}{2})^2 = \left(\frac{-1/2}{2}\right)^2 = \frac{1}{16}$. Ans. Check: $x^2 - \frac{1}{2}x + \frac{1}{16} = (x - \frac{1}{4})^2$ ✔

3. Solve $x^2 + 8x - 4 = 0$.
   Solution: $(\frac{b}{2})^2 = (\frac{8}{2})^2 = 16$
   $x^2 + 8x = 4$
   $x^2 + 8x + 16 = 4 + 16$
   $(x+4)^2 = 20$
   $x + 4 = \pm\sqrt{20}$
   $\therefore x = -4 \pm 2\sqrt{5}$. Ans.

4. Solve $x^2 - \frac{1}{2}x = \frac{1}{8}$.
   Solution: $(\frac{b}{2})^2 = \left(\frac{-1/2}{2}\right)^2 = \frac{1}{16}$
   $x^2 - \frac{1}{2}x = \frac{1}{8}$
   $x^2 - \frac{1}{2}x + \frac{1}{16} = \frac{1}{8} + \frac{1}{16}$
   $(x - \frac{1}{4})^2 = \frac{3}{16}$
   $x - \frac{1}{4} = \pm\sqrt{\frac{3}{16}}$
   $\therefore x = \frac{1}{4} \pm \frac{\sqrt{3}}{4}$. Ans.

5. Solve $2a^2 - 12a + 3 = 0$.
   Solution: $2a^2 - 12a = -3$
   Divide both sides by 2
   $a^2 - 6a = -\frac{3}{2}$
   $a^2 - 6a + 9 = -\frac{3}{2} + 9$
   $(a-3)^2 = \frac{15}{2}$
   $a - 3 = \pm\sqrt{\frac{15}{2}}$
   $\therefore a = 3 \pm \sqrt{\frac{15}{2}} = 3 \pm \frac{\sqrt{30}}{2}$. Ans.

EXERCISES

Write the value that makes a perfect-square when added to each expression.

1. x^2+6x **2.** x^2-6x **3.** x^2-10x **4.** x^2+10x

5. x^2+22x **6.** a^2-2a **7.** x^2-x **8.** x^2+38x

9. x^2-11x **10.** x^2+13x **11.** x^2+15x **12.** $p^2+\frac{1}{3}p$

13. $x^2+\frac{5}{6}x$ **14.** $y^2 \quad \frac{3}{8}y$ **15.** $4x^2-12x$ **16.** $4x^2+24x$

Complete the square.

17. $x^2+6x+\_\_\_\_=(x+\_\_\_\_)^2$ **18.** $x^2-10x+\_\_\_\_=(x-\_\_\_\_)^2$

19. $a^2-3a+\_\_\_\_=(a-\_\_\_\_)^2$ **20.** $m^2+\frac{3}{4}m+\_\_\_\_=(m+\_\_\_\_)^2$

21. $m^2-1.8m+\_\_\_\_=(m-\_\_\_\_)^2$ **22.** $4x^2+24x+\_\_\_\_=(2x+\_\_\_)^2$

Solve each equation by completing the square.

23. $x^2+6x=16$ **24.** $x^2-6x-16=0$ **25.** $x^2-10x-11=0$

26. $x^2+10x=11$ **27.** $x^2-22x+41=0$ **28.** $x^2+22x=-1$

29. $a^2-2a-11=0$ **30.** $x^2-x=\frac{3}{4}$ **31.** $x^2+38x+1=0$

32. $x^2-11x+\frac{1}{4}=0$ **33.** $x^2+13x+\frac{9}{4}=0$ **34.** $a^2+2a-12=0$

35. $a^2+10x-3=0$ **36.** $y^2+4y+20=0$ **37.** $y^2-14y+31=0$

38. $p^2-14p+50=0$ **39.** $p^2+2p-120=0$ **40.** $x^2-16x+60=0$

41. $x^2+6x-7=0$ **42.** $x^2-4x=50$ **43.** $x^2+10x=3$

44. $n^2+2n=7$ **45.** $n^2-14n=-31$ **46.** $y^2+8y=4$

47. $p^2+6p=6$ **48.** $y^2+y=3$ **49.** $x^2-3x=2$

50. $k^2+8k-2=0$ **51.** $k^2-4k-2=0$ **52.** $m^2-7m-6=0$

53. $x^2+\frac{1}{2}x-4=0$ **54.** $x^2-\frac{2}{3}x-1=0$ **55.** $2x^2+6x+3=0$

56. $3x^2+6x+2=0$ **57.** $\frac{1}{4}x^2-2x=3$ **58.** $\frac{2}{3}x^2-x=5$

59. Write the value that makes a perfect-square when added to the form x^2+bx.

60. Write the value that makes a perfect-square when added to the form $x^2+\frac{b}{a}x$.

61. Write the value that makes a perfect-square when added to the form ax^2+bx.

12-3 The Quadratic Formula

We learned that equations in the form $ax^2 + bx + c = 0$ ($a \neq 0$) are called **quadratic equations.**

Steps for solving a quadratic equation:

1. Solve the equation in the form of perfect-square. (See Section 12-1)
   If $x^2 = k$, then $x = \pm\sqrt{k}$.
2. Solve by factoring if it can be factored to the pattern: (See Section 8-5, 8-7)
   $(px + r)(qx + s) = 0$.
3. Solve by using the quadratic formula if factoring is not possible.

Quadratic Formula: $x = \dfrac{-b \pm \sqrt{b^2 - 4ac}}{2a}$

Quadratic formula can be used to solve any quadratic equation. But, we usually use it to solve the quadratic equation which cannot be solved by factoring.

Examples

1. Solve $2x^2 - 5x - 3 = 0$ by factoring.

Solution:

$$2x^2 - 5x - 3 = 0$$
$$(2x + 1)(x - 3) = 0$$

$2x + 1 = 0$, $2x = -1$, $x = -\frac{1}{2}$ | $x - 3 = 0$, $x = 3$

Ans: $x = -\frac{1}{2}$ or 3.

2. Solve $2x^2 - 5x - 3 = 0$ by formula.

Solution:

$$2x^2 - 5x - 3 = 0$$
$$a = 2,\ b = -5,\ c = -3$$
$$x = \frac{-b \pm \sqrt{b^2 - 4ac}}{2a}$$
$$= \frac{-(-5) \pm \sqrt{(-5)^2 - 4 \cdot 2 \cdot (-3)}}{2(2)}$$
$$= \frac{5 \pm \sqrt{25 + 24}}{4} = \frac{5 \pm 7}{4}$$
$$= \frac{5+7}{4} \text{ or } \frac{5-7}{4} = 3 \text{ or } -\tfrac{1}{2}. \text{ Ans.}$$

3. Solve $3x^2 + 2x - 4 = 0$.

Solution: $3x^2 + 2x - 4 = 0$

$$a = 3,\ b = 2,\ c = -4$$
$$x = \frac{-b \pm \sqrt{b^2 - 4ac}}{2a} = \frac{-2 \pm \sqrt{2^2 - 4 \cdot 3 \cdot (-4)}}{2 \cdot 3}$$
$$= \frac{-2 \pm \sqrt{4 + 48}}{6} = \frac{-2 \pm \sqrt{52}}{6} = \frac{-2 \pm 2\sqrt{13}}{6}$$
$$= \frac{-1 \pm \sqrt{13}}{3} \quad \text{Ans.}$$

4. Solve $n^2 - 7n + 11 = 0$.

Solution:

$$n^2 - 7n + 11 = 0$$
$$a = 1,\ b = -7,\ c = 11$$
$$n = \frac{-b \pm \sqrt{b^2 - 4ac}}{2a}$$
$$= \frac{-(-7) \pm \sqrt{(-7)^2 - 4 \cdot 1 \cdot 11}}{2(1)}$$
$$= \frac{7 \pm \sqrt{49 - 44}}{2} = \frac{7 \pm \sqrt{5}}{2}. \quad \text{Ans.}$$

EXERCISES

Solve each quadratic equation by using the appropriate method. Write irrational answers in simplest radical form.

1. $x^2 - 4 = 0$ **2.** $x^2 + 4 = 0$ **3.** $(x-2)^2 = 4$

4. $(x+2)^2 = 4$ **5.** $(x+2)^2 = 6$ **6.** $(y-4)^2 - 12 = 0$

7. $(t+5)^2 - 18 = 0$ **8.** $(x+2)^2 = -6$ **9.** $(y-4)^2 + 12 = 0$

10. $3x^2 - x - 4 = 0$ **11.** $6x^2 - 3x - 4 = 0$ **12.** $12x^2 + 11x - 15 = 0$

13. $3x^2 - 7x + 4 = 0$ **14.** $9x^2 - 9x - 4 = 0$ **15.** $x^2 - 2x - 4 = 0$

16. $y^2 - 7y + 10 = 0$ **17.** $x^2 - 2x + 4 = 0$ **18.** $y^2 - 7y - 8 = 0$

19. $3x^2 + x - 4 = 0$ **20.** $9x^2 - 8x - 4 = 0$ **21.** $t^2 + 4t + 8 = 0$

22. $t^2 + 4t - 9 = 0$ **23.** $3x^2 + 2x + 1 = 0$ **24.** $3x^2 - 2x - 2 = 0$

25. $x^2 + 0.8x + 0.15 = 0$ **26.** $x^2 + 0.8x - 0.15 = 0$ **27.** $2x^2 - 0.3x - 0.25 = 0$

28. $x^2 - \frac{1}{3}x - \frac{2}{3} = 0$ **29.** $x^2 + \frac{2}{3}x - \frac{1}{2} = 0$ **30.** $2x^2 + \frac{1}{2}x - \frac{1}{3} = 0$

Solve each equation. Write answers in decimal approximations to the nearest hundredth.

31. $a^2 - 2a - 11 = 0$ **32.** $x^2 + 38x + 1 = 0$ **33.** $y^2 + 2y - 12 = 0$

34. $a^2 + 10a - 3 = 0$ **35.** $y^2 - 14y + 31 = 0$ **36.** $p^2 + 6p - 6 = 0$

37. $2x^2 + x - 8 = 0$ **38.** $3x^2 - 2x - 3 = 0$ **39.** $2x^2 + 6x + 3 = 0$

40. If r_1 and r_2 are the two solutions of the quadratic equation $ax^2 + bx + c = 0$, find the sum of r_1 and r_2.

41. If r_1 and r_2 are the two solutions of the quadratic equation $ax^2 + bx + c = 0$, find the product of r_1 and r_2.

42. Find the quadratic equation whose two roots are 1 and -3.

43. Find the quadratic equation whose two roots are $\sqrt{6}$ and $-\sqrt{6}$.

12-4 Graphing Quadratic Functions – Parabolas

Quadratic Equation: An equation of the form $ax^2 + bx + c = 0$ $(a \neq 0)$.
Quadratic Function: A quadratic equation of the form $y = ax^2 + bx + c$ $(a \neq 0)$.
The graph of a quadratic function $y = ax^2 + bx + c$ $(a \neq 0)$ on the coordinate plane is a **parabola**. The central line of a parabola is called **the axis of symmetry**, or simply the **axis**. The point where the parabola crosses its axis is the **vertex**.

General Form of a parabola: $y - k = a(x - h)^2$. The vertex is (h, k).
Intercept form of a parabola: $y = a(x - p)(x - q)$. The x – intercepts are p and q.
The x – coordinate of the vertex is the midpoint between the two x – intercepts.
If $a > 0$ (positive), the parabola opens upward. Vertex is a minimum point.
If $a < 0$ (negative), the parabola opens downward. Vertex is a maximum point.
$y = x^2$ is the simplest quadratic function. Its vertex is located on the origin (0, 0).

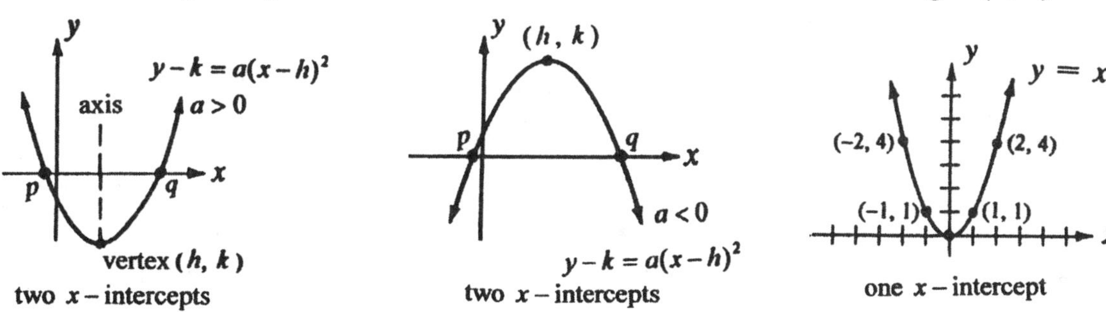

A parabola can have two, one, or no x – intercepts.
To find the x – intercepts of a parabola, we find the **zeros** (roots) of the related quadratic equation (let $y = 0$):

Solve $ax^2 + bx + c = 0$ by factoring , or by the formula: $x = \dfrac{-b \pm \sqrt{b^2 - 4ac}}{2a}$.

If $b^2 - 4ac > 0$, the equation has two roots. The parabola has two x – intercepts.
If $b^2 - 4ac = 0$, the equation has one root. The parabola has one x – intercept.
If $b^2 - 4ac < 0$, the equation has no real-number root. The parabola has no x – intercept.
The value of $b^2 - 4ac$ indicates the differences among the three cases. Therefore, we call the value of $b^2 - 4ac$ "**the discriminant**" of the quadratic equation.

Steps for graphing a quadratic function: $y = ax^2 + bx + c$ $(a \neq 0)$

1. Let $y = 0$. Find the zeros (roots). The roots are the x – intercepts.
2. Find the vertex (h, k) by matching the general form $y - k = a(x - h)^2$.
3. If $a > 0$, the graph opens upward. If $a < 0$, the graph opens downward.

The graph of a parabola is symmetrical. One half of the graph on one side of the axis is the **mirror image** of the other half. The x – coordinate of the vertex is $-\frac{b}{2a}$.

-----Continued-----

Examples

1. Graph $y = 2x^2$.

   Solution:

   One x – intercept = 0

   Vertex (0, 0)

   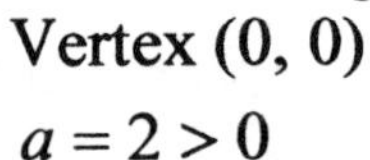

   $a = 2 > 0$

   Open upward

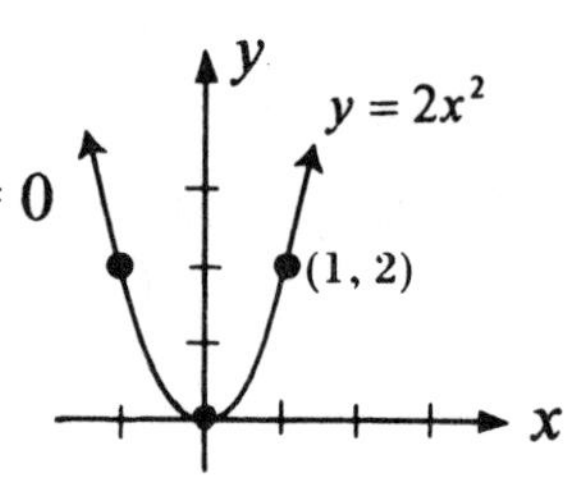

2. Graph $y = -2x^2$

   Solution:

   One x – intercept = 0

   Vertex (0, 0)

   $a = -2 < 0$

   Open downward

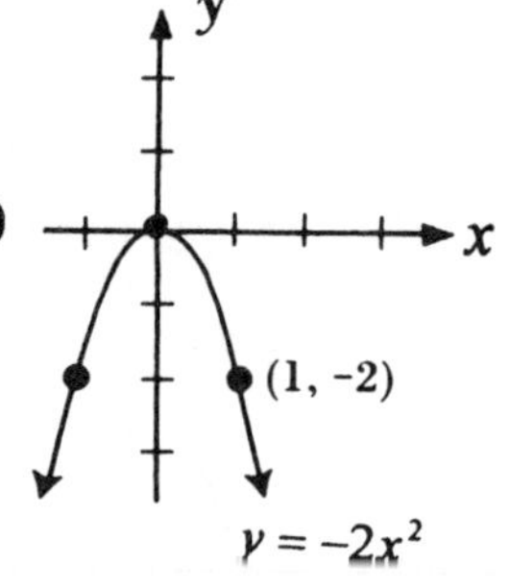

3. Graph $y = x^2 - 2$.

   Solution:

   $x^2 - 2 = 0$, $x^2 = 2$

   $x = \pm\sqrt{2} \approx \pm 1.4$

   $\therefore$ Two x – intercepts = 1.4 and –1.4

   $y + 2 = x^2$, $y - (-2) = (x - 0)^2$

   $\therefore$ Vertex (0, –2)

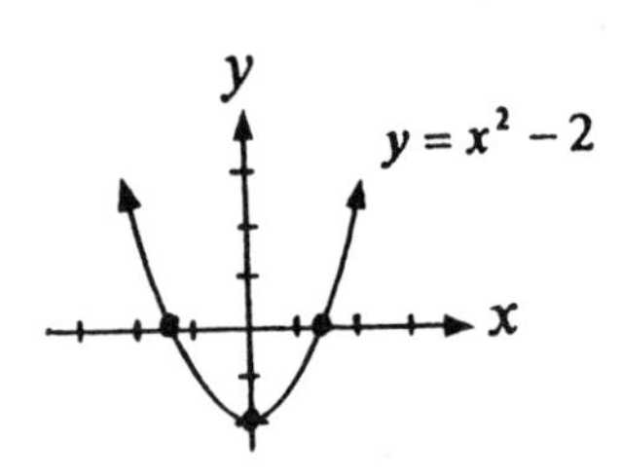

4. Graph $y = x^2 - 4x$.

   Solution:

   $x^2 - 4x = 0$, $x(x - 4) = 0$

   $x = 0$ or 4

   $\therefore$ Two x – intercepts = 0 and 4

   $y = (x^2 - 4x + 4) - 4$

   $y + 4 = x^2 - 4x + 4$

   $y - (-4) = (x - 2)^2$ $\therefore$ Vertex (2, –4)

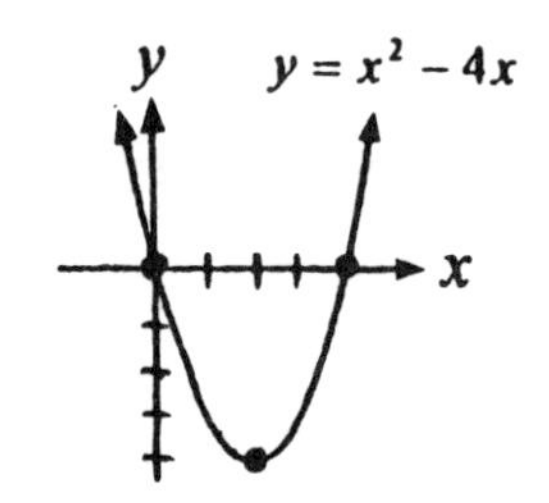

5. Graph $y = x^2 - 2x + 1$.

   Solution:

   $x^2 - 2x + 1 = 0$, $(x - 1)^2 = 0$

   $x = 1$

   $\therefore$ One x – intercept = 1

   $y = x^2 - 2x + 1$

   $y - 0 = (x - 1)^2$

   $\therefore$ Vertex (1, 0)

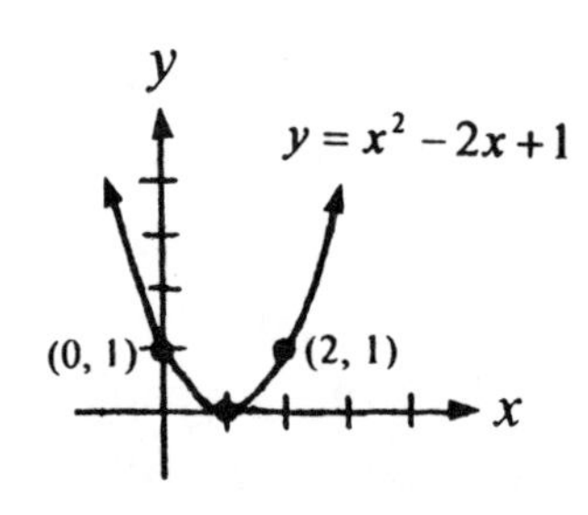

6. Graph $y = (x - 3)^2 + 1$

   Solution:

   $(x - 3)^2 + 1 = 0$, $(x - 3)^2 = -1$

   No real-number root

   $\therefore$ No x – intercept

   $y = (x - 3)^2 + 1$

   $y - 1 = (x - 3)^2$

   $\therefore$ Vertex (3, 1)

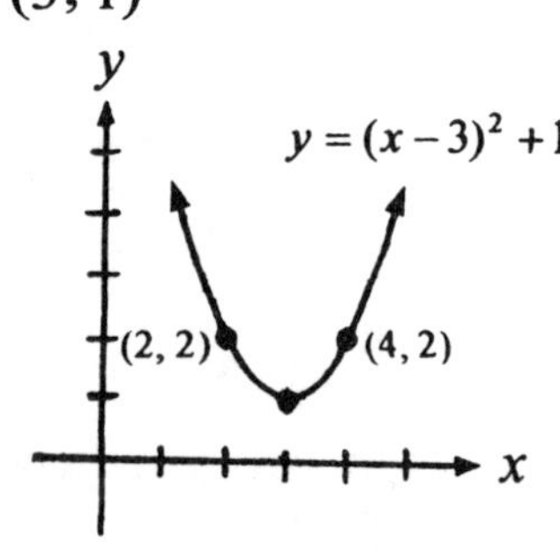

EXERCISES

Determine whether the graph (parabola) of each quadratic function opens upward or downward.

1. $y = -5x^2$ **2.** $y = 5x^2$ **3.** $y = x^2 - 2$

4. $y = x^2 + 2$ **5.** $y = -x^2 + 2$ **6.** $y = x^2 - 2x + 1$

7. $y = x^2 + 2x - 1$ **8.** $y = 4 - 5x - x^2$ **9.** $y = -12 + 3x^2$

Determine how many $x-$intercepts **each quadratic function has by using the discriminant** $b^2 - 4ac$**.**

10. $y = 2x^2 - 4x + 2$ **11.** $y = x^2 - 4x - 3$ **12.** $y = x^2 + 2x + 1$

13. $y = x^2 - 3x + 6$ **14.** $y = x^2 - 6x - 3$ **15.** $y = 2x^2$

16. $y = -2x^2$ **17.** $y = 2x^2 + 4$ **18.** $y = 2x^2 - 4$

19. $y = -2x^2 + 4x + 1$ **20.** $y = -3x^2 + 6x - 5$ **21.** $y = 3 - 6x + 3x^2$

Find the $x-$intercepts (zeros) **of each quadratic function. Round answers to the nearest tenth when necessary.**

22. $y = x^2 - x - 6$ **23.** $y = x^2 + x - 6$ **24.** $y = x^2 - 5x - 6$

25. $y = x^2 + 5x - 6$ **26.** $y = x^2 - 3x - 28$ **27.** $y = x^2 + 12x + 35$

28. $y = 2x^2 - 4x + 2$ **29.** $y = -2x^2 + 4x + 1$ **30.** $y = 3x^2 - 12$

31. $y = -2x^2 - 5$ **32.** $y = x^2 - 4x - 3$ **33.** $y = 3x^2 - 5x - 1$

Find the vertex $(h,\ k)$ **of each parabola by using the form** $y - k = a(x - h)^2$**.**

34. $y - 5 = 2(x - 7)^2$ **35.** $y - 1 = (x - 4)^2$ **36.** $y + 4 = 3(x - 2)^2$

37. $y + 1 = (x + 4)^2$ **38.** $y + 5 = 2(x - 7)^2$ **39.** $y - 4 = 3(x + 2)^2$

40. $y = 3(x - 4)^2 - 2$ **41.** $y = (x - 5)^2 - 1$ **42.** $y = 3(x + 4)^2 - 2$

43. $y = (x + 5)^2 - 1$ **44.** $y = (x + 7)^2 + 9$ **45.** $y = 5(x - 6)^2 + 12$

46. $y - 4 = -6(x + 12)^2$ **47.** $y = -(x - 1)^2 - 7$ **48.** $y = \frac{1}{2}(x - 10)^2 - 9$

49. $y = 3x^2 + 12$ **50.** $y + 7 = -5x^2$

-----Continued-----

Complete the square and rewrite each quadratic function to match the form $y-k=a(x-h)^2$. Then find the vertex (h, k) of each parabola.

51. $y=x^2+4x+4$
52. $y=x^2-4x+4$
53. $y=x^2+4x+9$
54. $y=x^2-4x-1$
55. $y=x^2-6x+7$
56. $y=x^2+6x+11$
57. $y=x^2-2x-4$
58. $y=x^2-10x+13$
59. $y=x^2+14x+58$
60. $y=x^2+10x+24$
61. $y=3x^2-12x+8$
62. $y=2x^2+16x+34$

Find the vertex and zeros ($x-$intercepts). **Then graph each parabola.**

63. $y=\frac{1}{2}x^2$
64. $y=-\frac{1}{2}x^2$
65. $y=x^2+1$
66. $y=-x^2+2$
67. $y=(x-1)^2$
68. $y=(x+1)^2$
69. $y=2(x-2)^2$
70. $y=(x-3)^2+1$
71. $y=2(x-3)^2+1$
72. $y=(x+3)(x+1)$
73. $y=(x+3)(x-1)$
74. $y=x^2+2x+2$

75. Find k so that the equation $9x^2+30x+k=0$ has one real-number root.
76. Find k so that the equation $5x^2-3x-k=0$ has two real-number roots.
77. Find k so that the equation $4kx^2+2x-5=0$ has no real-number root.
78. Compare the graph of $y=x^2+1$ to the graph of $y=x^2$, describe the transformation (shift) of he graph.
79. Compare the graph of $y=(x+1)^2$ to the graph of $y=x^2$, describe the transformation (shift) of the graph.
80. Compare the graph of $y=(x-2)^2-3$ to the graph of $y=x^2$, describe the transformation (shift) of the graph.
81. Determine how many $x-$intercepts the quadratic function has. (Hint: $D=b^2-4ac$)
a. $y=4x^2+4x+1$ **b.** $y=4x^2-5x+2$ **c.** $y=2x^2-7x-3$
82. Prove that the $x-$coordinate of a parabola $y=ax^2+bx+c$ ($a\neq 0$) is $-\frac{b}{2a}$.
(Hint: It is the midpoint between the two $x-$intercepts.)
83. Find the quadratic function that has zeros($x-$intercepts) at $x=1$ and $x=-3$ and passing the point $(0, -3)$.
84. Find the quadratic function that has zeros($x-$intercepts) at $x=\sqrt{2}$ and $x=-\sqrt{2}$ and passing the point (1, 1).
85. Find the quadratic function that has zero ($x-$intercept) at $x=1$ only and passing the point (2, 1).

On graduation day, a high school graduate said to his math teacher.
Student: Thank you very much for giving me a "D" in your class.
For your kindness, do you want me to do anything for you ?
Teacher: Yes. Please do me a favor.
Just don't tell anyone that I was your math teacher.

CHAPTER 12 EXERCISES

Write each expression as a perfect-square of a binomial.

1. $x^2 - 6x + 9$ **2.** $x^2 + 10x + 25$ **3.** $y^2 + 18y + 81$ **4.** $y^2 - 16y + 64$

5. $a^2 + 20x + 100$ **6.** $a^2 - 24a + 144$ **7.** $x^2 - \frac{2}{3}x + \frac{1}{9}$ **8.** $x + \frac{1}{3}x + \frac{1}{36}$

9. $4x^2 - 20x + 25$ **10.** $9x^2 + 36x + 36$

Complete the square.

11. $x^2 - 12x + \_\_\_ = (x - \_\_\_)^2$ **12.** $x^2 + 14x + \_\_\_ = (x + \_\_\_)^2$

13. $a^2 + 5a + \_\_\_ = (a + \_\_\_)^2$ **14.** $a^2 - a + \_\_\_ = (a - \_\_\_)^2$

15. $p^2 - 2.4p + \_\_\_ = (p - \_\_\_)^2$ **16.** $9x^2 - 6x + \_\_\_ = (3x - \_\_\_)^2$

Solve each quadratic equation by using the appropriate method. Write answers in the simplest radical form if necessary.

17. $x^2 = 1$ **18.** $y^2 = 2.25$ **19.** $y^2 = 28$

20. $x^2 = 48$ **21.** $x^2 = 2.56$ **22.** $a^2 = \frac{4}{25}$

23. $a^2 = \frac{25}{4}$ **24.** $p^2 = \frac{3}{4}$ **25.** $p^2 = \frac{4}{3}$

26. $(x-3)^2 = 9$ **27.** $(x-3)^2 = -9$ **28.** $(x+4)^2 = -16$

29. $(x+4)^2 = 16$ **30.** $(x-8)^2 = 18$ **31.** $(x-8)^2 = 20$

32. $4(x-1)^2 - 20 = 0$ **33.** $5(x-1)^2 - 40 = 0$ **34.** $9(x-3)^2 - 16 = 0$

35. $2(x+1)^2 - 3 = 0$ **36.** $y^2 - 2y + 1 = 8$ **37.** $x^2 - 14x + 49 = 12$

38. $\frac{1}{2}x^2 - \frac{9}{32} = 0$ **39.** $\frac{1}{3}x^2 - 2 = 0$ **40.** $4(x-5)^2 = \frac{3}{4}$

41. $x^2 + 3x - 4 = 0$ **42.** $x^2 - 2x - 35 = 0$ **43.** $y^2 - 3y - 88 = 0$

44. $y^2 - 18y + 80 = 0$ **45.** $t^2 - 8t - 9 = 0$ **46.** $2x^2 + 7x - 4 = 0$

47. $2x^2 + 3x - 5 = 0$ **48.** $x^2 - 2x - 4 = 0$ **49.** $x^2 - 2x - 7 = 0$

50. $x^2 - 6x + 1 = 0$ **51.** $x^2 + 4x - 14 = 0$ **52.** $3x^2 + x - 2 = 0$

53. $3x^2 + 2x - 1 = 0$ **54.** $4x^2 - 2x + 1 = 0$ **55.** $5x^2 - 8x + 2 = 0$

-----Continued-----

Solve each equation. Write answers in decimal approximation to the nearest hundredth.

56. $x^2 = 52$ **57.** $(x-4)^2 = 40$ **58.** $8(x+1)^2 = 64$

59. $4(x+5)^2 - 48 = 0$ **60.** $x^2 - x - 1 = 0$ **61.** $y^2 + 3y - 5 = 0$

62. $y^2 + 5y - 2 = 0$ **63.** $a^2 + 6a - 8 = 0$ **64.** $a^2 + 3a - 9 = 0$

65. $2x^2 - 3x - 7 = 0$ **66.** $3x^2 - x - 1 = 0$ **67.** $-4x^2 - 2x + 3 = 0$

68. $x^3 - 12x = 0$ **69.** $x^3 - 18x = 0$ **70.** $x^3 + 3x^2 + x = 0$

Find the vertex and zeros (x – intercepts) **of each quadratic equation (parabola). Round answers to the nearest tenth when necessary.**

71. $y = x^2 + 4$ **72.** $y = x^2 - 8$ **73.** $y + 9 = (x+2)^2$

74. $y - 1 = (x-2)^2$ **75.** $y - 10 = (x+7)^2$ **76.** $y - 3 = (x+6)^2$

77. $y = 5(x-4)^2 - 1$ **78.** $y = 12(x-11)^2 + 8$ **79.** $y + 15 = 9(x+3)^2$

80. $y = -(x-6)^2$ **81.** $y = 5x^2 + 6$ **82.** $y = -(x-4)^2 - 7$

83. $y = x^2 + 4x - 5$ **84.** $y = x^2 - 4x - 2$ **85.** $y = x^2 + 3x + 3$

Graph each quadratic equation (parabola).

86. $y = (x+1)^2 - 1$ **87.** $y = 2(x-3)^2 + 1$ **88.** $y = -(x-2)^2 + 4$ **89.** $y = x^2 + 2x - 3$

90. Determine how many x – intercepts the quadratic function $y = x^2 + 5x - 7$ has.

91. Determine how many x – intercepts the quadratic function $y = x^2 - 5x + 7$ has.

92. Find k so that the equation $2x^2 + kx + 8 = 0$ has one-real number root.

93. Find k so that the equation $kx^2 + 4x + 1 = 0$ has no real-real number root.

94. Write the equation of the axis of symmetry of the function $y = (x+1)^2 - 1$.

95. Write the equation of the axis of symmetry of the function $y = 2(x-3)^2 + 1$

96. Compare the graph of $y = x^2 - 4$ to the graph of $y = x^2$, describe the transformation (shift) of the graph.

97. Compare the graph of $y = (x-3)^2 + 4$ to the graph of $y = x^2$, describe the transformation (shift) of the graph.

98. Find the minimum value of $y = 2x^2 - 10x - 5$.

99. Find the maximum value of $y = -2x^2 + 8x + 10$.

100. Find the maximum profit if the equation of profit is $y = -2x^2 + 32x - 38$, where y is the profit and x is the number of products sold.

Proof of the Quadratic Formula

We learned to use the method of completing the square to solve quadratic equation of the form $ax^2 + bx + c = 0$ ($a \neq 0$). However, the steps of completing the square are very tedious and sometimes uneasy . Therefore, we have developed an efficient formula that can be used in all cases. Such a formula can be derived by applying the method of completing the square to the quadratic equation.
This formula is called **the quadratic formula**. We use it to find the roots (solutions) directly from the coefficients a, b, and c without repeating the tedious steps of completing the square. The formula has already absorbed the steps of completing the square.

The Quadratic Formula

If $ax^2 + bx + c = 0$, where $a \neq 0$, then

$$x = \frac{-b \pm \sqrt{b^2 - 4ac}}{2a}$$

Proof:

$$ax^2 + bx + c = 0$$

Divide both sides by a: $$x^2 + \frac{b}{a}x + \frac{c}{a} = 0$$

$$x^2 + \frac{b}{a}x = -\frac{c}{a}$$

Completing the square by adding $\left(\frac{b}{2a}\right)^2$ on both sides:

$$x^2 + \frac{b}{a}x + \left(\frac{b}{2a}\right)^2 = -\frac{c}{a} + \left(\frac{b}{2a}\right)^2$$

Square on the left and combine on the right:

$$\left(x + \frac{b}{2a}\right)^2 = \frac{b^2 - 4ac}{4a^2}$$

$$x + \frac{b}{2a} = \pm\sqrt{\frac{b^2 - 4ac}{4a^2}}$$

$$x = -\frac{b}{2a} \pm \frac{\sqrt{b^2 - 4ac}}{2a}$$

$$\therefore x = \frac{-b \pm \sqrt{b^2 - 4ac}}{2a}$$. The proof is complete.

The value of $b^2 - 4ac$ determines the nature of the roots (solutions).
We call $b^2 - 4ac$ "**the discriminant** D". (See Section 12 ~ 4)

Additional Example

1. Find the dimensions of the rectangular flower bed that can be built by the 12-feet fence to have as much area as possible ?

Solution:

Let $y =$ the area

$x =$ the length of one side

We have the equation:

$y = x(6-x)$

$y = -x^2 + 6x$

$y = -(x^2 - 6x)$

Completing the square:

$y = -(x-3)^2 + 9$

$y - 9 = -(x-3)^2$

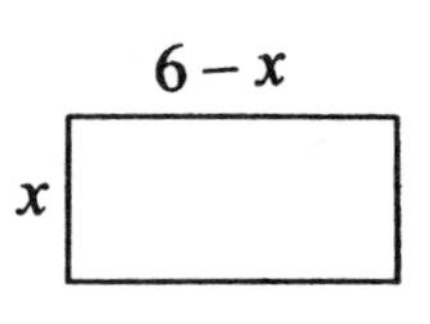

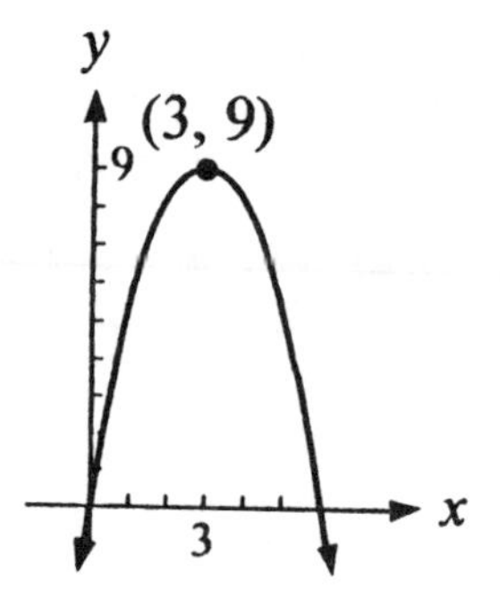

Ans: Dimensions = 3 × 3 (It is a square.)

Maximun area = 9 ft^2

2. Prove that it must be a square to have the greatest area that can be made by a perimeter of length p.

Proof:

Let $y =$ the area

$x =$ the length of one side

The equation is:

$\frac{p}{2} - x$

x

$y = x(\frac{p}{2} - x)$

$y = -x^2 + \frac{p}{2}x$

$y = -(x^2 - \frac{p}{2}x)$

Completing the square:

$y = -(x - \frac{p}{4})^2 + \frac{p^2}{16}$

$y - \frac{p^2}{16} = -(x - \frac{p}{4})^2$

It is a parabola (open downward) with vertex $(\frac{p}{4}, \frac{p^2}{16})$.

Max. area $y = \frac{p^2}{16}$ when $x = \frac{p}{4}$. It is a square.

CHAPTER THIRTEEN

Rational Expressions (Algebraic Fractions)

13-1 Simplifying Rational Expressions

Rational number: A number that is a quotient (ratio) of two integers.

2 and 4.6 are rational numbers because $2 = \frac{4}{2}$ and $4.6 = \frac{46}{10}$.

$\sqrt{2} \approx 1.414$ is not a rational number. It is an irrational number.

Rational Expression (algebraic fraction):

An algebraic expression that is a quotient (ratio) of two polynomials. It is also referred to as **algebraic fraction.**

$\frac{x+1}{2x}$, $\frac{2x^2+3x-5}{x^2+4}$, and $\frac{xy^2}{x^2+xy}$ are rational expressions.

All rules for operating with rational numbers that we have learned in Chapter 4 apply to rational expressions as well.

A rational expression is in simplest form if the numerator and the denominator have no common factors (except 1 and -1).

To simplify a simple rational expression, we reduce the fraction to lowest terms by using their greatest common factor (GCF) or the laws of exponents (see Section 8 ~ 1).

To simplify a rational expression, we factor the numerator and the denominator and then cancel any common factor.

To simplify a rational fraction, we also **restrict** the variables by **excluding** any values that make the denominator equal to zero. It is **undefined** if the denominator equals 0.

Important: If factors of the numerator and the denominator are opposites of one another, we take the negative of the numerator or the denominator. (See examples 4 and 5)

Examples (simplify each expression)

1. $\frac{16x^2}{24x} = \frac{16x^2/8x}{24x/8x} = \frac{2x}{3},\ x \neq 0$ **or:** $\frac{16x^2}{24x} = \frac{16}{24} \cdot x^{2-1} = \frac{2}{3}x,\ x \neq 0$

2. $\frac{18x^4y^2}{27xy^6} = \frac{18}{27} \cdot \frac{x^4}{x} \cdot \frac{y^2}{y^6} = \frac{2x^3}{3y^4}, x \neq 0, y \neq 0$ **3.** $\frac{3xy}{3x+3y} = \frac{\cancel{3}xy}{\cancel{3}(x+y)} = \frac{xy}{x+y}, x \neq -y$

4. $\frac{x-1}{1-x} = \frac{\cancel{x-1}}{-(\cancel{x-1})} = -1,\ x \neq 1$ **5.** $\frac{3-x}{4x-12} = \frac{3-x}{4(x-3)} = \frac{-\cancel{(x-3)}}{4\cancel{(x-3)}} = -\frac{1}{4},\ x \neq 3$

6. $\frac{x^2-4}{x+2} = \frac{\cancel{(x+2)}(x-2)}{\cancel{x+2}} = x-2,\ x \neq -2$

7. $\frac{x^2-5x+6}{x^2-x-2} = \frac{\cancel{(x-2)}(x-3)}{(x+1)\cancel{(x-2)}} = \frac{x-3}{x+1},\ x \neq -1,\ x \neq 2$

EXERCISES

Determine all values of variables for which the given rational expression is undefined.

1. $\frac{2}{x}$
2. $\frac{5}{x-1}$
3. $\frac{15}{a-5}$
4. $\frac{y-3}{y+4}$
5. $\frac{4x}{2x-6}$
6. $\frac{3}{(n-2)(n+1)}$
7. $\frac{n+4}{(n+5)(n-1)}$
8. $\frac{t-1}{t^2+2t}$
9. $\frac{20}{x^2+5}$
10. $\frac{x-8}{4}$
11. $\frac{p-3}{p^2-4p-5}$
12. $\frac{y+9}{y^3-y^2-2y}$

Simplify each rational expression. Give all restrictions on the variables.

13. $\frac{9x}{3x}$
14. $\frac{6x^5}{2x^2}$
15. $\frac{9x^2}{6x^5}$
16. $\frac{15a^2b^4}{25a^6b}$
17. $\frac{4x+2y}{2}$
18. $\frac{2}{4x+2y}$
19. $\frac{6}{9y-12}$
20. $\frac{9y-12}{6}$
21. $\frac{4(x+y)}{8(x-y)}$
22. $\frac{x-5}{5-x}$
23. $\frac{2a-6}{3-a}$
24. $\frac{a^2-4}{2-a}$
25. $\frac{a-2}{4-a^2}$
26. $\frac{x-8}{(x+8)(x-8)}$
27. $\frac{14+7x}{7x}$
28. $\frac{12+4y}{3+y}$
29. $\frac{2(x+2y)}{(x+2y)(x-y)}$
30. $\frac{3(x+y)(x-y)}{9(x+y)}$
31. $\frac{(a-2)^2}{a-2}$
32. $\frac{5(a-1)}{15(a-1)^2}$
33. $\frac{(a-b)^2}{(a+b)(a-b)}$
34. $\frac{x^2-8x+16}{x-4}$
35. $\frac{x^2-8x+15}{x-5}$
36. $\frac{3-n}{n^2-2n-3}$
37. $\frac{2a-12}{a^2-4a-12}$
38. $\frac{k^2+k-12}{5k+20}$
39. $\frac{10-2p}{p^2-4p-5}$
40. $\frac{x^2+5x-14}{x^2+3x-28}$

41. Determine all values of x for which the rational expression is undefined: $\frac{2x^2+8x}{x^2-3x-18}$
42. Determine all values of x for which the rational expression equals zero: $\frac{2x^2+8x}{x^2-3x-18}$
43. Solve the equation $ax^2-a^2=-bx-b^2$ for x in terms of a and b.
44. The area of a rectangle is given by $x^2+10x+21$. The length is given by $x+7$. Find the width in terms of x.

13-2 Adding and Subtracting Rational Expressions

1) To add or subtract rational expressions (algebraic fractions) having the same denominators, we combine their numerators.

$$\frac{b}{a}+\frac{c}{a}=\frac{b+c}{a} \quad ; \quad \frac{b}{a}-\frac{c}{a}=\frac{b-c}{a}$$

2) To add or subtract rational expressions (algebraic fractions) having different denominators, we rewrite each fraction having the least common denominator (**LCD**), and then combine the resulting fractions. **LCD** is the least common multiple (**LCM**) of their denominators.

$$\frac{c}{a}+\frac{d}{b}=\frac{bc}{ab}+\frac{ad}{ab}=\frac{bc+ad}{ab} \quad ; \quad \frac{c}{a}-\frac{d}{b}=\frac{bc}{ab}-\frac{ad}{ab}=\frac{bc-ad}{ab}$$

Examples

1. $\frac{3x}{5}+\frac{7x}{5}=\frac{3x+7x}{5}=\frac{10x}{5}=2x$

2. $\frac{x}{5}-\frac{3x}{2}=\frac{2x}{10}-\frac{15x}{10}=\frac{2x-15x}{10}=-\frac{13x}{10}$

3. $\frac{5a+3}{4}-\frac{a-1}{4}=\frac{5a+3-(a-1)}{4}=\frac{5a+3-a+1}{4}=\frac{4a+4}{4}=\frac{\cancel{4}(a+1)}{\cancel{4}}=a+1$

4. $\frac{1}{x}+\frac{1}{2x}=\frac{2}{2x}+\frac{1}{2x}=\frac{2+1}{2x}=\frac{3}{2x}, \ x\neq 0$

5. $\frac{1}{x^2}-\frac{2}{x^3}=\frac{x}{x^3}-\frac{2}{x^3}=\frac{x-2}{x^3}, x\neq 0$

6. $\frac{1}{2x^2}-\frac{5}{6x}=\frac{3}{6x^2}-\frac{5x}{6x^2}=\frac{3-5x}{6x^2}, x\neq 0$

7. $\frac{3}{x-1}+\frac{5}{x-1}=\frac{3+5}{x-1}=\frac{8}{x-1}, x\neq 1$

8. $\frac{7}{x-2}+\frac{2}{2-x}=\frac{7}{x-2}-\frac{2}{x-2}=\frac{5}{x-2}, x\neq 2$

9. $\frac{3}{x}-\frac{2}{x-2}=\frac{3(x-2)}{x(x-2)}-\frac{2x}{x(x-2)}=\frac{3(x-2)-2x}{x(x-2)}=\frac{3x-6-2x}{x(x-2)}=\frac{x-6}{x^2-2x}, x\neq 0, 2$

10. $\frac{2a}{a^2-4}-\frac{5}{a+2}=\frac{2a}{(a+2)(a-2)}-\frac{5(a-2)}{(a+2)(a-2)}=\frac{2a-5(a-2)}{(a+2)(a-2)}=\frac{2a-5a+10}{(a+2)(a-2)}$

$=\frac{-3a+10}{a^2-4}, \ a\neq -2, 2$

11. $\frac{-2}{x^2-5x+6}+\frac{2}{x-3}=\frac{-2}{(x-3)(x-2)}+\frac{2}{x-3}=\frac{-2}{(x-3)(x-2)}+\frac{2(x-2)}{(x-3)(x-2)}$

$=\frac{-2+2(x-2)}{(x-3)(x-2)}=\frac{-2+2x-4}{(x-3)(x-2)}=\frac{2x-6}{(x-3)((x-2)}=\frac{2\cancel{(x-3)}}{\cancel{(x-3)}(x-2)}=\frac{2}{x-2}, x\neq 3, 2$

EXERCISES

Find the LCM of each of the following. (See section 3 ~ 4)

1. 3, 5 **2.** 4, 9 **3.** 2, 6 **4.** 5, 15
5. 6, 10 **6.** 6, 8 **7.** $2, 5x$ **8.** $3,\ 4x$
9. $4,\ 6x$ **10.** $8,\ 10x$ **11.** $4a,\ 8ab$ **12.** $8a,\ 4ab$
13. $3x,\ 6x^2$ **14.** $x,\ y$ **15.** $7y^3,\ 28y$ **16.** $x+3,\ x-3$
17. $12x^2y,\ 16xy^2$ **18.** $a^2-9,\ a-3$ **19.** $5(a-2), 6(a-2)^2$ **20.** $x(x-1),\ x^2-1$

Find the LCD of the fractions.

21. $\frac{1}{3}, \frac{1}{5}$ **22.** $\frac{1}{5}, \frac{1}{15}$ **23.** $\frac{1}{2x}, \frac{1}{5x}$ **24.** $\frac{1}{8x}, \frac{1}{10x}$

25. $\frac{1}{4a}, \frac{1}{8ab}$ **26.** $\frac{1}{3x}, \frac{1}{6x^2}$ **27.** $\frac{1}{x+3}, \frac{1}{x-3}$ **28.** $\frac{1}{a^2-9}, \frac{1}{a-3}$

29. $x, \frac{1}{x}$ **30.** $\frac{1}{a}, 5$ **31.** $\frac{1}{x}, \frac{1}{y}$ **32.** $\frac{1}{x}, \frac{1}{x^2y}$

33. $\frac{1}{x}, \frac{1}{x-5}$ **34.** $\frac{1}{3c}, \frac{1}{2(c-2)}$ **35.** $\frac{1}{t+1}, \frac{1}{(t+1)^2}$ **36.** $\frac{1}{x-2}, \frac{1}{x+1}$

Simplify. Give all restrictions on the variables.

37. $\frac{x}{3}+\frac{x}{5}$ **38.** $\frac{4a}{5}-\frac{a}{15}$ **39.** $\frac{1}{2x}-\frac{3}{5x}$ **40.** $\frac{1}{8x}+\frac{9}{10x}$

41. $\frac{3}{4a}-\frac{1}{8ab}$ **42.** $\frac{2}{3x}-\frac{5}{6x^2}$ **43.** $\frac{4}{x+3}+\frac{3}{x-3}$ **44.** $\frac{8a}{a^2-9}-\frac{4}{a-3}$

45. $x+\frac{2}{x}$ **46.** $\frac{1}{a}-5$ **47.** $\frac{2}{x}+\frac{5}{y}$ **48.** $\frac{2}{x}+\frac{9}{x^2y}$

49. $\frac{2}{x}-\frac{1}{x-5}$ **50.** $\frac{2}{3c}-\frac{5}{2(c-2)}$ **51.** $\frac{3}{t+1}-\frac{3t}{(t+1)^2}$ **52.** $\frac{1}{2(x+1)}+\frac{2}{x+1}$

53. $\frac{3}{x-2}+\frac{7}{x+1}$ **54.** $\frac{2x}{x-4}-\frac{1}{3x-12}$ **55.** $\frac{3}{x^2-9}-\frac{1}{2x-6}$ **56.** $\frac{x}{x-2}+\frac{1}{2-x}$

57. $\frac{4n}{2n-1}+\frac{2}{1-2n}$ **58.** $\frac{a-2}{a-1}-\frac{a+1}{a+2}$ **59.** $\frac{4}{x^2+2x-3}+\frac{1}{x+3}$ **60.** $\frac{4}{y-3}-\frac{5y-12}{y^2-6y+9}$

61. $\frac{1}{b^2+4b+3}+\frac{1}{b^2-2b-15}$ **62.** $\frac{5}{3+c}-\frac{3}{3-c}+\frac{2c+7}{9-c^2}$ **63.** $\frac{x-1}{x^2+5x}-\frac{x+1}{x^2-25}+\frac{1}{x^2-5x}$

13-3 Multiplying and Dividing Rational Expressions

1) To multiply two rational expressions (algebraic fractions), we multiply their numerators and multiply their denominators. We can multiply first and then simplify, or simplify first and then multiply.

$$\frac{x^2}{3y}\cdot\frac{6}{5x}=\frac{\overset{2}{\cancel{6}}\overset{1}{\cancel{x^2}}}{{}_5\cancel{15}\cancel{x}y}=\frac{2x}{5y}\ ;\qquad \text{Or: } \frac{\overset{1}{\cancel{x^2}}}{{}_1\cancel{3}y}\cdot\frac{\overset{2}{\cancel{6}}}{5\cancel{x}}=\frac{2x}{5y}$$

2) To divide two rational expressions (algebraic fractions), we multiply the reciprocal of the divisor.

$$\frac{4x}{y}\div\frac{8x^2}{3y}=\frac{\overset{1}{\cancel{4}}\cancel{x}}{\cancel{y}}\cdot\frac{3\cancel{y}}{{}_2\cancel{8}\cancel{x^2}^{1}}=\frac{3}{2x}$$

3) To divide long polynomials, we follow the same ways of dividing real numbers. We must arrange the terms in both polynomials in order of decreasing degree of one variable. Keep space (using 0) on missing terms in degree (see example **10**).

Examples (Assume that no variable has a value for which the denominator is zero.)

1. $\dfrac{x^2}{4}\cdot\dfrac{x}{2}=\dfrac{x^2\cdot x}{4\cdot 2}=\dfrac{x^3}{8}$

2. $\dfrac{x^2}{4}\div\dfrac{x}{2}=\dfrac{\cancel{x^2}^{1}}{{}_2\cancel{4}}\cdot\dfrac{\cancel{2}^{1}}{\cancel{x}}=\dfrac{x}{2}$

3. $\dfrac{\cancel{2}x}{3\cancel{y}}\cdot\dfrac{x\cancel{y}}{\cancel{8}_4}=\dfrac{x^2}{12}$

4. $\dfrac{2x}{3y}\div\dfrac{xy}{8}=\dfrac{2\cancel{x}}{3y}\cdot\dfrac{8}{\cancel{x}y}=\dfrac{16}{3y^2}$

5. $\dfrac{a-3}{a}\cdot\dfrac{a^3}{a^2-9}=\dfrac{\cancel{(a-3)}}{\cancel{a}}\cdot\dfrac{\cancel{a^3}^{2}}{(a+3)\cancel{(a-3)}}=\dfrac{a^2}{a+3}$

6. $\dfrac{a^2-5a+6}{a^2}\div\dfrac{a^2-9}{a}=\dfrac{a^2-5a+6}{a^2}\cdot\dfrac{a}{a^2-9}=\dfrac{(a-2)\cancel{(a-3)}}{\cancel{a^2}^{1}}\cdot\dfrac{\cancel{a}}{(a+3)\cancel{(a-3)}}=\dfrac{a-2}{a(a+3)}$

7. $\dfrac{x^2-4x-5}{x}\cdot\dfrac{x^2+x}{x-5}=\dfrac{(x+1)\cancel{(x-5)}}{\cancel{x}}\cdot\dfrac{\cancel{x}(x+1)}{\cancel{x-5}}=(x+1)^2$

8. $\dfrac{x^2-4x-5}{x}\div\dfrac{x^2+x}{x-5}=\dfrac{x^2-4x-5}{x}\cdot\dfrac{x-5}{x^2+x}=\dfrac{\cancel{(x+1)}(x-5)}{x}\cdot\dfrac{x-5}{x\cancel{(x+1)}}=\dfrac{(x-5)^2}{x^2}$

9. Divide: $\dfrac{x^3+4x^2+6x+2}{x+3}$

Solution:

$$\begin{array}{r l}
 & x^2+\ x+3 \\
x+3 \big) & x^3+4x^2+6x+2 \\
 & \underline{-)\ x^3+3x^2\qquad\qquad} \\
 & \qquad x^2+6x \\
 & \qquad \underline{-)\ x^2+3x\quad} \\
 & \qquad\qquad 3x+2 \\
 & \qquad\qquad \underline{-)\ 3x+9} \\
 & \qquad\qquad\quad -7 \leftarrow \text{Remainder}
\end{array}$$

Ans: $\dfrac{x^3+4x^2+6x+2}{x+3}=x^2+x+3-\dfrac{7}{x+3}$.

10. Divide: $\dfrac{1+x+12x^3}{1+2x}$

Solution:

$$\begin{array}{r l}
 & 6x^2-3x+2 \\
2x+1 \big) & 12x^3+0\ +\ x+1 \\
 & \underline{-)\ 12x^3+6x^2\qquad\quad} \\
 & \qquad -6x^2+\ x \\
 & \qquad \underline{-)\ -6x^2-3x} \\
 & \qquad\qquad 4x+1 \\
 & \qquad\qquad \underline{-)\ 4x+2} \\
 & \qquad\qquad\quad -1 \leftarrow \text{Remainder}
\end{array}$$

Ans: $\dfrac{1+x+12x^3}{1+2x}=6x^2-3x+2-\dfrac{1}{2x+1}$.

EXERCISES

Simplify. Give all restrictions on the variables.

1. $\frac{3}{x^3}\cdot\frac{x^2}{6}$
2. $\frac{8}{x}\cdot\frac{x^3}{2}$
3. $\frac{3y}{2}\cdot\frac{6}{y^2}$
4. $\frac{5y^4}{4}\cdot\frac{2}{y}$
5. $\frac{6x^2}{y}\cdot\frac{y^4}{2x^5}$
6. $\frac{12a}{5b^3}\cdot\frac{15b}{4a^5}$
7. $\frac{7a^4}{b^2}\cdot\frac{3b^6}{14a}$
8. $\frac{cd^2}{3c}\cdot\frac{18}{d^7}$
9. $\left(\frac{2x}{y}\right)^3\cdot\frac{y}{2x}$
10. $\left(\frac{4x}{y}\right)^2\cdot\frac{y}{2x}$
11. $\frac{x+2}{x-5}\cdot\frac{x-5}{x-2}$
12. $\frac{2x-4y}{3y}\cdot\frac{6y^2}{x-2y}$
13. $\frac{a-3}{3-a}$
14. $\frac{5a-10b}{2b-a}$
15. $\frac{n^2-16}{2n}\cdot\frac{n}{n+4}$
16. $\frac{n^2-4n-5}{n^2-25}$
17. $\frac{3x}{10}\div\frac{x^2}{5}$
18. $\frac{x^3}{y^4}\div\frac{x}{y^2}$
19. $\frac{5n^2}{2p}\div\frac{n^4}{4p^3}$
20. $\frac{ab^3}{5}\div\frac{a^3b}{10}$
21. $4x\div\left(\frac{2x}{3}\right)^2$
22. $\left(\frac{3y}{2}\right)^3\div 15y^2$
23. $\frac{x+2}{x-5}\div\frac{x-2}{x-5}$
24. $\frac{2x-4y}{3y}\div\frac{x-2y}{6y^2}$
25. $\frac{3+3a}{8}\div\frac{1+a}{4a}$
26. $\frac{x^2-4}{3}\div\frac{x-2}{9}$
27. $\frac{2}{x-y}\div\frac{4}{y-x}$
28. $\frac{10-2a}{5}\div\frac{15-3a}{10}$

Simplify. Assume that no variable has a value for which the denominator is zero.

29. $\frac{x^2+2x-3}{x^2+x-2}\cdot\frac{4x+8}{x+4}$
30. $\frac{x^2-3x+2}{x+2}\cdot\frac{2x+4}{x^2+2x-3}$
31. $\frac{x^2-9}{15x}\cdot\frac{3x}{x+3}$
32. $\frac{x+1}{x^2-5x+6}\cdot\frac{x-2}{x^2-1}$
33. $\frac{m^2-2m-8}{m+5}\div\frac{m^2-4}{m^2-25}$
34. $\frac{x^2-y^2}{x^3}\div\frac{4x-4y}{x}$
35. $\frac{a^2-2a-15}{a^2+6a+8}\div\frac{a^2-6a+5}{a^2+3a-4}$
36. $\frac{3a+9}{5a+10}\cdot\frac{3a+12}{a+3}\div\frac{a^2-16}{a^2-4a-12}$

Divide polynomials.

37. $\frac{x^2-5x+6}{x+2}$
38. $\frac{x^2+5x-6}{x-2}$
39. $\frac{4a^2+4a-12}{2a-1}$
40. $\frac{6a^2-5a+10}{3a+2}$
41. $\frac{2x^3-2x+1}{2x+4}$
42. $\frac{2x^3-x^2-5}{2x-5}$
43. $\frac{a^4-4}{a-4}$
44. $\frac{2n^4-n^3-2n+5}{n^2+2n+1}$

13-4 Solving Rational Equations

In Section 5 –1 and 5 –2, we learned to solve an equation forming by two equivalent ratios.
If a rational equation consists of one fraction equal to another (it is a proportion), then the **cross products** are equal (see example 1).
To solve a rational equation consisting of two or more fractions on one side, we multiply both sides of the equation by their **least common denominator (LCD)**.

Example 1:

Solve $\frac{2}{x} = \frac{4}{5}$.

Solution:

$$2 \cdot 5 = 4 \cdot x$$
$$10 = 4x$$
$$\therefore x = \tfrac{10}{4} = 2\tfrac{1}{2}. \text{ Ans.}$$

Example 2:

Solve $\frac{x}{3} + \frac{x}{4} = 7$.

Solution: LCD = 12

$$12\left(\frac{x}{3} + \frac{x}{4}\right) = 12(7)$$
$$4x + 3x = 84$$
$$7x = 84$$
$$\therefore x = 12. \text{ Ans.}$$

Example 3:

Solve $\frac{3}{x} - \frac{1}{2x} = \frac{1}{3}$.

Solution: LCD = $6x$

$$6x\left(\frac{3}{x} - \frac{1}{2x}\right) = 6x\left(\frac{1}{3}\right)$$
$$18 - 3 = 2x$$
$$15 = 2x$$
$$\therefore x = \tfrac{15}{2} = 7\tfrac{1}{2}. \text{ Ans.}$$

To solve a rational equation, we always test each root in the original equation. A root is **not permissible** if it makes the denominator of the original equation equal to 0.

Example 4: Solve $\frac{18}{x^2 - 9} = \frac{x}{x-3}$.

Solution:

$$18(x-3) = x(x^2 - 9)$$
$$18(\cancel{x-3}) = x(x+3)(\cancel{x-3})$$
$$18 = x(x+3)$$
$$18 = x^2 + 3x$$
$$0 = x^2 + 3x - 18$$
$$x^2 + 3x - 18 = 0$$
$$(x+6)(x-3) = 0$$

$x = -6 \mid x = 3$ **(not permissible)** $\therefore x = -6$. Ans.

If the final statement is a "**true**" statement, the equation has roots for all real numbers.
If the final statement is a "**false**" statement, the equation has no root.

Example 5: Solve $\frac{4}{2x-2} = \frac{2}{x-1}$.

Solution:

$$4(x-1) = 2(2x-2)$$
$$4x - 4 = 4x - 4$$
$$0 = 0 \text{ (true)}$$

Ans: The equation has roots for all real numbers except 1.

Example 6: Solve $\frac{4}{2x+1} = \frac{2}{x}$.

Solution:

$$4x = 2(2x+1)$$
$$4x = 4x + 2$$
$$0 = 2 \text{ (false)}$$

Ans: No solution.

EXERCISES

Solve each rational equation (fractional equation).

1. $\frac{x}{3}=\frac{4}{5}$ **2.** $\frac{x}{6}=\frac{3}{5}$ **3.** $\frac{2}{5}=\frac{7}{3x}$ **4.** $\frac{5}{2x}=\frac{12}{7}$

5. $\frac{2y}{3}=\frac{7}{6}$ **6.** $\frac{12}{5}=\frac{4y}{9}$ **7.** $\frac{15}{x}=3$ **8.** $\frac{3x}{7}=-6$

9. $\frac{a-3}{8}=\frac{3}{4}$ **10.** $\frac{a-7}{7}=\frac{7-a}{3}$ **11.** $\frac{n-5}{4}=12$ **12.** $12=\frac{2n+4}{3}$

13. $\frac{4}{2x-1}=\frac{3}{x}$ **14.** $\frac{5n-2}{3}=\frac{5n+2}{4}$ **15.** $\frac{2+3x}{2-3x}=-5$ **16.** $\frac{12}{t+3}=\frac{24}{t+14}$

17. $\frac{2a-2}{a-4}=\frac{1}{4}$ **18.** $\frac{5}{3a+2}=\frac{5}{3a-2}$ **19.** $\frac{3}{5y+2}=\frac{4}{2+5y}$ **20.** $\frac{3x}{x+2}-\frac{1}{x}=0$

21. $\frac{8}{4x+5}=\frac{2}{x}$ **22.** $\frac{2}{4x-6}=\frac{1}{2x-3}$ **23.** $\frac{5}{y^2-16}=\frac{2}{y+4}$ **24.** $\frac{4}{y^2-4}=\frac{9}{y-2}$

25. $\frac{x}{5}+\frac{x}{4}=2$ **26.** $\frac{2x}{7}-\frac{x}{3}=2$ **27.** $\frac{2}{x}+\frac{1}{3x}=6$ **28.** $\frac{5}{2x}+\frac{4}{3x}=7$

29. $\frac{5}{x}-\frac{3}{4x}=\frac{1}{2}$ **30.** $\frac{3}{2x}-\frac{2}{3x}=\frac{5}{6}$ **31.** $\frac{9}{2y}-\frac{2}{3}=\frac{4}{y}$ **32.** $\frac{3y}{8}-y=\frac{5}{6}$

33. $\frac{4}{n+1}-\frac{1}{n}=1$ **34.** $\frac{3}{n-1}+\frac{2}{n}=\frac{3}{2}$ **35.** $\frac{12}{a(a-2)}-\frac{6}{a-2}=1$ **36.** $\frac{8}{x^2-4}=\frac{x}{x-2}$

37. $\frac{2c+5}{3}-\frac{c-4}{5}=\frac{24}{5}$ **38.** $\frac{3c+1}{10}+\frac{2c-1}{5}=2$ **39.** $\frac{m-4}{16}+\frac{m+4}{48}=\frac{3}{2}$

40. $\frac{1}{n-4}=\frac{8}{n^2-16}$ **41.** $\frac{4n-8}{n-2}=4$ **42.** $\frac{1}{a-3}+\frac{1}{a^2-9}=\frac{8}{a+3}$

43. Find the ratio of x to y: $\frac{x+y}{2}=\frac{x-y}{3}$.

44. Find the ratio of x to y: $\frac{2x+3y}{5}=\frac{3x+2y}{4}$.

45. Find the slope of the graph of the given equation: $\frac{2x-1}{3}+\frac{2y+1}{2}=\frac{5}{6}$.

13-5 Solving Word Problems by Rational Equations

Rational equations (fractional equations) are often used in solving word problems.

Examples:

1. One fifth of a number is 7 less than two thirds of the number. Find the number.

Solution:

Let $n =$ the number

The equation is:

$$\frac{n}{5} + 7 = \frac{2n}{3}$$

LCD = 15

$3n + 105 = 10n$

$105 = 7n \quad \therefore n = 15$

Ans: 15.

2. The sum of two numbers is 48 and their quotient is $\frac{3}{5}$. Find the numbers.

Solution:

Let $n = 1^{st}$ number

$48 - n = 2^{nd}$ number

The equation is: $\dfrac{n}{48 - n} = \dfrac{3}{5}$

$5n = 3(48 - n)$

$5n = 144 - 3n$

$8n = 144 \quad \therefore n = 18$

Ans: 18 and 30.

3. If two pounds of beef cost \$5.80, how much do 8 pounds cost ?

Solution:

Let $x =$ cost for 8 pounds

The equation is:

$$\frac{5.80}{2} = \frac{x}{8}$$

$5.80(8) = 2x$

$46.40 = 2x \quad \therefore x = 23.20$

Ans: \$23.20.

4. A bus uses 4.5 gallons of gasoline to travel 75 miles. How many gallons would the bus use to travel 95 miles ?

Solution:

Let $x =$ the number of gallons

The equation is: $\dfrac{75}{4.5} = \dfrac{95}{x}$

$75x = 4.5(95)$

$75x = 427.5 \quad \therefore x = 5.7$

Ans: 5.7 gallons.

5. John can finish a job in 5 hours. Steve can finish the same job in 4 hours. How many hours do they need to finish the job if they work together ?

Solution:

Let $x =$ number of hours needed to do the job together

John can do $\frac{1}{5}$ of the job per hour.

Steve can do $\frac{1}{4}$ of the job per hour.

The equation is: $\dfrac{x}{5} + \dfrac{x}{4} = 1$

$4x + 5x = 20, \quad 9x = 20$

$\therefore x = 2\frac{2}{9}$ Ans: $2\frac{2}{9}$ hours.

6. A 16-gallon salt-water solution contains 25% pure salt. How much water should be added to produce the solution to 20% salt ?

Solution:

Let $x =$ gallons of water added

Pure salt in the original solution:

$16 \times 25\% = 16 \times 0.25 = 4$ gal.

The equation is:

$$\frac{4}{16 + x} = 20\%, \quad \frac{4}{16 + x} = \frac{1}{5}$$

$20 = 16 + x \quad \therefore x = 4$

Ans: 4 gallons of water.

EXERCISES

1. One sixth of a number is 7 less than three fourths of the number. Find the number.
2. Three fourths of a number is 7 more than one sixth of the number. Find the number.
3. The sum of two numbers is 44 and their quotient is $\frac{3}{8}$. Find the numbers.
4. The sum of the reciprocals of two consecutive positive even integers is $\frac{5}{12}$. Find the integers.
5. The sum of a number and its reciprocal is $\frac{25}{12}$. Find the number.
6. The sum of two numbers is 12 and the sum of their reciprocals is $\frac{12}{35}$. Find the numbers.
7. If three pounds of beef cost $4.92, how much do 10 pounds cost ?
8. If five cans of dog food cost $3.45, how much do 12 cans cost ?
9. A car uses 6 gallons of gasoline to travel 96 miles. How many gallons would the car use to travel 160 miles ?
10. A car uses 6 gallons of gasoline to travel 96 miles. How many miles can the car travel on a full tank of 17 gallons ?
11. In a random survey taken in a city election, 25 of the 120 voters surveyed voted John for mayor. Based on the result of this survey, estimate how many of 24,000 total voters in the city should vote John for mayor ?
12. A map shows the length and the width of a rectangular field with the drawing 1.2 m by 0.8 m. 1 m represents 250 meters. Find the actual dimensions of the field.
13. Tom can paint the garage in 4 hours. Jack can paint the same garage in 2.4 hours. How long would it take them to paint the garage if they work together ?
14. Tom can paint the garage in 4 hours. Working together, Tom and Jack can paint the garage in 1.5 hours. How long would it take Jack to paint the garage alone ?
15. John can finish a job in 5 hours. Steve can finish the same job in 4 hours. After working together for 1 hour, Steve leaves. How long will it take John to complete the job ?
16. Nancy can ride 15 miles on her bicycle in the same time that it takes her to walk 6 miles. If her riding speed is 5 miles per hour faster than her walking speed, how fast does she walk ?
17. Nancy drives 20 miles to her school each day. If she drives 10 miles per hour faster, it takes her 4 minutes less to get to school. Find her new speed.
18. Flying in still air, an airplane travels 550 km/h. It can travel 1200 km with the wind in the same time that it travels 1000 km against the wind. Find the speed of the wind.
19. A boat travels 25 miles per hour in still water. It takes $3\frac{1}{3}$ hours to travel 40 miles up a river and then to return by the same route. What is the speed of the current in the river ?
20. A 12-gallon salt-water solution contains 25% pure salt. How much water should be added to produce the solution to 15% salt ?
21. A 18-gallon salt-water solution contains 15% pure salt. How much water should be added to produce the solution to 12% salt ?
22. A 18-gallon salt-water solution contains 15% pure salt. How much pure salt should be added to produce the solution to 20% salt ?

12-5 Inverse Variations

In Section 7-9, we learned that a linear equation in the form $y = kx$ is called an equation having a **direct variation** with x and y.
An equation in the form $y = \frac{k}{x}$ (where $x \neq 0$), or $xy = k$ is called an equation having an **inverse variation** with x and y. k is a nonzero constant. k is called the **constant of variation (constant of proportionality)**. For example, in the equation $y = \frac{3}{x}$, we say that y **varies inversely as** x, **or** y **is inversely proportional to** x. The graph of an equation having inverse variation is a **hyperbola**.
If (x_1, y_1) and (x_2, y_2) are two ordered pairs of an equation having an inverse variation defined by $y = \frac{k}{x}$, we have: $x_1y_1 = k$ and $x_2y_2 = k$. $\therefore x_1y_1 = x_2y_2$
In inverse variation, the variables that vary may involve powers (exponents).

Note that the equation $y = \frac{3}{x-1}$ does not show x and y as an inverse variation because the products of its ordered pairs x_1y_1 and x_2y_2 are not equal (not proportional).
However, the equation $y = \frac{3}{x-1}$ shows $(x-1)$ and y as an inverse variation $[\, y(x-1) = 3 \,]$.

Combined Variation: The combination of direct and inverse variations.

In science, there are many word problems which involve the concept of inverse variations. The gravitational force (F) attracted each other between any two objects in the universe is given by $F = G\frac{m_1m_2}{r^2}$ (read example on Page 41).

Examples

1. If y varies inversely as x, and if $y = 6$ when $x = 2$, find the constant of variation.
   Solution: Let $xy = k$ $\quad \therefore k = xy = 2 \cdot 6 = 12$. Ans.
2. If y varies inversely as x, and if $y = 6$ when $x = 2$, find y when $x = 3$.
   Solution: Let $xy = k$ $\quad \therefore k = xy = 2 \cdot 6 = 12$, $xy = 12$ is the equation.
   When $x = 3$, $y = \frac{12}{x} = \frac{12}{3} = 4$. Ans.
3. If (x_1, y_1) and (x_2, y_2) are ordered pairs of the same inverse variation, find y_1.
   $x_1 = 3$, $y_1 = ?$, $x_2 = 12$, $y_2 = 8$
   Solution: Let $x_1y_1 = x_2y_2$, $3y_1 = 12 \cdot 8$, $3y_1 = 96$ $\quad \therefore y_1 = 32$. Ans.
   Or: $k = x_2y_2 = 12 \cdot 8 = 96$ $\quad \therefore x_1y_1 = 96$, $y_1 = \frac{96}{x_1} = \frac{96}{3} = 32$. Ans.
4. The number of days needed to build a house varies inversely as the number of workers working on the job. It takes 140 days for 5 workers to finish the job. If the job has to be finished in 100 days, how many workers are needed ?
   Solution: Let $d_1 = 140$, $w_1 = 5$, $d_2 = 100$, $w_2 = ?$
   We have: $d_1w_1 = d_2w_2$, $140 \cdot 5 = 100 \cdot w_2$ $\quad \therefore w_2 = \frac{700}{100} = 7$ workers. Ans.

EXERCISES

Find the constant of variation and write an equation of inverse variation.

(Hint: "varies inversely as" and "is inversely proportional to" have the same meaning.

1. If y varies inversely as x, and if $y = 10$ when $x = 2$.
2. If y is inversely proportional to x, and if $y = 10$ when $x = 5$.
3. If n is inversely proportional to m, and if $n = 200$ when $m = 20$.
4. If m varies inversely as n, and if $m = 30$ when $n = 90$.
5. If y is inversely proportional to $(x+8)$, and if $y = 150$ when $x = 2$.

For each inverse variation described, find each missing value.

6. If y varies inversely as x, and if $y = 10$ when $x = 2$, find y when $x = 12$.
7. If y varies inversely as x, and if $y = 10$ when $x = 2$, find x when $y = 12$.
8. If y is inversely proportional to x, and if $y = 2$ when $x = 10$, find y when $x = 12$.
9. If n varies inversely as m, and if $n = 150$ when $m = 6$, find n when $m = 20$.
10. If n varies inversely as m, and if $n = 150$ when $m = 6$, find m when $n = 20$.

If (x_1, y_1) and (x_2, y_2) are ordered pairs of the same inverse variation, find each missing value.

11. $x_1 = 1$, $y_1 = ?$, $x_2 = 3$, $y_2 = 12$
12. $x_1 = 3$, $y_1 = ?$, $x_2 = 4$, $y_2 = 80$
13. $x_1 = 6$, $y_1 = ?$, $x_2 = 8$, $y_2 = 9.6$
14. $x_1 = 6$, $y_1 = 4$, $x_2 = 9$, $y_2 = ?$
15. $x_1 = ?$, $y_1 = 1.5$, $x_2 = 15$, $y_2 = 9$
16. $x_1 = 1$, $y_1 = 7$, $x_2 = ?$, $y_2 = 17.5$
17. $x_1 = \frac{4}{5}$, $y_1 = ?$, $x_2 = \frac{2}{5}$, $y_2 = \frac{1}{10}$
18. $x_1 = \frac{1}{4}$, $y_1 = \frac{1}{5}$, $x_2 = \frac{5}{3}$, $y_2 = ?$
19. $x_1 = \frac{7}{8}$, $y_1 = ?$, $x_2 = \frac{1}{2}$, $y_2 = \frac{1}{7}$
20. $x_1 = \frac{1}{5}$, $y_1 = 5$, $x_2 = 5$, $y_2 = ?$

21. Does the equation $xy = 5$ define an inverse variation ? Explain.
22. Are the ordered pairs (6, 4), (3, 8), (2, 12) in the same inverse variation ? Explain.
23. The number of days needed to finish a job varies inversely as the number of workers working on the job. It takes 40 hours for 3 workers to finish the job. If the job has to be finished in 15 hours, how many workers are needed ?
24. The time required to travel a given distance is inversely proportional to the speed of a car. If it takes 5.6 hours to travel from city A to city B at an average speed of 60 miles/h, how long will it take to make the same trip at an average speed of 70 miles/h.
25. The length of a rectangle of given area varies inversely as the width. A rectangle field has length $24\,m$ and width $18\,m$. Find the width of another rectangle field of equal area whose length is $12\,m$.
26. At a fixed water pressure, the speed of the water varies inversely as the diameter of the pipe. If the water flows at 30 miles/h through a pipe with a $2\,cm$ diameter, what would be the speed of the water through a pipe with a $1.5\,cm$?
27. If z varies directly as x and inversely as y, and if $z = 9$ when $x = 15$ and $y = 3$, find z when $x = 45$ and $y = 2$.

CHAPTER 13 EXERCISES

Simplify each rational expression. Give all restrictions on the variables.

1. $\dfrac{6x}{3x^2}$ 2. $\dfrac{12x^6}{4x^4}$ 3. $\dfrac{24a^3b^2}{32ab^5}$ 4. $\dfrac{8x-12}{4}$

5. $\dfrac{2-4x}{2x-1}$ 6. $\dfrac{n^2-9}{n-3}$ 7. $\dfrac{16(a+3)^4}{6(a+3)^2}$ 8. $\dfrac{x-4}{x^2+x-20}$

9. $\dfrac{a^2-25}{a^2-3a-10}$ 10. $\dfrac{t^2-4}{t^2+2t}$ 11. $\dfrac{y^2+2y-24}{y^2+3y-28}$ 12. $\dfrac{x^2+2x-3}{x^3+5x^2+6x}$

Simplify. Assume that no variable has a value for which the denominator is 0.

13. $\dfrac{x}{2}+\dfrac{x}{7}$ 14. $\dfrac{y}{2}-\dfrac{y}{7}$ 15. $\dfrac{1}{2x}-\dfrac{1}{7x}$ 16. $\dfrac{1}{2y}+\dfrac{1}{7y}$

17. $\dfrac{a}{5}\cdot\dfrac{10}{a^4}$ 18. $\dfrac{9}{x^2}\cdot\dfrac{x}{3}$ 19. $\dfrac{15n^2}{2}\div\dfrac{3n^5}{4}$ 20. $\dfrac{7}{n^3}\div\dfrac{21}{12n^6}$

21. $\dfrac{1}{x^2y}+\dfrac{1}{xy^2}$ 22. $\dfrac{2}{ab}+\dfrac{4}{a^2b}$ 23. $\dfrac{2}{ab}\cdot\dfrac{a^2b}{4}$ 24. $\dfrac{4}{x^2y}\div\dfrac{8}{xy^2}$

25. $\dfrac{1}{a+2}+\dfrac{2}{a+3}$ 26. $\dfrac{2}{n-2}-\dfrac{3}{n+3}$ 27. $\dfrac{2}{n-2}\cdot\dfrac{2-n}{4}$ 28. $\dfrac{12}{3-y}\div\dfrac{9}{y-3}$

29. $\dfrac{3}{a}+\dfrac{4}{a+2}$ 30. $\dfrac{3}{a-2}-\dfrac{4}{a+2}$ 31. $\dfrac{4}{a-2}\cdot\dfrac{a^2-4}{2}$ 32. $\dfrac{4}{a+2}\div\dfrac{2}{a^2-4}$

33. $\dfrac{5}{x^2-25}-\dfrac{4}{x+5}$ 34. $\dfrac{6}{x^2-25}+\dfrac{2}{x-5}$ 35. $\dfrac{12}{x^2-25}\cdot\dfrac{4x-20}{8}$ 36. $\dfrac{6}{2x-8}\div\dfrac{9}{x^2-16}$

37. $\dfrac{x^2+5x-14}{x-2}\cdot\dfrac{x}{x+7}$ 38. $\dfrac{x^2-36}{x+6}\div\dfrac{x-6}{x+6}$ 39. $\dfrac{a^2-b^2}{a}\div\dfrac{2a-2b}{a^4}$ 40. $\dfrac{m^2+2m-3}{m}\cdot\dfrac{m^2}{m-1}$

41. $\dfrac{1}{c+1}+\dfrac{2}{c-1}-\dfrac{4}{c^2-1}$ 42. $\dfrac{4}{x^2+2x-3}-\dfrac{1}{x-1}$ 43. $\dfrac{4}{x^2-49}\div\dfrac{x+2}{x+7}\cdot\dfrac{x^2-5x-14}{8}$

Divide polynomials.

44. $\dfrac{x^2-4x-5}{x+2}$ 45. $\dfrac{x^3-2x^2+4x-6}{x-4}$ 46. $\dfrac{4x^3-3x+4}{2x+1}$ 47. $\dfrac{x^3-a^3}{x+a}$

-----Continued-----

Solve each rational equation (Fractional equation).

48. $\frac{x}{5}=\frac{3}{4}$ **49.** $\frac{5}{x}=\frac{2}{9}$ **50.** $\frac{12}{5y}=\frac{2}{5}$ **51.** $\frac{3y}{4}=\frac{9}{2}$

52. $\frac{n-4}{3}=\frac{2}{7}$ **53.** $\frac{2n+1}{2}=\frac{n}{3}$ **54.** $\frac{3x+5}{4}=5$ **55.** $\frac{4x-2}{3x-6}=2$

56. $\frac{2}{k+3}=\frac{4}{k+13}$ **57.** $\frac{6}{2u-3}=\frac{3}{u+5}$ **58.** $\frac{10}{5x-4}=\frac{2}{x}$ **59.** $\frac{6n-12}{n-2}=6$

60. $\frac{4a+2}{4}=\frac{2a+1}{2}$ **61.** $\frac{9}{y^2-36}=\frac{3}{y-6}$ **62.** $\frac{32}{x^2-16}=\frac{x}{x-4}$ **63.** $\frac{50}{x^2-25}=\frac{x}{x+5}$

64. $\frac{x}{4}+\frac{x}{8}=3$ **65.** $\frac{y}{5}-\frac{y}{6}=1$ **66.** $\frac{2x}{3}-\frac{x}{6}=-1$ **67.** $\frac{9}{x}-\frac{3}{x}=\frac{1}{2}$

68. $\frac{1}{y}-\frac{5}{y}=-14$ **69.** $\frac{5}{x+2}-\frac{2}{x}=\frac{1}{3}$ **70.** $\frac{2c-5}{5}+\frac{c}{3}=\frac{8}{3}$ **71.** $\frac{3}{a-3}+2=\frac{a}{a-3}$

72. $\frac{14}{x}-\frac{2x-5}{2x}=\frac{5}{6}$ **73.** $\frac{5-a}{a-3}-\frac{1}{3-a}=0$ **74.** $\frac{8}{x-1}-\frac{16}{x^2-1}=3$ **75.** $\frac{x^2-9}{x+3}+x^2=3$

76. Two fifths of a number is 6 more than one tenth of the number. Find the number.

77. The sum of a number and its reciprocal is $\frac{58}{21}$. Find the number.

78. The sum of the reciprocals of two positive numbers is $\frac{1}{9}$, and one of the numbers is 3 times the other. Find the numbers.

79. If five pounds of beef cost $12.60, how much do 8 pounds cost ?

80. David can finish a job in 7 days. Coby can finish the same job in 5 days. How many days would it take them to finish the job if they work together ?

81. David can finish a job in 7 days. Coby can finish the same job in 5 days. After working together for 2 days, Coby leaves. How many days will it take David to complete the job ?

82. A 10-gallon alcohol-water solution contains 30% pure alcohol. How much water should be added to produce the solution to 20% alcohol ?

83. A 10-gallon alcohol-water solution contains 30% pure alcohol. How much pure alcohol should be added to produce the solution to 34% alcohol ?

84. If y is inversely proportional to x, and if $y=6$ when $x=1.5$, find y when $x=1.2$.

85. If (5, 14) and (7, y_2) are ordered pairs of the same inverse variation, find y_2.

86. The number of days needed to build a tower varies inversely as the number of workers working on the job. It takes 120 days for 8 workers to finish the jobs. If the job has to be finished in 80 days, how many workers are needed ?

87. If z varies directly as x and inversely as y^2, and if $z=6$ when $x=2$ and $y=4$, find z when $x=4$ and $y=2$.

88. Write the equation if f varies directly as the product of m_1 and m_2, and inversely as r^2.

CHAPTER FOURTEEN

Statistics and Probability

14-1 Mean, Median, Mode, and Range

In order to find a convenient way to describe and compare the information of entire collection of data, we need to organize the data and find a single number that is most significant to represent the data. Then we can analyze and make conclusion about the data. **Statistics** is the study of collecting, organizing, and analyzing numerical data.
The **mean**, **median**, **mode**, and **range** are important statistical numbers often used in analyzing numerical data. The mean, median, and mode are three kinds of averages.

Mean: It is the sum of the numbers divided by the number of items in the data. It is the arithmetic average and is most commonly used as an average.
Median: It is the middle number in the data when the data are arranged in an increasing order (from least to greatest). For an even number of items, the average of the two numbers in the middle is the median.
Mode: It is the number that appears most often in the data. There may be no mode or more than one mode.
Range: It is the difference between the highest and the lowest numbers.

The mean, median, or mode provides the information about the **central tendency** of the data. Which of these three measures is the best to represent the central tendency depends on the way in which you need to use the data. If the mean is inflated by a few very larger or very small numbers in the data, the median can be used as an average. Mode can be used as an average if we want the most frequent number in the data.
The range provides the information about the **spread** of the data.

Examples

1. The heartbeats per minute (pulse rates) for 9 students are listed below.

   66, 71, 72, 69, 70, 71, 72, 85, 72

   Find: **a)** the mean **b)** the median **c)** the mode **d)** the range.

   Solution: **a)** Divide the sum of the numbers by the number of students.

   The mean $= \frac{66+71+72+69+70+71+72+85+72}{9} = \frac{648}{9} = 72$. Ans.

   b) Arrange the numbers from least to greatest.

   66, 69, 70, 71, **71**, 72, 72, 72, 85. The median = 71. Ans.

   c) 72 appears most often. The mode = 72. Ans.

   d) $85 - 66 = 19$ The range = 19. Ans.

2. Find the median of the data 15, 19, 6, 17, 12, 12, 17, 11.

   Solution: Arrange the data from least to greatest: 6, 11, 12, **12**, **15**, 17, 17, 19

   The median $= \frac{12+15}{2} = 13.5$. Ans.

EXERCISES

Find the mean. Round to the nearest tenth.

1. 2, 3, 4, 5, 6

2. 5, 6, 7, 8, 9

3. 21, 25, 20, 10, 27, 20

4. 10, 8, 29, 6, 20, 17

5. 13, 11, 14, 16, 19, 12, 16

6. 15, 19, 13, 15, 29, 17, 35

7. 32, 28, 40, 20, 48, 28, 21, 28

8. 38, 29, 48, 29, 52, 46, 48, 28

9. 140, 250, 362, 250, 185

10. 214, 252, 301, 184, 252, 270

11. 2.2, 1.8, 4.5, 2.7, 5.7

12. 1.8, 4.2, 3.7, 2.6, 7.5, 4.7

13. 7.9, 5.1, 7.9, 3.4, 7.9, 4.5, 1.8

14. 6.9, 4.2, 5.9, 4.2, 6.9, 7.2, 1.5

15. 19, 29, 35, 11, 27, 29, 17, 28, 29, 22, 27, 56, 32, 37

Find the median, mode, and range.

16. 7, 8, 9, 12, 18, 18, 21

17. 9, 9, 12, 18, 22, 24, 32

18. 5, 9, 15, 7, 14, 12, 24, 31, 14

19. 18, 12, 20, 12, 18, 24, 18, 26, 12

20. 2, 12, 8, 18, 24, 6

21. 10, 20, 18, 10, 10, 20, 15, 20

22. 1.4, 1.2, 8.5, 6.2, 2.2, 1.4, 6.4

23. 1.8, 3.2, 1.8, 2.5, 1.2, 3.1, 1.8, 2.5, 4.5

24. 32, 26, 20, 42, 32, 54, 26, 26

25. 25, 46, 20, 27, 47, 25, 18, 27, 27, 25

26. John's math scores for 8 tests are 72, 85, 84, 82, 72, 85, 85, 98.
a) Find the mean of his scores rounded to the nearest tenth.
b) How many tests fell below the mean ? **c)** Find the median, mode, and range.

27. Gary's mean score in 5 math tests is 82. His four test scores are 76, 80, 84, 81. What is his other score ?

28. Choose the measure (mean, median, or mode) that best describes the average of the ages of a group of 11 people. 10, 15, 11, 14, 12, 14, 86, 11, 10, 9, 13

29. The test scores of 40 students are listed in the table below.
a) Find the mean of the scores. **b)** Which measure can best describe the students' abilities ?

Scores	72	76	78	84	86	90
Numbers of students	3	4	20	6	5	2

30. The numbers of hours spent on homework per week of 10 students are listed below.
12, 14.5, 10, 18.5, 11.5, 12, 20, 14.5, 3.5, 12
Find the mean, median, mode, and range.

31. What are the statistical numbers providing the information about the central tendency of the data ?

32. What is the statistical number providing the information about the spread of the data ?

14-2 Statistical Graphs

In statistics, tables and graphs are often used to display and represent numerical data. Data represented in table or graph are easily to read and understand.
There are many different kinds of statistical tables and graphs we use in statistics, such as **bar graph**, **line graph**, **circle graph**, **picture graph**, **Stem-and-Leaf plot**, **Box-and-Whisker plot**, and others. Three major graphs are shown below.
We can estimate data from graphs.

Examples

1. The bar graph below shows the population of a town between 1930 and 2000.

1. What data unit is used on the horizontal axis ?
2. What data unit is used on the vertical axis ?
3. In which year did the population increase the most ?
4. Estimate the approximate population in 1980.
5. Find the approximate total increase in population during the years from 1930 to 2000.
6. In which year did the population stay the same as the previous year.

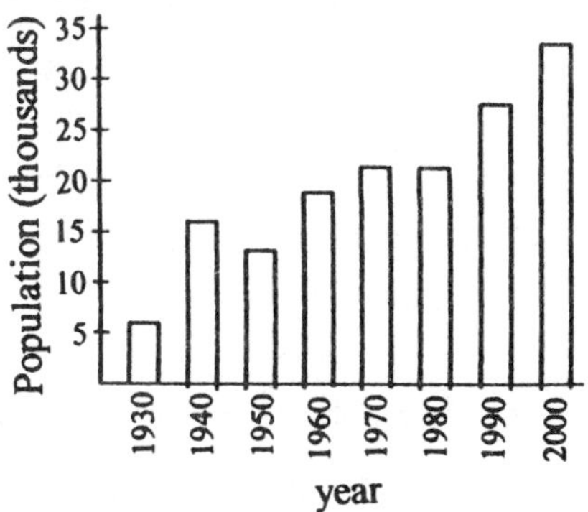

Solution: **1.** The year **2.** Thousands of persons **3.** 1940
4. 22,000 **5.** 34,000 – 6,000 = 28,000 persons **6.** 1980

2. Make a line graph to illustrate the given data.

Population of a town (in Thousands)

1930	1940	1950	1960	1970	1980	1990	2000
6	16	13	17	22	22	28	34

Solution:

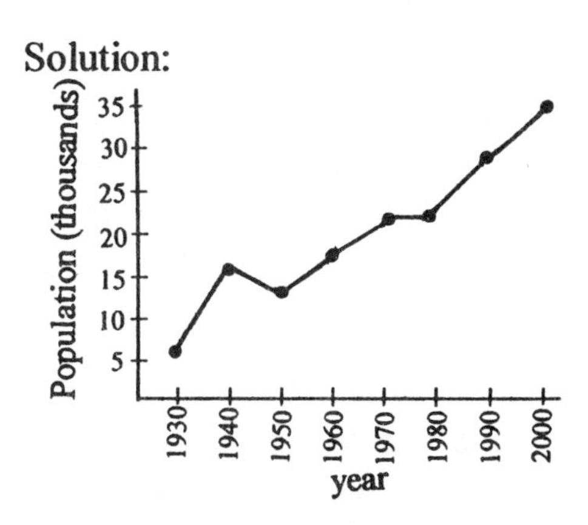

3. The circle graph below shows Ronald's budget to spend his money. His monthly income is \$2,700. Find how much money does he plan to spend on food, clothes, housing, girl friends, savings and others.

Solution:

Food: $\$2,700 \times 25\% = \$675$
Clothes: $\$2,700 \times 8\% = \$216$
Housing: $\$2,700 \times 17\% = \$459$
Girl friends: $\$2,700 \times 3\frac{2}{3}\% = \$99$
Saving: $\$2,700 \times 35\frac{1}{3}\% = \$954$
Others: $\$2,700 \times 11\% = \$297$

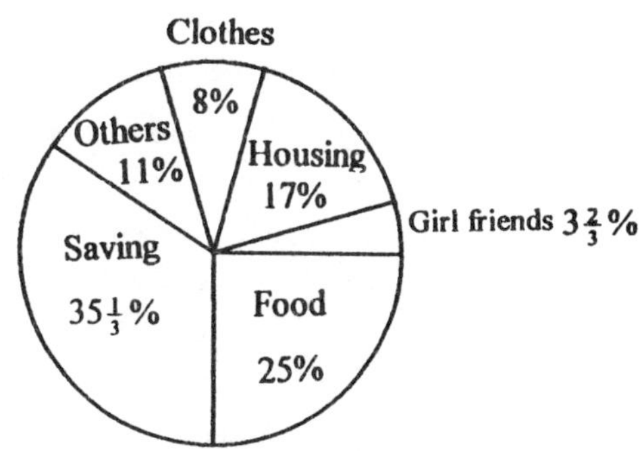

EXERCISES

The bar graph below shows the numbers of hours spent on homework per week for 5 students.

1. Who spent most time doing homework ?
2. Who is the laziest student ?
3. How many more hours did Ron spend doing homework than David ?
4. Who were the students who spent the same number of hours doing homework ?
5. How many more hours did Ron spend on homework than Jim ? What percent more would that be ?
6. How many less hours did David spend on homework than Ron ? What percent less would that be ?

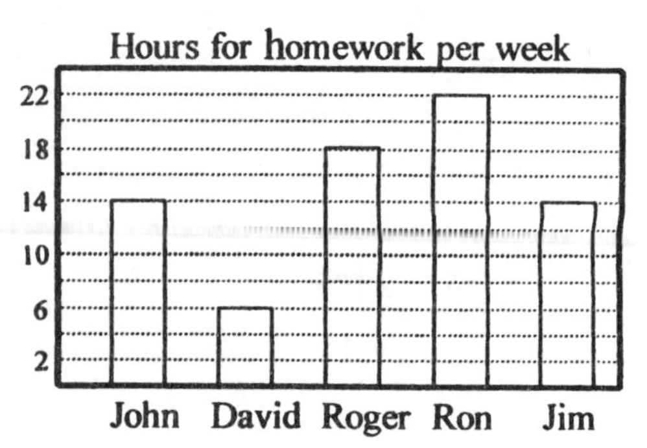

The line graph below shows the Dow Jones industrial average for stock over several weeks.

7. Estimate what was the Dow Jones industrial average on September 9 ?
8. Estimate what was the Dow Jones industrial average on September 23 ?
9. How many points did the Dow Jones industrials average climb from October 7 to 14 ?
10. What percent did the Dow Jones industrial average increase from October 7 to 14 ?

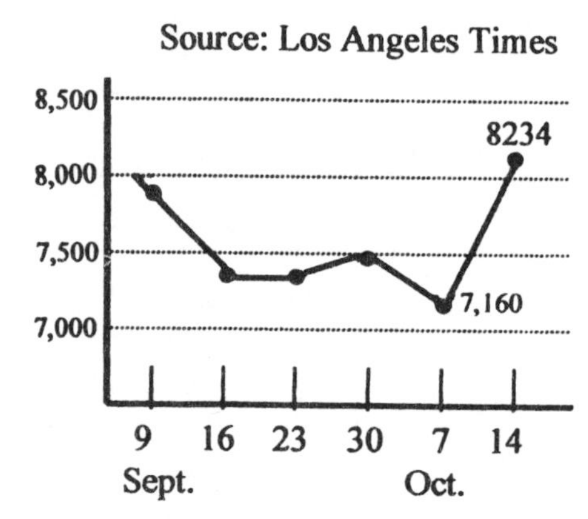

The circle graph below shows an annual sales of a book publishing company.

11. Where does the company sell the most copies ?
12. Where does the company sell the least copies ?
13. How many copies does the company sell to the chain stores in a year ?
14. How many copies does the company sell to the college stores in a year ?
15. How many copies does the company sell to the libraries in a year ?
16. How many copies does the company sell to the schools in a year ?

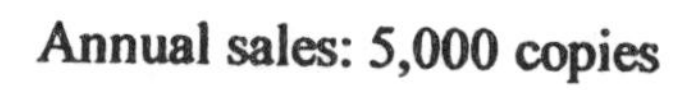

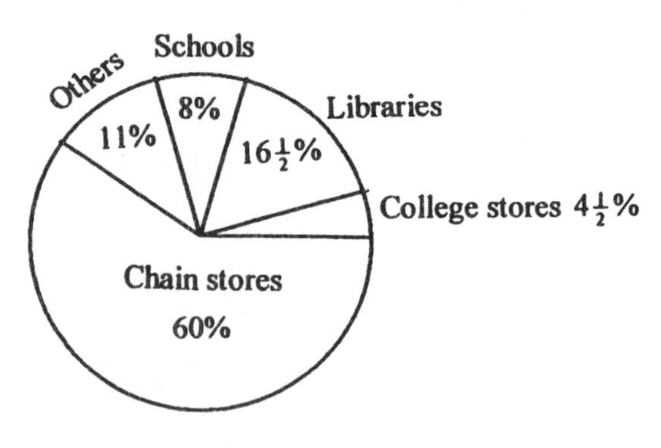

14-3 Histogram and Normal Distribution

In statistics, we use a table called **frequency distribution** to show information about the frequency of occurrences of statistical data.
Histogram is a statistical graph to describe a frequency distribution.
In a histogram, data are grouped into convenient intervals. A boundary datum in a histogram is included in the interval to its left. If the midpoints at the tops of the bars are connected, a smooth **"bell-shaped"** curve called **normal curve** can result from the plotting of a larger collection of data. The area under the normal curve represents the probability from **normal distribution**. The area under the curve and above the $x-$axis is 1. Therefore, the sum of the probabilities of all possible outcomes is 1 (or 100%).
The probability of each region under a normal curve is shown below. We can find the probability of a given region under the normal curve.

Frequency Distribution

Test scores	numbers of students
45 ~ 55	3
55 ~ 65	13
65 ~ 75	21
75 ~ 85	27
85 ~ 95	15
95 ~100	7

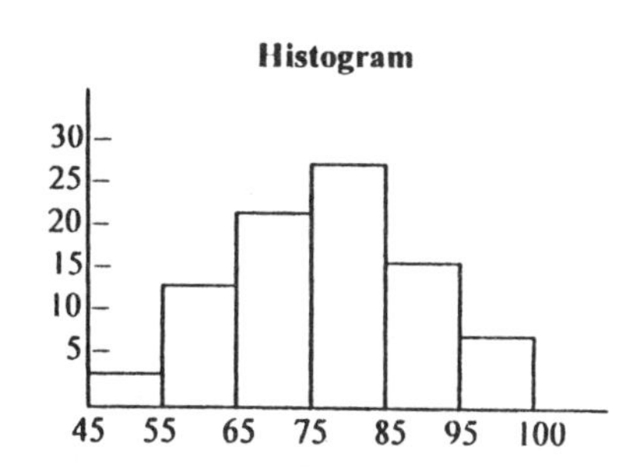

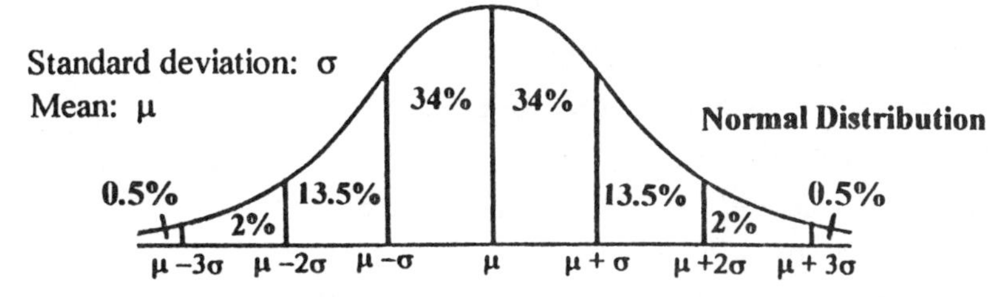

Examples

1. The heights of the students in a high school are normally distributed as shown in the graph. Find the percentage of the students in each the following groups.
   a. Height $>172\,cm$ **b.** Height $<172\,cm$ **c.** Height $>179\,cm$ **d.** Height $<186\,cm$
   Solution:
   a. 50% **b.** 50%
   c. $13.5\% + 2\% + 0.5\% = 16\%$
   d. $50\% + 34\% + 13.5\% = 97.5\%$

2. The sales of a math book each month are normally distributed as shown in the graph.
   a. What is the probability that a month will sell more than 600 copies in a month ?
   b. What is the probability that a month will sell less than 300 copies in a month ?
   Solution:
   a. $13.5\% + 2\% + 0.5\% = 16\%$
   b. $2\% + 0.5\% = 2.5\%$

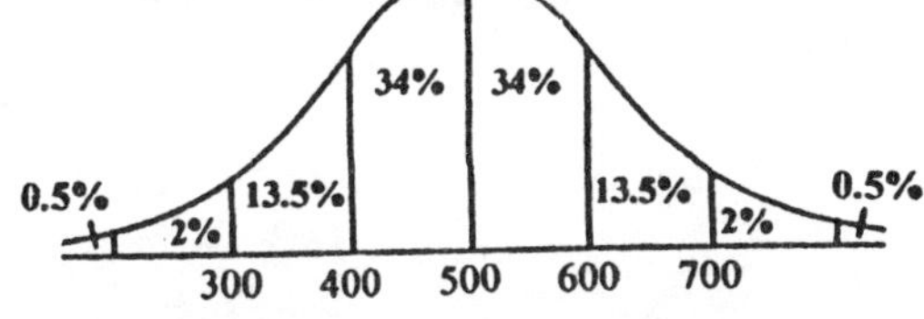

EXERCISES

1. The test scores of 50 students are listed below.

12	25	41	74	52	70	32	49	54	63
26	44	47	33	55	16	42	24	89	34
44	79	58	21	38	54	39	57	42	51
52	59	29	35	69	43	56	63	30	44
37	40	45	52	51	18	61	75	88	31

a. Make a frequency distribution

Test scores	Tally	Frequency
10 ~ 20		
20 ~ 30		
30 ~ 40	卌 ////	9
40 ~ 50		
50 ~ 60		
60 ~ 70		
70 ~ 80		
80 ~ 90		

b. Complete the following histogram

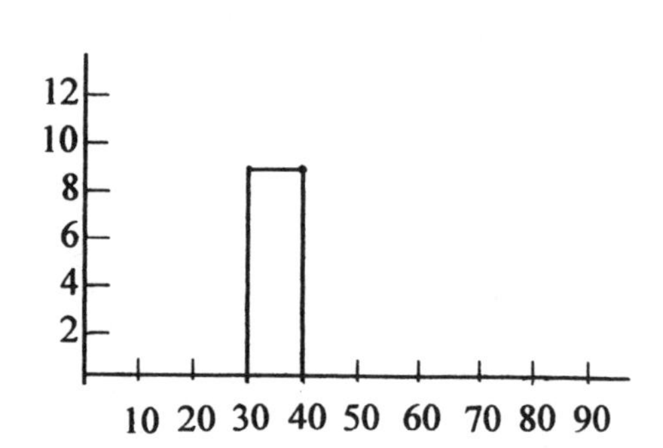

2. The time spent by shoppers in the mall is normally distributed as shown in the graph.
   a. What percent of the shoppers in the mall will spend between 58 and 82 minutes ?
   b. What is the probability that a shopper will spend more than 58 minutes ?
   c. What is the probability that a shopper will spend less than 58 minutes ?
   d. What percent of the shoppers in the mall will spend between 46 and 94 minutes ?

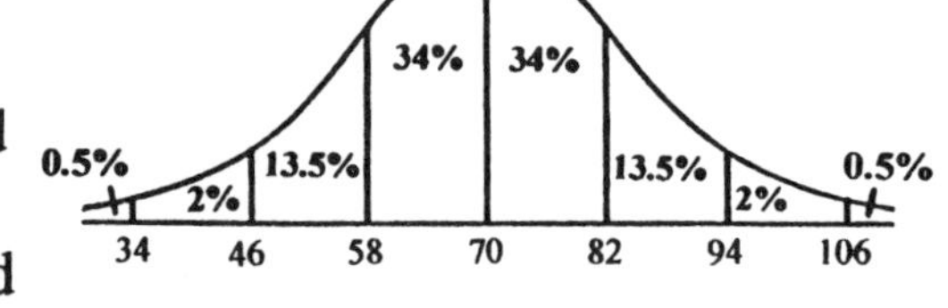

3. In the experiment, the probability of operating life of a new brand battery is normally distributed as shown in the graph.
   a. What is the probability that a battery will last more than 25 hours ?
   b. What is the probability that a battery will last less than 25 hours ?

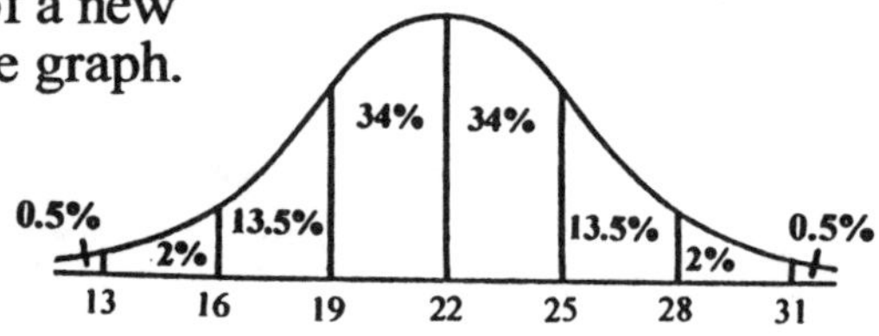

4. The scores on a math standardized test gave a normal distribution as shown in the graph. What is the probability that a student's score is more than 500 ?

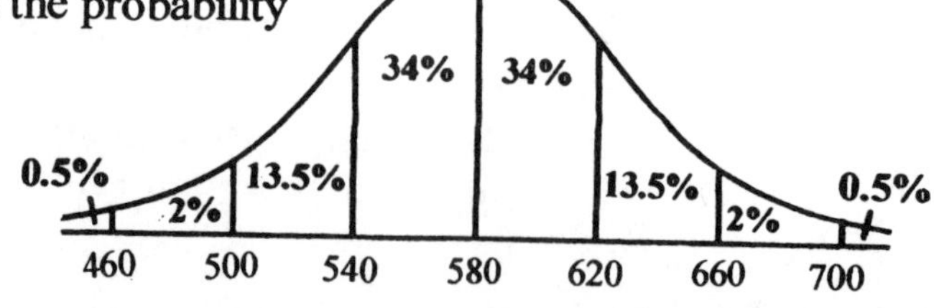

14-4 Stem-and-Leaf Plots & Box-and-Whisker Plots

A stem-and-leaf plot is a statistical plot used to organize numerical data. We can use the stem-and-leave plot to convert data into a bar graph or compare two sets of data.

Examples

1. Draw a stem-and-leaf plot of the data.

13, 4, 23, 34, 40, 52, 55, 52, 9, 38
60, 36, 22, 68, 15, 39, 29, 15, 41

Solution:
Identify the lowest number 4.
Identify the highest number 68.
In the plot, use the tens digits as the stems and the ones digits as the leaves.
Draw a vertical line between the stems and the leaves.
Arrange each row of leaves from smallest to largest.
A legend is used if the decimal point is not included in the stems.

Stems	Leaves
0	4, 9
1	3, 5, 5
2	2, 3, 9
3	4, 6, 8, 9
4	0, 1
5	2, 2, 5
6	0, 8

1|3 = 13

2. Draw a stem-and-leaf plot of the data below.

8.9, 4.6, 11.2, 10.2, 8.3, 8.4, 10.4, 8.8
4.2, 8.1, 10.8, 5.7, 10.4, 5.3, 11, 6.4

Solution:
Identify the lowest number 4.2.
Identify the highest number 11.2.
In the plot, use the whole number as the stems and the tenths as the leaves.
Arrange each row of leaves from smallest to largest.

Stems	Leaves
4.	2, 6
5.	3, 7
6.	4
7.	
8.	1, 3, 4, 8, 9
9.	
10.	2, 4, 4, 8
11.	0, 2

A **box-and-whisker plot** is also a statistical plot used to organize numerical data. In the box-and-whisker plot, the data are divided into four sections (2 whiskers and 2 boxes). Each section contains 25% of data. The lines to the left and right of the boxes are the whiskers of the plot.
We can use the box-and-whisker plot to identify the three **quartiles** (Q_1, Q_2, Q_3) between the two **outliers** (The minimum and the maximum).
Q_1 is the 1st **(lower) quartile**. Q_2 is the 2nd **quartile** (It is the median.). Q_3 is the 3rd **(upper) quartile**.

3. Draw a box-and-whisker plot for the test scores of 12 students.

75, 86, 65, 95, 85, 68, 92, 90, 85, 75, 82, 80

Solution:
Arrange the numbers from least to greatest.
65, 68, 75, 75, 80, **82, 85**, 85, 86, 90, 92, 95
The median = 83.5
The median of the numbers below 83.5. It is 75.
The median of the numbers above 83.5. It is 88.

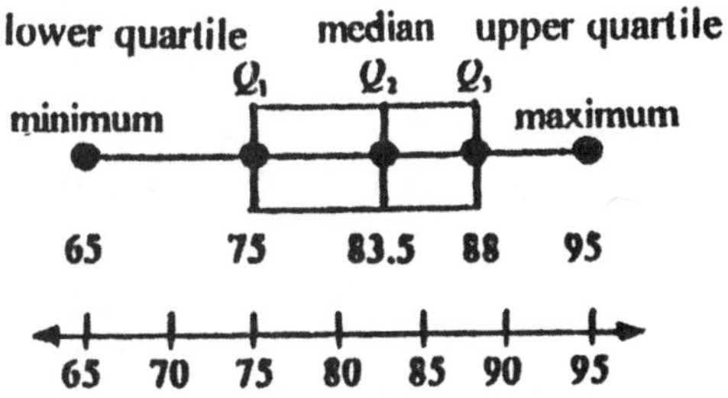

lower quartile Q_1 median Q_2 upper quartile Q_3
minimum maximum
65 75 83.5 88 95
65 70 75 80 85 90 95

EXERCISES

The test scores of 20 students are listed below.

74 52 70 49 66 63 48 79 81 54
89 94 68 75 78 81 79 90 76 82

1. Draw a stem-and-leaf plot for the data.
2. Draw a box-and-whisker plot for the data.
3. What is the mean, median, mode, and range of the data ?
4. What numbers are the outliers of the data ?
5. What are the quartiles Q_1, Q_2, and Q_3 of the data ?
6. The top 25% of the students should get A's. What is the cutoff score for an A ?
7. What is the range of the scores of the lowest 25% of the students ?
8. What percent of the students scored less than 81 ?
9. What percent of the students scored more than 64.5 ?
10. Draw a histogram (bar graph) for the data.

The box-and-whisker plot below shows the scores on a test.

11. What is the median score on the test ?
12. What percent of the students scored less than 66 ?
13. What percent of the students scored more than 82 ?
14. Find the lower and upper quartiles from the plot.
15. What is the mean score on the test ?
16. How many students were tested ?

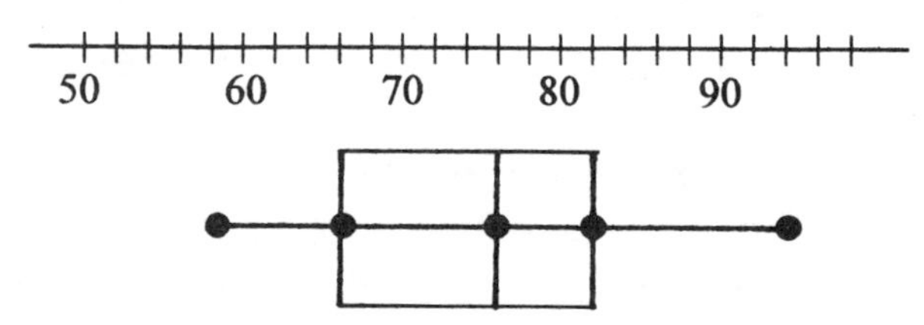

The box –and-whisker plot shows the monthly sales in copies of a book.

17. What is the median sale of the book ?
18. There are four sections in the plot. Which section contains more data ?
19. Between what two readings is the middle 50% of the copies of the book sold ?
20. Find the lower and upper quartiles from the plot.

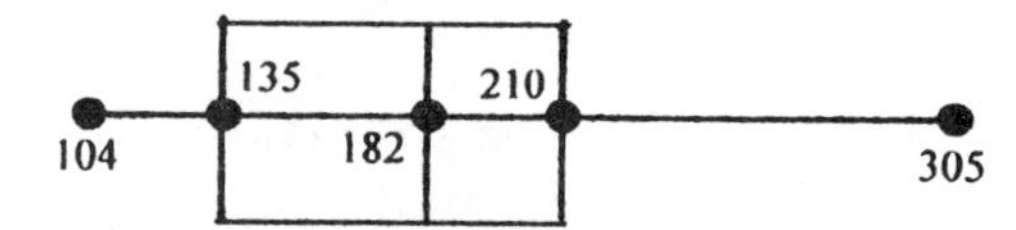

A three-year old boy was taken to church by his mother for the first time. The boy was extremely quiet during the sermon.
Mother: Why were you so quiet in church ?
Boy: I don't want to be nailed onto the big letter " T ".

14-5 Sets and Venn Diagrams

Set: Any particular group of elements or members. We use braces, { }, to name a set.

U: The **universal** set. It is the set consisting of all the elements.

$A \cap B$: The **intersection** of two sets. The set consists of the elements that are common to both sets A and B.

$A \cup$ B: The **union** of two sets. The set consists of all members of both sets A and B.

ϕ: The **empty** set, or null set. The set consists of no member.

The set {0} contains exactly one member "0". The set {0} is not an empty set.

$\in$: It indicates " **it is a member of** ". $\notin$: It indicates " **it is not a member of** ".

$A \subset B$ or $A \subseteq B$: If every member in set A is also a member of set B, then A is a **subset** of B, B is a **superset** of A. Every set is the subset of itself. It is agreed that the empty set is also a subset of every set.

If A is a subset of B and $A \neq B$, we write $A \subset B$, then A is a **proper subset** of B. If A is a subset of B and $A = B$, we write $A \subseteq B$.

$A \not\subset B$: Set A is **not a subset** of set B.

A' or A^c: The **complement** of set A. A' is the set consisting of all the elements not in A.

$A \cup A' = U$; $A \cap A' = \phi$

Venn Diagram: A diagram used to indicate relations between sets.

1. $n(A \cup B) = n(A) + n(B) - n(A \cap B)$
2. $n(A \cup B \cup C) = n(A) + n(B) + n(C) - n(A \cap B) - n(B \cap C) - n(A \cap C) + n(A \cap B \cap C)$

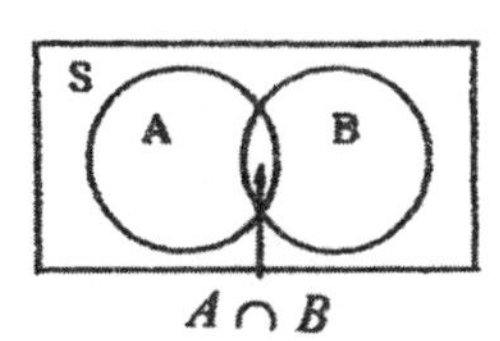

$A \cap B$

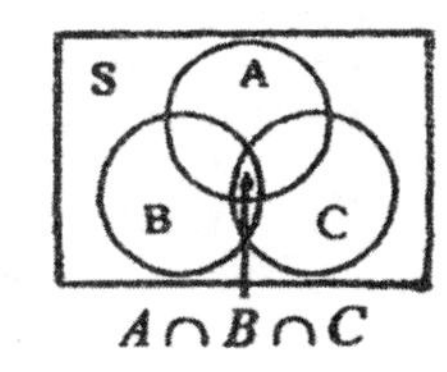

$A \cap B \cap C$

Examples

1. If $A = \{1, 3, 5, 7\}$, $B = \{2, 4, 6, 8\}$, $U = \{1, 2, 3, 4, 5, 6, 7, 8\}$, we have:

$3 \in \{1, 3, 5, 7\}$, $6 \notin \{1, 3, 5, 7\}$, $A \neq B$, $\{1, 3, 5, 7\} = \{1, 5, 7, 3\}$

$A \subset U$, $B \subset U$, $A \not\subset B$, $B \subset B$, $\phi \subset A$, $\phi \subset B$, $A' = B$, $A \cup A' = U$

$A \cup B = U$, $A \cap B = \phi$, $A \cap U = \{1, 3, 5, 7\}$, $U \cap B = \{2, 4, 6, 8\}$

2. In a high school class, there are 25 students who went to see movie A, 18 students went to see move B. These figures include 9 students went to see both movies. How many students are in this class ?

Solution:

$25 + 18 - 9 = 34$ students.

16 9 9
A B

3. In a high school class, there are 17 students in the basketball team, 18 in the volley ball team, 20 in the baseball team. Of these, there are 9 students in both the basket ball and baseball teams, 6 in both volleyball and baseball teams, and 5 in both volleyball and basketball teams. These figures include 2 students who are in all three teams. How many students are in these class ?

Solution:

$17 + 18 + 20 - 5 - 6 - 9 + 2 = 37$ students.

Basketball 5 3 9 Valleyball
7 2 4
7
Baseball

EXERCISES

Specify each of the following sets, referring to the Venn Diagram.

1. $A \cap B$ **2.** $A \cup B$ **3.** $B \cap C$ **4.** $B \cup C$

5. $(A \cup B) \cap C$ **6.** $(A \cap B) \cap C$ **7.** $A \cap \phi$ **8.** $A \cup \phi$

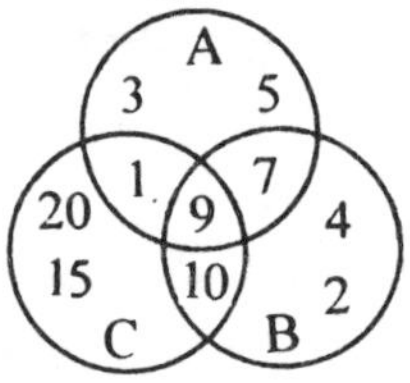

A 3 5 1 7 20 9 4 15 10 2 C B

Insert a symbol of sets to make a true statement.

9. 5 ___ {0, 5, 8, 11} **10.** 4 ___ {0, 5, 8, 11} **11.** {5, 8} ___ {0, 5, 8, 11}

12. {5} ___ {0, 5, 8, 11} **13.** {0, 5, 8, 11} ___ {5, 11, 8, 0} **14.** {3, 9} ___ {0, 5, 8, 11}

15. ϕ ___ {0, 5, 8, 11} **16.** {0, 5, 8, 11) ___ {5, 8, 0, 12} **17.** ϕ ___ {0}

Specify each set of numbers.

18. {the even numbers between 1 and 10} **19.** {the odd numbers between 6 and 16}

20. {the positive integers less than 4} **21.** {the positive integers less than –1}

22. {the real number that are neither positive nor negative}

23. {the whole numbers less than 6}

24. {the real numbers less than 1 and the real numbers greater than 5}

25. {the real numbers greater than or equal to –4 and less than 2 }

26. List all the subsets of {1, 2, 3}.

27. What is the total number of subsets of {1, 2, 3, 4) ?

28. What is the total number of subsets of {1, 2, 3, 4. 5} ?

29. If a set contains n numbers, what is the total number of subsets of n ?

30. Draw a Venn Diagram to illustrate $A \supset B$.

31. Draw a Venn Diagram to illustrate $C \subset B$ and $B \subset A$.

32. What is the relation between M an N if $M \subset N$ and $N \subset M$?

33. There are 34 students in a class, 25 students went to see movie A, 18 students went to see movie B. These figures include the students who went to see both movie A and movie B. How many students went to see both movie A and movie B ?

34. There are 50 students in a class, 24 were registered in Algebra, 32 students were registered in biology, and 8 were in both courses. How many were registered in neither course ?

35. In a survey of 50 investors in the stock market, 17 owned shares in company A, 18 owned shares in company B, 20 owned shares in company C, 5 owned shares in both company A and B, 6 owned shares in both company B and C, 9 owned shares in both company A and C, 2 owned shares in all three companies.

- **a.** How many owned only company A shares ?
- **b.** How many owned shares in both company A and B , but no company C ?
- **c.** How many did not own shares in any of the three companies ?

36. Write the set consisting of the possible outcomes from tossing two coins. Use H for "heads" and T for "tails".

37. One-third of all Matts are Patts. Half of all Patts are Natts. No Matt is a Natt. All Natts are Patts. There are 25 Natts and 45 Matts. How many Patts are neither Matts nor Natts ?

14-6 Probability

In mathematics, we want to find the possibilities of uncertainty. For example, we want to know the chance of winning a state lottery, or the chance to have a life of more than 90 years.
There are countless of uncertainty in our everyday life. To measure the uncertainties, we use the term "the probability". For example, the probability of tossing a fair coin turned up head is one-half, or 50%.
The main use for the theory of probability is to make decision by performing a survey or experiment that will allow us to study, predict, and make conclusions from data.
In the study of probability, the various results in a random experiment are called the **outcomes**. The collection of all possible outcomes is called the **sample space**.
Any subset of the sample space is called an **event**.

There are two types of probability, theoretical probability and experimental probability.
Theoretical Probability: It is simply called the probability.

If there are n equally likely outcomes and an event A for which there are k of these outcomes, then the probability that the event A will occur is given by:

$$P(A) = \frac{k}{n}$$

The probability of tossing a fair coin turned up head is $P(H) = \frac{1}{2}$.

The probability of tossing a fair coin turned up tail is $P(T) = \frac{1}{2}$.

The probability of rolling a die turned up an even number is $P(E) = \frac{3}{6} = \frac{1}{2}$.

The probability of an event must be between 0 and 1, or between 0% to 100%.
The statement $P(A) = 0$ means that event A cannot occur (0%).
The statement $P(A) = 1$ means that event A must occur (100%).
If p is the probability of an event, $1 - p$ is the probability of an event not occurring.

Experimental Probability: In real-life situations, sometimes it is not possible to find the theoretical probability of an event. Therefore, we find the experimental probability by performing an experiment or a survey. We can use experimental probability to predict future occurrences of events.

If a sufficient number of trials is conducted n times, and an event A occurs e of these times, then the experimental probability that the event A will occur in another trial is given by:

$$P(A) = \frac{e}{n}$$

A school with 2,000 students has 40 students who are left-handed this year. Then the probability of a student chosen at random will be left-handed is $\frac{40}{2000} = \frac{1}{50}$.

If there is no other information available, we can best estimate that the probability of a student chosen at random next year will be left-handed is $\frac{1}{50}$.

-----Continued----

A **tree diagram** is helpful in listing all the possible outcomes (the sample space).
The sample space for tossing two coins is shown below. Use H for heads and T for tails.
There are four possible outcomes (four simple events) {(H, H), (H, T), (T, H), (T, T)}.

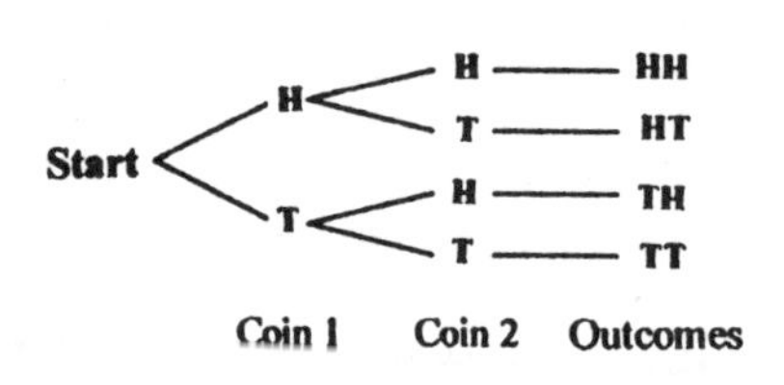

The probability of two heads is $\frac{1}{4}$.
The probability of two tails is $\frac{1}{4}$.
The probability of "one is heads and one is tails" is $\frac{1}{4}+\frac{1}{4}=\frac{1}{2}$.
The probability of "at least one is heads" is $\frac{3}{4}$.

Basic Counting Principal (Principal of Multiplication):

If we choose one from m ways, then we choose another one from n ways.
Then, the number of possible choices is: $m \times n$

Examples

1. You have 5 shirts and 6 pants to choose. In how many ways you can choose, one of each kind ?
   Solution: $5 \times 6 = 30$ ways. Ans.
2. In how many different ways can a 5-question true- false test be answered if each question must be answered ?
   Solution: $2 \times 2 \times 2 \times 2 \times 2 = 32$ ways. Ans.

Independent Events: In an experiment, A and B are two independent events if the occurrence of one event does not affect the occurrence of the other.

Dependent Events: In an experiment, A and B are two dependent events if the occurrence of one event affect the occurrence of the other.

Probability of Independent Events:

If A and B are two independent events, then the probability that both A and B occur is $P(A \cap B) = P(A) \times P(B)$. Or, $P(A \text{ and } B) = P(A) \times P(B)$.

3. A bag contains 5 red balls and 4 white balls. Two balls are drawn at random from the bag. The first ball drawn is put back into the bag before the second ball is drawn. What is the probability that the two balls drawn are both red ?
   Solution:
   $$P(\text{red}) = \tfrac{5}{9} \;;\; P(\text{red and red}) = \tfrac{5}{9} \times \tfrac{5}{9} = \tfrac{25}{81}. \text{ Ans.}$$

Probability of Dependent Events:

If A and B are two dependent events, then the probability that both A and B occur is $P(A \cap B) = P(A) \times P(B \mid A)$. Or, $P(A \text{ and } B) = P(A) \times P(B \mid A)$.
$P(B \mid A)$ is called the **conditional probability** of B given that A has occurred.

4. A bag contains 5 red balls and 4 white balls. Two balls are drawn at random from the bag. The first ball drawn is not put back into the bag before the second ball is drawn. What is the probability that the two balls drawn are both red ?
   Solution:
   $$P(\text{red}) = \tfrac{5}{9} \;;\; P(\text{red} \mid \text{red}) = \tfrac{4}{8} \;;\; P(\text{red and red}) = \tfrac{5}{9} \times \tfrac{4}{8} = \tfrac{20}{72} = \tfrac{5}{18}. \text{ Ans.}$$

EXERCISES

One card is drawn at random from a 52-card bridge deck. Find the probability of each event.

1. It is a 10. **2.** It is a club. **3.** It is the ace of heart. **4.** It is a black heart.
5. It is a 4, 5, or 6. **6.** It is a red card. **7.** It is a 5 or 6. **8.** It is an even number.
9. It is not a spade. **10.** It is an ace or a face card.

One card is drawn from number cards 1, 2, 3, ..., 50. Find the probability of the card drawn turned up:

11. an even number. **12.** a multiple of 5. **13.** a multiple of 7
14. a multiple of both 5 and 7. **15.** a multiple of 5 or 7.

A die is tossed. Find the probability of each event.

16. It is a 3. **17.** It is a 7. **18.** It is an even number. **19.** It is an odd number.
20. It is greater than 4. **21.** It is greater than or equal to 2. **22.** It is less than 5.
23. It is between 1 and 5.

What is the probability that after a spin the arrow will stop on:

24. region 7 ? **25.** region 1 ? **26.** region 5 or 1 ?
27. an odd-numbered region ? **28.** a region with a number less than 6 ?

29. A box contains 12 red balls, 10 white balls, and 8 green balls. A ball is drawn at random. What is the probability that the ball drawn is red ?

30. John throws a dart and hit the bull's eye 12 times out of 50 throws.
  a. What is the experimental probability that he will hit the bull's eye in the next throw ?
  b. What is the experimental probability that he will not hit the bull's eye in the next throw ?

31. John tosses a fair coin 20 times and 12 times turned up heads.
  a. What is the experimental probability that it will turn up head in the next toss ?
  b. What is theoretical probability that it will turn up head in the next toss ?

32. In a basketball game, John makes the basket on 6 free throws out of 15.
  a. What is the experimental probability that he will make a basket on the next free throw ?
  b. How many times would you expect John to make the basket on the next 5 free throws ?
  c. If John actually makes the baskets 4 times out of next 5 free throws, what is the experimental probability that John will make a basket on the next free throw ?

33. Draw a tree diagram to show the results of tossing three coins.

34. Three coins are tossed. How many outcomes are possible ?

35. Four coins are tossed. How many outcomes are possible ?

36. Two dice are rolled. How many outcomes are possible ?

37. Three dice are rolled. How many outcomes are possible ?

38. Two coins are tossed. Find the probability that both are heads.

39. Two coins are tossed. Find the probability that one comes up head and one comes up tail.

40. Three coins are tossed. Find the probability that at least one comes up heads.

-----Continued-----

41. In the basketball game, the probability that you will make the basket on free throws is $\frac{3}{5}$. How many times can you expect to make the basket if you makes 125 free throws ?

42. In the basketball game, the probability that you will make the basket on free throws is 0.25. How many of 120 free throws would probable make the basket ?

43. If you take a 5-question true-false test by just guessing at each answer. Find the probability that your score will be 100% ?

44. In a survey of city mayor election, 34% of voters preferred Candidate A, 20% preferred Candidate B, and 46% preferred Candidate C. Based on this survey, how many of 3,900 voters would probably prefer:

a. Candidate A **b.** Candidate B **c.** Candidate C

45. The probability of a person being colorblind is $\frac{1}{40}$. There are 2,400 students in a high school. How many students would probably be colorblind ?

46. The probability of a person being left-handed is $\frac{1}{30}$. There are 2,400 students in a high school. How many students would probably be left-handed ?

47. The probability of a person being colorblind is $\frac{1}{40}$, and the probability of a person being left-handed is $\frac{1}{30}$. There are 2,400 students in a high school. How many students would probably be both colorblind and left-handed ?

48. Spin two spinners shown below.

a. What is the probability of getting a 7 and a D.

b. What is the probability of getting a 4 or a 5, followed by A ?

c. What is the probability of getting a 2, followed by a A or B ?

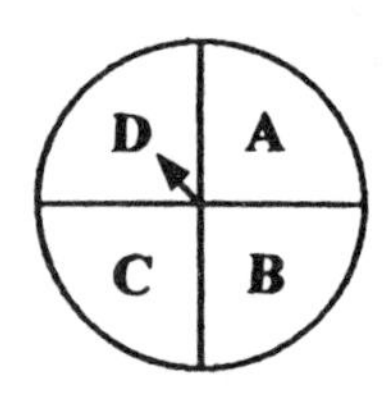

49. Two cards are drawn at random from a 52-card bridge deck with the first card replaced before the second card is drawn. Find the probability of each event.

a. Both cards are red. **b.** Both cards are clubs. **c.** Both cards are the ace of hearts.
d. The first card is a red card and the second card is a black card.

50. Two cards are drawn at random from a 52-card bridge deck without replacement. Find the probability of each event.

a. Both cards are red. **b.** Both cards are clubs. **c.** Both cards are the ace of hearts.
d. The first card is a red card and the second card is a black card.

51. You roll two fair dice. If the total of the numbers equals 10 or more, you win. What is the probability that you will win ?

52. Scientists tag 100 salmons in a river. Later, they caught 50 salmons. Of these, 2 have tags. Estimate how many salmons are in the river ?

53. One card is drawn at random from a 52-card bridge deck. Find the probability that it is a 10 or a face card ?

54. One card is drawn at random from a 52-card bridge deck. Find the probability that it is a 10 or a red card ?

55. One card is drawn at random from a 52-card bridge deck. Find the probability that it is a 10 and a red card ?

CHAPTER 14 EXERCISES

Find the mean, median, mode, and range. Round to the nearest tenth.

1. 3, 4, 4, 5, 5, 5, 6, 6, 6, 7, 8, 9

2. 2, 2, 3, 3, 7, 7, 9, 9, 9, 11, 12

3. 24, 7, 9, 17, 15, 10, 25, 8, 21

4. 2.8, 2.4, 3.1, 4.5, 2.3, 2.4, 2.6, 2.4

5. 0.4, 1.5, 0.8, 3.6, 3.6, 1.5, 1.5, 5.2

6. 41, 30, 15, 52, 8, 20, 37, 32, 26, 49

7. What are the statistical numbers providing the information about the central tendency of the data ?

8. What is the statistical number providing the information about the spread of the data ?

9. The test scores of 20 students listed in the table below.

a. Find the mean of the scores. **b.** How many students' scores fell below the mean ?

Scores	70	72	78	80	84	90
Numbers of students	2	3	8	4	2	1

10. Roger's mean score in 5 math tests is 79. His four test scores are 70, 78, 82, 83. What is his other score ?

11. The test scores of 40 students are listed below.

31, 65, 41, 69, 65, 48, 69, 75, 67, 62
61, 82, 69, 52, 74, 58, 49, 53, 76, 42
71, 56, 63, 43, 54, 61, 51, 64, 84, 42
42, 57, 56, 36, 58, 66, 53, 43, 62, 34

a. List the data by using equal intervals in a frequency table.

Intervals: 30~40, 40~50, 50~60, 60~70, 70~80, 80~90

b. Draw a histogram.

12. The picture graph below shows an Airline flights canceled during six months.

January	✈✈✈✈✈✈
February	✈✈✈✈
March	✈✈✈
April	✈✈✈✈✈
May	✈✈
June	✈✈✈✈

✈: 10 flights canceled

a. How many flights were canceled in January ?

b. How many flights were canceled in June ?

c. What is the mean (average) of flights canceled per month ?

13. In an random survey, the lengths of the adult gramineous snake are normally distributed as shown in the graph.

a. What is the probability that the length of an adult snake will be longer than 20 inches ?

b. What is the probability that the length of an adult snake will be shorter than 17 inches ?

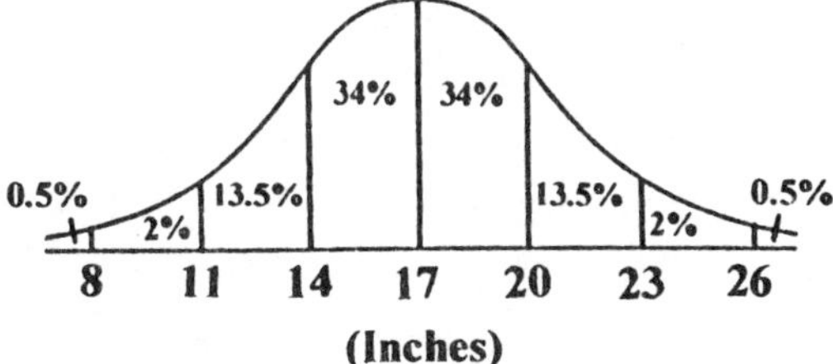

-----Continued-----

14. The bar graph below shows the number of a product sold during a 6-month period.

a. What is the total number of the product sold ?

b. What is the range of the product sold in 6 months ?

c. Which month were the month that sold the most number of the product ?

d. Which month were the month that sold the product two times as many as were sold in February ?

e. What percent of the product sold during the 6-month period were sold in April ?

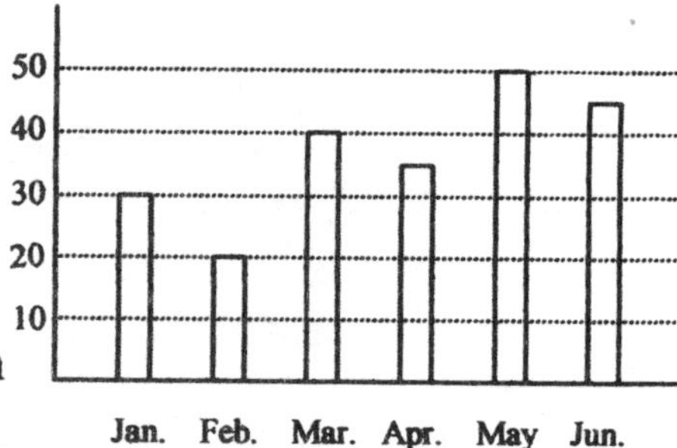

15. In the math seminar, 500 people were asked to judge the instructor's performance as poor, fair, good, or excellent. The results are shown in the circle graph below.

a. How many people rated the instructor's performance as good or excellent ?

b. How many people rated the instructor's performance as fair or poor ?

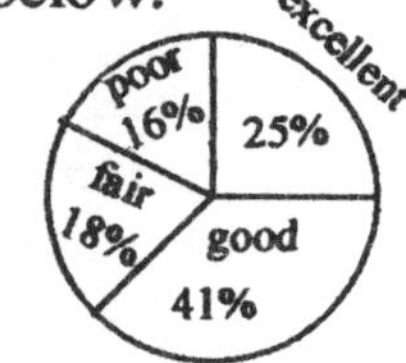

16. The line graph below shows the closing price per share of a company's stock during a week.

a. What is the stock price per share on Tuesday ?

b. What is the stock price per share on Friday ?

c. Which day had the greatest increase in the value of this stock over that of the previous day ?

d. What percent did the price of the stock per share increase during this week ?

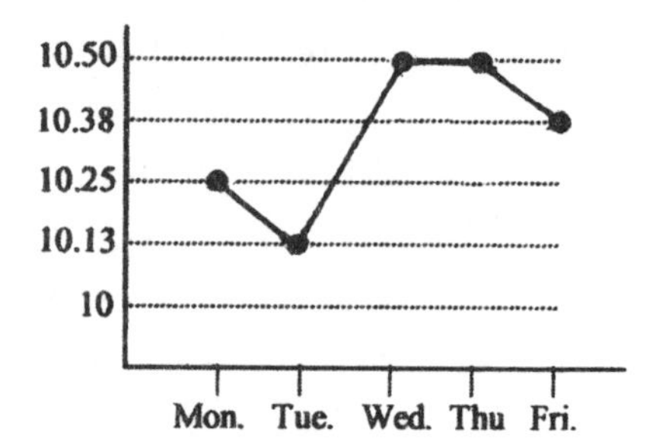

17. The box-and-whisker plot shows the scores on a test.

a. What is the median score on the test ?

b. What percent of the students scored less than 56 ?

c. What percent of the students scored more than 72 ?

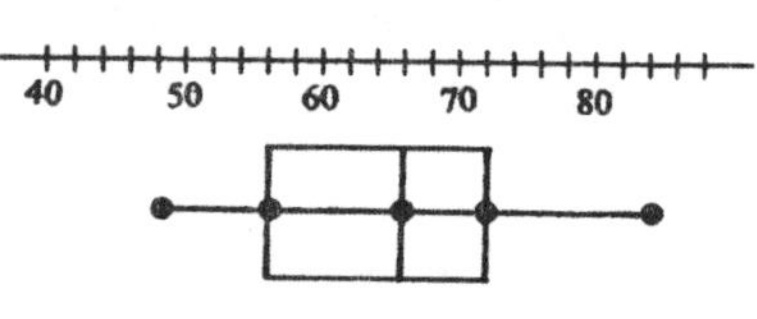

18. In a high school class, there are 23 students who took Algebra class, 17 students took Biology class. These figure include 5 students took both Algebra and Biology classes. How many students are in this class ?

19. A player spins the spinner twice. (Hint: Draw a tree diagram to list all possible outcomes)

a. What is the probability of getting two 5's ?

b. What is the probability of getting a sum of 4 ?

c. What is the probability of getting a sum of 5 ?

d. What is the probability of getting a sum of 10 ?

e. What is the probability of getting a sum of 5 or 10 ?

20. A box contains 10 red balls, 5 white balls, and 3 green balls. A ball is drawn at random. What is the probability that the ball drawn is white ?

21. A bag contains 8 red balls and 4 white balls. Two balls are drawn at random from the bag. What is the probability that the two balls drawn are both red and

a. with replacement ? b. without replacement ?

ANSWERS FOR ALL EXERCISES

CHAPTER 1 – 1 Numbers page 11-12

1. 0 → real number, rational number, whole number, integer.
$-\frac{3}{7}$ → real number, rational number, fraction.
2.4··· → real number, irrational number, decimal.
π → real number, irrational number, decimal.
$3.\overline{12}$ → real number, rational number, decimal.
5 → real number, rational number, whole number, integer, prime number.

2. –5 **3.** –2 **4.** 0 **5.** 4.5 **6.** –14 **7.** –3 **8.** 6 **9.** –22.5 **10.** –5 **11.** 0 **12.** 12.5 **13.** 30
14. 1 **15.** 1.5 **16.** –2.5 **17.** –4 **18.** 20 **19.** 1 **20.** –7.5 **21.** 4 **22.** 4 **23.** 8 **24.** 8
25. 4 **26.** 6 **27.** 17 **28.** 3 **29.**10.7 **30.** 2 **31.** 7, 12, 18, 24 **32.** 3, 13, 20,
33. 2 < 5 **34.** 8 > 3 **35.** –2 < 5 **36.** –2 > –5 **37.** 2 > –5 **38.** –8 < –3 **39.** 2.5 < 3.5
40. –5.3 > –5.4 **41.** –4.8 < –4.7 **42.** –38 > –60 **43.** –9 = –9
44. 0 **45.** 10 **46.** 70 **47.** 60 **48.** 340 **49.** 350 **50.** 3,410 **51.** 5,800 **52.** 6,560
53. 37,200 37,000 40,000 **54.** 24,400 24,000 20,000 **55.** 31,400 31,000 30,000
56. 60,300 60,000 60,000 **57.** 67,500 67,000 70,000 **58.** 370,000 370,000 370,000
59. 500,000 500,000 500,000 **60.** 2.6 2.64 **61.** 3.0 3.03 **62.** 32.4 32.37
63. 323.9 323.91 **64.** 245.9 245.95 **65.** 248.0 247.99 **66.** 259.0 259.00
67. \$0.99 **68.** \$4.42 **69.** \$5.40 **70.** \$19.99 **71.** \$56.89 **72.** \$126.00 **73.** 1,300,000,000
74. 1 thousand 234 **75.** 2 thousand 347 **76.** 45 thousand 786 **77.** 345 thousand 625
78. 6 tenths **79.** 627 thousandths **80.** 204 and 5 thousandths
81. one thousand two hundred thirty-four
82. two thousand three hundred forty-seven
83. forty-five thousand seven hundred eighty-six
84. three hundred forty-five thousand six hundred twenty-five
85. two million four hundred twenty-five thousand two hundred fifty-six
86. twenty-three million four hundred thirty-five thousand two hundred fifty-three
87. one hundredth **88.** two hundredths
89. two hundred seventy-eight thousandths
90. sixty-three and five hundredths
91. twenty-five and one hundred twenty-four thousandths
92. fifty-eight and three thousand six hundred seventy-four ten-thousandths
93. four hundred six and forty-six thousandths
94. nine hundred five and seventy-eight thousandths
95. four thousand and twelve ten-thousandths
96. one thousand and one hundred-thousandth
97. five thousand and twenty-four hundred-thousandths

CHAPTER 1 – 2 Adding Integers page 14

1. –4 **2.** –8 **3.** –4 **4.** –8 **5.** 85 **6.** 23 **7.** –85 **8.** –23 **9.** 10 **10.** 18 **11.** –59
12. –40 **13.** –7 **14.** –147 **15.** –171 **16.** 0 **17.** –42 **18.** –60 **19.** 300 **20.** 125 **21.** 0
22. 91 **23.** 8 **24.** 4 **25.** 8 **26.** 4 **27.** 36 **28.** 96 **29.** –37 **30.** 7 **31.** 4 **32.** 6 **33.** 7
34. 3 **35.** 7 **36.** 2 **37.** –4 **38.** 1 **39.** 5 **40.** 522 books **41.** 134 pounds **42.** –5
43. \$24,850 **44.** $-11°C$ **45.** –80 feet (80 feet below sea level) **46.** 250 feet
47. He gained \$2 per share of stock.

CHAPTER 1 – 3 Subtracting Integers page 16

1. 1 **2.** 2 **3.** 3 **4.** 4 **5.** 5 **6.** 99 **7.** 100 **8.** 4 **9.** –4 **10.** 8 **11.** 8 **12.** –4 **13.** 4
14. –23 **15.** –85 **16.** 23 **17.** 85 **18.** 85 **19.** –23 **20.** 36 **21.** –22 **22.** –11 **23.** 0
24. –147 **25.** –7 **26.** 31 **27.** –40 **28.** –126 **29.** 60 **30.** 340 **31.** –365 **32.** –42
33. 227 **34.** 0 **35.** –4 **36.** 4 **37.** –22 **38.** –20 **39.** –6 **40.** 241 **41.** –3 **42.** –6
43. 8 **44.** 7 **45.** 1 **46.** 1 **47.** –4 **48.** 0 **49.** 1 **50.** 1 **51.** 238 books **52.** 132 pounds
53. $-17°C$ **54.** –130 feet (130 feet below sea level) **55.** 375 feet **56.** 13 **57.** \$81 **58.** 39
59. $25°\ C$ (It increases $25°\ C$.) **60.** –239 volts (It decreases 239 volts.)

CHAPTER 1 – 4 Multiplying Integers page 18

1. 20 **2.** –20 **3.** 20 **4.** –20 **5.** –4 **6.** 4 **7.** 0 **8.** 0 **9.** 24 **10.** 25 **11.** –6 **12.** –250
13. –3 **14.** –9 **15.** 12 **16.** –125 **17.** –35 **18.** –68 **19.** 62 **20.** –1 **21.** 12 **22.** 6
23. –12 **24.** 20 **25.** 24 **26.** 30 **27.** –96 **28.** –5 **29.** 14 **30.** –21 **31.** 13 **32.** –34
33. –18 **34.** 30 **35.** –86 **36.** –8 **37.** 8 **38.** –12 **39.** 2 **40.** –2 **41.** –11 **42.** $480
43. 225 balls **44.** 1,600 meters **45.** 770 passengers **46.** 600 nails **47.** $32 **48.** 6,790 miles
49. 15,000 books **50.** $240 **51.** $10 **52.** 126 $in.^2$

CHAPTER 1 – 5 Dividing Integers page 20

1. –5 **2.** 5 **3.** –5 **4.** 0 **5.** undefined **6.** –49 **7.** 49 **8.** –11 **9.** 12 **10.** undefined
11. –3 **12.** –8 **13.** –24 **14.** 200 **15.** –3 **16.** 18 **17.** 4 **18.** –37 **19.** –6 **20.** –10
21. 15 **22.** 48 **23.** –6 **24.** 3 **25.** –3 **26.** –6 **27.** 9 **28.** 5 **29.** 36 **30.** –3 **31.** –1
32. –6 **33.** 3 **34.** 2 **35.** $120 **36.** 9 balls **37.** $6 **38.** 120 books **39.** 480 books
40. 18 inches **41.** 26 pieces **42.** 72 times **43.** 73 times per minute **44.** 87 **45.** 71°
46. 10 bags **47.** 33 boxes **48.** 14 hours

CHAPTER 1 – 6 Powers and Exponents page 22

1. 10^3 **2.** 9^2 **3.** 7^4 **4.** 4^4 **5.** 1^7 **6.** a^2 **7.** p^3 **8.** x^4 **9.** x^2y^3 **10.** x^3y^2 **11.** 8^3
12. 8^2 **13.** 5^2a^2 **14.** $(-a)^2y^2$ **15.** 9^5 **16.** $(-3)^2$ **17.** -3^2 **18.** $(-6)^3$ **19.** $-x^2(-y)^2$
20. $2(-a)^3y$ **21.** $1\cdot1\cdot1\cdot1\cdot1$ **22.** $3\cdot3\cdot3\cdot3$ **23.** $4\cdot4\cdot4$ **24.** $6\cdot6\cdot6\cdot6$ **25.** $4\cdot4\cdot4\cdot4\cdot4\cdot4$
26. $10\cdot10$ **27.** $10\cdot10\cdot10$ **28.** $a\cdot a\cdot a$ **29.** $p\cdot p\cdot p\cdot p$ **30.** $3\cdot3\cdot3\cdot3\cdot3\cdot3$ **31.** $-4\cdot-4\cdot-4$
32. $-4\cdot4\cdot4$ **33.** $-3\cdot3\cdot3\cdot3$ **34.** $-3\cdot-3\cdot-3\cdot-3$ **35.** $-10\cdot10\cdot10\cdot10$ **36.** $-a\cdot a\cdot a$
37. $-a\cdot-a\cdot-a$ **38.** $3\cdot x\cdot x\cdot x$ **39.** $3x\cdot3x\cdot3x$ **40.** $-3x\cdot3x$ **41.** 125 **42.** 243 **43.** 32
44. 25 **45.** 64 **46.** 81 **47.** 729 **48.** 1 **49.** 1 **50.** 81 **51.** –8 **52.** 64 **53.** –8 **54.** –216
55. –36 **56.** 1 **57.** –1 **58.** –1 **59.** a^{16} **60.** $-a^{73}$ **61.** 1 **62.** 1 **63.** 1 **64.** undefined
65. 1 **66.** –27 **67.** 4 **68.** 16 **69.** 16 **70.** 36 **71.** –24 **72.** –32 **73.** –972 **74.** 72
75. 144 **76.** –16 **77.** –64 **78.** –64 **79.** –256 **80.** 256 **81.** 144 m^2 **82.** 1,728 m^3
83. 400 feet **84.** 2×10^5 miles **85.** 4×10^8 meters **86.** 1×10^8 miles **87.** a) a^{12} b) a^2 c) 1
d) 3^{42} e) 3^{18} f) 1

CHAPTER 1 – 7 Order of Operations page 24

1. 8 **2.** 16 **3.** 24 **4.** 18 **5.** 38 **6.** 20 **7.** 2 **8.** 6 **9.** 1 **10.** 8 **11.** 2 **12.** 6 **13.** 2
14. 4 **15.** 27 **16.** 72 **17.** 5 **18.** 0 **19.** 4 **20.** 0 **21.** 1,600 **22.** 1,000 **23.** 800
24. 8,100,000 **25.** 675 **26.** 17 **27.** –4 **28.** 16 **29.** 15 **30.** 52 **31.** 5 **32.** 5 **33.** 5
34. –60 **35.** 4 **36.** 24 **37.** 5 **38.** 4 **39.** 3 **40.** 7 **41.** 9 **42.** 6. **43.** 13. **44.** 8 **45.** 0
46. 1 **47.** 91 **48.** –4 **49.** –1 **50.** 2 **51.** 48 **52.** 1 **53.** 4 **54.** 39 **55.** 2 **56.** 13
57. 30 **58.** 35 **59.** –7 **60.** 17 **61.** 3 **62.** –216 **63.** 250 **64.** 1,020 **65.** –2 **66.** 16
67. –60 **68.** 126 **69.** –15 **70.** –72 **71.** 5 **72.** –18 **73.** –18 **74.** –7 **75.** 5 **76.** 25
77. –1 **78.** –23 **79.** –21 **80.** 3 **81.** 3 **82.** –13 **83.** –3 **84.** –82 **85.** –16 **86.** 24
87. –120 **88.** 120 **89.** 0 **90.** –7 **91.** $31 **92.** 20 *cm*

In a medical school, a professor asked a student about his major.
Professor : Why do you think a dentist can make more money than a heart surgeon ?
Student: A patient has 32 teeth, but only one heart.

CHAPTER 1 Numbers EXERCISES page 25 – 26

1. 20 → real number, rational number, whole number, integer, composite number.
6. 29 → real number, rational number, whole number, integer, prime number.
9. $-\frac{1}{2}$ → real number, rational number, fraction.
10. 2.15 → real number, rational number, decimal.
13. –3 → real number, rational number, integer.
14. $2.\overline{15}$ → real number, rational number, decimal.
20. 1.732··· → real number, irrational number, decimal.
21. 7 **22.** –7 **23.** $12\frac{1}{2}$ or 12.5 **24.** $13\frac{1}{2}$ or 13.5 **25.** $-13\frac{1}{2}$ or –13.5 **26.** 9 **27.** 18
28. 6 **29.** 3 **30.** 10 **31.** –4.5 or $-4\frac{1}{2}$ **32.** 13 **33.** 1.5 or $1\frac{1}{2}$
34. $-7 < 5$ **35.** $7 > -5$ **36.** $-7 < -5$ **37.** $4.5 < 4.9$ **38.** $-4.5 > -4.9$ **39.** $-1 > -2$
40. $-2.5 < 3$ **41.** $-2 = -2$ **42.** $4.1 > 3.9$ **43.** $0.5 = \frac{1}{2}$
44. 12,350 12,300 12,000 **45.** 54,520 54,500 55,000 **46.** 23,550 23,500 24,000
47. 44,450 44,500 44,000 **48.** 555,550 555,500 556,000 **49.** 3.5 3.55
50. 3.5 3.53 **51.** 45.4 45.36 **52.** 235.3 235.33 **53.** 245.4 245.36
54. 12 thousand 345 ; twelve thousand three hundred forty-five
55. 325 thousand 423 ; three hundred twenty-five thousand four hundred twenty-three
56. 24 hundredths ; twenty-four hundredths
57. 214 and 252 thousandths ; two hundred fourteen and two hundred fifty-two thousandths
58. 31 million 324 thousand 534 ;
thirty-one million three hundred twenty-four thousand five hundred thirty-four
59. 2,365,000,000 **60.** 2,370,000,000 **61.** 2,400,000,000 **62.** \$0.95 **63.** \$19.89 **64.** \$42.79
65. –6 **66.** 6 **67.** 20 **68.** –20 **69.** –4 **70.** –10 **71.** –2 **72.** 10 **73.** 30 **74.** –10
75. –10 **76.** 6 **77.** 16. **78.** –29 **79.** –13 **80.** –29 **81.** 12 **82.** 0 **83.** –9 **84.** 31
85. –8 **86.** 8 **87.** 8 **88.** 12 **89.** –14 **90.** –2 **91.** –24 **92.** –12 **93.** –18 **94.** –24
95. –8 **96.** 1 **97.** 12 **98.** –2 **99.** 6 **100.** 1,000 **101.** –2 **102.** 11 **103.** –32 **104.** 4
105. 4 **106.** –3 **107.** 2 **108.** –2 **109.** –10 **110.** –11 **111.** 177 **112.** –112
113. –1 **114.** 5 **115.** –1 **116.** 1 **117.** 4 **118.** 5 **119.** 3 **120.** –24 **121.** 19 **122.** –14
123. –4 **124.** –6 **125.** –17 **126.** 1 **127.** –73 **128.** 69 **129.** –37 **130.** 14
131. 40 **132.** 78 **133.** \$3 **134.** 270 cents **135.** 432 cents **136.** \$1

CHAPTER 2 – 1 Decimals page 30

1. tenths **2.** hundredths **3.** ones **4.** thousandths **5.** tens **6.** hundreds **7.** ten-thousandths
8. thousands **9.** 3 and 5 hundredths **10.** 5 thousandths **11.** 2 and 54 thousandths
12. 235 ten-thousandths **13.** 24 and 15 ten-thousandths **14.** 324 and 12 hundredths
15. 3 thousand 452 and 8,246 ten-thousandths **16.** two and one hundredth
17. twenty-three and two hundredths **18.** three hundred twenty-five thousandths
19. three hundred fifty-four and forty-eight thousandths
20. nine thousand seven hundred eighty-two and forty-six ten-thousandths
21. six thousand and one thousand four hundred twenty-three ten-thousandths
22. 3.0, 3.02, 3.015 **23.** 32.1, 32.14, 32.144 **24.** 0.3, 0.25, 0.254 **25.** 7.4, 7.35, 7.351
26. 72.1, 72.14, 72.144 **27.** 0.1, 0.10, 0.102 **28.** 1 **29.** 3 **30.** 6 **31.** 12. **32.** 0 **33.** 0
34. 423 **35.** 243 **36.** $0.3 < 0.5$ **37.** $-0.3 > -0.5$ **38.** $0.7 > 0.09$ **39.** $0.07 < 0.9$
40. $-0.02 < -0.01$ **41.** $7.01 < 7.05$ **42.** $-7.01 > -7.05$ **43.** $2.01 < 2.011$ **44.** $23.68 > 23.67$
45. $-28.34 > -29.43$ **46.** $34.431 < 34.432$ **47.** $0.00010 < 0.00012$
48. terminating **49.** terminating **50.** repeating **51.** repeating **52.** nonterminating
53. nonterminating **54.** terminating **55.** repeating **56.** terminating **57.** nonterminating
58. repeating **59.** repeating
60. $0.\overline{1}$ **61.** $0.0\overline{5}$ **62.** $0.\overline{02}$ **63.** $3.\overline{54}$ **64.** $0.0\overline{12}$ **65.** $0.\overline{234}$
66. 0.25, 0.22, 0.2, 0.12, 0.025, 0.02 **67.** 0.22, 0.2, 0.12, –0.02, –0.025, –0.25
68. 3.43, 3.34, 3.30, 1.61, 1.6, 0.16 **69.** 3.43, 0.16, –1.6, –1.61, –3.30, –3.34
70. –0.12, –0.10, –0.012, 0.12, 0.21 **71.** –8.54, –8.45, –6.45, 8.45, 8.54, 9.54
72. –4.39, –3.039, 3.04, 4.039, 4.04 **73.** –2.39, –2.04, –2.039, 2.04, 2.39, 2.40
74. 2.01368 **75.** *a*) The highest score is 3.910. *b*) The lowest score is 3.091.

CHAPTER 2 – 2 Adding and Subtracting Decimals page 32

1. 4.2 **2.** 6.8 **3.** 2.15 **4.** 6.86 **5.** 15.328 **6.** 3.1 **7.** 3.9 **8.** 1.52 **9.** 10.88 **10.** 2.733
11. 0.69 **12.** 10.75 **13.** –3.2 **14.** –5.2 **15.** –2.304 **16.** –9 **17.** –12.6 **18.** –10.2
19. –10.2 **20.** 10.2 **21.** 100.123 **22.** 99.877 **23.** 199.862 **24.** –199.658 **25.** –200.342
26. 76.4 **27.** –297.86 **28.** –700.45 **29.** –399.01 **30.** –499.761 **31.** 3.5 **32.** 0.52
33. 2.1 **34.** 5.9 **35.** –4.98 **36.** 1.69 **37.** 12.04 **38.** 54.71 **39.** 149.845 **40.** 125.425
41. 117.761 **42.** –3.75 **43.** –7.19 **44.** 320.411
45. 5.9 **46.** –1.1 **47.** 15.8 **48.** –11.2 **49.** 9.1 **50.** –13.4 **51.** –7.5 **52.** –27 **53.** 3.5
54. –36.9 **55.** 28.1 **56.** 16.8 **57.** 24.6 **58.** –8.8 **59.** 16.8 **60.** 19.2
61. $289.46 **62.** 6.43 inches **63.** 131.5 pounds **64.** 144.9 pounds **65.** He gained $0.13
66. $d = 12.97$ **67.** $d = 23.4$ **68.** $d = 210.49$ **69.** 114.37 seconds **70.** $59.40 **71.** 30.92
72. –40.71 volts

CHAPTER 2 – 3 Multiplying Decimals page 34

1. 0.02 **2.** –0.12 **3.** 0.3 **4.** 0.72 **5.** 2.66 **6.** 4.14 **7.** 0.00035 **8.** 0.009 **9.** 1.5
10. 0.8398 **11.** 0.0432 **12.** 369 **13.** 0.5024 **14.** 5.992 **15.** 3.87816 **16.** 4.08 **17.** –11.5
18. 4.8 **19.** 4.9 **20.** 4.59 **21.** 23 **22.** 320 **23.** 12 **24.** 0.4 **25.** 10 **26.** 320 **27.** 3,200
28. 213 **29.** 40 **30.** 0.35 **31.** –1.4 **32.** 21.99 **33.** –21.96 **34.** 65.6 **35.** 0.8 **36.** 3.3
37. 5.7 **38.** –25 **39.** 5 **40.** 3.82 **41.** 14.6 **42.** –7.62 **43.** 6.3 **44.** –6.72 **45.** 10.5
46. –22.4 **47.** –25.2 **48.** 13.44 **49.** 11.1 **50.** 6.9 **51.** 0.7 **52.** 11.7 **53.** –0.8 **54.** –1
55. 31.5 **56.** –28.8 **57.** 7 **58.** 9.72 **59.** –55.86 **60.** –37.632 **61.** 9 **62.** 4.41 **63.** 10.24
64. 23.8 **65.** 28.36 **66.** 0.739 **67.** –9.6 **68.** –1.12 **69.** 8.76 **70.** 18.9
71. $230 **72.** 96.12 $in.^2$ **73.** $41.44 **74.** $39.95 **75.** 94.022 **76.** $28.14 **77.** $1.04
78. $2,448 **79.** $12.77 **80.** perimeter = 33.44 inches, area = 66.3552 $in.^2$ **81.** $1,980
82. $200

CHAPTER 2 – 4 Dividing Decimals page 36

1. 1.25 **2.** 0.625 **3.** 0.3125 **4.** 12.5 **5.** 6.25 **6.** 3.125 **7.** 125 **8.** 62.5 **9.** 31.25
10. 12.5 **11.** 12,500 **12.** 5.25 **13.** 52.5 **14.** 5,250 **15.** 525 **16.** 5.08 **17.** 508
18. 5.08 **19.** 24.3 **20.** 2,430 **21.** 125 **22.** 1,250 **23.** 508 **24.** 5,080 **25.** 508,000
26. 83.3 **27.** –833.3 **28.** –83.3 **29.** 3.6 **30.** 35.7 **31.** 0.3 **32.** –227.3 **33.** –18.8
34. 188.1 **35.** –1,880.8 **36.** 96.67 **37.** –966.67 **38.** 96.67 **39.** 4.14 **40.** –41.43
41. 0.32 **42.** 263.64 **43.** –0.03 **44.** –6.83 **45.** 682.56
46. 3.5 **47.** –1.5 **48.** 3.33 **49.** 7.5 **50.** 46.67 **51.** 3.25 **52.** 2.5 **53.** –16.6 **54.** 9.63
55. 5.33 **56.** 0.5 **57.** –1.55 **58.** 6.71 **59.** 6.25 **60.** 7.86
61. 0.7 **62.** 1.43 **63.** –1.31 **64.** 3.73 **65.** –3.43 **66.** –2.1 **67.** 6.75 **68.** 5.67
69. –1.25 **70.** 6.3 **71.** 1.61 **72.** 1.47 **73.** 0.56 **74.** 10.92 **75.** –22.54 **76.** 1.4 **77.** 2.86
78. 274.29 miles per hour. **79.** –2.31 **80.** $2.14 **81.** 9.6 miles per gallon **82.** $11.88

CHAPTER 2 – 5 Scientific Notation page 38

1. 3.2×10^2 **2.** 3.42×10^2 **3.** 1×10^3 **4.** 3×10^3 **5.** 3.58×10^3 **6.** 9×10^3 **7.** 1×10^4 **8.** 3.2×10^4
9. 7.65×10^4 **10.** 5.0123×10^4 **11.** 2.34×10^5 **12.** 2.345×10^6 **13.** 3.05×10^5 **14.** 2.22×10^6
15. 3.24432×10^5 **16.** 7.77×10^5 **17.** 7.777×10^5 **18.** 4.02×10^7 **19.** 2.022×10^7 **20.** 1.93×10^7
21. 7×10^8 **22.** 7.12×10^8 **23.** 9.01×10^8 **24.** $8.015\times^8$ **25.** 4.0253×10^7 **26.** 3.2045×10^2
27. 4.58267×10^3 **28.** 5.634612×10^4 **29.** 3.2254123×10^4 **30.** 2.4314218×10^5 **31.** 1×10^{-6}
32. 1.23×10^{-4} **33.** 7×10^{-7} **34.** 4.2×10^{-7} **35.** 4.36×10^{-6} **36.** 1.4258×10^{-1} **37.** 5.328×10^{-2}
38. 4.567×10^{-3} **39.** 3.68×10^{-4} **40.** 2.83456×10^{-1} **41.** 100 **42.** 120 **43.** 2,400 **44.** 40,000,000
45. 460,000 **46.** 40,600,000 **47.** 3,520,000 **48.** 1,101 **49.** 402,000,000 **50.** 500,000 **51.** 14.8
52. 424,300 **53.** 999,000 **54.** 234.5 **55.** 20.34 **56.** 0.1 **57.** 0.01 **58.** 0.002 **59.** 0.024
60. 0.000005 **61.** 0.000076 **62.** 0.02145 **63.** 0.000000524 **64.** 0.0000081 **65.** 0.0205
66. 0.0000007 **67.** 0.0032345 **68.** 0.00000802 **69.** 0.00001111 **70.** 0.000000001
71. 6×10^5 **72.** 4.2×10^5 **73.** 2×10^5 **74.** 3.8×10^5 **75.** 8×10^9 **76.** 8×10^1 **77.** 3×10^{16}
78. 8.4×10^8 **79.** 4.2×10^{-3} **80.** 8×10^{-13} **81.** 7.5×10^{15} **82.** 1.378×10^{15} **83.** 2×10^4
84. 5×10^{-5} **85.** 3.5×10^5 **86.** 3.99×10^8 **87.** 5,900,000,000,000 miles
88. 9.82×10^8 feet **89.** 1.18×10^{11} miles **90.** 2.5075×10^{13} miles **91.** 1.18×10^{19} miles

CHAPTER 2 Decimals EXERCISES page 39 – 40

1. tenths **2.** thousandths **3.** ones **4.** ten-thousandths **5.** hundreds
6. tens **7.** three and twenty-one hundredths **8.** four and one hundredth
9. five hundred forty-two and five thousandths **10.** forty-two and twelve thousandths
11. five thousand twenty-three and forty-five thousandths
12. three thousand and one thousand two hundred forty-three ten-thousandths
13. $0.4 > 0.3$ **14.** $-0.4 < 0.3$ **15.** $-0.4 < -0.3$ **16.** $0.05 < 0.5$ **17.** $3.02 > 3.01$
18. $3.02 < 3.021$ **19.** $54.453 < 54.54$ **20.** $-53.4 > -54.3$
21. repeating **22.** terminating **23.** repeating **24.** repeating **25.** terminating
26. nonterminating **27.** repeating **28.** terminating **29.** repeating **30.** terminating
31. repeating **32.** nonterminating
33. 0.023, 0.23, 0.32, 3.23, 3.32 **34.** –2.32, –0.3, –0.03, 0.2, 3.01
35. 6.13, 5.03, 5.02, 1.78, 0.61, 0.50 **36.** 3.52, 3.25, 2.2, –2.2, –4.2, –4.24
37. 0.5 **38.** 0.1 **39.** –0.1 **40.** 2.48 **41** 3.54 **42.** –11 **43.** 1.66 **44.** 2.136 **45.** 4.5
46. –4.5 **47.** –2.182 **48.** –4.14 **49.** 199.99 **50.** –398.676 **51.** –98.8 **52.** 6.1 **53.** 1
54. 1.4 **55.** –0.2 **56.** –0.4
57. 0.0002 **58.** –0.6 **59.** 0.00072 **60.** 9.17 **61.** 940 **62.** 12.8 **63.** –134.88 **64.** 1,050
65. 2,600 **66.** 3,400 **67.** 15.33 **68.** –1.15 **69.** 2 **70.** 1,500 **71.** 3.51 **72.** 0.06
73. –12.285 **74.** 5 **75.** 102.33 **76.** 4,200 **77.** 11.7 **78.** –9.76 **79.** 2.9 **80.** 46
81. 10.7 **82.** 2.4 **83.** 4
84. 7.1 **85.** 4.8 **86.** 21.7 **87.** –3.2 **88.** 16.5 **89.** 3.25 **90.** –0.35 **91.** 29 **92.** –4.15
93. 1.91 **94.** 6.4 **95.** 1.84 **96.** –1.49 **97.** 17.73 **98.** –3.70
99. 5.62×10^{6} **100.** 5.62×10^{-4} **101.** 1.2×10^{-5} **102.** 7.68×10^{7} **103.** 2.04×10^{5}
104. 4.532×10^{-3} **105.** 4.532×10^{-1} **106.** 1.453272×10^{4} **107.** 1.2412343×10^{5}
108. 1.4×10^{-4} **109.** 12 **110.** 310 **111.** 230 **112.** 4,500 **113.** 10,000 **114.** 503,000
115. 0.12 **116.** 0.04 **117.** 0.0042 **118.** 0.0000342
119. 19 cm **120.** 87.75 m^2 **121.** \$2.95 **122.** \$0.35 **123.** $V = 1{,}481.544\ m^3$
124. $A \approx 50.24\ cm^2$ **125.** 29.04 seconds **126.** 2×10^{11} stars **127.** 1.18×10^{11} miles
128. 2.53×10^{13} miles **129.** 9.119×10^{-28} grams **130.** 1.673×10^{-24} grams
131. 1.675×10^{-24} grams **132.** 500 seconds

CHAPTER 3 – 1 Factors and Multiples page 44

1. True **2.** False **3.** True **4.** True **5.** False **6.** False **7.** True **8.** False **9.** True
10. True **11.** True **12.** True **13.** False **14.** False **15.** True
Hint (16 ~ 35): 1. Test each whole number as a divisor, starting with 1.
2. Stop at the divisor if the quotient is also a factor which begins to repeat.
16. 1, 2, 5, 10 **17.** 1, 3, 9 **18.** 1, 2, 4, 8 **19.** 1, 7 **20.** 1, 2, 3, 6 **21.** 1, 5 **22.** 1, 2, 4
23. 1, 3 **24.** 1, 2 **25.** 1 **26.** 1, 3, 5, 15 **27.** 1, 2, 3, 6, 9, 18 **28.** 1, 2, 4, 5, 10, 20
29. 1, 2, 3, 4, 6, 8, 12, 16, 24, 48 **30.** 1, 3, 5, 15, 25, 75 **31.** 1, 2, 4, 5, 8, 10, 16, 20, 40, 80
32. 1, 2, 3, 5, 6, 9, 10, 15, 18, 30, 45, 90 **33.** 1, 5, 19, 95 **34.** 1, 3, 9, 11, 33, 99
35. 1, 2, 4, 7, 8, 14, 16, 28, 56, 112
36. 0, 1, 2, 3, 4 **37.** 0, 2, 4, 6, 8 **38.** 0, 3, 6, 9, 12 **39.** 0, 4, 8, 12, 16 **40.** 0, 5, 10, 15, 20
41. 0, 6, 12, 18, 24 **42.** 0, 7, 14, 21, 28 **43.** 0, 8, 16, 24, 32 **44.** 0, 9, 18, 27, 36
45. 0, 10, 20, 30, 40 **46.** 0, 11, 22, 33, 44 **47.** 0, 13, 26, 39, 52 **48.** 0, 15, 30, 45, 60
49. 0, 21, 42, 63, 84 **50.** 0, 25, 50, 75, 100 **51.** 0, 30, 60, 90, 120 **52.** 0, 40, 80, 120, 160
53. 0, 75, 150, 225, 300 **54.** 0, 92, 184, 276, 368 **55.** 0, 101, 202, 303, 404
56. Odd **57.** Odd **58.** Even **59.** Odd **60.** Even **61.** Odd **62.** Odd **63.** Odd **64.** Even
65. Odd
66. 0 **67.** 4 **68.** Yes **69.** No **70.** None. 0 is not a factor of any number.
71. Any nonzero integer **72.** Yes **73.** Yes **74.** 1 **75.** 1 **76.** 0 **77.** 0 **78.** 12
79. $6 = 1 + 2 + 3$ (it is called a perfect number.) **80.** 528 **81.** 1, 2, 4, 5, 10, 20, 25, 50, 100
82. $16 = 1 \times 16 = 2 \times 2 \times 2 \times 2 = 2 \times 2 \times 4 = 2 \times 8 = 4 \times 4$
83. $x+1$ **84.** $(x+1)(y+1)$ **85.** $(x+1)(y+1)(z+1)$

CHAPTER 3 – 2 Divisibility Tests pages 46

1. No **2.** Yes **3.** Yes **4.** Yes **5.** No **6.** Yes **7.** Yes
8. Yes **9.** No **10.** No **11.** Yes **12.** NO **13.** Yes **14.** Yes
15. No **16.** Yes **17.** No **18.** Yes **19.** No **20.** Yes **21.** No
22. Yes **23.** No **24.** Yes **25.** Yes **26.** Yes **27.** No **28.** Yes
29. No **30.** No **31.** Yes **32.** No **33.** Yes **34.** No **35.** No
36. Yes **37.** Yes **38.** Yes **39.** No **40.** Yes **41.** No **42.** No
43. Yes **44.** No **45.** Yes **46.** No **47.** Yes **48.** Yes **49.** Yes
50. Yes **51.** No **52.** Yes **53.** Yes **54.** No **55.** Yes **56.** No
57. Yes **58.** No **59.** Yes **60.** Yes **61.** Yes **62.** No **63.** Yes

64. $140 \times = 2 \times 70$, $142 = 2 \times 71$, $144 = 2 \times 72$, $146 = 2 \times 73$, $148 = 2 \times 74$.
65. $225 = 3 \times 75$, $255 = 3 \times 85$, $285 = 3 \times 95$. **66.** 534
67. A number is divisible by 50 if its last two digits are 00 or 50.
68. Yes **69.** No **70.** 9,996 **71.** 10,008 **72.** 6 pounds

CHAPTER 3 – 3 Prime Factorization page 48

1. prime **2.** composite **3.** composite **4.** prime **5.** prime **6.** composite
7. composite **8.** prime **9.** composite **10.** prime **11.** prime **12.** composite
13. composite **14.** composite **15.** composite **16.** composite **17.** prime **18.** Composite
19. composite **20.** Composite

Hint (21 ~ 30): 1. Test each whole number as a divisor, starting with 1.
2. Stop at the divisor if the quotient is also a factor which begin to repeat.
Or, stop at the divisor less than or equal to the square root of the number.

21. 1, 2, 4, 8 **22.** 1, 3, 9 **23.** 1, 3, 5, 15 **24.** 1, 2, 3, 6, 9, 18 **25.** 1, 2, 4, 5, 10, 20
26. 1, 2, 3, 4, 6, 8, 12, 16, 24, 48 **27.** 1, 2, 3, 5, 6, 9, 10, 15, 18, 30, 45, 90
28. 1, 3, 9, 11, 33, 99 **29.** 1, 2, 4, 7, 8, 14, 16, 28, 56, 112 **30.** 1, 13, 169
31. 2 **32.** 3 **33.** 3, 5 **34.** 2, 3 **35.** 2, 5 **36.** 2, 3 **37.** 2, 3, 5 **38.** 3, 11 **39.** 2, 7 **40.** 13
41. 5, 7 **42.** 5 **43.** 2, 3, 5 **44.** 2, 3 **45.** 47 **46.** 2, 5 **47.** 5, 11 **48.** 2, 23 **49.** 2, 19 **50.** 3, 19
51. $8 = 2^3$ **52.** $9 = 3^2$ **53.** $15 = 3 \times 5$ **54.** $18 = 2 \times 3^2$ **55.** $20 = 2^2 \times 5$ **56.** $48 = 2^4 \times 3$
57. $90 = 2 \times 3^2 \times 5$ **58.** $99 = 3^2 \times 11$ **59.** $112 = 2^4 \times 7$ **60.** $169 = 13^2$ **61.** $12 = 2^2 \times 3$
62. $50 = 2 \times 5^2$ **63.** $28 = 2^2 \times 7$ **64.** $39 = 3 \times 13$ **65.** $40 = 2^3 \times 5$ **66.** $63 = 3^2 \times 7$
67. $84 = 2^2 \times 3 \times 7$ **68.** $51 = 3 \times 17$ **69.** $70 = 2 \times 5 \times 7$ **70.** $196 = 2^2 \times 7^2$
71. 2 **72.** 7 **73.** 2 **74.** No. The number 0 has many factors.
75. No. There are many prime factorizations of a number if the number 1 is defined as a prime number.
76. $12 = 1 \times 2^2 \times 3 = 1 \times 1 \times 2^2 \times 3 = 1 \times 1 \times 1 \times 2^2 \times 3 = 1 \times 1 \times 1 \times \cdots \times 2^2 \times 3$
There are many prime factorizations of a number. The Fundamental Theorem of Arithmetic would be false if the number 1 is defined as a prime number.
77. Because the sum of two prime numbers greater than 2 is an even number and all even numbers greater than 2 are not prime numbers.
78. $1{,}575 = 3^2 \times 5^2 \times 7$ **79.** $3{,}213 = 3^3 \times 7 \times 17$ **80.** $6{,}825 = 3 \times 5^2 \times 7 \times 13$
81. $10{,}780 = 2^2 \times 5 \times 7^2 \times 11$ **82.** $40{,}425 = 3 \times 5^2 \times 7^2 \times 11$

CHAPTER 3 – 4 Greatest Common Factor and Least Common Multiple pages 51

1. 1 **2.** 1 **3.** 2 **4.** 1 **5.** 4 **6.** 3 **7.** 1 **8.** 3 **9.** 1 **10.** 4 **11.** 1 **12.** 5 **13.** 6 **14.** 16
15. 4 **16.** 5 **17.** 8 **18.** 19 **19.** 14 **20.** 3 **21.** 3 **22.** 70 **23.** 72 **24.** 61 **25.** 1 **26.** 1
27. 2 **28.** 2 **29.** 6 **30.** 8 **31.** 2 **32.** $2x$ **33.** $2x$ **34.** $2b$ **35.** $4ab$ **36.** $7y$ **37.** cd
38. $3v^2$ **39.** $7mn$ **40.** $15r^2s$ **41.** 3 **42.** 6 **43.** 15 **44.** 12 **45.** 36 **46.** 30 **47.** 60
48. 18 **49.** 84 **50.** 70 **51.** 20 **52.** 28 **53.** 60 **54.** 144 **55.** 84 **56.** 60 **57.** 57
58. 990 **59.** 6 **60.** 60 **61.** 24 **62.** 30 **63.** 180 **64.** 42 **65.** $6x$ **66.** $14x$ **67.** $12x^2$
68. $35a^2b$ **69.** $36u^2v$ **70.** $117x^2y^3$
71. GCF = $6xy^2$, LCM = $1{,}260x^3y^3$ **72.** GCF = ab, LCM = $84a^3b^2$
73. GCF = $2 \times 3^2 \times 5^2$, LCM = $2^2 \times 3^3 \times 5^2 \times 7$ **74.** GCF = 13, LCM = 2,210
75. GCF = 133, LCM = 175,560 (Hint: By prime factorization.) **76.** GCF = 6, LCM = 69,300
77. 30 **78.** 7 (It is the GCF.) **79.** 300 seconds (It is the LCM.) **80.** 2:00 PM **81.** 6 in., 29 pcs
82. 10 lights.(Hint: $1890 \div \text{LCM} + 1$) **83.** 40 **84.** 1 **85.** 7 and 105, or 21 and 35

CHAPTER 3 – 5 Sequences and Number Patterns page 54

1. 19, 22, 25, 28, 31 **2.** 20, 23, 26, 29, 32 **3.** 22, 26, 30, 34, 38
4. 54, 62, 70, 78, 86 **5.** 11, 9, 7, 5, 3 **6.** 14, 11, 8, 5, 2
7. 21, 25, 29, 33, 37 **8.** 36, 32, 28, 24, 20 **9.** 28, 33, 38, 43, 48
10. 51, 58, 65, 72, 79 **11.** 4.0, 4.5, 5.0, 5.5, 6.0 **12.** 9.0, 10.5, 12, 13.5, 15
13. 3, 0, –3, –6, –9 **14.** –23, –27, –31, –35, –39 **15.** –6, –3, 0, 3, 6
16. 22, 29, 37, 46, 56 **17.** 26, 37, 50, 65, 82 **18.** 30, 42, 56, 72, 90
19. 27, 38, 51, 66, 83 **20.** 13, 21, 34, 55, 89 **21.** 28, 45, 73, 118, 191
22. 32, 64, 128, 256, 512 **23.** 29, 47, 76, 123, 199 **24.** 21, 34, 55, 89, 144
25. 36, 49, 64, 81, 100 **26.** 34, 45, 58, 73, 90 **27.** 12.5, 18, 24.5, 32, 40.5
28. 21.5, 28, 35.5, 44, 53.5 **29.** 6, 12, 19, 27, 36 **30.** –24, –30, –37, –45, –54
31. Yes **32.** No **33.** No **34.** 11th day **35.** 8:50 **36.** 15 games (Hint: 8 + 4 + 2 + 1 = 15.)
37. 36 handshakes (Hint: 0, 1, 3, 6, 10, 15, 21, 28, **36**)
38. 2 pieces (1 cut), 4 pieces (2 cuts), 7 pieces (3 cuts), 11 pieces (4 cuts), 16 pieces (5 cuts).
39. No **40.** Yes

CHAPTER 3 Number Theory EXERCISES page 55 – 56

1. 1, 2, 3, 4, 6, 12 **2.** 1, 2, 7, 14 **3.** 1, 17 **4.** 1, 19 **5.** 1, 5, 25 **6.** 1, 5, 7, 35 **7.** 1, 7, 49
8. 1, 3, 9, 27, 81 **9.** 1, 3, 31, 93 **10.** 1, 2, 4, 31, 62, 124
11. 0, 1, 2, 3, 4 **12.** 0, 4, 8, 12, 16 **13.** 0, 12, 24, 36, 48 **14.** 0, 16, 32, 48, 64
15. 0, 20, 40, 60, 80 **16.** 0, 24, 48, 72, 96 **17.** 0, 31, 62, 93, 124 **18.** 0, 42, 84, 126, 168
19. 0, 60, 120, 180, 240 **20.** 0, 91, 182, 273, 364, 455
21. Yes(2, 4), No(3, 5) **22.** Yes(3, 5), No(2, 4) **23.** Yes(2,3,4), No(5) **24.** Yes(3), No(2,4,5)
25. Yes(3,5), No(2,4) **26.** Yes(2,4,5), No(3) **27.** Yes(2,3,4), No(5) **28.** Yes(2,4), No(3,5)
29. Yes(2,3,5), No(4) **30.** Yes(3,5), No(2,4)
31. Yes(6,8), No(9,10,11) **32.** Yes(6,9), No(8,10,11) **33.** Yes(6,8,9), No(10,11)
34. Yes(8,11), No(6,9,10) **35.** Yes(6,9,10), No(8,11) **36.** Yes(6,8), No(9,10,11)
37. Yes(10,11), No(6,8,9) **38.** Yes(6,10), No(8,9,11) **39.** Yes(6,8,9,10), No(11)
40. Yes(8,11), No(6,9,10)
41. Neither **42.** Neither **43.** Prime **44.** Prime **45.** Prime **46.** Composite **47.** Composite
48. Prime **49.** Composite **50.** Composite (671=61 × 11)
51. 2, 3 **52.** 2, 7 **53.** 17 **54.** 19 **55.** 5 **56.** 5, 7 **57.** 7 **58.** 3 **59.** 3, 31 **60.** 2, 31
61. 2×5 **62.** 2×7 **63.** $2^3\times3$ **64.** $2^2\times11$ **65.** $3^2\times5$ **66.** $2^3\times3^2$ **67.** $2^2\times23$
68. 11^2 **69.** $2^4\times3^2$ **70.** $3^2\times5^2$
71. 1 ; 5 **72.** 2 ; 8 **73.** 2 ; 20 **74.** 5 ; 15 **75.** 6 ; 12 **76.** 3 ; 18 **77.** 1 ; 76 **78.** 8 ; 24
79. 10 ; 20 **80.** 1 ; 285 **81.** 5 ; 140 **82.** 17 ; 51 **83.** 6 ; 330 **84.** 6 ; 378 **85.** 1 ; 4,800
86. 13 ; 455 **87.** 60 ; 120 **88.** 2 ; 12 **89.** 1 ; 120 **90.** 2 ; 24 **91.** 3 ; 60 **92.** 5 ; 180
93. 1 ; 468 **94.** 5 ; 600 **95.** 2 ; 720 **96.** 1 ; $12x$ **97.** $2x$; $12x^2$ **98.** $2a$; $8ab$
99. $5y$; $10y^3$ **100.** $2uv$; $24u^3v^2$ **101.** 18 ; 72 **102.** 12 ; 144 **103.** 20 ; 400
104. 2,025 ; 20,250 **105.** 288 ; 4,320
106. 15, 17, 19, 21, 23 **107.** 28, 33, 38, 43, 48 **108.** 37, 44, 51, 58, 65
109. 15, 13, 11, 9, 7 **110.** 20, 17, 14, 11, 8 **111.** 13, 15.5, 18, 20.5, 23
112. 6.5, 5, 3.5, 2, 0.5 **113.** 12.5, 12, 11.5, 11, 10.5 **114.** 8, 8.5, 9, 9.5, 10
115. –4, –2, 0, 2, 4 **116.** –5, –7, –9, –11, –13 **117.** 16, 22, 29, 37, 46
118. 63, 84, 108, 135, 165 **119.** 25, 36, 49, 64, 81 **120.** 37, 60, 97, 157, 254
121. 16, 26, 42, 68, 110 **122.** 33, 45, 59, 75, 93 **123.** 14, 25, 38, 53, 70
124. 30, 34, 39, 43, 48 **125.** 18.5, 25, 32.5, 41, 50.5
126. None. 0 is not a factor of any number. **127.** Any nonzero integer **128.** 1 **129.** 0
130. 936 **131.** 1, 2, 3, 4, 5, 6, 8, 10, 12, 15, 20, 24, 25, 30, 40, 60, 120
132. 1,000 **133.** 99,990 **134.** 6 students **135.** $4{,}860 = 2^2\times3^5\times5$
136. $5{,}544 = 2^3\times3^2\times7\times11$ **137.** GCF = 20, LCM = 1,540 **138.** GCF = 10, LCM = 1,540
139. 63 **140.** 12 lights **141.** 6 students **142.** 31 games **143.** 55 handshakes
144. 22 pieces **145.** Yes

CHAPTER 4 – 1 Introduction to Fractions page 61~62

1. Proper **2.** Proper **3.** Improper **4.** Proper **5.** Improper **6.** Undefined **7.** Improper
8. Proper **9.** Proper **10.** Improper
11. Equal **12.** Equal **13.** Not equal **14.** Equal **15.** Not equal **16.** Equal **17.** Not equal
18. Equal **19.** Not equal **20.** Equal
21. $\frac{1}{2}$ **22.** $\frac{1}{2}$ **23.** $-\frac{2}{5}$ **24.** $\frac{1}{7}$ **25.** $\frac{1}{8}$ **26.** $\frac{1}{2}$ **27.** $5\frac{3}{4}$ **28.** $-4\frac{2}{3}$ **29.** $6\frac{1}{3}$ **30.** $-5\frac{1}{2}$
31. $1\frac{1}{3}$ **32.** 5 **33.** $2\frac{2}{3}$ **34.** 1 **35.** $-3\frac{1}{3}$ **36.** $7\frac{1}{2}$ **37.** 4 **38.** $-1\frac{9}{11}$ **39.** $4\frac{7}{9}$ **40.** $2\frac{6}{11}$
41. $\frac{8}{3}$ **42.** $\frac{10}{3}$ **43.** $-\frac{22}{5}$ **44.** $\frac{31}{4}$ **45.** $\frac{22}{3}$ **46.** $-\frac{32}{3}$ **47.** $\frac{39}{4}$ **48.** $\frac{53}{8}$ **49.** $\frac{26}{3}$ **50.** $-\frac{97}{49}$
51. $\frac{1}{2}=\frac{4}{8}$ **52.** $\frac{1}{2}>\frac{3}{8}$ **53.** $\frac{2}{3}>\frac{4}{7}$ **54.** $\frac{3}{5}>\frac{4}{7}$ **55.** $\frac{1}{3}>\frac{1}{6}$ **56.** $\frac{1}{2}<\frac{6}{10}$ **57.** $\frac{4}{5}=\frac{16}{20}$
58. $\frac{3}{14}>\frac{5}{28}$ **59.** $\frac{20}{30}=\frac{2}{3}$ **60.** $\frac{15}{45}<\frac{7}{15}$
61. one-fourth **62.** three-fourths **63.** four fifths **64.** one-eighth **65.** seven-eighths
66. five-halves **67.** Seven-elevenths **68.** eleven-sevenths **69.** ten-twelfths
70. twenty-four-fiftyths **71.** five-wholes **72.** one and two-thirds **73.** four and one-fifth
74. six and five-ninths **75.** ten and ten-thirteenths
76. $\frac{1}{3}, \frac{1}{2}, \frac{3}{5}, \frac{2}{3}, \frac{5}{6}$ **77.** $\frac{3}{8}, \frac{1}{2}, \frac{5}{8}, \frac{11}{16}, \frac{3}{4}$ **78.** $\frac{7}{12}, \frac{2}{3}, \frac{3}{4}, \frac{19}{24}, \frac{5}{6}$
79. $-\frac{7}{10}, -\frac{17}{30}, \frac{2}{5}, \frac{8}{15}, \frac{2}{3}$ **80.** $-\frac{1}{2}, -\frac{1}{3}, \frac{5}{9}, \frac{13}{18}, \frac{5}{6}$
81. $2x$ **82.** $3x^2$ **83.** $6x^2y$ **84.** $16a$ **85.** $\frac{3b^2}{2}$ **86.** $\frac{4cd}{3}$
87. $\frac{2m}{3n}$ **88.** $\frac{2z}{3y}$ **89.** $\frac{3x}{2y}$ **90.** $\frac{4n^2}{3m}$
91. $\frac{3}{9},\frac{1}{9}$ **92.** $\frac{2}{8},\frac{1}{8}$ **93.** $\frac{8}{12},\frac{1}{12}$ **94.** $\frac{6}{8},\frac{5}{8}$ **95.** $\frac{8}{20},\frac{15}{20}$ **96.** $\frac{6}{16},\frac{5}{16}$
97. $\frac{4}{18},\frac{3}{18}$ **98.** $\frac{35}{42},\frac{24}{42}$ **99.** $-\frac{4}{14},\frac{5}{14}$ **100.** $\frac{21}{36},\frac{14}{36}$ **101.** $-\frac{18}{22},\frac{1}{22}$ **102.** $\frac{16}{60},\frac{39}{60}$
103. $\frac{11}{30},-\frac{20}{30}$ **104.** $\frac{44}{99},\frac{72}{99}$ **105.** $\frac{18}{75},\frac{13}{75}$ **106.** $\frac{33}{132},\frac{60}{132}$ **107.** $-\frac{7}{48},\frac{44}{48}$ **108.** $\frac{169}{195},\frac{45}{195}$
109. $\frac{36}{168},\frac{21}{168}$ **110.** $\frac{80}{504},\frac{27}{504}$
111. $\frac{6}{12},\frac{4}{12},\frac{3}{12}$ **112.** $-\frac{10}{40},\frac{8}{40},\frac{5}{40}$ **113.** $\frac{40}{60},-\frac{45}{60},\frac{48}{60}$ **114.** $\frac{10}{24},\frac{9}{24},\frac{6}{24}$ **115.** $-\frac{24}{360},\frac{60}{360},-\frac{285}{360}$
116. $\frac{3}{4}$ **117.** $2\frac{2}{5}$ **118.** $\frac{1}{5}$ **119.** $1\frac{1}{6}$ **120.** $\frac{13}{14}$ **121.** $-\frac{1}{5}$ **122.** 6 **123.** $1\frac{2}{7}$
124. $1\frac{1}{3}$ **125.** 1
126. $\frac{y}{xy}, \frac{x}{xy}$ **127.** $\frac{2b}{ab}, \frac{a}{ab}$ **128.** $\frac{1}{2a}, \frac{4}{2a}$ **129.** $\frac{4y}{6xy}, \frac{9x}{6xy}$ **130.** $\frac{yz}{xyz}, \frac{2xz}{xyz}, \frac{3xy}{xyz}$

CHAPTER 4 – 2 Fractions and Decimals page 64

1. 0.25 **2.** 0.2 **3.** $0.1\overline{6}$ **4.** -0.4 **5.** 0.375 **6.** $-0.\overline{2}$ **7.** $0.8\overline{3}$ **8.** $0.\overline{4}$ **9.** $-0.\overline{6}$
10. 0.9 **11.** $0.\overline{8}$ **12.** $-0.2\overline{7}$ **13.** $1.\overline{3}$ **14.** $0.41\overline{6}$ **15.** 1.6 **16.** 0.625 **17.** $2.\overline{3}$
18. $0.47\overline{2}$ **19.** $0.3\overline{7}$ **20.** $0.\overline{01}$ **21.** $-0.\overline{13}$ **22.** $0.1\overline{4}$ **23.** $0.\overline{1}$ **24.** $0.1\overline{2}$ **25.** 1.5
26. $-2.\overline{1}$ **27.** $4.8\overline{3}$ **28.** 5.7 **29.** $5.\overline{7}$ **30.** $2.\overline{714285}$ **31.** $1.\overline{6}$ **32.** $1.08\overline{3}$ **33.** 3.24
34. 4.05 **35.** 7.048
36. $\frac{1}{50}$ **37.** $\frac{1}{4}$ **38.** $-\frac{1}{25}$ **39.** $\frac{1}{200}$ **40.** $\frac{1}{8}$ **41.** $\frac{5}{8}$ **42.** $-\frac{3}{4}$ **43.** $\frac{9}{20}$ **44.** $-\frac{31}{50}$
45. $\frac{9}{40}$ **46.** $\frac{3}{8}$ **47.** $\frac{1}{20}$ **48.** $\frac{1}{1000}$ **49.** $-\frac{9}{25}$ **50.** $\frac{2}{125}$ **51.** $\frac{7}{8}$ **52.** $\frac{1}{125}$ **53.** $2\frac{7}{20}$
54. $-4\frac{9}{20}$ **55.** $10\frac{1}{100}$ **56.** $-9\frac{1}{4}$ **57.** $1\frac{1}{250}$ **58.** $6\frac{1}{40}$ **59.** $5\frac{1}{8}$ **60.** $8\frac{2}{125}$
61. $\frac{13}{99}$ **62.** $\frac{4}{33}$ **63.** $\frac{7}{33}$ **64.** $\frac{26}{111}$ **65.** $\frac{2}{165}$ **66.** $\frac{7}{330}$ **67.** $\frac{1}{15}$ **68.** $\frac{1}{165}$ **69.** $-\frac{1}{150}$
70. $\frac{1}{825}$ **71.** $\frac{10}{99}$ **72.** $\frac{11}{90}$ **73.** $\frac{1}{6}$ **74.** $\frac{37}{300}$ **75.** $\frac{61}{495}$ **76.** $\frac{11}{900}$ **77.** $\frac{1}{300}$ **78.** $\frac{1}{75}$
79. $\frac{179}{330}$ **80.** $\frac{122}{225}$ **81.** $-\frac{4111}{33300}$ **82.** $\frac{37}{30000}$ **83.** $1\frac{4}{33}$ **84.** $1\frac{1}{2}$ **85.** $2\frac{17}{90}$ **86.** $-2\frac{25}{99}$
87. $3\frac{2}{5}$ **88.** $4\frac{2}{165}$ **89.** $5\frac{56}{225}$ **90.** $1\frac{1}{12}$

CHAPTER 4 – 3 Adding and Subtracting Fractions page 67 ~ 68

1. $\frac{3}{5}$ **2.** $1\frac{2}{5}$ **3.** $\frac{3}{5}$ **4.** $-\frac{2}{5}$ **5.** $-\frac{4}{5}$ **6.** $-\frac{2}{5}$ **7.** $1\frac{4}{7}$ **8.** $-\frac{1}{4}$ **9.** $\frac{1}{2}$ **10.** $-1\frac{5}{16}$
11. 2 **12.** $2\frac{2}{3}$ **13.** 0 **14.** $1\frac{1}{2}$ **15.** $-\frac{1}{3}$ **16.** $1\frac{1}{3}$ **17.** $-\frac{5}{8}$ **18.** $-\frac{7}{8}$ **19.** -1 **20.** 2
21. 6 **22.** $13\frac{1}{5}$ **23.** $2\frac{2}{5}$ **24.** $4\frac{1}{7}$ **25.** $5\frac{5}{7}$ **26.** $11\frac{4}{7}$ **27.** $-3\frac{4}{5}$ **28.** $2\frac{4}{9}$ **29.** $-5\frac{1}{6}$
30. $3\frac{3}{4}$ **31.** $\frac{5}{x}$ **32.** $-\frac{5}{a}$ **33.** $\frac{x-2y}{x^2}$ **34.** $\frac{2a+b}{c}$ **35.** $-\frac{1}{a}$
36. $\frac{5}{6}$ **37.** $\frac{11}{12}$ **38.** $\frac{1}{12}$ **39.** $-\frac{1}{3}$ **40.** $1\frac{3}{8}$ **41.** $-1\frac{3}{20}$ **42.** $-\frac{1}{21}$ **43.** $-\frac{16}{45}$ **44.** $1\frac{11}{18}$
45. $-\frac{5}{16}$ **46.** $5\frac{5}{6}$ **47.** $10\frac{1}{2}$ **48.** $3\frac{3}{35}$ **49.** $3\frac{3}{8}$ **50.** $3\frac{1}{20}$ **51.** $5\frac{11}{14}$ **52.** $2\frac{3}{10}$ **53.** $\frac{3}{4}$
54. $5\frac{3}{10}$ **55.** $2\frac{7}{8}$ **56.** $10\frac{7}{8}$ **57.** $9\frac{1}{8}$ **58.** $7\frac{2}{5}$ **59.** $8\frac{4}{9}$ **60.** $21\frac{5}{9}$
61. $\frac{5x+4y}{20}$ **62.** $\frac{2a-3b}{6}$ **63.** $\frac{4a-7b}{14}$ **64.** $\frac{8m+10n}{40}$ **65.** $\frac{4x-y}{10}$
66. He gained $\$1\frac{1}{8}$. **67.** $8\frac{1}{10}$ **68.** $3\frac{19}{60}$ **69.** $24\frac{2}{3}$ **70.** $70\frac{5}{7}$ in. **71.** $\frac{1}{8}$ **72.** $-13\frac{3}{20}$
73. $-3\frac{5}{12}$ **74.** $53\frac{5}{6}$ feet **75.** $-22\frac{1}{4}$ pounds (He lost $22\frac{1}{4}$ pounds.) **76.** $159\frac{1}{30}$ pounds
77. $7\frac{11}{16}$ inches **78.** $3\frac{47}{56}$ hours **79.** $\frac{1}{12}$ of the gas was left. **80.** $1\frac{7}{12}$

CHAPTER 4 – 4 Multiplying Fractions page 71~72

1. $\frac{15}{28}$ **2.** $\frac{2}{35}$ **3.** $\frac{18}{35}$ **4.** $\frac{8}{45}$ **5.** $-\frac{2}{9}$ **6.** $-\frac{4}{7}$ **7.** -2 **8.** $-\frac{1}{4}$ **9.** $\frac{2}{5}$ **10.** $\frac{2}{3}$ **11.** $-1\frac{5}{9}$
12. $1\frac{2}{13}$ **13.** $3\frac{1}{15}$ **14.** $4\frac{1}{4}$ **15.** $\frac{7}{8}$ **16.** -2 **17.** 3 **18.** $\frac{5}{8}$ **19.** $1\frac{1}{18}$ **20.** -2 **21.** $8\frac{1}{2}$
22. $23\frac{4}{5}$ **23.** $-14\frac{3}{5}$ **24.** $-10\frac{22}{25}$ **25.** $24\frac{1}{2}$ **26.** $4\frac{1}{5}$ **27.** $20\frac{1}{4}$ **28.** $59\frac{1}{3}$ **29.** $-53\frac{1}{4}$
30. 31 **31.** $5\frac{5}{6}$ **32.** -4 **33.** $20\frac{1}{4}$ **34.** $8\frac{2}{5}$ **35.** -42 **36.** $2\frac{1}{2}$ **37.** 9 **38.** 116
39. $43\frac{3}{5}$ **40.** $13\frac{8}{9}$
41. $\frac{x}{10}$ **42.** $\frac{y}{10}$ **43.** $\frac{mn}{14}$ **44.** $\frac{ab}{10}$ **45.** $\frac{22d}{7}$ **46.** $\frac{21t}{22}$ **47.** $\frac{6xy}{25}$ **48.** $3\frac{1}{3}$ **49.** $\frac{1}{9}$ **50.** $4p$
51. $\frac{8}{15}$ **52.** $\frac{37}{56}$ **53.** $-\frac{8}{15}$ **54.** $-4\frac{1}{2}$ **55.** $1\frac{1}{4}$ **56.** $-\frac{1}{2}$ **57.** $4\frac{9}{20}$ **58.** 8 **59.** $\frac{7}{24}$ **60.** $\frac{1}{12}$
61. $\frac{3}{8}$ **62.** $\frac{3}{8}$ **63.** 6 **64.** $3\frac{7}{12}$ **65.** $\frac{4}{15}$ **66.** $\frac{2}{5}$ **67.** 27 books **68.** 24 books **69.** $5\frac{1}{3}$ $ft.^2$
70. $7\frac{2}{3}$ m^2 **71.** 231 $in.^3$ **72.** 26 students **73.** 2,130 boys and 1,065 girls **74.** $98.06^\circ F$
75. 137.5 miles **76.** 112.2 $ft.^2$ **77.** 50 hours. He will earn \$325. **78.** 360 pounds
79. $-\frac{24}{95}$ **80.** 108

CHAPTER 4 – 5 Dividing Fractions page 75~76

1. $\frac{7}{2}$ **2.** $\frac{2}{7}$ **3.** $\frac{8}{3}$ **4.** $\frac{5}{9}$ **5.** $\frac{4}{13}$ **6.** $\frac{16}{3}$ **7.** $\frac{4}{13}$ **8.** $\frac{3}{17}$ **9.** None **10.** 1
11. $1\frac{1}{20}$ **12.** $2\frac{4}{5}$ **13.** $\frac{5}{14}$ **14.** $\frac{5}{18}$ **15.** $-\frac{8}{9}$ **16.** $-1\frac{2}{7}$ **17.** -2 **18.** $-\frac{1}{4}$ **19.** $\frac{2}{5}$ **20.** $\frac{2}{3}$
21. $-\frac{5}{21}$ **22.** $\frac{4}{43}$ **23.** 6 **24.** $4\frac{2}{5}$ **25.** $\frac{5}{44}$ **26.** $\frac{5}{7}$ **27.** $\frac{11}{13}$ **28.** $-1\frac{3}{31}$ **29.** $-\frac{3}{4}$ **30.** $1\frac{13}{18}$
31. $\frac{5}{42}$ **32.** $-\frac{4}{81}$ **33.** $\frac{1}{4}$ **34.** $\frac{6}{35}$ **35.** $-\frac{1}{8}$ **36.** $57\frac{3}{5}$ **37.** 18 **38.** 5 **39.** $10\frac{9}{10}$ **40.** $\frac{5}{9}$
41. $\frac{5x}{8}$ **42.** $\frac{5y}{18}$ **43.** $\frac{2m}{7}$ **44.** $\frac{a}{10}$ **45.** $\frac{18d}{77}$ **46.** $\frac{24t}{77}$ **47.** $\frac{x}{8}$ **48.** $\frac{10}{3x^2}$ **49.** $9y$ **50.** $\frac{p}{2}$
51. $\frac{1}{6}$ **52.** $1\frac{11}{14}$ **53.** $-\frac{8}{15}$ **54.** $5\frac{1}{2}$ **55.** $5\frac{3}{5}$ **56.** $-\frac{1}{2}$ **57.** $4\frac{11}{12}$ **58.** $1\frac{1}{8}$ **59.** $\frac{6}{7}$ **60.** $\frac{1}{6}$
61. $\frac{1}{2}$ **62.** 16 **63.** $5\frac{3}{8}$ **64.** 10 miles per gallon **65.** $\frac{1}{9}$ of the whole pizza **66.** 27 pieces
67. $\frac{2}{3}$ of the day **68.** 2 *ft.* **69.** 20 bars **70.** 32 bags **71.** 25 gallons **72.** $3\frac{1}{2}$ pounds
73. $\frac{2}{5}$ mile **74.** $2\frac{2}{9}$ hours **75.** 40 students **76.** \$400 **77.** 96 meters **78.** $-\frac{5}{6}$ **79.** 3
80. $-3\frac{1}{8}$ **81.** $x=10$ **82.** 2.4 hours

CHAPTER 4 Fractions EXERCISES page 77 – 78

1. Not equal **2.** Equal **3.** Equal **4.** Not equal **5.** Not equal **6.** Equal **7.** Equal
8. Equal **9.** Not equal **10.** Not equal

11. $\frac{3}{4}$ **12.** $\frac{3}{5}$ **13.** $\frac{3}{5}$ **14.** $1\frac{2}{3}$ **15.** 1 **16.** 0 **17.** -3 **18.** $6\frac{3}{4}$ **19.** $-4\frac{1}{13}$ **20.** $7\frac{1}{2}$

21. $-\frac{1}{3}$ **22.** $\frac{1}{3}$ **23.** $7\frac{3}{4}$ **24.** $5\frac{1}{4}$ **25.** $-5\frac{1}{3}$ **26.** $9\frac{1}{7}$ **27.** $\frac{x}{5}$ **28.** $2y$ **29.** $\frac{3a}{2}$ **30.** $\frac{3n}{4m}$

31. $\frac{6}{8}, \frac{1}{8}$ **32.** $\frac{24}{30}, \frac{25}{30}$ **33.** $-\frac{12}{21}, \frac{14}{21}$ **34.** $\frac{9}{21}, \frac{5}{21}$ **35.** $\frac{3}{33}, -\frac{4}{33}$ **36.** $\frac{28}{36}, \frac{9}{36}$

37. $\frac{55}{66}, \frac{54}{66}$ **38.** $\frac{20}{24}, -\frac{9}{24}$ **39.** $\frac{10}{24}, \frac{21}{24}$ **40.** $\frac{15}{36}, \frac{10}{36}$ **41.** $-\frac{24}{336}, -\frac{14}{336}$

42. $\frac{20}{60}, \frac{15}{60}, \frac{24}{60}$ **43.** $-\frac{24}{30}, \frac{25}{30}, -\frac{4}{30}$ **44.** $\frac{21}{24}, \frac{10}{24}, \frac{4}{24}$ **45.** $\frac{10}{180}, \frac{105}{180}, \frac{54}{180}$

46. 0.6 **47.** 0.8 **48.** $0.\overline{5}$ **49.** $0.\overline{7}$ **50.** 0.3 **51.** -0.625 **52.** -2.25 **53.** 3.6

54. $-0.\overline{13}$ **55.** $0.1\overline{2}$ **56.** $0.2\overline{3}$ **57.** $0.3\overline{2}$ **58.** $-3.\overline{6}$ **59.** 6.25 **60.** -12.875

61. $\frac{1}{20}$ **62.** $\frac{3}{20}$ **63.** $-\frac{1}{25}$ **64.** $\frac{9}{20}$ **65.** $\frac{13}{20}$ **66.** $1\frac{1}{10}$ **67.** $2\frac{1}{100}$ **68.** $-9\frac{1}{2}$

69. $12\frac{11}{20}$ **70.** $4\frac{1}{200}$ **71.** $-10\frac{1}{1000}$ **72.** $8\frac{1}{8}$ **73.** $\frac{5}{99}$ **74.** $\frac{5}{33}$ **75.** $-\frac{2}{45}$ **76.** $\frac{41}{90}$

77. $\frac{73}{198}$ **78.** $1\frac{1}{9}$ **79.** $2\frac{1}{90}$ **80.** $-4\frac{61}{495}$

81. $1\frac{5}{12}$ **82.** $-\frac{1}{12}$ **83.** $-\frac{1}{2}$ **84.** $\frac{8}{9}$ **85.** $1\frac{2}{7}$ **86.** $\frac{4}{5}$ **87.** $-1\frac{2}{7}$ **88.** $\frac{4}{5}$ **89.** $-\frac{9}{40}$

90. $\frac{7}{10}$ **91.** $8\frac{1}{8}$ **92.** $1\frac{1}{8}$ **93.** 22 **94.** $1\frac{1}{2}$ **95.** $\frac{4}{33}$ **96.** $-1\frac{1}{2}$ **97.** $-\frac{32}{363}$ **98.** $\frac{22}{31}$

99. $\frac{3}{5}$ **100.** $15\frac{3}{4}$ **101.** $\frac{1}{12}$ **102.** $\frac{7}{12}$ **103.** $\frac{4}{5}$ **104.** $\frac{5}{14}$ **105.** $\frac{5}{6}$ **106.** $1\frac{3}{10}$ **107.** $\frac{2}{3}$

108. $-25\frac{1}{5}$ **109.** $-2\frac{3}{4}$ **110.** 4 **111.** $7\frac{5}{11}$ **112.** Perimeter = $12\frac{1}{3}$ *in.* , Area = $7\frac{1}{2}$ $in.^2$

113. $101\frac{17}{27}$ $in.^3$ **114.** $\frac{1}{10}$ **115.** $\frac{8}{15}$ **116.** $\frac{2}{3}$ **117.** Donna ate $\frac{4}{15}$, $\frac{2}{5}$ was left **118.** 60 pieces

CHAPTER 5 – 1 Ratios and Rates page 82

1. $\frac{5}{8}$ **2.** $\frac{8}{5}$ **3.** $\frac{2}{3}$ **4.** $\frac{3}{2}$ **5.** $\frac{6}{5}$ **6.** $\frac{5}{6}$ **7.** $\frac{3}{4}$ **8.** $\frac{9}{5}$ **9.** $\frac{4}{1}$ **10.** $\frac{1}{4}$ **11.** $\frac{1}{5}$ **12.** $\frac{5}{1}$ **13.** $\frac{7}{9}$
14. $\frac{9}{7}$ **15.** $\frac{3}{4}$ **16.** $\frac{4}{3}$ **17.** $\frac{3}{5}$ **18.** $\frac{5}{3}$ **19.** $\frac{13}{5}$ **20.** $\frac{63}{200}$ **21.** $\frac{9}{80}$ **22.** $\frac{25}{6}$ **23.** $\frac{1}{2}$ **24.** $\frac{21}{2}$
25. $\frac{4}{3}$ **26.** $\frac{3}{4}$ **27.** $\frac{2}{1}$ **28.** $\frac{9}{64}$ **29.** $\frac{7}{4}$ **30.** $\frac{4}{9}$
31. $\frac{3}{1}$ **32.** $\frac{1}{3}$ **33.** $\frac{4}{7}$ **34.** $\frac{2}{1}$ **35.** $\frac{12}{5}$ **36.** $\frac{8}{3}$ **37.** $\frac{10}{3}$ **38.** $\frac{3}{2}$ **39.** $\frac{8}{3}$ **40.** $\frac{1}{4}$ **41.** $\frac{5}{12}$
42. $\frac{3}{5}$ **43.** *a.* $\frac{5}{2}$, *b.* $\frac{2}{5}$, *c.* $\frac{5}{7}$ **44.** *a.* $\frac{5}{2}$, *b.* $\frac{2}{5}$, *c.* $\frac{5}{7}$, *d.* $\frac{2}{7}$ **45.** $\frac{2}{5}$
46. He can finish $\frac{1}{3}$ of the job in 2 hours. **47.** $\frac{7}{25}$ **48.** 1 out of 16 apples is overripe.
49. 1 out of 3 students wears glasses. **50.** $\frac{1}{6}$ **51.** 62.5 miles per gallon.
52. average speed = 62.6 mph. Drive 438.2 miles in 7 hours. **53.** \$0.18, 100 eggs
54. B-brand is a better buy. **55.** \$29.70

CHAPTER 5 – 2 Proportions page 84

1. Yes **2.** No **3.** Yes **4.** Yes **5.** No **6.** No **7.** Yes **8.** No **9.** No **10.** No **11.** Yes
12. Yes **13.** Yes **14.** Yes **15.** No **16.** Yes **17.** No **18.** Yes **19.** No **20.** Yes
21. $n = 1$ **22.** $n = 6$ **23.** $x = 15$ **24.** $x = 4$ **25.** $y = 6$ **26.** $y = 21$ **27.** $k = 3$ **28.** $k = 3$
29. $n = 42$ **30.** $n = 5$ **31.** $p = 18$ **32.** $r = 15$ **33.** $x = 2.8$ **34.** $x = 6$ **35.** $y = 5.6$
36. $m = 6.4$ **37.** $n = 4.16$ **38.** $p = 22.5$ **39.** $n = 1.2$ **40.** $x = 0.2$
(41 ~ 50 Write a proportion and use cross-products to solve it.)
41. \$14 **42.** 5 pounds **43.** 15 hours **44.** 25 cm **45.** 6.8 in. **46.** 50 meters **47.** 512 miles
48. 5.4 gallons **49.** \$2,460 **50.** 3.96 in.

CHAPTER 5 – 3 Percents page 86

1. 25% **2.** 50% **3.** 75% **4.** 100% **5.** 125% **6.** 200% **7.** 50% **8.** 40% **9.** 100% **10.** 30% **11.** 90% **12.** 100% **13.** 4% **14.** 16% **15.** 40% **16.** 60% **17.** $33\frac{1}{3}$% **18.** $66\frac{2}{3}$% **19.** $133\frac{1}{3}$% **20.** 375% **21.** 100% **22.** 94% **23.** 46% **24.** 180% **25.** 1% **26.** 37% **27.** 25% **28.** 240% **29.** 175% **30.** 35%

31. $\frac{1}{50}$ **32.** $\frac{1}{25}$ **33.** $\frac{1}{20}$ **34.** $\frac{9}{100}$ **35.** $\frac{1}{10}$ **36.** $\frac{1}{2}$ **37.** $\frac{3}{4}$ **38.** $\frac{9}{10}$ **39.** $\frac{3}{25}$ **40.** $\frac{3}{20}$

41. $\frac{9}{50}$ **42.** $\frac{1}{4}$ **43.** $\frac{11}{20}$ **44.** $\frac{49}{50}$ **45.** $\frac{99}{100}$ **46.** $1\frac{1}{50}$ **47.** $1\frac{1}{5}$ **48.** $1\frac{1}{4}$ **49.** $1\frac{3}{10}$ **50.** $3\frac{3}{4}$

51. 4 **52.** 5 **53.** 2 **54.** 8 **55.** 15. **56.** 50 **57.** 60 **58.** 11.25 **59.** 10 **60.** 150 **61.** 36 **62.** 360 **63.** 30 **64.** 450 **65.** 20% **66.** 50% **67.** 80% **68.** 20% **69.** 6.75 **70.** 13.75 **71.** 40.5 **72.** 8.8 **73.** 60% **74.** 2.5% **75.** 100% **76.** 11% **77.** 6% **78.** 6.25% **79.** 100% **80.** 100%

CHAPTER 5 – 4 Percents and Decimals page 88

1. 0.02 **2.** 0.04 **3.** 0.05 **4.** 0.09 **5.** 0.1 **6.** 0.5 **7.** 0.75 **8.** 0.9 **9.** 0.12 **10.** 0.15 **11.** 0.18 **12.** 0.25 **13.** 0.55 **14.** 0.98 **15.** 0.99 **16.** 1.02 **17.** 0.005 **18.** 0.0005 **19.** 0.00005 **20.** 0.0045 **21.** 0.00045 **22.** 0.824 **23.** 0.798 **24.** 0.0235 **25.** 1.5 **26.** 11.2 **27.** 0.0999 **28.** 0.999 **29.** 10.01 **30.** 2.49

31. 20% **32.** 2% **33.** 0.2% **34.** 90% **35.** 10% **36.** 100% **37.** 1% **38.** 99% **39.** 101% **40.** 9.9% **41.** 12.5% **42.** 125% **43.** 450% **44.** 83.5% **45.** 0.35% **46.** 0.01% **47.** 7.1% **48.** 705% **49.** 1412% **50.** 0.29% **51.** 2.19% **52.** 2120% **53.** 1025% **54.** 210% **55.** 33.3% **56.** 333% **57.** 3.3% **58.** 100.1% **59.** 87.5% **60.** 8.75%

61. 4 **62.** 5 **63.** 4 **64.** 10 **65.** 13.5 **66.** 40 **67.** 22.5 **68.** 11.25 **69.** 120 **70.** 96 **71.** 8.8% **72.** 9.9% **73.** $7.43 **74.** $0.16 **75.** $0.22 **76.** $360 **77.** $375 **78.** $7,000 **79.** $47,040 **80.** 36 cases

CHAPTER 5 – 5 Percents and Fractions page 90

1. 25% **2.** 75% **3.** 80% **4.** 87.5% **5.** 10% **6.** 30% **7.** 4% **8.** 180% **9.** 240% **10.** 260% **11.** 640% **12.** 237.5% **13.** 187.5% **14.** 312.5% **15.** 1% **16.** 6% **17.** 23% **18.** 1.5% **19.** 125% **20.** 160%

21. 41.7% **22.** 55.6% **23.** 83.3% **24.** 42.9% **25.** 6.7% **26.** 22.2% **27.** 30.8% **28.** 133.3% **29.** 266.7% **30.** 108.3%

31. $41\frac{2}{3}$% **32.** $55\frac{5}{9}$% **33.** $83\frac{1}{3}$% **34.** $42\frac{6}{7}$% **35.** $6\frac{2}{3}$% **36.** $22\frac{2}{9}$% **37.** $30\frac{10}{13}$% **38.** $133\frac{1}{3}$% **39.** $266\frac{2}{3}$% **40.** $108\frac{1}{3}$%

41. $\frac{1}{50}$ **42.** $\frac{1}{80}$ **43.** $\frac{1}{20}$ **44.** $\frac{9}{100}$ **45.** $\frac{1}{10}$ **46.** $\frac{11}{200}$ **47.** $\frac{3}{4}$ **48.** $\frac{3}{200}$ **49.** $\frac{3}{25}$ **50.** $\frac{7}{200}$ **51.** $\frac{13}{300}$ **52.** $\frac{1}{4}$ **53.** $\frac{9}{10}$ **54.** $\frac{2}{3}$ **55.** $\frac{99}{100}$ **56.** $1\frac{1}{50}$ **57.** $\frac{1}{200}$ **58.** $\frac{1}{2000}$ **59.** $\frac{9}{2000}$ **60.** $10\frac{1}{100}$

61. 1.5 **62.** 1.4 **63.** 1.4 **64.** 16.2 **65.** 3.2 **66.** 20.5 **67.** 2 **68.** 3.19 **69.** 25% **70.** 80% **71.** 20% **72.** 15%

CHAPTER 5 – 6 Percent and Applications page 94

1. 20% **2.** 30% **3.** 25% **4.** 35% **5.** 25% **6.** 60% **7.** 55% **8.** 85% **9.** 35% **10.** 5% **11.** 42% **12.** $33\frac{1}{3}$% **13.** 600 **14.** 160 **15.** 400 **16.** 240 **17.** 20 **18.** 7.5 **19.** 320 **20.** 25 **21.** 3,000 **22.** 1,000 **23.** 90 **24.** 60

25. 7.5% **26.** $90 **27.** 30% **28.** $180 **29.** 70% **30.** 50 questions **31.** 20% **32.** 25% **33.** 2,875 students **34.** $15.92 **35.** $4,500 **36.** $180,000 **37.** $1,200 **38.** $2,192 **39.** $2016.03 **40.** $2024.10

CHAPTER 5 Ratios, Proportions, and Percent EXERCISES page 95 – 96

1. $\frac{7}{4}$ **2.** $\frac{3}{1}$ **3.** $\frac{5}{1}$ **4.** $\frac{1}{4}$ **5.** $\frac{5}{3}$ **6.** $\frac{5}{2}$ **7.** $\frac{5}{8}$ **8.** $\frac{3}{1}$ **9.** $\frac{7}{10}$ **10.** $\frac{2}{3}$
11. $\frac{10}{13}$ **12.** $\frac{1}{6}$ **13.** $\frac{9}{2}$ **14.** $\frac{3}{2}$ **15.** $\frac{2}{1}$ **16.** $\frac{2x}{5}$ **17.** $\frac{2}{5}$ **18.** $\frac{8}{3}$ **19.** $\frac{7}{3}$ **20.** $\frac{5}{3}$
21. yes **22.** no **23.** no **24.** yes **25.** yes **26.** no **27.** yes **28.** no **29.** no **30.** yes
31. $n=1$ **32.** $n=16$ **33.** $n=8.25$ **34.** $x=15$ **35.** $y=1$ **36.** $m=6$ **37.** $k=4$
38. $x=63$ **39.** $p=2.5$ **40.** $y=1.3$
41. 20% **42.** 100% **43.** 80% **44.** 2% **45.** 22% **46.** 12% **47.** 100% **48.** 87.5%
49. 7% **50.** 32.5%
51. $\frac{1}{5}$ **52.** $\frac{2}{25}$ **53.** $\frac{2}{5}$ **54.** $\frac{99}{100}$ **55.** $\frac{1}{100}$ **56.** $1\frac{1}{100}$ **57.** $1\frac{99}{100}$ **58.** $\frac{11}{20}$ **59.** $1\frac{1}{5}$ **60.** $3\frac{1}{4}$
61. 0.08 **62.** 0.11 **63.** 0.045 **64.** 0.99 **65.** 0.01 **66.** 0.0109 **67.** 0.001 **68.** 0.0002
69. 2.25 **70.** 0.225
71. 8% **72.** 11% **73.** 4.5% **74.** 99% **75.** 1% **76.** 1.09% **77.** 0.1% **78.** 0.02%
79. 225% **80.** 22.5%
81. 25% **82.** 60% **83.** 160% **84.** 450% **85.** 87.5% **86.** 57.5% **87.** $66\frac{2}{3}$% **88.** $3\frac{1}{3}$%
89. $83\frac{1}{3}$% **90.** $55\frac{5}{9}$%
91. $\frac{1}{4}$ **92.** $\frac{3}{5}$ **93.** $1\frac{3}{5}$ **94.** $4\frac{1}{2}$ **95.** $\frac{7}{8}$ **96.** $\frac{23}{40}$ **97.** $\frac{2}{3}$ **98.** $\frac{1}{30}$ **99.** $\frac{5}{6}$ **100.** $\frac{5}{9}$
101. 1.2 **102.** 7 **103.** 37.5 **104.** 18 **105.** 0.675 **106.** 8.75 **107.** 12.5% **108.** 8.75%
109. 16% **110.** 100% **111.** 7.2% **112.** 1.3% **113.** 3.4 **114.** 39 **115.** 34 **116.** 3.12
117. $\frac{3}{1}$ **118.** 75% **119.** 72% **120.** 16 **121.** 555 miles **122.** \$1.46 **123.** 7.5% **124.** \$240
125. 56% **126.** 25% **127.** 16.7% **128.** \$84 **129.** \$2,375 **130.** \$6,125 **131.** \$5121.10

CHAPTER 6 – 1 Variable and Expressions page 102

1. 3 **2.** 5 **3.** 7 **4.** 1 **5.** 3 **6.** 5 **7.** 12 **8.** 14 **9.** 16 **10.** –6 **11.** 0 **12.** 4
13. –6 **14.** –10 **15.** –14 **16.** –2 **17.** –6 **18.** –10 **19.** 2 **20.** 8 **21.** –4 **22.** –10
23. 0 **24.** 2 **25.** 5 **26.** $6\frac{1}{2}$ or 6.5 **27.** 7 **28.** 0 **29.** $-3\frac{1}{2}$ or –3.5 **30.** $-1\frac{1}{2}$ or –1.5
31. 4 **32.** $1\frac{1}{2}$ or 1.5 **33.** $-6\frac{1}{2}$ or –6.5 **34.** 13 **35.** 7 **36.** –6.5 **37.** 14 **38.** 19
39. –3 **40.** 3 **41.** 16 **42.** 36 **43.** \$40 **44.** \$44 **45.** 225 square meters **46.** 1024 feet

CHAPTER 6 – 2 Adding and Subtracting Expressions page 104

1. $9x$ **2.** $-3x$ **3.** $-9x$ **4.** $3x$ **5.** $-9x+5$ **6.** $3x-5$ **7.** $9x+5$ **8.** $9x-5$
9. $3x+5$ **10.** $9x+5$ **11.** $15-4x$ **12.** $9x+4y$ **13.** $9x+12y$ **14.** $5x-4y$
15. $5x+4y$ **16.** $-4a+19$ **17.** $a-8b$ **18.** $9u-9w$ **19.** $3w-4u$ **20.** $9w-u$
21. $9m+13$ **22.** $13+2x$ **23.** $-14+y$ **24.** $-n-7$ **25.** 7 **26.** $\frac{5}{6}x+\frac{2}{3}y$ **27.** $\frac{1}{6}x+\frac{1}{6}y$
28. $\frac{5}{12}x+\frac{11}{20}y$ **29.** $\frac{4}{5}x+\frac{1}{3}y$ **30.** $-\frac{2}{5}x+y$ **31.** $4.7x-2.3y$ **32.** $-2.3x-4.7y$ **33.** $-2.3x+4$
34. $6x^2-1$ **35.** $-2y^2+2$ **36.** $4x^2-7$ **37.** $9x^2-8x$ **38.** $-8x^2-10$ **39.** $-3n^2-4$
40. $-5x^2+x+4$ **41.** $2a^2-9$ **42.** $-2k^2+8$ **43.** $-5y$ **44.** $-a-5b+9$ **45.** $3m-5n-8$
46. $-p-2q-4$ **47.** $-2x^2+2x-2$ **48.** $8x^2-6x+10$ **49.** $2x+2y$ or $2(x+y)$
50. $4x+4y$ or $4(x+y)$

CHAPTER 6 – 3 Multiplying and Dividing Expressions page 106

1. x^2 **2.** $6x^2$ **3.** $-24x^4$ **4.** $-4x^{12}$ **5.** $7.2n^{10}$ **6.** 1 **7.** $-\frac{2}{3}$ **8.** $\frac{3}{2}x^2$ **9.** $\frac{1}{4}x^2$ **10.** $0.8n^2$
11. $18x$ **12.** $3a$ **13.** $-25n$ **14.** $-4x+5$ **15.** $5-4x$ **16.** $2x^2+x$ **17.** $-3a^2+4a-6$
18. x^2-6 **19.** $-3x^2+3$ **20.** $6x+10$ **21.** $-10+6x$ **22.** $-12a-20$ **23.** $15x^2-30x$
24. $18a^2-42a$ **25.** $8a^2+20a$ **26.** k^2-k **27.** $2k^2-4k$ **28.** $3xy-27xz$ **29.** $8ab-4ac$
30. $10ab-35ac$ **31.** $-2xy+2x^2$ **32.** $3x^2-12x-18$ **33.** $4y^2+12y-8$
34. $-10a^2-15a+35$ **35.** $-6n^2+6n-10$ **36.** $6y^3-3y^2+18y$ **37.** $2x^3-6x^2-18x$
38. $-8x^2-4xy+4xz$ **39.** $15x^2+6xy-21xz$ **40.** $-a^2+ab-ac$ **41.** $13x^2-8$
42. $4x^2-8$ **43.** $-5a^2+6$
44. $3x-1$ **45.** $-4x+1$ **46.** $-3x^2+x$ **47.** $4x^2-3x$ **48.** $5a^2-2a$ **49.** a^2-a
50. $4x^2-2x$ **51.** $2a^3-a$ **52.** $4x^2+5x+3$ **53.** a^2-2a-1 **54.** $3a-b$ **55.** $8xy-4$
56. $\frac{4}{3}x+2$ **57.** $\frac{4}{5}x-\frac{3}{5}$ **58.** $2a-\frac{4}{3}b$ **59.** $\frac{2}{3}-\frac{1}{3}x$ **60.** *a*) $2n^2+2n$ *b*) x^2+2xy

CHAPTER 6 – 4 Solving One–Step Equations page 109 ~ 110

1. $x=12$ **2.** $x=6$ **3.** $x=3$ **4.** $x=27$ **5.** $n=6$ **6.** $n=18$ **7.** $n=72.$ **8.** $n=2$
9. $y=2$ **10.** $y=-6$ **11.** $y=-8$ **12.** $y=-\frac{1}{2}$ **13.** $a=45$ **14.** $a=15$ **15.** $a=2$
16. $a=450$ **17.** $x=-1$ **18.** $x=-2$ **19.** $a=-\frac{1}{2}$ **20.** $x=4$ **21.** $x=-6$ **22.** $x=-12$
23. $x=-3$ **24.** $x=-27$ **25.** $x=10$ **26.** $x=14$ **27.** $x=-24$ **28.** $x=-6$ **29.** $x=-14$
30. $x=-10$ **31.** $x=24$ **32.** $x=6$ **33.** $x=-14$ **34.** $x=10$ **35.** $x=-6$ **36.** $x=-24$
37. $m=3$ **38.** $m=6$ **39.** $m=6.75$ **40.** $m=3$ **41.** $n=2.5$ **42.** $n=7.5$ **43.** $n=-0.5$
44. $n=-12.5$ **45.** $y=1$ **46.** $y=-8$ **47.** $y=-11.5$ **48.** $x=-1.84$ **49.** $x=3\frac{1}{2}$
50. $x=4\frac{1}{2}$ **51.** $x=-3\frac{1}{2}$ **52.** $x=-4\frac{1}{2}$ **53.** $y=1\frac{1}{3}$ **54.** $y=3\frac{1}{3}$ **55.** $y=\frac{1}{2}$ **56.** $y=\frac{2}{9}$
57. $a=-\frac{3}{5}$ **58.** $a=2\frac{1}{5}$ **59.** $a=\frac{3}{5}$ **60.** $a=-2\frac{1}{5}$ **61.** $x=-14.4$ **62.** $x=-12.8$
63. $t=2$ **64.** $t=1152$ **65.** $k=10$ **66.** $x=-\frac{1}{12}$ **67.** $k=-115.2$ **68.** $k=-0.2$
69. $x=-\frac{3}{5}$ **70.** $x=-1\frac{7}{8}$ **71.** $x=-1\frac{3}{5}$ **72.** $x=\frac{5}{8}$

CHAPTER 6 – 5 Solving Two–Step Equations page 113 ~ 114

1. $x=4$ **2.** $x=9$ **3.** $x=-5$ **4.** $x=-3$ **5.** $a=4$ **6.** $y=3$ **7.** $y=5$ **8.** $a=15$
9. $x=-3$ **10.** $x=-2$ **11.** $y=3$ **12.** $y=-3\frac{1}{3}$ **13.** $a=-12$ **14.** $a=-15$ **15.** $m=2$
16. $m=-2$ **17.** $x=6\frac{2}{5}$ **18.** $x=5\frac{3}{5}$ **19.** $x=-5\frac{3}{5}$ **20.** $x=6\frac{2}{5}$ **21.** $y=-10$ **22.** $y=2$
23. $n=2$ **24.** $n=-2$ **25.** $x=4$ **26.** $x=-1$ **27.** $y=-\frac{1}{2}$ **28.** $y=\frac{1}{2}$ **29.** $p=-2$
30. $p=2$ **31.** $c=1\frac{1}{2}$ **32.** $c=-3\frac{3}{4}$ **33.** $c=-4$ **34.** $c=4$ **35.** $p=4\frac{1}{2}$ **36.** $p=-4\frac{1}{2}$
37. $x=2$ **38.** $x=4$ **39.** $y=4$ **40.** $y=-4$ **41.** $y=0.64$ **42.** $y=3.2$ **43.** $k=-3.2$
44. $k=0.8$ **45.** $c=0.6$ **46.** $c=-0.8$ **47.** $n=0.6$ **48.** $n=-1.2$ **49.** $x=1.2$
50. $x=-8.4$ **51.** $x=8.4$ **52.** $x=-8.4$ **53.** $x=-10$ **54.** $x=2.5$ **55.** $y=-4.8$
56. $y=1.4$ **57.** $n=8$ **58.** $n=2$ **59.** $m=0.6$ **60.** $m=6.8$ **61.** $x=4$ **62.** $x=12$
63. $y=9$ **64.** $y=33$ **65.** $k=75$ **66.** $k=15$ **67.** $b=42$ **68.** $b=-42$ **69.** $x=14.4$
70. $x=34.4$ **71.** $x=12$ **72.** $x=60$ **73.** $y=-24$ **74.** $y=-4\frac{4}{5}$ **75.** $y=20$
76. $y=-8\frac{4}{7}$ **77.** $x=\frac{2}{5}$ **78.** $x=1\frac{1}{3}$ **79.** $x=\frac{11}{18}$ **80.** $x=\frac{5}{32}$ **81.** $n=-\frac{3}{4}$ **82.** $n=-\frac{1}{4}$
83. $n=\frac{1}{5}$ **84.** $n=-\frac{2}{5}$ **85.** $x=43$ **86.** $y=-17$ **87.** $d=-17$

CHAPTER 6 – 6 Solving Multistep Equations page 116

1. $x=8$ **2.** $x=-3$ **3.** $x=-6$ **4.** $x=3$ **5.** $y=-1$ **6.** $y=1$ **7.** $x=3$ **8.** $x=-2.5$
9. $y=3$ **10.** $y=-3$ **11.** $m=-5$ **12.** $m=2$ **13.** $a=-4\frac{1}{2}$ **14.** $x=-2$ **15.** $x=-2\frac{1}{9}$
16. $a=-2$ **17.** $x=9$ **18.** $n=-1\frac{1}{2}$ **19.** $x=2$ **20.** $x=-14$ **21.** $x=-2$ **22.** $y=2\frac{3}{4}$
23. $y=\frac{3}{4}$ **24.** $y=2$ **25.** $m=2\frac{4}{5}$ **26.** $m=1\frac{3}{5}$ **27.** $m=1\frac{1}{7}$ **28.** $n=-1$ **29.** $n=1$
30. $n=1$ **31.** $x=-5$ **32.** $y=3$ **33.** $z=-3$ **34.** $c=2$ **35.** $c=-2$ **36.** $c=1$
37. $k=10$ **38.** $k=8$ **39.** $k=-8$ **40.** $b=-21$ **41.** $b=11$ **42.** $b=7$ **43.** $p=8$
44. $p=12$ **45.** $p=55$ **46.** $x=5$ **47.** $x=-5$ **48.** $x=2$ **49.** $y=1$ **50.** $y=-1$
51. $y=-1$ **52.** $x=-11$ **53.** $x=11$ **54.** $x=-\frac{2}{3}$ **55.** $y=1$ **56.** $y=-1$ **57.** $y=1\frac{3}{7}$
58. $x=7$ **59.** $x=-17$ **60.** $x=-7$ **61.** $x=-7$ **62.** $y=1$ **63.** $a=-1\frac{1}{3}$ **64.** $x=-2$
65. $y=-8$

A teacher is testing a three-year old boy.
Teacher: What is the result of 7×6 ?
Boy: 42.
Teacher: What a smart boy !
How about 6×7 ?
Boy: 24.

CHAPTER 6 – 7 Translating Words into Symbols page 119

1. $n+5$ **2.** $5+n$ **3.** $n-5$ **4.** $5-n$ **5.** $5>n$ **6.** $n<5$ **7.** $2x+5$ **8.** $\frac{1}{2}n-5$
9. $5n-10$ **10.** $5(n+10)$ **11.** $5\le m$ **12.** $5\ge m$ **13.** $2(x+y)$ **14.** $2x+y$ **15.** $5(n+2)$
16. $5m$ **17.** $2(x+5)$ **18.** $2x+5$ **19.** $5-4k$ **20.** $5-k$ **21.** $5(x+2)$ **22.** $5x+2$
23. $5(x-2)$ **24.** $5x-2$ **25.** $5x-2$ **26.** $5x+2$ **27.** $4k+5$ **28.** $4k(5)$ **29.** $4k-5$
30. $4k/5$ **31.** $5>\frac{k}{3}$ **32.** $5\le\frac{k}{2}$ **33.** $5+2m$ **34.** $5+\frac{m}{2}$ **35.** $3n+5$ **36.** $3n-5$
37. $5+3n$ **38.** $5-3n$ **39.** $5(n+3)$ **40.** $\frac{1}{2}(\frac{n}{3})$
41. $g+15$ **42.** $n,\ n+1,\ n+2,\ n+3$ **43.** $n,\ n+2,\ n+4,\ n+6$
44. $n,\ n+2,\ n+4,\ n+6$ **45.** $200+20x$ **46.** $110+0.25x$ **47.** $2x+24$ **48.** $12x$
49. $n-35$ **50.** $19.5n$
51. $n+21=28$ **52.** $n-8=15$ **53.** $n-12=30$ **54.** $8-n=15$ **55.** $x+5=25$
56. $5n=45$ **57.** $2x+5=13$ **58.** $10-x=29$ **59.** $2(n+5)=-35$ **60.** $\frac{1}{3}x=12$
61. $6(n+2)=14$ **62.** $3x-4=8$ **63.** $\frac{1}{2}(\frac{x}{3})=10$ **64.** $2n-2=6$ **65.** $4n+12=2n$
66. $9x=2(x-7)$ **67.** $5(x-2)=15$ **68.** $\frac{1}{3}x-4=\frac{3}{8}$
69. $2x+24=54$ **70.** $12x=180$ **71.** $2n+1=31$ **72.** $3n+6=48$ **73.** $3n+15=60$
74. $240=80x$ **75.** $120=50+0.25x$ **76.** $3x+8=92$ **77.** $6x=30$ **78.** $4x+12=36$
79. $3x=36$ **80.** $2x+8=27$

CHAPTER 6 – 8 Solving Word Problems in One Variable page 125 ~ 126

1. $x=9$ **2.** $x=-9$ **3.** $x=35$ **4.** $x=-35$ **5.** $n=-6$ **6.** $n=-6$ **7.** $x=12$
8. $x=-18$ **9.** $x=36$ **10.** $x=36$ **11.** $x=-4$ **12.** $x=-4$ **13.** $x=-3\frac{1}{2}$ **14.** $k=18$
15. $x=-6$ **16.** $x=2$ **17.** $x=-\frac{6}{7}$ **18.** $x=6$ **19.** $y=5$ **20.** $x=-9$ **21.** $a=6$
22. $p=-5$ **23.** $p=4\frac{2}{3}$ **24.** $x=-3$ **25.** $x=-28$ **26.** $x=2\frac{1}{3}$ **27.** $x=-46$ **28.** $x=2\frac{8}{9}$
29. $y=3$ **30.** $n=3$ **31.** $n=2$ **32.** $x=-3$ **33.** $a=-4\frac{1}{2}$ **34.** $x=-6\frac{1}{2}$ **35.** $y=-6$
36. 7 **37.** 23 **38.** 42 **39.** –7 **40.** 20 **41.** 9 **42.** 4 **43.** –22.5 **44.** 36 **45.** $7\frac{1}{3}$ **46.** $\frac{1}{3}$
47. 60 **48.** –6 **49.** –2 **50.** $x=5$ **51.** 27 and 28 **52.** 56 and 58 **53.** 57 and 59
54. 22, 23, and 24 **55.** 72, 74, and 76 **56.** 40, 42, 44, and 46 **57.** 35, 37, 39, and 41
58. 50, 55, 60, and 65 **59.** 12 **60.** Roger has \$42, David has \$21 **61.** 21 **62.** 4.5
63. 27 books **64.** 36 books **65.** length = 15 inches, width = 9 inches **66.** length = 32 in.
67. \$20 **68.** 5 gallons of water should be added. **69.** 120 grams of water should be added.
70. $9.5\,cm$ and $17.5\,cm$
71. the largest piece = 18 in., the median piece = 13 in., the shortest piece = 8 in.
72. Roger has \$44, Carol has \$22, Maria has \$11
73. 24 one-dollar bills and 30 five-dollar bills **74.** \$120 **75.** 45 candies

CHAPTER 6 – 9 Literal Equations and Formulas page 128

1. $y=-x+5$ **2.** $x=-y+5$ **3.** $x=y+5$ **4.** $y=x-5$ **5.** $y=-2x+10$
6. $x=-\frac{1}{2}y+5$ **7.** $y=2x+10$ **8.** $x=\frac{1}{2}y+5$ **9.** $y=\frac{1}{2}x+\frac{5}{4}$ **10.** $b=c-2a$
11. $a=\dfrac{c-b}{2}$ **12.** $a=\dfrac{c}{2}-b$ **13.** $a=-b$ **14.** $x=1-\dfrac{2}{a}$ **15.** $a=\dfrac{2}{x-1}$
16. $a-b=0$ **17. a.** $r=\dfrac{C}{2\pi}$, **b.** $r\approx 0.64$ meters **18. a.** $h=\dfrac{2A}{b}$, **b.** $h=8$ feet
19. a. $l=\dfrac{p}{2}-w$, **b.** $l=28\ cm$ **20. a.** $r=\dfrac{d}{t}$ **b.** $r=650$ miles per hour
21. $h=\dfrac{3V}{b}$ **22.** $h=\dfrac{A}{2\pi r}-r$

CHAPTER 6 Equations EXERCISES page 129~130

1. 7 **2.** −17 **3.** 17 **4.** −17 **5.** −5 **6.** −11 **7.** 21 **8.** −21 **9.** −9 **10.** −3.5
11. 0.5 **12.** 7.5 **13.** 66 **14.** 25 **15.** −324
16. $6x$ **17.** $3x$ **18.** $-3x$ **19.** $10x-2$ **20.** $2x+2$ **21.** $-2x+2$ **22.** $2x+12$
23. $-2x+2$ **24.** $-2x+12$ **25.** $8x-1$ **26.** $-2x+15$ **27.** $-2x+1$ **28.** $-2x+15$
29. $8x-1$ **30.** $8x-15$ **31.** $-2x$ **32.** $5x-5y$ **33.** $-3x+y$ **34.** $y-\frac{3}{4}$ **35.** $\frac{3}{5}x-\frac{1}{3}y$
36. $\frac{1}{5}x+y$ **37.** $\frac{1}{15}x+\frac{13}{15}y$ **38.** $5.7x+1.5y$ **39.** $8.1n-4$ **40.** $-2x^2-9$ **41.** $-2x^2+3$
42. $7x^2-6$ **43.** $3x-4y-5$ **44.** $2q$ **45.** $-2x^2+3x+5$ **46.** $8x^2$ **47.** $-8x^3$
48. $-4.5a^7$ **49.** $8x^2-32x$ **50.** $-15x+20y-25$ **51.** $-2a^2+4ab-10a$ **52.** $-8x+18$
53. $-2x^2-x+12$ **54.** $2p^2+3p-1$ **55.** $x+2$ **56.** $x-2$ **57.** $3a-2$ **58.** $2x^2+x+3$
59. a^2-3a-1 **60.** $3x^2-2y$
61. $x=39$ **62.** $y=10$ **63.** $a=90$ **64.** $p=-6\frac{2}{5}$ **65.** $x=7\frac{1}{4}$ **66.** $y=8\frac{3}{4}$ **67.** $p=-\frac{2}{5}$
68. $x=2\frac{2}{3}$ **69.** $x=21\frac{3}{5}$ **70.** $x=2$ **71.** $y=2$ **72.** $p=2\frac{1}{4}$ **73.** $x=-3\frac{2}{3}$ **74.** $y=\frac{3}{8}$
75. $x=-54$ **76.** $y=2x-6$ **77.** $x=\frac{1}{2}y+3$ **78.** $y=\frac{1}{2}x-2$ **79.** $y=-\frac{4}{5}x-5$
80. $b=-a+\frac{3}{2}c$

81. 65 **82.** −15 **83.** 19 **84.** 13.5 **85.** 20 **86.** 6 **87.** −7.2 **88.** 67 and 68 **89.** 86 and 88
90. 80, 85, and 90 **91.** 32 inches and 43 inches **92.** 20, 25, and 30 inches
93. length = 13 feet, width = 6 feet **94.** 5 gallons of water added
95. leg = 29 *cm*, base = 15 *cm* **96.** $115 **97.** 54 one-dollar bills and 26 five-dollar bills
98. $J=R-500$

CHAPTER 7 −1 Number line and Coordinate Plane page 139 ~ 140

1. −5 **2.** −2 **3.** 0 **4.** 4.5 or $4\frac{1}{2}$ **5.** −14 **6.** −3 **7.** 6 **8.** −22.5 or $-22\frac{1}{2}$ **9.** −5 **10.** 0
11. 12.5 or $12\frac{1}{2}$ **12.** 30 **13.** 2.5 or $2\frac{1}{2}$ **14.** −6.5 or $-6\frac{1}{2}$ **15.** −7
16. −6 **17.** −2.5 or $-2\frac{1}{2}$ **18.** 20 **19.** −8 **20.** 15 **21.** 2 **22.** −4 **23.** −4 **24.** 5
25. 12.5 or $12\frac{1}{2}$ **26.** B = 4, C = 9, D = 15, E = 21

27. $x=2$ **28.** $y=-1$

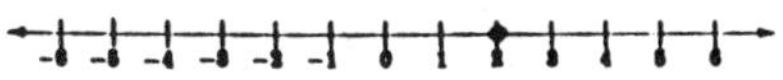

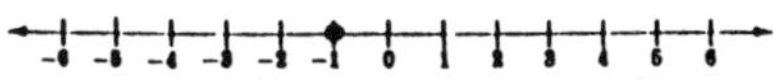

29. $a=-2.5$ 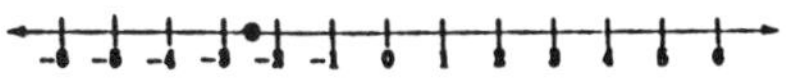 **30.** $x=9$

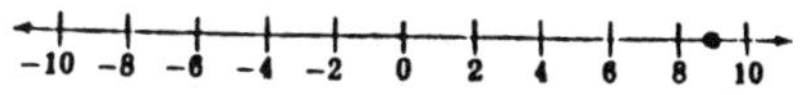

−10 −8 −6 −4 −2 0 2 4 6 8 10

31. $A(2,4)$ **32.** $B(-2,3)$
33. $C(-4,-4)$ **34.** $D(4,-1)$
35. $E(0,2)$ **36.** $F(3,-5)$
37. $G(-3,0)$ **38.** $K(1\frac{1}{2},0)$
39. $P(0,-2\frac{1}{2})$ **40.** $Q(-5,2\frac{1}{2})$

41 ~ 45

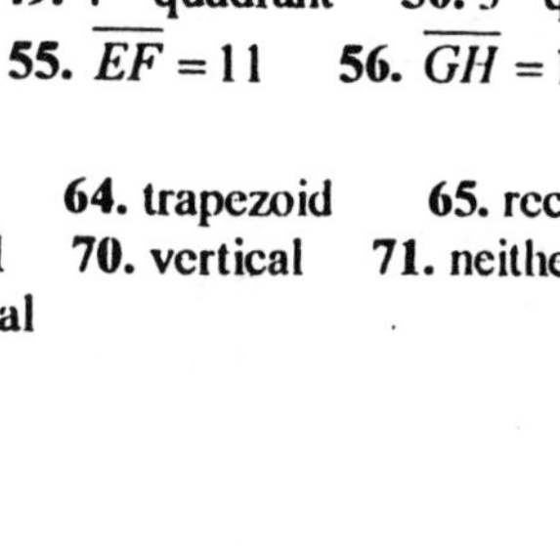

P(−3, 4) Q(3, 2½) A(−4, 0) B(2, −2) M(−1, −5) x y

46. 2nd quadrant **47.** 3rd quadrant **48.** 1st quadrant **49.** 4th quadrant **50.** 3rd quadrant
51. $\overline{AB}=5$ **52.** $\overline{MN}=4$ **53.** $\overline{PT}=6$ **54.** $\overline{CD}=5$ **55.** $\overline{EF}=11$ **56.** $\overline{GH}=11$
57. $\overline{JK}=4$ **58.** $\overline{ST}=8$ **59.** $\overline{UV}=9$ **60.** $\overline{WZ}=15$
61. rectangle **62.** right triangle **63.** parallelogram **64.** trapezoid **65.** rectangle
66. vertical **67.** vertical **68.** horizontal **69.** vertical **70.** vertical **71.** neither
72. horizontal **73.** vertical **74.** neither **75.** horizontal
76. Translates (moves) the figure 2 units to the right.
77. Translates (moves) the figure 2 units to the left.
78. Translates (moves) the figure 2 units upward.
79. Translates (moves) the figure 2 units downward.
80. Reflects the figure over the y-axis.
81. Reflects the figure over the x-axis.
82. Yes **83.** Yes **84.** No
85. Yes **86.** 4 by 5 **87.** 6.75

CHAPTER 7 –2 Linear Equations page 142

1. Yes **2.** No **3.** Yes **4.** Yes **5.** No **6.** Yes **7.** Yes **8.** No **9.** Yes **10.** No
11. Yes **12.** No **13.** Yes **14.** Yes **15.** No **16.** Yes **17.** No **18.** No **19.** No **20.** Yes
21. N0 **22.** Yes **23.** Yes **24.** Yes **25.** Yes **26.** Yes **27.** No **28.** Yes **29.** Yes **30.** Yes
31. $3x-y=-5$ **32.** $4x+y=7$ **33.** $3x+4y=6$ **34.** $5x-4y=-9$ **35.** $2x+4y=8$
36. $5x+3y=20$ **37.** $4x+7y=1$ **38.** $2x-4y=-7$ **39.** $2x-5y=4$ **40.** $6x-4y=5$
41. (–2, –4), (0, 0), (2, 4) **42.** (–2, –3), (0, 1), (2, 5) **43.** (–2, –5), (0, –1), (2, 3)
44. (–2, 5), (0, 1), (2, –3) **45.** (–2, 3), (0, –1), (2, –5) **46.** (–2, 7), (0, 5), (2, 3)
47. (–2, –3), (0, –5), (2, –7) **48.** (–2, 6), (0, 4), (2, 2) **49.** (–2, 8), (0, 4), (2, 0)
50. (–2, –8), (0, –4), (2, 0)
51. $c=1.50n$ **52.** $c=p-0.30p$, or $c=0.70p$ **53.** $c=p+0.075p$, or $c=1.075p$
54. for one day: $c=20+0.20m$ **55.** $c=11.50n+5$
for n days: $c=20n+0.20m$ $c=\$396$ for 34 copies

CHAPTER 7 – 3 Equations and Functions Page 144

1. $y=-x+5$ **2.** $y=x-5$ **3.** $y=x-5$ **4.** $y=-x-5$ **5.** $y=-2x+6$
6. $y=2x+6$ **7.** $y=x+\frac{7}{2}$ **8.** $y=2x-5$ **9.** $y=2x-\frac{8}{3}$ **10.** $y=\frac{4}{5}x-4$
11. $y=\frac{1}{5}x-2$ **12.** $y=3x-1$ **13.** $y=2x+24$ **14.** $y=3x-15$ **15.** $y=\frac{1}{3}x-4$
16. (–2, –4), (0, 0), (2, 4) **17.** (–1, 1), (0, 3), (2, 7) **18.** (–2, –7), (1, –1), (3, 3)
19. (–2, 0), (0, 1), $(1, 1\frac{1}{2})$ **20.** (–3, –3), (0, –1), $(1, -\frac{1}{3})$
21. $f(x)=2x-4$ **22.** $f(x)=-2x+5$ **23.** $f(x)=-8x+7$ **24.** $f(x)=8x-5$
25. $f(x)=-x+4$ **26.** $f(x)=x-4$ **27.** $f(x)=-2x+4$ **28.** $f(x)=-2x-4$
29. $f(x)=-x+\frac{7}{2}$ **30.** $f(x)=2x-\frac{3}{2}$
31. a) $f(-1)=-9$ **b)** $f(0)=-6$ **c)** $f(1)=-3$ **d)** $f(2)=0$
32. a) $f(-2)=12$ **b)** $f(\frac{1}{2})=-\frac{1}{2}$ **c)** $f(3)=-13$ **d)** $f(5)=-23$
33. a) $p(-\frac{1}{2})=1\frac{1}{4}$ **b)** $p(0)=0$ **c)** $p(3)=3$ **d)** $p(6)=24$
34. a) $q(-2)=4$ **b)** $q(0)=-2$ **c)** $q(1)=1$ **d)** $q(\frac{1}{2})=-1$
35. a) $f(-3)=-17$ **b)** $f(-\frac{1}{2})=-7$ **c)** $f(1)=-1$ **d)** $f(1\frac{1}{4})=0$
36. $x=4$ **37.** $x=\frac{3}{4}$ **38.** $f(1)+g(2)=5$ **39.** $f(2)\cdot g(3)=-5$
40. $f[g(2)]=1$, $g[f(2)]=-5$. (Hint: we call it "function of function") **41.** $n=3$

Amazing arrangement of numbers with the same sum in all directions: (vertical, horizontal, diagonal)

1) 1 to 9 in " 3 × 3 matrix "

4	3	8
9	5	1
2	7	6

2) 1 to 25 in " 5 × 5 matrix "

11	4	25	2	23
10	12	19	18	6
21	17	13	9	5
20	8	7	14	16
3	24	1	22	15

2) You may try to arrange numbers from 1 to 49 in " 7 × 7 matrix ", from 1 to 81 in " 9 × 9 matrix ", ……… and so on.
3) Put the median in the center of the matrix. Put consecutive integers on the main diagonal.
4) The sum of the numbers on the diagonal is the sum of all directions.
5) Try all other directions to get the same sum.

CHAPTER 7 – 4 Relations and Functions page 147 ~ 148

1. R = {2, 3, 4, 5, 6} **2.** R = {−2, −1, 0, 1, 2} **3.** R = {1, 3, 5, 7, 9}
4. R = {−1, 1, 3, 5, 7} **5.** R = {3, 4, 5, 6, 7} **6.** R = {−7, −6, −5, −4, −3}
7. R = {−9, −5, −3, 1, 5} **8.** R = {−13, −7, 8, 14, 17} **9.** R = {2, 3, 6, 11}
10. R = { −1, 0, 1, 2}
11. D = {0, 1, 2, 3}, R = {1, 3, 4, 6}, a function
12. D = {−2, −1, 2, 3}, R = {1, 3, 5}, a function
13. D = {−3, −2, 3}, R = {1, 3, 4}, not a function
14. D = {−4, −1, 0, 4}, R = {1, 2, 3}, a function
15. D = {1, 3}, R = {3, 4, 5, 7}, not a function
16. D = {3, 4, 5}, R = {1, 3, 4, 5}, not a function
17. D = {−7, −6, 4, 5}, R = {2, 3, 7, 8}, a function
18. D = {2, 3}, R = {−2, −1, 1, 2}, not a function
19. D = {0, 1, 2, 3}, R = {0, 1, 3, 4}, not a function
20. D = {0, 1, 2, 3, 4}, R = {1, 3, 5, 6}, a function
21. D = all real numbers, R = all real numbers **22.** D = $x \geq 2$, R = all real numbers
23. Yes, the relation is a function because each value of x is assigned to exactly one value of y.
24. No, the relation is not a function because $x = -5$ is assigned to two values of y, 2 and −1.
25. Yes, y is a function of x because each value of x can find exactly one value of y.
26. No, y is not a function of x because each values of x can find two values of y.
Hint: (3, 2), (3, −2), (9, 4), (9, −4), ...
27. No, y is not a function of x because each value of x can find two values of y.
Hint: (1, 2), (1, −2), (2, 3), (2, −3), ...
28. Yes, y is a function of x because each value of x can find exactly one value of y.
29. $d(h) = 60\,h$ **30.** $c(n) = 19.50\,n + 2.50$
31. Yes, it is a function. **32.** No, it is not a function. **33.** Yes, it is a function.
34. Yes, it is a function. **35.** No, it is not a function. **36.** Yes, it is a function.
37. Yes, it is a function. **38.** No, it is not a function. **39.** Yes, it is a function.
40. Yes, it is a function. **41.** Yes, it is a function. **42.** No, it is not a function.
43. No, it is not a function. **44.** Yes, it is a function. **45.** No, it is not a function.

A kid said to his mother.
Kid: Did you wear contact lenses this morning ?
Mother: No. Why do you ask ?
Kid: I heard you yell to father this morning:
"After 20 years of marriage, I have finally seen what kind of man you are !".

CHAPTER 7 – 5 Graphing Linear Equations page 151 ~ 152

1~5 graphs are left to the students (see examples 1, 2, 3 on page 153 and 154).

1. (0, 0), (1, 2), (2, 4), (3, 6), (4, 8) **2.** (–1, 2), (0, 0), (1, –2), (2, –4), (3, –6)

3. (–2, –5), (–1, –3), (0, –1), (1, 1), (2, 3) **4.** (–4, –10), (–2, –4), (0, 2), (2, 8), (3, 11)

5. (–2, –2), $(-1, -1\frac{1}{2})$, (0, –1), (2, 0), $(5, 1\frac{1}{2})$

6~35 graphs are left to the students (see example 4 on page 154).

6. x - intercept = –1, y - intercept = 1
7. x - intercept = 2, y - intercept = –2
8. x - intercept = 2, y - intercept = –4
9. x - intercept = 4, y - intercept = 4
10. x - intercept = 2, y - intercept = 4
11. x - intercept = 2, y - intercept = –6
12. x - intercept = 2, y - intercept = 6
13. x - intercept = $\frac{1}{2}$, y - intercept = 2
14. x - intercept = $-\frac{1}{2}$, y - intercept = $\frac{1}{2}$
15. x - intercept = $\frac{1}{2}$, x - intercept = $-\frac{1}{2}$
16. x - intercept = –8, y - intercept = 4
17. x - intercept = –9, y - intercept = –3
18. x - intercept = 4, y - intercept = –2
19. x - intercept = 4, y - intercept = 2
20. x - intercept = 5, y - intercept = –5
21. x - intercept = 5, y - intercept = 5
22. x - intercept = 2, y - intercept = 4
23. x - intercept = 2, y - intercept = –4
24. x - intercept = 9, y - intercept = 3
25. x - intercept = 9, y - intercept = –3
26. x - intercept = $\frac{1}{2}$, y - intercept = –2
27. x - intercept = $-\frac{1}{2}$, y - intercept = 2
28. x - intercept = $-\frac{1}{2}$, y - intercept = 1
29. x - intercept = –5, y - intercept = 2
30. x - intercept = 0, y - intercept = 0
31. x - intercept = 0, y - intercept = 0
32. x - intercept = 6, y - intercept = –9
33. x - intercept = 9, y - intercept = –6
34. x - intercept = $1\frac{1}{2}$, y - intercept = 2
35. x - intercept = 6, y - intercept = $4\frac{1}{2}$

36~45 graphs are left to the students (see example 5).

46. The graph is the $y-axis$. **47.** The graph is the $x-axis$. **48.** $k = 3$ **49.** $k = -8$

50. $k = -3$ **51.** $k = 2\frac{1}{2}$ **52.** $y = x + 5$ **53.** $y = \frac{1}{2}x$ **54.** $y = 2x - 1$

55. $y = 2x + 4$

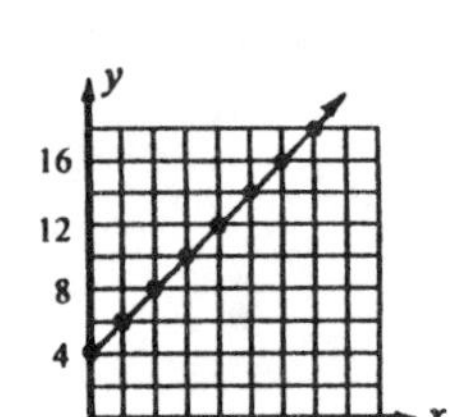

56. $d = 60\,t$

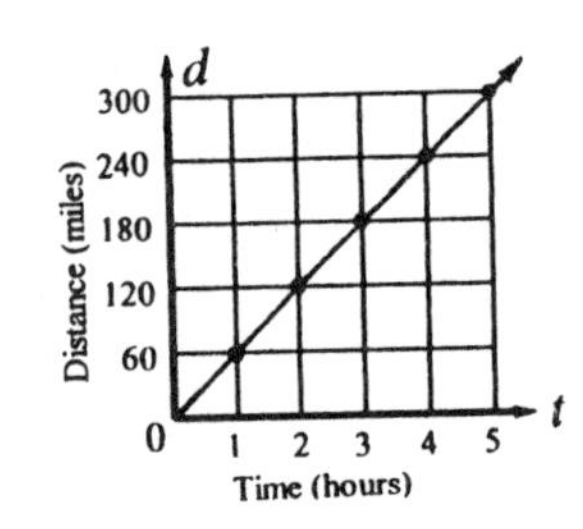

57. $d = 50\,t$

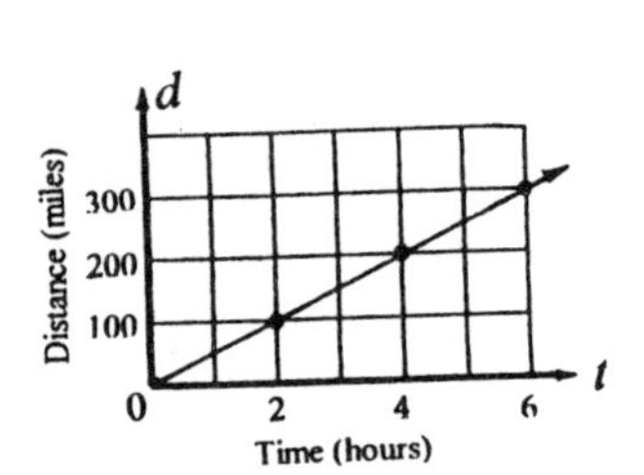

> A student said to a priest.
> Student: I don't believe that heaven is as good as you said. You have never been there.
> Priest: Have you ever heard of anyone who came back here because he didn't like it there.

CHAPTER 7 – 6 Slope of a Line page 155 ~ 156

1. slope $= \frac{4}{4} = 1$ **2.** slope $= \frac{-2}{3} = -\frac{2}{3}$ **3.** slope $= \frac{1}{2}$ **4.** slope $= -\frac{5}{8}$ **5.** Slope is undefined.
6. slope $= 0$

7. 3 **8.** $\frac{1}{3}$ **9.** –2 **10.** $\frac{1}{3}$ **11.** $-\frac{1}{3}$ **12.** $\frac{5}{2}$ **13.** $\frac{2}{5}$ **14.** undefined **15.** 0 **16.** $-\frac{1}{6}$

17. $\frac{3}{2}$ **18.** –2 **19.** 0.5 **20.** 3

21. –1 **22.** $\frac{1}{3}$ **23.** –3 **24.** 0 **25.** 1 **26.** 1 **27.** undefined **28.** –2 **29.** –1 **30.** –2

31. 2 **32.** –1 **33.** $\frac{6}{5}$ **34.** $-\frac{7}{9}$ **35.** $-\frac{14}{15}$ **36.** 1 **37.** 1 **38.** 2 **39.** $\frac{1}{5}$ **40.** 1

41.

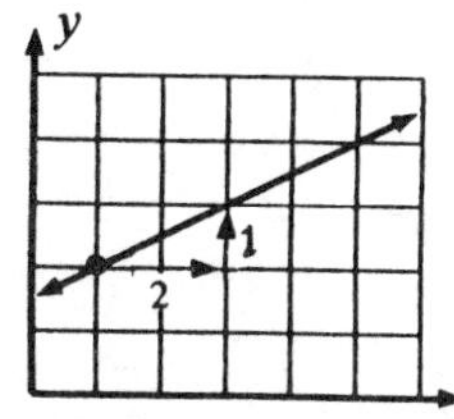

42.

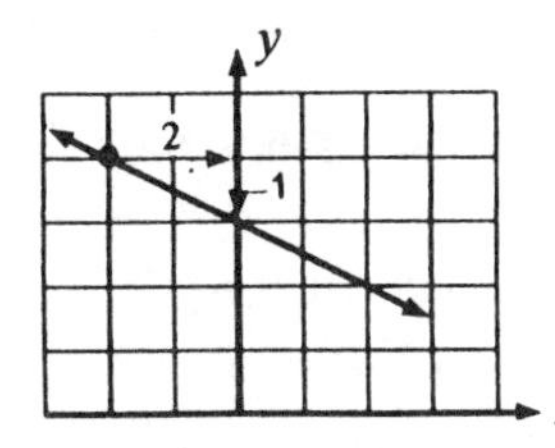

43 ~ 50 graphs are left to the students.

51. 2 ; –5 **52.** –2 ; 5 **53.** –7 ; –8 **54.** 7 ; –8 **55.** $\frac{2}{3}$; 4 **56.** $-\frac{2}{3}$; 4 **57.** $-\frac{1}{2}$; $-\frac{3}{4}$

58. $\frac{1}{2}$; $\frac{3}{4}$ **59.** –4 ; 0 **60.** –1 ; 5 **61.** 1 ; –10 **62.** 7 ; 0 **63.** 1 ; 0 **64.** –1 ; 0 **65.** 0 ; 5

66. 0 ; –2 **67.** –8 ; 7 **68.** 2 ; 1 **69.** $-\frac{1}{3}$; –2 **70.** $-\frac{4}{5}$; 8 **71.** 2 ; 15 **72.** –2 ; 15

73. –2 ; 6 **74.** 2 ; –6 **75.** 1 ; $\frac{3}{2}$ **76.** –2 ; $\frac{1}{2}$ **77.** $-\frac{1}{2}$; $\frac{5}{4}$ **78.** $\frac{1}{2}$; $\frac{3}{4}$ **79.** $\frac{3}{2}$; –3

80. $-\frac{3}{2}$; $\frac{7}{2}$ **81.**

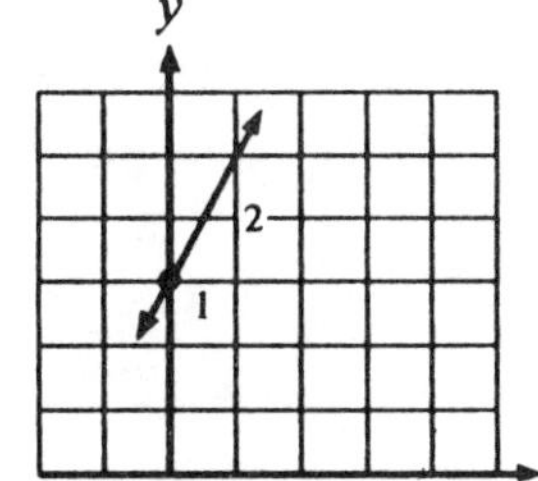

82.

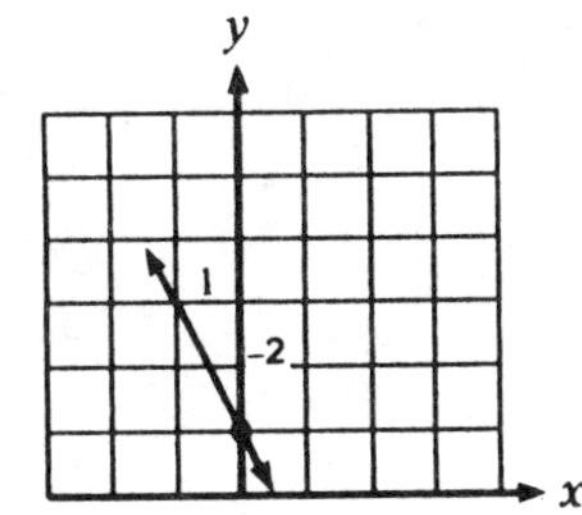

83 ~ 90 graphs are left to the students.

91. undefined (The slope of a vertical line does not exit.) **92.** 0
93. undefined (The value does not exit.) **94.** $a = -8$ **95.** $a = -3$
96. Yes. The slope between (1, 3) and (2, 5) equals the slope between (2, 5) and (4, 9).
97. No. **98.** $y = 3$ **99.** $x = -10$ **100.** $m = -\frac{a}{b}$; y – intercept $= \frac{c}{b}$ **101.** 6.5
102. $-3°F$ per hour, $66°F$ at 12 P.M.

CHAPTER 7 – 7 Finding the Equation of a Line page 159 ~ 160

1. $3x + y = 5$ **2.** $3x - y = 5$ **3.** $3x + y = -5$ **4.** $6x - y = 9$ **5.** $x - 2y = -14$
6. $x + 2y = 14$ **7.** $2x + 3y = 24$ **8.** $3x - 4y = -32$ **9.** $2x - 3y = 12$ **10.** $3x + 4y = -16$
11. $y = 2x + 1$ **12.** $y = 2x - 8$ **13.** $y = -2x + 10$ **14.** $y = -2x - 2$ **15.** $y = -2x + 2$
16. $y = 2x + 2$ **17.** $y = -2x - 10$ **18.** $y = 7x - 1$ **19.** $y = 6x + 38$ **20.** $y = -x + 13$
21. $y = x - 6$ **22.** $y = 5$ **23.** $y = 0$ **24.** $x = 5$ **25.** $x = 0$ **26.** $y = \frac{1}{2}x - 4$
27. $y = -\frac{3}{5}x + 9$ **28.** $y = -\frac{2}{3}x + \frac{1}{3}$ **29.** $y = \frac{4}{5}x + 9$ **30.** $y = \frac{5}{4}x - \frac{45}{4}$ **31.** $x - y = -1$
32. $x - y = 1$ **33.** $x - 2y = -8$ **34.** $3x - 2y = 12$ **35.** $x - 2y = -10$ **36.** $2x + y = 0$
37. $3x - y = 15$ **38.** $11x - 3y = -21$ **39.** $13x + y = -45$ **40.** $2x - 13y = 14$
41. $3x - y = 4.5$ **42.** $2x - y = 1.5$ **43.** $y = \frac{1}{6}x + \frac{1}{3}$ **44.** $5x - 3y = 1$ **45.** $2x - 24y = -9$
46. $y - 4 = 3(x - 2)$ **47.** $y - 4 = -3(x + 2)$ **48.** $y + 2 = 4(x + 1)$ **49.** $y + 5 = -5(x - 7)$
50. $y - 5 = \frac{3}{4}(x + 1)$ **51.** $y + 7 = -\frac{1}{5}(x - 4)$ **52.** $y - 5 = -2(x - 4)$ or $y + 3 = -2(x - 8)$
53. $y + 2 = -\frac{4}{5}(x - 5)$ or $y - 2 = -\frac{4}{5}x$ **54.** $y - 5 = -x$ or $y = -(x - 5)$
55. $y = \frac{1}{3}x$ or $y + 1 = \frac{1}{3}(x + 3)$
56. $10x - y = 34$ **57.** $10x - y = 46$ **58.** $5x - y = 20$ **59.** $5x + y = 20$ **60.** $2x - 3y = -12$
61. $2x - 3y = 12$ **62.** $y = -2$ **63.** $x = 3$ **64.** $x = -2$ **65.** $y = 3$ **66.** $a = -8$
67. $a = 1$ **68.** $a = 11$ **69.** $a = 6.5$ **70.** $a = -2$

CHAPTER 7 – 8 Parallel and Perpendicular Lines page 162

1. $y=-x+5$ **2.** $y=x-5$ **3.** $y=-x-5$ **4.** $y=x+5$ **5.** $y=-2x+6$ **6.** $y=2x+6$
7. $y=x+\frac{7}{2}$ **8.** $y=-\frac{3}{2}x-\frac{5}{2}$ **9.** $-\frac{1}{2}$ **10.** $-\frac{5}{4}$ **11.** 5 **12.** $\frac{1}{3}$ **13.** 1 **14.** -1
15. $-\frac{1}{2}$ **16.** $\frac{1}{2}$ **17.** undefined **18.** 0 **19.** $-\frac{1}{6}$ **20.** -3
21. parallel **22.** neither **23.** perpendicular **24.** parallel **25.** neither **26.** perpendicular
27. perpendicular **28.** parallel **29.** parallel **30.** perpendicular **31.** parallel **32.** neither
33. perpendicular **34.** perpendicular **35.** neither **36.** $\frac{5}{4}$ **37.** undefined **38.** undefined
39. 0 **40.** 0 **41.** $y=-\frac{1}{7}x+15$ **42.** $2x+5y=10$ **43.** $5x-2y=54$

CHAPTER 7 – 9 Direct Variations page 164

1. $k=5$; $y=5x$ **2.** $k=2$; $y=2x$ **3.** $k=\frac{5}{2}$; $y=\frac{5}{2}x$ **4.** $k=0.5$; $y=0.5x$
5. $k=10$; $n=10m$ **6.** $k=\frac{6}{5}$; $p=\frac{6}{5}n$ **7.** $k=\frac{1}{2}$; $s=\frac{1}{2}t$ **8.** $k=\frac{1}{3}$; $m=\frac{1}{3}n$
9. $k=3$; $y=3(x-4)$ **10.** $k=15$; $y=15(x+8)$ **11.** $y=60$ **12.** $x=\frac{12}{5}$ or $2\frac{2}{5}$
13. $y=\frac{12}{5}$ or $2\frac{2}{5}$ **14.** $n=500$ **15.** $m=\frac{4}{5}$ **16.** $y_1=4$ **17.** $y_1=60$ **18.** $y_1=7.2$ **19.** $y_2=6$
20. $x_1=2.5$ **21.** $x_2=2.5$ **22.** $y_1=\frac{1}{5}$ **23.** $y_2=\frac{4}{3}$ or $1\frac{1}{3}$ **24.** $y_1=\frac{1}{4}$ **25.** $y_2=125$
26. No. If $xy=5$, then $y=\frac{5}{x}$. y is decreased when x is increases. It is an inverse variation.
27. Yes. The ratios are equal. $\frac{4}{6}=\frac{2}{3}$, $\frac{6}{9}=\frac{2}{3}$, $\frac{18}{27}=\frac{2}{3}$ **28.** $D=10L$ **29.** 14.5 cm **30.** $441

CHAPTER 7 Functions and Graphs EXERCISES page 165 ~ 166

1. 10 **2.** 8 **3.** 6 **4.** 3 **5.** 1 **6.** 8 **7.** yes **8.** yes **9.** no **10.** no **11.** yes **12.** yes
13. yes **14.** no **15.** Yes **16.** $2x-y=-5$ **17.** $2x+y=5$ **18.** $2x+y=5$
19. $2x-y=5$ **20.** $4x-5y=7$ **21.** $3x-8y=9$ **22.** $2x-6y=7$ **23.** $2x+6y=3$
24. $y=-2x+4$ **25.** $y=2x+4$ **26.** $y=2x-4$ **27.** $y=2x+4$ **28.** $y=-2x+\frac{5}{2}$ **29.** $y=2x-\frac{5}{2}$
30. $y=\frac{5}{3}x-2$ **31.** $y=-\frac{3}{5}x+\frac{6}{5}$ **32.** $f(1)=-5$ **33.** $f(0)=-7$ **34.** $f(-1)=-9$ **35.** $f(2)=-3$
36. $f(1)=5$ **37.** $f(0)=7$ **38.** $f(2)=3$ **39.** $f(-\frac{5}{4})=9\frac{1}{2}$ **40.** $p(2)=3$ **41.** $p(0)=1$
42. $p(5)=21$ **43.** $p(\frac{1}{2})=\frac{3}{4}$ **44.** $f(2)=6$ **45.** $f(-2)=-14$ **46.** $f(\frac{1}{5})=-3$ **47.** $f(-1\frac{1}{5})=-10$
48. R = {–7, –4, –1, 2, 5} **49.** R = {–10, –6, –2, 2, 6} **50.** R = { –2, –1, 7} **51.** R = {–2, 0, 6}
52. R = {–4, –3, –1, 1} **53.** R = {–1, 1, 3, 4} **54.** D = {0, 1, 2, 3}, R = {–2, –1, 1, 4}, a function
55. D = {–2, 0, 2}, R = {1, 2, 3}, not a function
56. D = {–2, –1, 1}, R = {–2, 0, 1, 2}, not a function
57. D = {–3, –1, 2, 3}, R = {–3, –2, 1, 3}, a function
58. D = {0, 1, 2, 3, 4}, R = {–2, –1, 0, 1, 2}, a function
59. D = {1, 2, 8}, R = {2, 3, 9}, a function **60.** D = {–4, 1}, R = {–2, 2, 7}, not a function
61 ~ 75 graphs are left to the students (see Section 7 – 5, example 4.)
61. x – intercept 3, y – intercept –3
62. x – intercept 3, y – intercept 3
63. x – intercept –3, y – intercept –3
64. x – intercept –1, y – intercept 2
65. x – intercept 6, y – intercept 3
66. x – intercept 6, y – intercept –3
67. x – intercept 6, y – intercept –4
68. x – intercept 6, y – intercept 4
69. x – intercept "none", y – intercept –2
70. x – intercept 1, y – intercept "none"
71. x – intercept 0, y – intercept 0
Need a third point to graph it, say (1, 2).
72. x – intercept 0, y – intercept 0
Need a third point to graph it, say (1, –2).
73. x – intercept 6, y – intercept –4
74. x – intercept 8, y – intercept 12
75. x – intercept 8, y – intercept –6 **76.** $\frac{5}{2}$ **77.** $-\frac{2}{5}$ **78.** 0 **79.** undefined **80.** 12 **81.** –4
82. $-\frac{6}{5}$ **83.** 5 **84.** –1 **85.** $-\frac{5}{3}$ **86.** $\frac{5}{3}$ **87.** undefined **88.** 0 **89.** undefined **90.** $\frac{2}{3}$ **91.** –4
92. –4 **93.** 1.8 **94.** 1.6 **95.** –5 **96.** –4 ; 9 **97.** 4 ; 9 **98.** $\frac{4}{5}$; –15 **99.** $-\frac{4}{5}$; 15 **100.** –8 ; 10
101. 8 ; –10 **102.** 8 ; 0 **103.** 0 ; –8 **104.** $-\frac{1}{2}$; $\frac{4}{3}$ **105.** $\frac{1}{2}$; $\frac{4}{3}$ **106.** 8 ; 14 **107.** 2 ; $-\frac{7}{5}$
108. $5x-y=-7$ **109.** $4x+y=-2$ **110.** $3x-2y=-24$ **111.** $12x+15y=-10$ **112.** $7x-y=11$
113. $x+2y=-6$ **114.** $3x-4y=-2$ **115.** $y=9$ **116.** $x=5$ **117.** $2x-y=7$ **118.** $2x-y=-5$
119. $27x+y=22$ **120.** $3x+2y=7$ **121.** perpendicular **122.** neither
123. parallel **124.** perpendicular **125.** $y=8$ **126.** $n=55$ **127.** $y_2=\frac{4}{15}$

CHAPTER 8 – 1 Laws of Exponents page 173 ~ 174

1. 8 **2.** 9 **3.** −8 **4.** −8 **5.** −16 **6.** 16 **7.** 5^{4n} **8.** 243 **9.** 25 **10.** −25 **11.** −27
12. 81 **13.** −81 **14.** 81 **15.** 1 **16.** 1 **17.** −1 **18.** 1 **19.** undefined **20.** 729
21. 2^7 **22.** 2^{12} **23.** a^7 **24.** a^{12} **25.** m^{15} **26.** m^{50} **27.** a^{n+2} **28.** a^{2n} **29.** 10^8 **30.** 10^{15}
31. x^{13} **32.** $-a^6$ **33.** a^6 **34.** a^5 **35.** a^{-5}, or $\frac{1}{a^5}$ **36.** 3^0, or 1 **37.** a^0, or 1 **38.** x^{a+b} **39.** a^{x+y}
40. a^{xy} **41.** 32. **42.** $\frac{1}{32}$ **43.** a^5 **44.** a^{-5}, or $\frac{1}{a^5}$ **45.** 100,000 **46.** 8 **47.** 125 **48.** $\frac{1}{8}$
49. $\frac{1}{125}$ **50.** x^{-4}, or $\frac{1}{x^4}$ **51.** x^4 **52.** x^{-6}, or $\frac{1}{x^6}$ **53.** 1 **54.** 1 **55.** x^{12-n} **56.** x^{n-12} **57.** $\frac{a^4}{b^4}$
58. x^{2-n} **59.** $\frac{a^8}{b^{12}}$ **60.** $\frac{x^{5n}}{y^{5n}}$ **61.** $7x$ **62.** $-x$ **63.** $12x^2$ **64.** $6x^7$ **65.** $18a^5$ **66.** x^6y^8
67. $18a^8b^7$ **68.** $-4x^2y^5$ **69.** $2x^6y^6$ **70.** $10p^4q^9$ **71.** $-x^3y^5$ **72.** $2a^7$ **73.** $-24m^2n^9$
74. $-7m^7p$ **75.** $8x^{12}$ **76.** $-\frac{1}{32}a^{20}$ **77.** $16a^8$ **78.** $375x^9$ **79.** $-27x^{12}y^3$ **80.** p^9q^{13}
81. $48xy^{14}$ **82.** $972a^{24}$ **83.** $16x^{4n}$ **84.** $72a^{5n}$ **85.** $\frac{9}{4}x^{16}$
86. $3\frac{1}{3}$ **87.** $\frac{3}{10}$ **88.** 200 **89.** $\frac{9}{25}$ **90.** $\frac{25}{9}$ **91.** $2x$ **92.** 2 **93.** $2x$ **94.** $\frac{1}{2}$ **95.** $\frac{1}{2x}$
96. $\frac{a^3}{5}$ **97.** $4m$ **98.** $\frac{xy^2}{2}$ **99.** $-\frac{2ab^3}{5}$ **100.** $\frac{p^3}{q^3}$ **101.** y^3 **102.** $\frac{a^3}{7b^3}$ **103.** $\frac{n^2}{m^4}$
104. $-\frac{12}{k^5}$ **105.** $-3x^2y^2$ **106.** $4a^5$ **107.** $9x^2$ **108.** $\frac{1}{32x^3}$ **109.** $\frac{1}{2m^2}$ **110.** $\frac{1}{3}$
111. $-\frac{m^4}{5}$ **112.** $-2a^4$ **113.** $-\frac{1}{x^2}$ **114.** $\frac{3}{4r^2}$ **115.** $-8x$ **116.** $2y^2$ **117.** $4a$
118. 4 **119.** $-a^3$ **120.** $\frac{1}{m^2}$ **121.** $\frac{16}{81}$ **122.** $\frac{81}{16}$ **123.** $\frac{4a^4}{b^6}$ **124.** $\frac{1}{x^6y^3}$ **125.** x^6y^3
126. $\frac{27b^3}{8a^3}$ **127.** $\frac{1}{a^3}$ **128.** a^8 **129.** $\frac{x^2}{2}$ **130.** $\frac{x^7}{y^5}$ **131.** $3x^2$ **132.** $10a^2$ **133.** $7x^2$
134. surface area $=18a^2$, volume $= 4a^3$ **135.** surface area $= 6a^2$, volume $= a^3$

CHAPTER 8 – 2 Multiplying Polynomials page 176

1. $10x$ **2.** $2x$ **3.** $-2x$ **4.** $-4-6x$ **5.** $-4+6x$ **6.** $4-6x$ **7.** $-x+4$ **8.** $4x-8$
9. $4x+8$ **10.** $-2a^2+6a-13$ **11.** x^2 **12.** $6x^2$ **13.** $6x^5$ **14.** $-24a^4$ **15.** $-10a^6$
16. $5x^4$ **17.** $-12x^9$ **18.** $2x^2y^2$ **19.** $18x^3y^3$ **20.** $-20a^5b^2$ **21.** $5x-20$ **22.** $5x+20$
23. a^2+9a **24.** a^2-9a **25.** $-a^2+9a$ **26.** $6x-12x^2$ **27.** $5x-6x^2$ **28.** $-2m^2+8m^3$
29. $2x^2y-3xy^2$ **30.** $6a^2b-8ab^2$ **31.** $-2a^3b^2-6a^3b^3$ **32.** $3x^3y^4+6x^4y^3$ **33.** x^3-3x^2+5x
34. $2x^4-8x^3-18x^2$ **35.** $2a^3-4a^2+12a$ **36.** $3a^5-6a^4+3a^3$ **37.** $2x^4+x^3$ **38.** x^3+2x^2-3x
39. $4a^3+2a^2-2a$ **40.** $16a^3b^3-12a^2b^2$ **41.** $2a^2-13a$ **42.** $-8a^2+17a$ **43.** $8x^2-18x$
44. $-20x^2+37x$ **45.** $2x^2+x$ **46.** x^2-23x **47.** $9p^3+3p^2$ **48.** $8a^3+9a^2-4a$ **49.** $-2n^2+4n$
50. $-8n^2+26n$ **51.** x^2+5x+6 **52.** x^2+x-6 **53.** x^2-5x+6 **54.** $2x^2+13x+20$
55. $2x^2-3x-20$ **56.** $2x^2+3x-20$ **57.** $9a^2-6a-8$ **58.** $9a^2+6a-8$ **59.** $6n^3-8n^2-4n+6$
60. $6a^3-10a^2+11a+5$ **61.** $x=7$ **62.** $x=-14$ **63.** $x=26$ **64.** $x=3$
65. base = 6 *in.*; each leg = 14 *in.* **66.** length = 6 *cm*; width = 4 *cm* **67.** 14 *cm* by 7 *cm*

CHAPTER 8-3 Multiplying Binomials page 178

1. x^2-2x-8 **2.** $x^2+2x-48$ **3.** $x^2-12x+35$ **4.** $x^2+13x+36$ **5.** $6x^2+7x-20$
6. $4a^2+35a+24$ **7.** $15a^2-53a+42$ **8.** $4a^2+20a+25$ **9.** $4a^2-20a+25$ **10.** $4a^2-25$
11. x^2+2x-3 **12.** x^2-2x-3 **13.** $a^2+12a+35$ **14.** $a^2+2a-35$ **15.** $2a^2-19a+35$
16. $2a^2-9a-35$ **17.** $4a^2+24a+35$ **18.** x^2-4 **19.** x^2+4x+4 **20.** x^2-4x+4
21. $6x^2-xy-y^2$ **22.** $6x^2-5xy+y^2$ **23.** $a^2+ab-2b^2$ **24.** a^2-4b^2 **25.** p^2-q^2
26. $p^2-pq-2q^2$ **27.** $4a^2+20a+25$ **28.** $4a^2-20a+25$ **29.** $4a^2-25$
30. $25-20a+4a^2$
31. $x^2+10x+25$ **32.** $x^2-10x+25$ **33.** x^2-25 **34.** a^2-36 **35.** $a^2-12a+36$
36. $a^2+12a+36$ **37.** $16-8x+x^2$ **38.** $16-x^2$ **39.** $9-x^2$ **40.** $4x^2-12x+9$
41. $16-4x^2$ **42.** $9-6x+x^2$ **43.** $x^2+14x+49$ **44.** $x^2-14x+49$ **45.** x^2-49
46. $4x^2+28x+49$ **47.** $4x^2-28x+49$ **48.** $4x^2-49$ **49.** $9x^2-6xy+y^2$
50. $9x^2+6xy+y^2$ **51.** $a^2-16a+64$ **52.** $n^2+18n+81$ **53.** $25y^2-60y+36$
54. $x^2+8xy+16y^2$ **55.** $4x^2-20xy+25y^2$ **56.** $16a^2+8a+1$ **57.** $16a^2-1$
58. $25x^2-20xy+4y^2$ **59.** $a^2+6ab+9b^2$ **60.** $4c^2-12cd+9d^2$
61. $4x^2-36y^2$ **62.** $16x^2+64xy+64y^2$ **63.** $4a^2-20ab+25b^2$ **64.** $p^2-6pq+9q^2$
65. $20a^2-41ab+20b^2$ **66.** $15x^2-2x-24$ **67.** $k^2-15k-100$ **68.** $6a^2-21a-12$
69. c^2-4d^2 **70.** $4c^2+16cd+16d^2$ **71.** $4x^2-32x+64$ **72.** $4x^2-32xy+64y^2$
73. $14s^2+32s-30$ **74.** a^4+6a^2+9 **75.** $x^{2n}-2x^ny^n+y^{2n}$ **76.** a^2-b^4
77. a^6-b^6 **78.** x^4-y^2 **79.** $a^2b^2-9c^2$ **80.** $a^4-2a^2b+b^2$

CHAPTER 8 – 4 Factoring Polynomials page 180

1. 4 **2.** 4 **3.** 1 **4.** 4 **5.** 5 **6.** 3 **7.** 2 **8.** $9x$ **9.** $6x$ **10.** xy **11.** ab^2 **12.** $3a$ **13.** $4xy^2$
14. cd **15.** $4y^2$ **16.** $2(x+6)$ **17.** $4(x-3)$ **18.** $4(x^2-3)$ **19.** $3x(2x-5)$ **20.** $6a^2(2a+3)$
21. $3a(3a+7)$ **22.** $2m(5m-1)$ **23.** $2(3x^2+5y^2)$ **24.** $7(p+2q)$ **25.** $2y(3x^2+5y)$
26. $2xy(1+2xy)$ **27.** $3ab(2a+3)$ **28.** $2ab(2a^2-3)$ **29.** $5mn(1+4m^2)$ **30.** $2x^2(8x^2-9)$
31. $2x^3(2x^4-1)$ **32.** $2(x^2+2x+3)$ **33.** $2x(x^2+2x+3)$ **34.** $3p^2(1+2p+3p^2)$
35. $4m(m^3-2m+1)$ **36.** $ab(b^2+3ab+a^2)$ **37.** $xy(1+xy+x^2y^2)$ **38.** $2xy(2+3x+4y)$
39. $xy(1-4y+y^2)$ **40.** $xy^2z(x+z)$ **41.** $3x^2(2x^2-xy+3y^2)$ **42.** $2ab^2c(3a-7c)$
43. $x^2(x^n+1)$ **44.** $x^n(x^2+1)$ **45.** $2x^{2n}(3+4x^n)$ **46.** $(x+3)(x+2)$ **47.** $(x+3)(x-2)$
48. $(x-3)(x-2)$ **49.** $(x-3)(x-2)$ **50.** $(x-3)(x+2)$ **51.** $(q-1)(p-5)$ **52.** $(q-1)(p+5)$
53. $4(a+5)(a-1)$ **54.** $(2x+3)(9x+1)$ **55.** $2(x-3y)(2x-1)$ **56.** $(2a-3b)(5a+6)$
57. $(x-2)(2x-3)$ **58.** $(x-2)(2x+3)$ **59.** $(p-s)(2p-s)$ **60.** $(m-n)(m+2n)$
61. $(x-y)(4+a)$ **62.** $(y-1)(x+2)$ **63.** $2(y-2)(x+1)$ **64.** $(x-y)(2-a)$ **65.** $(a-3)(4+a^2)$
66. $2(2a-1)(b+1)$ **67.** $2(2a-1)(b-1)$ **68.** $(x+2a)(x+y)$ **69.** $(s+t)(s+1)$
70. $(m+n)(m-2)$ **71.** $(b+4)(4a-1)$ **72.** $(3a-1)(b+2)$ **73.** $(x+y)(x-2)$ **74.** $(x-4)(x^2-3)$
75. $(p+3)(p^2-2)$ **76.** $(2+b)(x+y)$ **77.** $(x-2y)(x+2z)$ **78.** $(a-b)(x-y)$
79. $(m+n)(m+4)$ **80.** $(x-y)(2a-b)$ **81.** $(2x+y)(c+4n)$ **82.** $(x+2)(x^2-18)$

In a biology class, a student said to his teacher.
Student: I just killed five mosquitoes, three were males and two were females.
Teacher: How did you know which were males and which were females ?
Student: Three were killed on a beer bottle. Two were killed on a mirror.

CHAPTER 8 – 5 Factoring Quadratic Trinomials page 182

1. $(x+2)(x+1)$ **2.** $(x-2)(x-1)$ **3.** $(x+5)(x+1)$ **4.** $(x-5)(x-1)$ **5.** $(x+7)(x+1)$ **6.** $(x-7)(x-1)$ **7.** $(x-8)(x+1)$ **8.** *prime* **9.** $(x+3)(x-1)$ **10.** $(x-3)(x+1)$ **11.** *prime* **12.** $(x+3)(x+3)$, or $(x+3)^2$ **13.** $(a+7)(a+2)$ **14.** $(y+2)(y+3)$ **15.** $(a-7)(a-2)$ **16.** $(y-2)(y-3)$ **17.** $-(y-6)(y+1)$ **18.** $-(y+6)(y-1)$ **19.** *prime* **20.** $(c-7)(c-2)$ **21.** $(s-5)(s-8)$ **22.** $(n+4)(n+7)$ **23.** $(n-4)(n-7)$ **24.** $(x-4)(x+3)$ **25.** $(p+30)(p-1)$ **26.** *prime* **27.** $(a-15)(a+2)$ **28.** *prime* **29.** *prime* **30.** $(x+8)(x-7)$ **31.** $(x+2y)(x+y)$ **32.** $(x-2y)(x-y)$ **33.** $(x-y)(x-y)$, or $(x-y)^2$ **34.** $(a+3b)(a+2b)$ **35.** $(a-6b)(a+b)$ **36.** $(x+4y)(x+3y)$ **37.** $(x-3y)(x-7y)$ **38.** $(m+21n)(m+2n)$ **39.** *prime* **40.** *prime* **41.** $(p-25q)(p+2q)$ **42.** $(r-7s)^2$ **43.** $(x+6y)(x+4y)$ **44.** $(x+12y)(x-2y)$ **45.** $(a-6b)(a-7b)$ **46.** $(2x+1)(x+1)$ **47.** $(5x+2)(x-1)$ **48.** $(2x-1)(x+3)$ **49.** $(3x+2)(x-3)$ **50.** $-(3x-1)(2x+3)$ **51.** *prime* **52.** $-(3n-1)(n+2)$ **53.** $(2x-1)(4x+3)$ **54.** $(5a+1)(2a-3)$ **55.** $(4c-7)(2c+1)$ **56.** *prime* **57.** $(3x+1)(4x-5)$ **58.** $(3x+2y)(x-3y)$ **59.** $(3x-4y)(x+2y)$ **60.** $(2a+b)(a+2b)$ **61.** $(5a-6b)(a+b)$ **62.** $(2x+3y)(x-y)$ **63.** $(3c-4)(6c+5)$ **64.** $(11y-8)(2y+3)$ **65.** $(2x^2+3)(2x^2-1)$ **66.** $(2x^2-7)(x^2+2)$ **67.** $(x^n+3)(x^n+2)$ **68.** $(x^{2n}-3)(x^{2n}+5)$ **69.** $(x^2+5)(x^2+3)$ **70.** $(3x^2+4)(x^2-5)$ **71.** $x(x+2)(x+1)$ **72.** $2x(x-2)(x-1)$ **73.** $4x(x-4)(x+1)$ **74.** $a(a-7)(a-2)$ **75.** $2xy(x+2y)(x+y)$ **76.** $n^2(n-3)(n+1)$ **77.** $ab(a-6b)(a+b)$ **78.** $3ab(a+3b)(a+2b)$ **79.** $x(2x^2-1)(2x^2+3)$ **80.** $x^2(x^3+2)(x^3-3)$

CHAPTER 8 – 6 Factoring Special Quadratic Patterns page 184

1. $x^2+8x+16$ **2.** $x^2-8x+16$ **3.** x^2-16 **4.** $9x^2+30x+25$ **5.** $9x^2-30x+25$ **6.** $9x^2-25$ **7.** $4x^2-12xy+9y^2$ **8.** $4x^2+12xy+9y^2$ **9.** $4x^2-9y^2$ **10.** $4a^2+28ab+49b^2$ **11.** $4a^2-28ab+49b^2$ **12.** $4a^2-49b^2$ **13.** $a^2b^2+6ab+9$ **14.** $4x^2y^2-4xyz+z^2$ **15.** $9x^2-30xy+25y^2$ **16.** x^4-4y^2 **17.** $a^2b^2-c^4$ **18.** p^4-q^4 **19.** $4p^6+16p^3+16$ **20.** $4a^4-12a^2b+9b^2$ **21.** Yes. $(x+2)^2$ **22.** Yes. $(x-2)^2$ **23.** No. $(x+2)(x+1)$ **24.** Yes. $(a+4)^2$ **25.** No. $(a-8)(a-2)$ **26.** Yes. $(a-4)^2$ **27.** Yes. $(p+9)^2$ **28.** Yes. $(y+11)^2$ **29.** Yes. $(3x+4)^2$ **30.** No. $(3x+2)(2x-3)$ **31.** Yes. $(2x-3)^2$ **32.** No. $2(x-3)(x+1)$ **33.** Yes. $(2a+3b)^2$ **34.** Yes. $(3x-5y)^2$ **35.** Yes. $(5x+4y)^2$ **36.** Yes. $(a+4)(a-4)$ **37.** No. *prime* **38.** Yes. $(x+2)(x-2)$ **39.** No. *prime* **40.** Yes. $(2x+3)(2x-3)$ **41.** Yes. $(2n+1)(2n-1)$ **42.** Yes. $(1+4x)(1-4x)$ **43.** Yes. $(x^2-y)(x^2+y)$ **44.** Yes. $(3+c)(3-c)$ **45.** No. *prime* **46.** No. *prime* **47.** Yes. $(7+2a)(7-2a)$ **48.** Yes. $(ab+c^2)(ab-c^2)$ **49.** Yes. $(2xy+3z)(2xy-3z)$ **50.** Yes. $(a^n+b^n)(a^n-b^n)$ **51.** $(x+8)^2$ **52.** $(x-9)^2$ **53.** $(2a+7)(2a-7)$ **54.** $(3n+1)(3n-1)$ **55.** $(1+3n)(1-3n)$ **56.** $(5n-6)^2$ **57.** $(3x-2y)^2$ **58.** $(2p+5q)^2$ **59.** $(6x+7y)(6x-7y)$ **60.** $(3a+2b)(3a-2b)$ **61.** $(ab+2c)(ab-2c)$ **62.** $x(x+5)^2$ **63.** $x^2(x-3)^2$ **64.** $2x(1+2x)(1-2x)$ **65.** $2c(3+c)(3-c)$ **66.** $3a(2a-3)^2$ **67.** $(x+y-3)(x-y-3)$ **68.** $(a+3b+2)(a-3b+2)$ **69.** $(x+y+3)(x-y-3)$ **70.** $(m+n-3)(m-n+3)$ **71.** $(x^3-4)^2$ **72.** $(xy-7)^2$ **73.** $(x^2+y^2)(x+y)(x-y)$ **74.** $(a^2+4)(a+2)(a-2)$ **75.** $(a^n+4)(a^n-4)$ **76.** $(x^{2n}+y^{2n})(x^n+y^n)(x^n-y^n)$ **77.** $(2x^n+3y^{2n})(2x^n-3y^{2n})$ **78.** $(5+x^2)(x+2)(x-2)$ **79.** $(x+3)(x-3)(x^2+4)$ **80.** $(x^4+y^4)(x^2+y^2)(x+y)(x-y)$

CHAPTER 8 – 7 Solving Equations by Factoring page 186

1. $x = -2, 1$ **2.** $x = 10, -8$ **3.** $x = 4$ (a double root) **4.** $x = 12$ (a double root) **5.** $x = -2, 1$
6. $x = 2, -1$ **7.** $x = -1, 1$ **8.** $x = \frac{1}{2}, 2$ **9.** $x = 0, 15$ **10.** $x = 0, 5, -2\frac{2}{3}$
11. $x = 0, -1\frac{1}{2}$ **12.** $x = 0, 1\frac{1}{2}$ **13.** $x = 0, 4$ **14.** $x = 0, x = \frac{2}{3}$ **15.** $x = 0, 7$ **16.** $x = 0, \frac{1}{7}$
17. $x = 0$ (a double root), 5 **18.** $a = 0$ (a double root), $\frac{1}{2}$ **19.** $x = 0, 5$
20. $x = 0$ (a double root), 5 **21.** $y = 0, 12$ **22.** $x = 0$ (a double root), 2 **23.** $p = 0, 1$
24. $n = 0$ (a double root), 1, –1 **25.** $x = 0, 10$ **26.** $a = 5, -7$ **27.** $p = -5, 5$
28. $x = 9$ (a double root) **29.** $n = -2, -12$ **30.** $x = -2, 6$ **31.** $x = -2, \frac{1}{2}$ **32.** $x = 0, 2, -4$
33. $y = 0, -\frac{1}{2}, 4$ **34.** $x = 0, 2, -\frac{1}{2}$ **35.** $x = -2, -1$ **36.** $x = 2, 1$ **37.** $n = 6, -1$
38. $x = -8$ (a double root) **39.** $x = 2$ (a double root) **40.** $x = 8, 2$ **41.** $x = -1\frac{1}{2}, -1$
42. $x = 1\frac{1}{2}, -4$ **43.** $x = \frac{1}{2}, 1\frac{1}{2}$ **44.** $x = -\frac{1}{3}, 2\frac{1}{2}$ **45.** $x = -\frac{3}{4}, 5$ **46.** $x = \frac{1}{6}, 2$ **47.** $u - 3, 3$
48. $a = -1\frac{1}{2}, 1\frac{1}{2}$ **49.** $n = 0, 1, -1$ **50.** $x = 4$ (a double root) **51.** $a = -\frac{1}{8}, 1$ **52.** $y = \frac{1}{2}, -\frac{2}{3}$
53. $x = 0, 3, -2$ **54.** $x = 0, -3, \frac{1}{2}$ **55.** $y = 0$ (a double root), 4(a double root)
56. $p = 0, 3, -3, 2, -2$ **57.** $x = -4, 5$ **58.** $x = 1, -5$ **59.** $x = 7, -2$ **60.** $x = -\frac{1}{3}, 2$

CHAPTER 8 – 8 Solving Word Problems by Factoring page 188

1. $x = -1\frac{1}{2}, 1\frac{1}{3}$ **2.** $x = -1\frac{2}{3}, 6$ **3.** $x = \frac{3}{5}, -\frac{1}{2}$ **4.** $x = 3\frac{1}{2}, 1$ **5.** $x = \frac{2}{3}, -3$ **6.** $x = \frac{3}{4}, \frac{1}{2}$
7. $x = -1, 4$ **8.** $x = -3, 2$ **9.** $x = -2, -1$ **10.** $t = 3, 8$ **11.** $t = 2, 6$ **12.** $t = 3, 7$
13. –9, or 8 **14.** –13, or 12 **15.** –12, or 11 **16.** 11, or –10 **17.** 11 and 13 **18.** 12 and 14
19. –15 and –13 **20.** –10 and –8, or 8 and 10 **21.** 11, 12, 13
22. –16, –14, –12, or 12, 14, 16 **23.** 7, 8, 9 **24.** 7 *m* by 12 *m* **25.** length 20 *ft*, width 12 *ft*
26. $9cm \times 9cm \times 6cm = 486cm^3$ **27.** base 6 *cm*, height 18 *cm* **28.** base 11 *m*, height 32 *m*
29. base 6 *m*, height 9 *m* **30.** 1.5 *m* **31.** 3 and 7 seconds **32.** 10 seconds. (Hint: let $h = 0$)

CHAPTER 8 EXERCISES Polynomial and Factoring page 189 ~ 190

1. 64 **2.** –64 **3.** –64 **4.** 16 **5.** 4^{10} **6.** 4^7 **7.** 1 **8.** 1 **9.** –1 **10.** 1 **11.** x^{12} **12.** x^7
13. a^{20} **14.** a^9 **15.** a^3 **16.** a^{-3}, or $\frac{1}{a^3}$ **17.** 64 **18.** $\frac{1}{64}$ **19.** a^{-3}, or $\frac{1}{a^3}$ **20.** x^3 **21.** $11x$
22. $-3x$ **23.** $28x^2$ **24.** $6a^6$ **25.** $8x^3$ **26.** $9x^2$ **27.** $6a^3b^2$ **28.** x^3y^5 **29.** $72x^8$ **30.** $\frac{4}{9}a^6$
31. n^6 **32.** $-108p^5$ **33.** $\frac{4}{25}$ **34.** $\frac{25}{4}$, or $6\frac{1}{4}$ **35.** 5×10^3, or 5,000 **36.** $\frac{x}{2}$ **37.** $\frac{1}{2x}$ **38.** $\frac{1}{5}a^2b^2$
39. $4x^2y^{-4}$, or $\frac{4x^2}{y^4}$ **40.** $\frac{81n^8}{p^8}$ **41.** $4x + 24$ **42.** $-5x + 15$ **43.** $2x^2 - 6x$ **44.** $-3a^2 + 4a$
45. $2p^2 - 4p^3$ **46.** $-2x + 22$ **47.** $5x^2 - 8x$ **48.** $9a^2 - 16$ **49.** $x^2 + 2x - 8$ **50.** $2a^2 + 7a - 30$
51. $x^2 + 4x - 21$ **52.** $a^2 + 4a - 45$ **53.** $4a^2 + 12a + 9$ **54.** $4a^2 - 12a + 9$ **55.** $4a^2 - 9$
56. $8x^2 - 22x + 5$ **57.** $x^2 - 100$ **58.** $c^2 - 9d^2$ **59.** $4x^2 - 4xy + y^2$ **60.** $9x^2 + 12xy + 4y^2$
61. $12n^2 + 2n - 30$ **62.** $9a^2 - 24ab + 16b^2$ **63.** $6p^2 - 21p + 15$ **64.** $9x^2 + 30x + 25$
65. $9x^2 - 25$ **66.** $a^4 - b^2$ **67.** $a^4 + 2a^2b + b^2$ **68.** $a^2 - 2ab^2 + b^4$ **69.** $x^{2n} + 2x^ny^n + y^{2n}$
70. $x^6 - y^4$ **71.** $2(x-9)$ **72.** $4x(x-3)$ **73.** $2xy(3x + 7y^2)$ **74.** $2x^3(6x^2 + 5)$
75. $3x(x-3)(x+1)$ **76.** $2xy(x+2y)(x+y)$ **77.** $(x-2)(2x+3)$ **78.** $(x-3)(x+2)$
79. $(y+3)(x-2)$ **80.** $(x+2y)(x-3)$ **81.** $(x-2)(x-5)$ **82.** $(x-2)(x+5)$ **83.** *prime*
84. $(a+9)(a-5)$ **85.** $(p-2)(p-3)$ **86.** $(p+1)(p-6)$ **87.** $(y-5)(y-8)$ **88.** $(x+40)(x-1)$
89. $-(x+4)(x-6)$ **90.** $(x+y)(x-2y)$ **91.** $(a+b)(a-3b)$ **92.** $(p-10q)(p+2q)$
93. $(x+8y)(x-6y)$ **94.** $(2x+5)(2x-3)$ **95.** $(3x-1)(4x-3)$ **96.** *prime* **97.** $(2a^2+1)(a^2-3)$
98. $(4x^2-5)(x^2+2)$ **99.** $x(2x+3)(2x-1)$ **100.** $3y(3y-2)(y+5)$ **101.** $(x+4)^2$ **102.** $(2x-3)^2$
103. $(a-9)^2$ **104.** $(2y+3)(2y-3)$ **105.** $(2a-b)^2$ **106.** $(4x-5y)^2$ **107.** $(3x+2y)(3x-2y)$
108. $(2ab+c)(2ab-c)$ **109.** $(5p-2q)^2$ **110.** $(2x-5y)^2$ **111.** $(x+2)^2(x-2)^2$ **112.** $(4n-5)^2$
113. $(x+2y+3)(x-2y+3)$ **114.** $(a+b-2)(a-b+2)$ **115.** $(a^{2n}+3)(a^{2n}-3)$ **116.** $(a^3-5)^2$
117. $(x^2+4y)^2$ **118.** $3x(2x+3)(2x-3)$ **119.** $(a+4)(a+3)(a-3)$ **120.** $(x+2)(x-2)(2x-3)$
121. $x = -6, 12$ **122.** $x = 2, -\frac{1}{2}$ **123.** $n = 0, 4, 2\frac{1}{2}$ **124.** $x = 7, -2$ **125.** $a = -7, -3$
126. $x = -\frac{1}{2}, 3$ **127.** $x = \frac{2}{5}, -2$ **128.** $x = 7$ (a double root) **129.** $x = -1\frac{1}{2}$ (a double root)
130. $x = 1\frac{1}{3}, -1\frac{1}{3}$ **131.** $x = 0, 6, -3$ **132.** $n = 0, 3, -3$ **133.** $x = -3, 5$ **134.** $x = -1, 1\frac{2}{3}$
135. $x = 1, 1\frac{1}{2}$ **136.** $x^2 + 2x - 3 = 0$ **137.** $3x^2 + 4x - 4 = 0$ **138.** 12, or –11 **139.** 13 and 14
140. 8, 9, 10 **141.** 7*cm* by 14*cm* **142.** 7*cm* by 15*cm* **143.** 5 seconds **144.** 100 feet

CHAPTER 9 – 1 Inequalities and Absolute Values page 196

1. $6 > 4$ **2.** $3 < 8$ **3.** $-3 < 5$ **4.** $-3 > -5$ **5.** $-7 < -4$ **6.** $4.5 < 6$ **7.** $-12 = -12$
8. $-4.2 > -4.5$ **9.** $-18 > -24$ **10.** $-32 < -15$
11. False **12.** True **13.** False **14.** True **15.** True (Hint: It is true since $-5 = -5$ is true.)
16. True **17.** False **18.** False **19.** True **20.** True (Hint: It is true since $3 = 3$ is true.)
21. $10 < 13$ **22.** $2 < 5$ **23.** $24 < 36$ **24.** $1.5 < 2.25$ **25.** $-24 > -36$ **26.** $-1.5 > -2.25$
27. $10 > -12$ **28.** $-10 < 15$ **29.** $-2 < 3$ **30.** $2 < 3$ **31.** $18 < 27$ **32.** $-2 > -3$ **33.** $x > 2$
34. $x < -2$ **35.** $x \le 8$ **36.** $x \ge -8$ **37.** $x > 9$ **38.** $x \ge -9$ **39.** $x \le -\frac{3}{2}$ **40.** $x \ge \frac{3}{2}$
41. $8 < 15$ **42.** $-8 > -15$ **43.** $-7 < -2$ **44.** $-9 < -7 < 2$ **45.** $-7 < -2 < 3$
46. $-10 < -7 < -2$ **47.** $x > 15$ **48.** $x \le 15$ **49.** $-8 < n < 2$ **50.** $4 < n \le 9$
51. $|n| < 5$ **52.** $|x| > 7$ **53.** $|x| \ge 7$ **54.** $|n| \le 5$

55. -4 -1 0 4 **56.** -2 0 4 **57.** -3 0 4

58. -5 0 3 **59.** -4 0 2 **60.** -3 0 3

61 ~ 62 graphs are left to the students

CHAPTER 9 – 2 Solving Inequalities in One Variable page 198

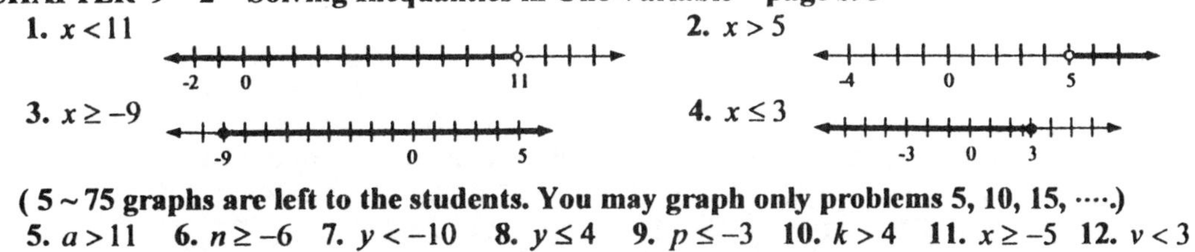

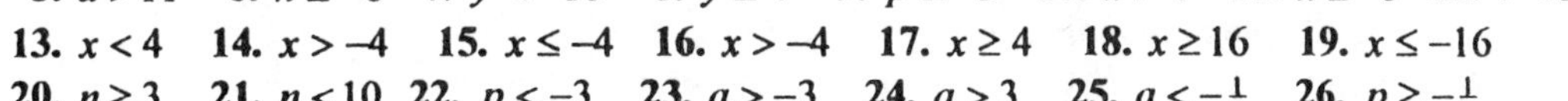

(5 ~ 75 graphs are left to the students. You may graph only problems 5, 10, 15, ...)

5. $a > 11$ **6.** $n \ge -6$ **7.** $y < -10$ **8.** $y \le 4$ **9.** $p \le -3$ **10.** $k > 4$ **11.** $x \ge -5$ **12.** $v < 3$
13. $x < 4$ **14.** $x > -4$ **15.** $x \le -4$ **16.** $x > -4$ **17.** $x \ge 4$ **18.** $x \ge 16$ **19.** $x \le -16$
20. $n \ge 3$ **21.** $n < 10$ **22.** $p < -3$ **23.** $a > -3$ **24.** $a > 3$ **25.** $a < -\frac{1}{3}$ **26.** $p \ge -\frac{1}{3}$
27. $y \le -1$ **28.** $x < -10$ **29.** $x > 20$ **30.** $x \le -12$ **31.** $x \le -6$ **32.** $x < -20$
33. $x \ge -9$ **34.** $p > -20$ **35.** $p > 18$ **36.** $a < -40$ **37.** $k \le 21$ **38.** $k \ge 2\frac{1}{2}$ **39.** $x > 2$
40. $x > -4$ **41.** $x \le -\frac{1}{3}$ **42.** $x \ge -\frac{2}{3}$ **43.** $x > 4$ **44.** $x > -1$ **45.** $x < 1$ **46.** $x < 4$
47. $x \ge -4$ **48.** $x \le 4$ **49.** $x > -18$ **50.** $x < -32$ **51.** $x \le 75$
52. All real numbers **53.** All real numbers **54.** No solution (The solution set is ϕ.)
55. No solution (The solution set is ϕ.) **56.** $x < -1\frac{1}{5}$ **57.** $x > -2\frac{1}{2}$ **58.** $n > 4$ **59.** $n \le -21$
60. $p \ge 0$ **61.** $x < 2$ **62.** $x > -\frac{1}{2}$ **63.** $x \le \frac{4}{5}$ **64.** $x \le 3$ **65.** $x > -2$ **66.** $x < -48$
67. $y > 6$ **68.** $x \ge -6$ **69.** $x \ge 7\frac{1}{2}$ **70.** $x \le \frac{3}{5}$ **71.** $x < -10$ **72.** $x < -4\frac{2}{3}$ **73.** $n < -5$
74. $p \ge 3\frac{1}{3}$ **75.** $y > 14$
76. All real numbers of c **77.** All positive real numbers of c, or all real numbers of $c > 0$
78. All negative real numbers of c, or all real numbers of $c < 0$.
79. False. (Hint: $-1 > -2$, but $(-1)^2 < (-2)^2$) **80.** True. (Hint: $-1 > -2$ and $(-1)^3 > (-2)^3$)
81. True. (Hint: They are positive numbers.)
82. All positive real numbers of y, or all real numbers of $y > 0$. (Hint: $3^0 = 2^0$ and $3^{-2} < 2^{-2}$)

> A middleaged single lady told her friend.
> Lady: I prefer to marry an archaeologist rather than a doctor or a lawyer.
> Friend: Why do you want to marry an archaeologist ?
> Lady: If my husband is an archaeologist, the more I get older, the more he is interested in me.

CHAPTER 9 – 3 Solving Combined Inequalities page 201 ~ 202

1. B **2.** D **3.** G **4.** C **5.** H **6.** F (all real numbers) **7.** A **8.** E **9.** E
10. F (all real numbers) **11.** B **12.** ϕ (empty, no graph)
13. $-2 < x < 5$ **14.** $x \ge 8$ **15.** $1 < x \le 4$ **16.** $-3 < x \le 7$ **17.** $-5 \le x \le -1$ **18.** $x \le 9$
19. $-6 \le n \le 4$ **20.** $-12 < a \le -4$
21 ~ 30 graphs are left to the students
21. $x < -5$ **22.** $x < 1$ **23.** $x \le 2$ **24.** $x \ge 6$ **25.** $x < 6$ **26.** $x < -6$ **27.** $x > -2\frac{1}{2}$
28. $x \ge 2\frac{2}{5}$ **29.** $x \le -18$ **30.** $a < 1$

31. **32.**

33 ~ 39 graphs are left to the students. **40.** No graph (ϕ, empty)
41 ~ 76 graphs are left to he students. The graphs are very useful to write the right answers.
41. $7 < x < 14$ **42.** $-7 \le x < 4$ **43.** $-4 < a \le 3$ **44.** $3 < a \le 6$ **45.** $-3 < n \le 5$
46. $-3 \le n < 2$ **47.** $-8 < x < 4$ **48.** No solution (ϕ, empty) **49.** $-2 < x < 9$
50. $-4 \le y \le 1$ **51.** $-1 \le y < 2$ **52.** $3 < b < 5$ **53.** $2 < p < 4$ **54.** $-3 \le x < 4$
55. $-2 < x \le 2$ **56.** $-2 < x < 3$ **57.** $-3 < x < 2$ **58.** No solution (ϕ, empty)
59. $x > -2$ **60.** $x < -5$ or $x > 2$ **61.** All real numbers of x **62.** $x > -5$
63. $x > -3$ **64.** $x \le -3$ or $x \ge 5$ **65.** No solution (ϕ, empty) **66.** $3 \le x \le 5$
67. $-2 \le n \le 4$ **68.** $n > -2$ **69.** $-2 \le p < 2$ **70.** $p > 3$ **71.** All real numbers of p
72. $p \ge -2$ **73.** $-2 < x \le 1.5$ **74.** $x \le -3$ or $x \ge 1\frac{2}{3}$ **75.** $x < -4.5$ or $x > 3.4$
76. $-18 \le x \le 12$

CHAPTER 9 – 4 Solving Word Problems Involving Inequalities page 204

1. $x < 13$ **2.** $x < 5$ **3.** $x \ge 46$ **4.** $n \le 43$ **5.** $n \le 10$ **6.** $n \ge 12$ **7.** $m > 420$
8. $m < 700$ **9.** $w > 20$ **10.** $w > 3.5$ **11.** $x \ge 85$ **12.** $x \ge 90$
13. At most 300 miles (300 miles or less) **14.** At most 7 days (7 days or less)
15. 180 miles or less **16.** 91 (or more) **17.** 95 (or more) **18.** 47 and 48 **19.** 49 and 50
20. 48 and 49 **21.** 42, 43, 44 **22.** {1, 3, 5} or {3, 5, 7} or {5, 7, 9}
23. At least 19 *cm* by 7 *cm* **24.** At least 15 *cm* by 9 *cm* **25.** At most 15 *cm* by 9 *cm*
26. At least 13 *in.* by 11 *in.* **27.** At most 20 quarters and 4 dimes in the box
28. At most 13.5 grams (or less) of salt added

CHAPTER 9 – 5 Operations with Absolute Values page 206

1. 15 **2.** 15 **3.** 7.2 **4.** $\frac{4}{5}$ **5.** 1 **6.** 1 **7.** 8 **8.** 0 **9.** 3 **10.** 3 **11.** 9 **12.** 8 **13.** 0 **14.** 18 **15.** 8
16. True **17.** True **18.** False **19.** False **20.** True **21.** False **22.** True **23.** False
24. True **25.** False **26.** True **27.** False **28.** True **29.** True **30.** True
31. D = all real numbers, R = all non-negative real numbers (or all real numbers of $y \ge 0$)
32. D = all real numbers, R = all real numbers of $y \ge 8$ **33.** D = all real numbers, R = all non-negative real numbers
34. D = all real numbers, R = all real numbers of $y \ge -8$
35. D = all real numbers, R = all non-positive real numbers (or all real numbers of $y \le 0$)
36. D = all real numbers, R = all non-negative real numbers **37.** D = all real numbers, R = all real numbers of $y \le 8$
38. D = all real numbers, R = all real numbers of $y \le -8$ **39.** D = all real numbers, R = all non-negative real numbers
40. D = all real numbers, R = all non-positive real numbers **41.** D = all real numbers, R = all non-positive real numbers
42. D = all real numbers, R = all non-negative real numbers **43.** D = all real numbers, R = all real numbers of $y \ge -1$
44. D = all real numbers, R = all real numbers of $y \ge 1$ **45.** D = all real numbers, R = all real numbers of $y \le 1$

46. $|x| = 3$
$x = 3$ or $x = -3$

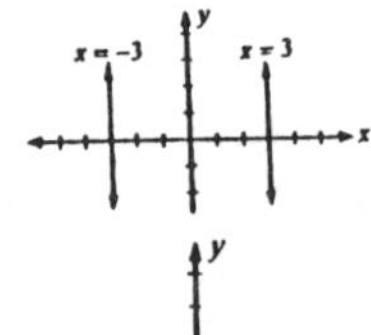

47. $y = -|x|$
For $x > 0$, $y = -x$
For $x < 0$, $y = -(-x)$
$y = x$
For $x = 0$, $y = 0$

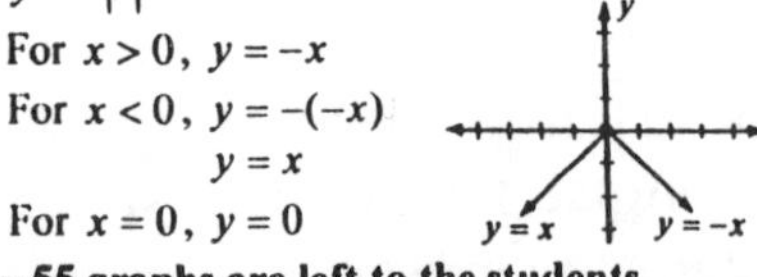

48. $y = |x - 3|$
For $x - 3 > 0$, $y = x - 3$
For $x - 3 < 0$, $y = -(x - 3)$, $y = -x + 3$
For $x - 3 = 0$ (or $x = 3$), $y = 0$

y = -x + 3 y y = x - 3 (5, 2) x

49 ~ 55 graphs are left to the students.
56. True **57.** False (Hint: It is false when $x < 0$) **58.** $x \ge 0$
59. It is true when a and b are both positive, both negative, or both 0.
60. False (Hint: $|0| = 0$. 0 is a real number. But, 0 is neither positive nor negative.)

CHAPTER 9 – 6 Absolute-Value Equations and Inequalities page 208

1. $x = 10$ or -10 **2.** $x < -10$ or $x > 10$ **3.** $-10 < x < 10$ **4.** $x \le -10$ or $x \ge 10$
5. $-10 \le x \le 10$ **6.** $x = 14$ or -6 **7.** $x = 6$ or -14 **8.** $x \le -6$ or $x \ge 14$ **9.** $-14 \le x \le 6$
10. $x = 4$ or $-1\frac{1}{5}$ **11.** $x = 1\frac{1}{5}$ or -4 **12.** $-2 < x < \frac{1}{3}$ **13.** $x < -\frac{1}{3}$ or $x > 2$
14. $y < -4$ or $y > -2$ **15.** $-6 \le y \le 0$ **16.** $x = 6$ or -6 **17.** $x = 3$ or -7 **18.** $x < -2$ or $x > 6$
19. $1 < a < 1\frac{2}{3}$ **20.** $a = -2$ or -4 **21.** $n = \frac{1}{2}$ or $-\frac{1}{6}$ **22.** $p \le -1$ or $p \ge 2\frac{1}{3}$ **23.** $x \le 0$ or $x \ge 8$
24. $-8 \le y \le 48$ **25.** $x < -2$ or $x > 1$ **26.** $-1 < x < 2$ **27.** $1 \le x \le 2$ **28.** $x < -2$ or $x > 2$
29. $-3 < x < 3$ **30.** $-6 < x < 10$ **31.** $x < -\frac{1}{2}$ or $x > 3$ **32.** $x \le -1$ or $x \ge 5$ **33.** $x \le -5$ or $x \ge 9$
34. $-8 \le x \le 16$ **35.** $x =$ all real numbers except 0 **36.** No solution **37.** $x =$ all real numbers
38. $x = 0$ only **39.** $x =$ all real numbers except 4 **40.** $x =$ all real numbers except -2
41. $n = 4$ only **42.** No solution **43.** $-5 \le x \le -1$ or $1 \le x \le 5$ **44.** $-4 \le x < -3$ or $1 < x \le 2$
45. $-2 < x < 2$ or $2 < x < 6$ **46.** $|p - 46| \le 5$ **47.** $41\% \le p \le 51\%$ **48.** $|t - 98.6| \le 1$
49. $a - b \le x \le a + b$ **50.** $x \le -a - b$ or $x \ge -a + b$

CHAPTER 9 EXERCISES Inequalities and Absolute Values page 209 ~ 210

1. False **2.** True **3.** True **4.** False **5.** True(Hint: Since $-1 = -1$) **6.** True **7.** True
8. False **9.** False **10.** True
11. $9 > 1$ **12.** $-9 < -1$ **13.** $-3 < 2 < 7$ **14.** $-8 < -4 < 1$ **15.** $x < -1$ **16.** $n \ge 15$
17. $0 < x < 12$ **18.** $n < -3$ or $n > 3$ **19.** $|x| \ge 25$ **20.** $|x - a| \le b$
21. $x \ge 21$ **22.** $x \le 31$ **23.** $x \le -21$ **24.** $x \ge -31$ **25.** $x > 5$ **26.** $x < -5$ **27.** $x < -3$
28. $x > 1$ **29.** $x \ge 30$ **30.** $x \le -30$ **31.** $a > 50$ **32.** $a > -100$ **33.** $x \le 5$ **34.** $x \ge 3$
35. All real numbers **36.** No solution **37.** $x < 31\frac{1}{4}$ **38.** $x < -56\frac{1}{4}$ **39.** $x \ge -4$
40. $x \ge 3$ **41.** $n < -\frac{7}{8}$ **42.** $-1 < a < 6$ **43.** $4 \le a < 15$ **44.** $3 < x < 8$ **45.** $-4 < x < 3$
46. $-5 \le n < -1$ **47.** $-2 < x < 3$ **48.** $x < -5$ **or** $x > 3$ **49.** All real numbers
50. No solution

51. $|x| = 2$
$x = 2$ or $x = -2$

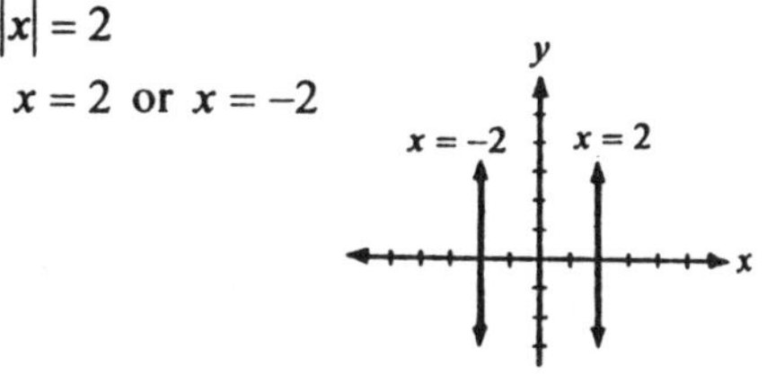

52. $|y| = 2$
$y = 2$ or $y = -2$

y
y = 2
x
y = −2

53 ~ 60 graphs are left to the students.
61. No solution **62.** $x = 15$ or -15 **63.** $x = 5$ or -19 **64.** $-19 < x < 5$
65. $x < -19$ or $x > 5$ **66.** $x = 12$ or -4 **67.** $-4 < x < 12$ **68.** $x < -4$ or $x > 12$
69. $x = -1$ or -3 **70.** $x \le \frac{5}{7}$ or $x \ge 1$ **71.** $\frac{5}{7} \le x \le 1$ **72.** $-1\frac{1}{5} \le x \le 2$ **73.** $k \le \frac{3}{5}$ or $k \ge 1$
74. $a < -6$ or $a > 5$ **75.** $\frac{1}{3} < x < 3\frac{1}{3}$ **76.** $x < -3$ or $x > 3$ **77.** $-2 \le x \le 2$
78. $-1 < x < 5\frac{1}{2}$ **79.** $x \le -16$ or $x \ge 24$ **80.** $-5 < x < 0$ or $2 < x < 5$
81. All negative real numbers of c, or all real numbers of $c < 0$.
82. False. (Hint: $-2 < -1$, but $(-2)^2 > (-1)^2$)
83. D = all real numbers, R = all non-negative real numbers (or all real numbers of $y \ge 0$)
84. D = all real numbers, R = all real numbers of $y \ge 5$
85. D = all real numbers, R = all non-positive real numbers (or all real numbers of $y \le 0$)
86. $x = a + b$ or $x = a - b$ **87.** $a - b < x < a + b$ **88.** 89 **89.** 150 miles or less
90. 72, 74 and 76 **91.** 91 feet by 84 feet **92.** 4.42 grams
93. $|p - 5| \le 0.2$ (Hint: $4.8\% \le p \le 5.2\%$)

CHAPTER 10 – 1 Solving Systems by Graphing page 216

1. x – intercept = 3
   y – intercept = –3

2. x – intercept = 4
   y – intercept = 2

3. x – intercept = 4
   y – intercept = –2

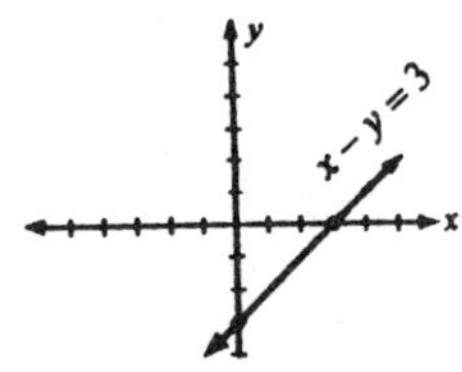

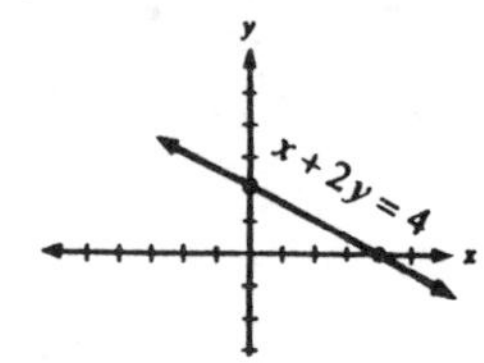

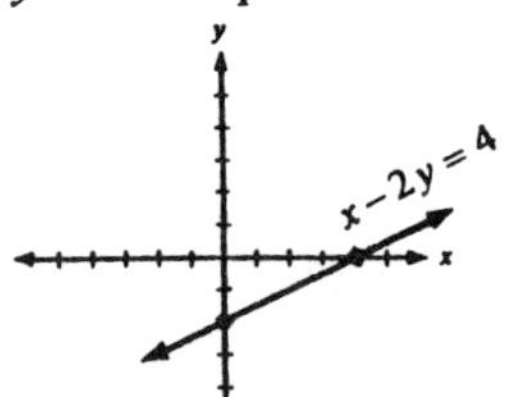

4 ~ 12 graphs are left to the students.

4. x – intercept = 3
   y – intercept = –6

5. x – intercept = 8
   y – intercept = $3\frac{1}{5}$

6. x – intercept = 5
   y – intercept = 4

7. x – intercept = $2\frac{1}{2}$
   y – intercept = –2

8. x – intercept = $1\frac{3}{4}$
   y – intercept = $-3\frac{1}{2}$

9. x – intercept = $2\frac{1}{4}$
   y – intercept = $1\frac{1}{2}$

10. x – intercept = $-\frac{1}{2}$
    y – intercept = 2

11. x – intercept = $-\frac{1}{2}$
    y – intercept = 1

12. x – intercept = –8
    y – intercept = 4

13. $x - y > 3$

14. $x - y \leq 3$

15. $x + 2y < 4$

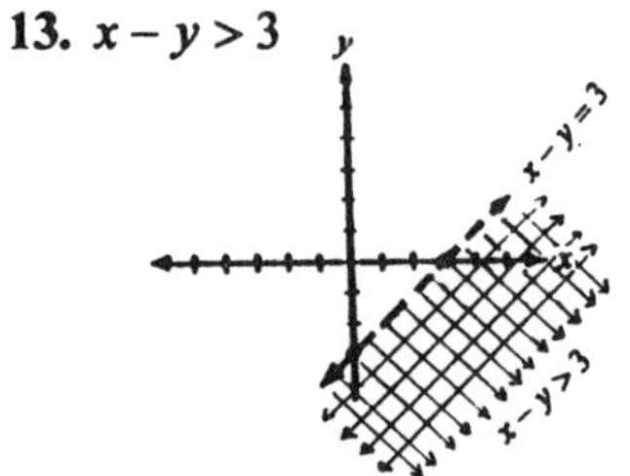

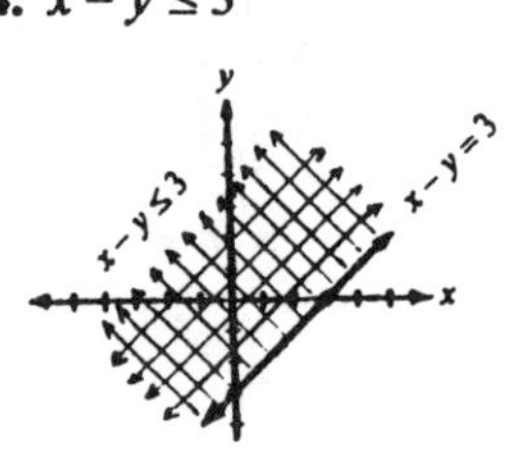

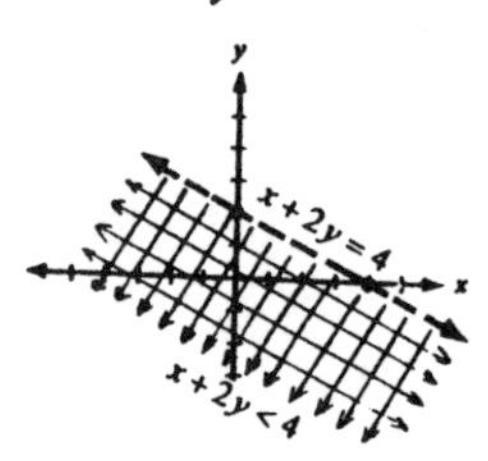

16 ~ 24 graphs are left to the students.

25 ~ 36 graphs are left to the students.

25. $x = 2$, $y = 1$ 26. $x = 1$, $y = 3$ 27. $x = 3$, $y = 2$ 28. $x = 4$, $y = -2$ 29. $a = -2$, $b = 1$ 30. $x = 5$, $y = -3$

31. Coincide (unlimited number of solutions) 32. Parallel (no solution)

33. Parallel (no solution) 34. Coincide (unlimited number of solutions)

35. $s = \frac{1}{2}$, $t = 4$ 36. $x = 2$, $y = -\frac{1}{2}$

37. $x + y \geq 2$
    $x - y \leq 1$

38. $2x - y \leq 4$
    $x \leq 3$

39. $2x - y \geq 4$
    $3x + 4y < 12$

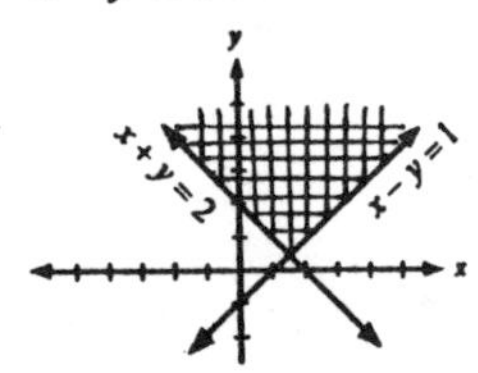

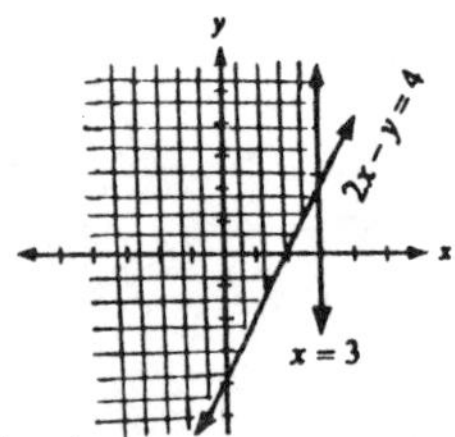

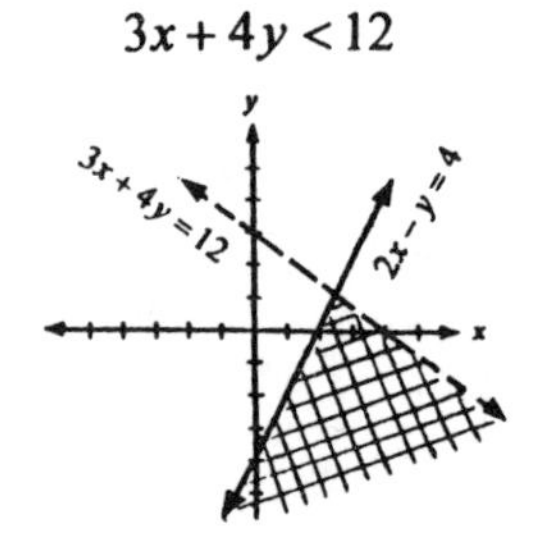

40 ~ 45 graphs are left to the students.

CHAPTER 10 – 2 Solving Systems Algebraically page 219 ~ 220

1. $x=2$, $y=2$ **2.** $x=2$, $y=4$ **3.** $x=1$, $y=-3$ **4.** $x=-2$, $y=2$ **5.** $x=6$, $y=3$ **6.** $x=2$, $y=-4$ **7.** $x=7$, $y=12$ **8.** $x=5$, $y=-8$

9. $x=1$, $y=1$ **10.** $x=3$, $y=-2$ **11.** $x=2$, $y=1$ **12.** $x=1$, $y=3$ **13.** $x=3$, $y=2$ **14.** $x=4$, $y=-2$ **15.** $a=-2$, $b=1$

16. $x=5$, $y=-3$ **17.** Unlimited number of solutions (they coincide.) **18.** No solution (they are parallel.)

19. Unlimited number of solutions (they coincide.) **20.** No solution (they are parallel.)

21. $x=\frac{1}{2}$, $y=4$ **22.** $x=2$, $y=-\frac{1}{2}$ **23.** $x=\frac{2}{3}$, $y=\frac{1}{3}$ **24.** $x=-\frac{1}{2}$, $y=\frac{3}{4}$

25. $x=9$, $y=-1$ **26.** $x=2$, $y=9$ **27.** $a=7$, $b=5$ **28.** $m=4$, $n=-4$ **29.** $p=-\frac{3}{4}$, $q=-2$ **30.** $x=1$, $y=-\frac{2}{3}$ **31.** $x=2\frac{2}{3}$, $y=3$

32. $c=2$, $d=-2\frac{3}{4}$ **33.** $a=-5$, $b=-4$ **34.** No Solution (They are parallel.) **35.** $x=5$, $y=-1\frac{3}{7}$ **36.** $s=\frac{1}{2}$, $t=\frac{1}{7}$

37. $x=3$, $y=-2$ **38.** $x=-1$, $y=4$ **39.** $x=4$, $y=6$ **40.** Unlimited number of solutions (They coincide.)

41. No solution (They are parallel.) **42.** $x=\frac{1}{3}$, $y=\frac{3}{4}$ **43.** Unlimited number of solutions (They coincide) **44.** $x=-6$, $y=7$

45. $x=3$, $y=-8$ **46.** $a=\frac{1}{2}$, $b=\frac{1}{3}$ **47.** $a=\frac{2}{3}$, $b=-4$ **48.** $p=20$, $q=40$ **49.** $x=1.2$, $y=1.5$ **50.** $x=6$, $y=8$

51. $x=4$, $y=-2$ **52.** $x=3$, $y=1$ **53.** $x=-5$, $y=7$ **54.** $x=-3$, $y=-4$ **55.** $x=7$, $y=8$ **56.** $x=9$, $y=10$ **57.** $x=1$, $y=-6$

58. $a=-1$, $b=2$ **59.** $a=4$, $b=-5$ **60.** $c=10$, $d=-4$ **61.** $m=5$, $n=12$ **62.** $m=-3$, $n=-6$ **63.** $x=1$, $y=-6$ **64.** $x=\frac{1}{3}$, $y=8$

65. $x=\frac{1}{5}$, $y=-\frac{3}{4}$ **66.** $x=10$, $y=5$ **67.** $x=20$, $y=-50$ **68.** $x=0.3$, $y=0.5$ **69.** $x=\frac{2}{3}$, $y=2$ **70.** $x=\frac{3}{5}$, $y=-3$ **71.** $a=-3$, $b=-\frac{3}{4}$

72. $x=2$, $y=3$ **73.** $x=5$, $y=-4$ **74.** $x=-6$, $y=8$

CHAPTER 10 – 3 Solving Word Problems by using Systems page 224

1. John \$104, David 52 **2.** 19 one-dollar bills, 45 five-dollar bills **3.** 20 quarters, 4 dimes

4. Airplane speed = 800 km/hr., Wind speed = 60km/hr. **5.** 28 and 36 **6.** 300 miles

7. 19 *cm* by 7 *cm* **8.** 12.5 *in.* by 10 *in.* **9.** 20 dozens of hamburgers, 25 dozens of bagels

10. Hamburger \$1.90 and bagel \$0.90 **11.** 12 boxes of hamburgers, 18 boxes of bagels

12. 150 seconds **13.** 28 students, 120 pencils **14.** \$0.60 each pen, 0.35 each pencil

15. 1,680 boys and 1,512 girls

CHAPTER 10 EXERCISES Systems of Equations and Inequalities page 225 ~ 226

1. x – intercept = 4, y – intercept = 4 **4.** x – intercept = 2, y – intercept = –4 **7.** x – intercept = $\frac{3}{4}$, y – intercept = $-\frac{1}{2}$ **10.** x – intercept = –4, y – intercept = $2\frac{2}{5}$

1 ~ 12 graphs are left to the students

13. $x=4$, $y=-1$ **14.** $x=4$, $y=-2$ **15.** No solution (They are parallel) **16.** Unlimited number of solutions (They coincide)

17. $a=2$, $b=1$ **18.** $x=\frac{1}{2}$, $y=-5\frac{1}{2}$ **19.** $x=1$, $y=-1$ **20.** $x=1$, $y=2$ **21.** $m=-3$, $n=4$ **22.** $x=-2$, $y=1$

23. $x=\frac{1}{2}$, $y=-4$ **24.** $x=-1\frac{1}{2}$, $y=2$

25. $\begin{cases} 2x+y\ge 4 \\ x-y\le 2 \end{cases}$

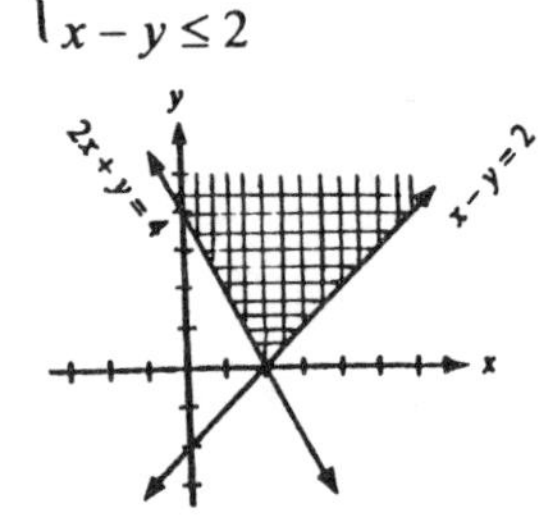

28. $\begin{cases} 2x+y\le 4 \\ x-y\ge 2 \end{cases}$

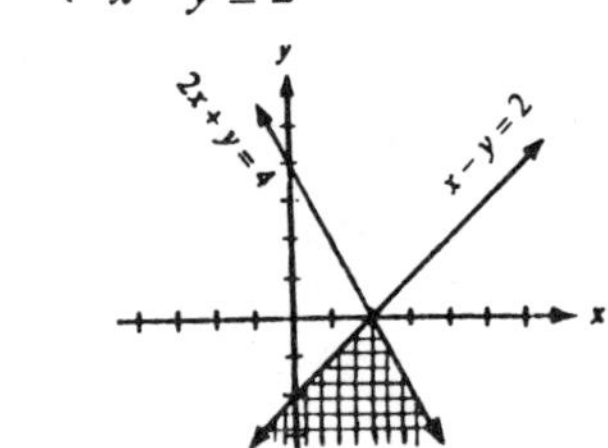

31. 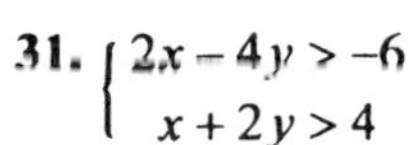
$\begin{cases} 2x-4y>-6 \\ x+2y>4 \end{cases}$

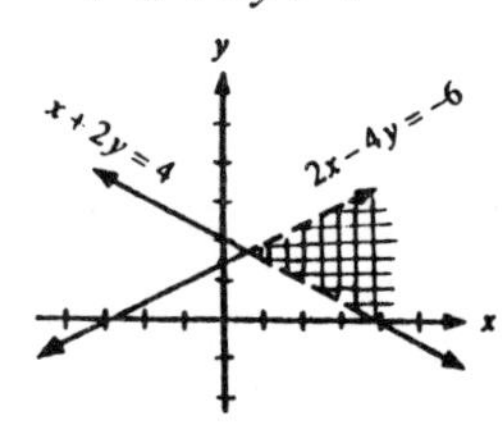

37. $\begin{cases} 2x+y\ge 4 \\ x-y\le 2 \\ x\ge 2 \end{cases}$

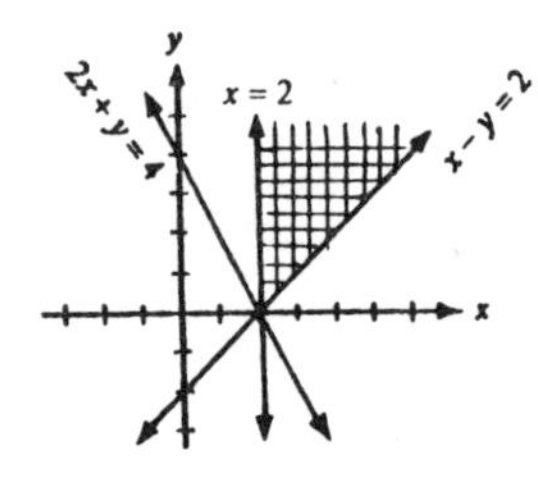

38. 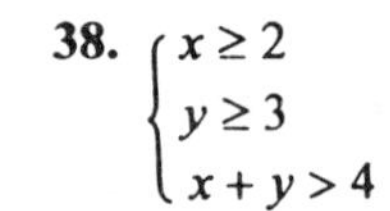
$\begin{cases} x\ge 2 \\ y\ge 3 \\ x+y>4 \end{cases}$

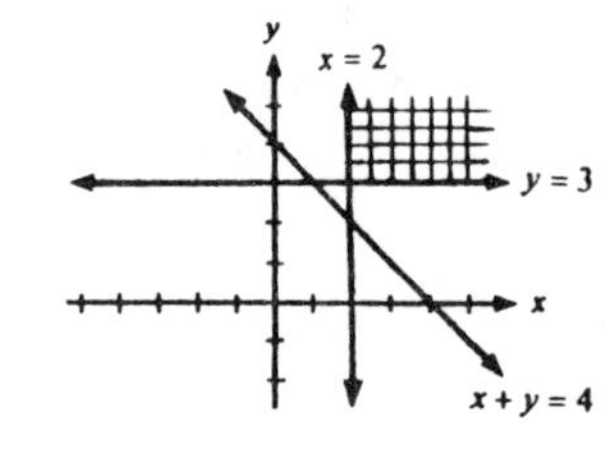

39. 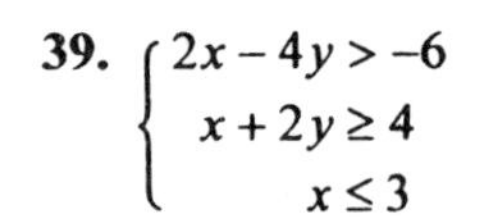
$\begin{cases} 2x-4y>-6 \\ x+2y\ge 4 \\ x\le 3 \end{cases}$

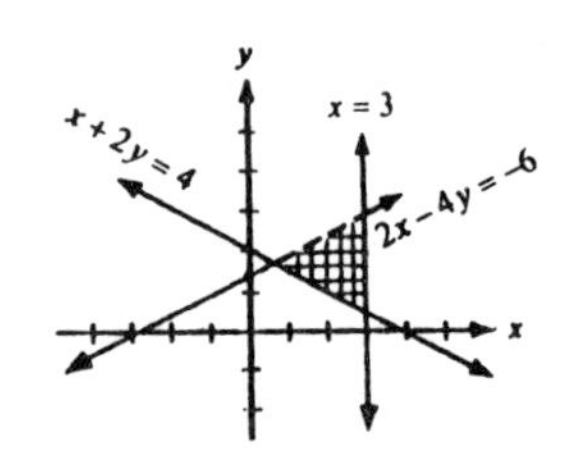

40. $x=9$, $y=-2$ **41.** $x=-8$, $y=7$ **42.** $x=\frac{2}{3}$, $y=7$ **43.** $x=12$, $y=13$ **44.** $x=-9$, $y=-1$ **45.** $a=\frac{1}{2}$, $b=\frac{1}{3}$

46. $x=0.6$, $y=0.8$ **47.** $x=-0.2$, $y=0.5$ **48.** $x=0.5$, $y=0.4$ **49.** $a=5$, $b=10$ **50.** $x=4$, $y=2$ **51.** $x=-3$, $y=2$

52. 7 and 17 **53.** 49 and 70 **54.** 25 one-dollar bills, 18 five-dollar bills

55. 54 nickels, 19 quarters **56.** 30 cm by 18 cm **57.** 40 hamburgers, 35 bagels

58. 33 students, 240 pencils **59.** \$4,500 in bank, \$2,500 in stocks

60. Speed of the airplane = 875 km/h , Speed of the wind = 125 km/h
(Hint: 3 h 18 min = $3\frac{18}{60}=3.3$ h , 4 h 24 min = $4\frac{24}{60}=4.4$ h)

61. 120 seconds

62 ~ 64 graphs are left to the students

62. $w+t\le 10$, $3w+t\ge 14$, $w\ge 0$, $t\ge 0$ **63.** $x+y>30$, $x+y<55$, $x>0$, $y>0$ **64.** Min.C = \$1,200, $x=80$, $y=120$

CHAPTER 11 – 1 Square Roots and Radical page 230

1. 4 **2.** 5 **3.** –6 **4.** 7 **5.** 8 **6.** –9 **7.** ±12 **8.** 100 **9.** –14 **10.** ±20 **11.** –30
12. 25 **13.** 28 **14.** 36 **15.** ±48 **16.** 105 **17.** 0.3 **18.** –0.4 **19.** –0.5 **20.** ±0.9
21. 1.2 **22.** 0.04 **23.** 0.05 **24.** 0.11 **25.** 0.12 **26.** ±0.13 **27.** 0.02 **28.** 0.006
29. $1\frac{1}{2}$ **30.** $-\frac{5}{7}$ **31.** $\pm\frac{1}{6}$ **32.** $\frac{9}{20}$ **33.** $\frac{10}{11}$ **34.** $\pm\frac{1}{2}$ **35.** $-\frac{2}{3}$ **36.** $\frac{1}{5}$
37. 3 **38.** 4 **39.** –2 **40.** 3 **41.** 2 **42.** ±3.162 **43.** 3.464 **44.** –6.325 **45.** 8.660
46. 5.196 **47.** –5.657 **48.** 11.180 **49.** ±0.1 **50.** 0.2 **51.** 0.346 **52.** ±18.028
53. 4 **54.** –4 **55.** 41.833 **56.** 31.177 **57.** 0.949 **58.** 1.871 **59.** ±55.426
60. 25.456 **61.** 0.775 **62.** ±1.354 **63.** –2.598 **64.** 2.789

CHAPTER 11 – 2 Simplifying Radicals page 232

1. $2\sqrt{3}$ **2.** $3\sqrt{2}$ **3.** $-2\sqrt{6}$ **4.** $\pm 3\sqrt{3}$ **5.** $4\sqrt{2}$ **6.** $2\sqrt{7}$ **7.** $4\sqrt{5}$ **8.** $4\sqrt{3}$ **9.** $5\sqrt{6}$
10. $\pm 6\sqrt{3}$ **11.** $2\sqrt{30}$ **12.** $-9\sqrt{2}$ **13.** $6\sqrt{11}$ **14.** $30\sqrt{2}$ **15.** $40\sqrt{2}$ **16.** $25\sqrt{3}$
17. $30\sqrt{2}$ **18.** $-8\sqrt{11}$ **19.** $\pm 35\sqrt{5}$ **20.** $110\sqrt{3}$ **21.** $6\sqrt{66}$ **22.** $162\sqrt{2}$ **23.** $160\sqrt{3}$
24. $300\sqrt{3}$ **25.** $-3\sqrt[3]{2}$ **26.** $-3\sqrt[3]{2}$ **27.** $\pm 2\sqrt[3]{2}$ **28.** $3\sqrt[3]{3}$ **29.** $-3\sqrt[3]{3}$ **30.** 3 **31.** 3
32. –3 **33.** 2 **34.** $\pm\frac{2\sqrt{2}}{7}$ **35.** $\frac{\sqrt{6}}{2}$ **36.** $-\sqrt{30}$ **37.** $\frac{\sqrt{3}}{2}$ **38.** $\frac{11\sqrt{2}}{4}$ **39.** $2\sqrt{6}$ **40.** $\frac{\sqrt{85}}{5}$
41. $2a$ **42.** $10x^2$ **43.** $2\sqrt{7}a^2$ **44.** $x-4$ **45.** $a+b$ **46.** $-4\sqrt{2}\,b$ **47.** $3a\sqrt{2a}$
48. $2x^2\sqrt{6x}$ **49.** $2\sqrt{6}\,ab^2$ **50.** $\frac{x^2y^2\sqrt{y}}{2z}$ **51.** $\pm\frac{2ab\sqrt{2a}}{c^2}$ **52.** $4m^3n^2\sqrt{5mn}$ **53.** $a^2b\sqrt[3]{b^2}$
54. a^3b^2 **55.** $-3x$ **56.** $-2x$ **57.** $2|a|$ **58.** $10x^2$ (Hint: x^2 is always nonnegative.)
59. $\pm 2\sqrt{7}a^2$ (Hint: a^2 is always nonnegative.) **60.** $|x-4|$ **61.** $|a+b|$ **62.** $-4\sqrt{2}|b|$
63. $3a\sqrt{2a}$ (Hint: a must be nonnegative.) **64.** $2x^2\sqrt{6x}$ (Hint: x must be nonnegative.)
65. $2\sqrt{6}\,|a|\,b^2$ **66.** $\frac{x^2y^2\sqrt{y}}{2|z|}$, $z \neq 0$ (Hint: y must be nonnegative.)
67. $\pm\frac{2a|b|\sqrt{2a}}{c^2}$, $c \neq 0$ (Hint: a must be nonnegative.)
68. $4m^3n^2\sqrt{5mn}$ (Hint: Both m and n must be nonnegative.)
69. $a^2b\sqrt[3]{b^2}$ (Hint: b could be negative.)
70. $\sqrt[4]{a^{12}b^8} = \sqrt[4]{a^{12}}\cdot\sqrt[4]{b^8} = \sqrt[4]{(a^3)^4}\cdot\sqrt[4]{(b^2)^4} = |a^3|\,b^2$ **71.** $-3x$ **72.** $-2x$

In advanced algebra, the following formulas are useful for problems **57 ~ 72**:

$$\sqrt[n]{a^n} = |a| \text{ , if } n \text{ is even.}$$

$$\sqrt[n]{a^n} = a \text{ , if } n \text{ is odd.}$$

CHAPTER 11 – 3 Simplifying Radical Expressions page 234

1. $7\sqrt{3}$ **2.** $3\sqrt{3}$ **3.** $-3\sqrt{3}$ **4.** $8\sqrt{6}$ **5.** $17\sqrt{5}$ **6.** $-5\sqrt{2}$ **7.** $5\sqrt{5}$ **8.** $3\sqrt[3]{4}$
9. $-15\sqrt{2}$ **10.** $13\sqrt{6}$ **11.** $7\sqrt{7}$ **12.** $3\sqrt{11}-\sqrt{13}$ **13.** $5\sqrt{2}$ **14.** $3\sqrt[3]{2}$ **15.** $16\sqrt{10}$
16. 108 **17.** $6\sqrt{2}$ **18.** $12\sqrt{6}$ **19.** 80 **20.** 320 **21.** $5-\sqrt{15}$ **22.** $6\sqrt{6}+6$
23. $13+2\sqrt{2}$ **24.** $52-19\sqrt{5}$ **25.** $9+4\sqrt{5}$ **26.** $9-4\sqrt{5}$ **27.** 1
28. $\frac{\sqrt{6}}{6}$ **29.** $\frac{18\sqrt{77}}{77}$ **30.** $\frac{\sqrt{5}}{5}$ **31.** $\frac{6\sqrt{10}}{5}$ **32.** $\frac{8\sqrt{6}}{3}$ **33.** $\frac{\sqrt{30}}{3}$ **34.** 1 **35.** $\frac{\sqrt{6}}{2}$ **36.** $\frac{\sqrt[3]{12}}{2}$
37. $2\sqrt{x}$ **38.** $2\sqrt{3x}$ **39.** $7\sqrt{x}$ **40.** $3\sqrt{2x}$ **41.** $-4\sqrt{5x}$ **42.** $-4\sqrt{3x}$ **43.** $18\sqrt{5x}$
44. $2x\sqrt{x}$ **45.** $7x\sqrt{2}$ **46.** $4x\sqrt{2x}$ **47.** $\frac{\sqrt{5x}}{4}$ **48.** $\frac{(a-x)\sqrt{ax}}{ax}$ **49.** $2\sqrt{6}\,m$ **50.** xy^4
51. $-20a^2b$ **52.** $4a-6a^2$ **53.** $2\sqrt{6}\,x+4x\sqrt{3x}$ **54.** $36x$ **55.** $192x\sqrt{3x}$ **56.** $3\sqrt[3]{7x}$
57. 0 **58.** $\frac{\sqrt{2x}}{2}$ **59.** $\frac{\sqrt{3a}}{3}$ **60.** $\frac{\sqrt[3]{9a}}{3}$

CHAPTER 11 – 4 Solving Equations involving Radicals page 236

1. $x = \pm 7$ **2.** $x = \pm 9$ **3.** $x = \pm 11$ **4.** $x = \pm 2\sqrt{2}$ **5.** $a = \pm 2\sqrt{6}$ **6.** $x = 4\sqrt{3}$
7. No real-number solution **8.** No real-number solution **9.** No real-number solution
10. $x = \pm 8$ **11.** $x = \pm 4$ **12.** No real-number solution **13.** $x = \pm 12$ **14.** $m = \pm 5$
15. No real-number solution **16.** $a = \pm 10$ **17.** $x = \pm 4\sqrt{2}$ **18.** $y = \pm 6\sqrt{2}$ **19.** $x = \pm 2\sqrt{7}$
20. $x = \pm 5\sqrt{6}$ **21.** $x = \pm 5$ **22.** $x = \pm 4$ **23.** $x = \pm 1\frac{1}{2}$ **24.** $x = \pm \frac{2}{3}$ **25.** $x = 6$ or -2
26. $x = 1$ or -9 **27.** $x = 1$ or -5 **28.** $x = 8$ or -4
29. $x = 2$ **30.** $x = -2$ **31.** $x = \frac{3}{4}$ **32.** $x = \pm 2$ **33.** $x = \pm \frac{2}{3}$

29 ~ 33 by using the following rules: If $x^n = a$, then $x = \pm \sqrt[n]{a}$ if n is even.
$x = \sqrt[n]{a}$ if n is odd.

34. $x = 36$ **35.** $x = 64$ **36.** No real-number solution **37.** $x = 8$ **38.** $x = 12$ **39.** $x = 16$
40. $x = 4$ **41.** $a = 4$ **42.** $x = 81$ **43.** $x = 77$ **44.** $x = 8$ **45.** $x = 11$ **46.** $x = 9$
47. $x = 7$ **48.** $p = 2$ **49.** $x = 3$ **50.** $x = 2\sqrt{2}$ **51.** $x = 5$ **52.** $x = \frac{2}{5}$ **53.** $x = 9$ or -1
54. $x = -8$ or 2 **55.** $a = 64$ **56.** $a = 68$ **57.** $a = 10$
58. 18 **59.** 2 or 1 (Hint: $\sqrt{3n-2} - n = 0$) **60.** 7.07 m **61.** Length = 22 cm, width = 11 cm
62. 8 seconds

CHAPTER 11 – 5 The Pythagorean Theorem page 238

1. 7 **2.** 8 **3.** 11 **4.** 4.24 **5.** 4.90 **6.** 12 **7.** 9.90 **8.** 8.66 **9.** 12 **10.** 18 **11.** 80
12. 50 **13.** 10 **14.** 8 **15.** 6 **16.** $s = 5$ **17.** $x = 9.90$ **18.** $m = 5.66$ **19.** $c = 8.66$
20. $c = 10.39$ **21.** $b = 7.94$ **22.** $a = 17$ **23.** $c = 2.60$ **24.** $c = 7.28$ **25.** No **26.** Yes
27. Yes **28.** No **29.** Yes **30.** Yes **31.** Yes **32.** No **33.** Yes **34.** Yes **35.** 141.42 feet
36. 72.11 meters **37.** 11.31 cm **38.** Hypotenuse = $\sqrt{2}\, s$ **39.** Area = 15.60 $in.^2$ **40.** $h = \frac{\sqrt{3}}{2} s$

CHAPTER 11 – 6 Distance and Midpoint Formulas page 240

1. $\overline{AB} = 6$ **2.** $\overline{CD} = 10$ **3.** $\overline{MN} = 6$ **4.** $\overline{AC} = 10$ **5.** $\overline{EF} = 10$ **6.** $\overline{GH} = 2$ **7.** $\overline{PQ} \approx 8.25$
8. $\overline{ST} \approx 8.25$ **9.** $\overline{PR} \approx 6.71$ **10.** $\overline{AB} \approx 7.21$ **11.** $\overline{AC} \approx 11.66$ **12.** $\overline{PQ} \approx 4.24$ **13.** $d \approx 17.72$
14. $d \approx 5.66$ **15.** $d \approx 5.83$ **16.** $d = 9$ **17.** $d = 4$ **18.** $d \approx 17.69$ **19.** $d = 10$ **20.** $d = 5$
21. $d \approx 2.03$ **22.** $d \approx 3.22$ **23.** $d \approx 6.56$ **24.** $d \approx 5.29$ **25.** $M(5, 0)$ **26.** $M(-3, 0)$
27. $M(0, -1)$ **28.** $M(5, 5)$ **29.** $M(-2.5, 2.5)$ **30.** $M(5, 2)$ **31.** $M(1.5, 2.5)$ **32.** $M(-7, 8)$
33. $M(-2, -5.5)$ **34.** $M(-6, -3.5)$ **35.** $M(3, \frac{5}{6})$ **36.** $M(4, -\frac{\sqrt{3}}{2})$ **37.** $Q(8, 0)$ **38.** $Q(-8, 0)$
39. $Q(8, 7)$ **40.** $Q(0, 4)$ **41.** $Q(7, 8)$ **42.** $Q(-7, 10)$ **43.** $Q(-7, 0)$ **44.** $Q(4, 1)$ **45.** $Q(6, -2\sqrt{3})$
46 ~ 53 Hint: Find the distance of each side, then apply the **Pythagorean Theorem**.
46. Yes **47.** Yes **48.** Yes **49.** No **50.** No **51.** No **52.** Yes **53.** No

A dog owner said to his friend. Dog owner: My dog is very good. He picks up my newspaper in front of my house and gives it to me every morning. Friend: My dog is the best. He picks up different newspapers from my neighbors and gives them to me every morning.

CHAPTER 11 EXERCISES Radicals page 241 ~ 242

1. 3 **2.** 1 **3.** ±11 **4.** No real-number solution **5.** No real-number solution **6.** 40 **7.** 1.9 **8.** 1.1 **9.** ±0.03 **10.** 0.25 **11.** 3.87 **12.** –4.47 **13.** $1\frac{1}{3}$ **14.** $\frac{1}{5}$ **15.** $\frac{5}{9}$ **16.** $\frac{2}{3}$ **17.** 25 **18.** ±24.25 **19.** –26.29 **20.** 21.47 **21.** 0.77 **22.** 1.29 **23.** 0.94 **24.** 1.06 **25.** 5 **26.** 5 **27.** –5 **28.** No real-number solution **29.** $2\sqrt{5}$ **30.** $10\sqrt{2}$ **31.** $5\sqrt{3}$ **32.** $42\sqrt{2}$ **33.** $\frac{\sqrt{15}}{5}$ **34.** $\frac{\sqrt{15}}{3}$ **35.** $\frac{2\sqrt{2}}{3}$ **36.** $\frac{3\sqrt{2}}{4}$ **37.** $5\sqrt[3]{2}$ **38.** $3\sqrt[4]{5}$ **39.** $2\sqrt[5]{2}$ **40.** $2\sqrt[4]{4}$ **41.** $3x$ **42.** $4x^2$ **43.** $8x\sqrt{x}$ **44.** $3x\sqrt{2x}$ **45.** $a-5$ **46.** $2a+1$ **47.** $2x\sqrt[3]{3x^2}$ **48.** $-2\sqrt[3]{2}\,n$ **49.** $2mn\sqrt[4]{2n}$ **50.** $-2n\sqrt[5]{m^4}$ **51.** $\dfrac{xy^3}{3z^2}$, $z \neq 0$ **52.** $\dfrac{3a^2\sqrt{2b}}{2c^3}$, $\neq 0$ **53.** $3|x|$ **54.** $4x^2$ **55.** $8x\sqrt{x}$ **56.** $3x\sqrt{2x}$ **57.** $|a-5|$ **58.** $|2a+1|$ **59.** $2x\sqrt[3]{3x^2}$ **60.** $-2\sqrt[3]{2}\,n$ **61.** $2|m|n\sqrt[4]{2n}$ **62.** $-2n\sqrt[5]{m^4}$ **63.** $\frac{|xy^3|}{3z^2}, z \neq 0$ **64.** $\frac{3a^2\sqrt{2b}}{2|c^3|}, c \neq 0$ **65.** $5\sqrt{5}$ **66.** $2\sqrt{6}$ **67.** $\sqrt{3}$ **68.** $-\sqrt[3]{2}$ **69.** $3\sqrt{10}$ **70.** $50\sqrt{6}$ **71.** 63 **72.** 189 **73.** $12-48\sqrt{3}$ **74.** $1-\sqrt{5}$ **75.** $14-4\sqrt{6}$ **76.** 10 **77.** $\frac{8\sqrt{5}}{15}$ **78.** $\frac{\sqrt{6}}{4}$ **79.** $\frac{\sqrt[3]{10}}{2}$ **80.** $-5\sqrt{5x}$ **81.** $7\sqrt{5x}$ **82.** $6\sqrt{2}\,a^3b$ **83.** $\frac{\sqrt{b}}{2}, b \neq 0$ **84.** $\frac{\sqrt{10y}}{2}, x \neq 0$ **85.** $\frac{\sqrt[3]{20y}}{2}$ **86.** $x = \pm 5$ **87.** $x = 25$ **88.** No real-number solution **89.** No real-number solution **90.** $x = 3$ **91.** $x = -3$ **92.** $x = 12$ **93.** $x = \pm\frac{2}{3}$ **94.** $x = -\frac{2}{3}$ **95.** $x = 4$ **96.** $x = -1$ or -7 **97.** $x = 7$ **98.** $x = \frac{3}{7}$ **99.** $x = 27$ **100.** $x = 3$ **101.** $20\,cm$ **102.** $6\sqrt{5}$ feet **103.** 127.28 feet **104.** $\overline{AB} \approx 10.44$, M(2.5, 2) **105.** $d = \sqrt{17}$, M(7.5, $3\sqrt{2}$) **106.** $B(-2, -14)$ **107.** Yes **108.** $\frac{\sqrt{x}+2}{x-4}$ **109.** $2-\sqrt{3}$ **110.** $5-2\sqrt{6}$

CHAPTER 12 – 1 Quadratic Equations involving Perfect Squares page 246

1. $x^2-8x+16$ **2.** $x^2+8x+16$ **3.** $x^2+24x+144$ **4.** $x^2-24x+144$ **5.** $4y^2-12y+9$ **6.** $4y^2+12y+9$ **7.** $x^2+x+\frac{1}{4}$ **8.** $x^2-x+\frac{1}{4}$ **9.** $4x^2-3x+\frac{9}{16}$ **10.** $\frac{1}{4}a^2-\frac{1}{5}a+\frac{1}{25}$ **11.** $(x-1)^2$ **12.** $(x+1)^2$ **13.** $(x+2)^2$ **14.** $(x-2)^2$ **15.** $(x+4)^2$ **16.** $(x-4)^2$ **17.** $(y-10)^2$ **18.** $(y+10)^2$ **19.** $(a+\frac{1}{8})^2$ **20.** $(2x-\frac{1}{2})^2$ **21.** $x = \pm 4$ **22.** No real-number solution **23.** $x = \pm 12$ **24.** $x = \pm 3\sqrt{6}$ **25.** $x = \pm 10\sqrt{2}$ **26.** $a = \pm\frac{4}{9}$ **27.** $a = \pm 2\sqrt{5}$ **28.** $y = \pm 2\sqrt{6}$ **29.** $y = \pm\frac{2}{5}$ **30.** $x = \pm 2\sqrt{2}$ **31.** $x = \pm\sqrt{23}$ **32.** No real-number solution **33.** $x = 2$ or -6 **34.** $x = 2 \pm 3\sqrt{6}$ **35.** $a = -5 \pm 2\sqrt{7}$ **36.** $a = 5 \pm 2\sqrt{7}$ **37.** No real-number solution **38.** $y = 10$ or 2 **39.** $x = -7 \pm 3\sqrt{2}$ **40.** $x = -1$ or -5 **41.** $m = 9 \pm \sqrt{6}$ **42.** $n = -1 \pm 2\sqrt{2}$ **43.** $x = 1$ or -7 **44.** $x = -6$ or -10 **45.** $x = 8$ or 6 **46.** $a = 10$ or -12 **47.** $x = \pm\frac{3}{5}$ **48.** $x = \pm 2\sqrt{6}$ **49.** $x = 8\frac{1}{2}$ or $3\frac{1}{2}$ **50.** $x = -5 \pm \frac{\sqrt{7}}{8}$ **51.** $m = 0$ or ± 7 **52.** $a = 0$ or $\pm 3\sqrt{2}$ **53.** $8.94\,m$ **54.** $r = 4\ cm$ **55.** 9 seconds **56.** 8 seconds to reach 272 feet (Hint: $h = 272$), 9 seconds to reach the ground (Hint: $h = 0$)

CHAPTER 12 – 2 Completing the Square page 248

1. 9 **2.** 9 **3.** 25 **4.** 25 **5.** 121 **6.** 1 **7.** $\frac{1}{4}$ **8.** 361 **9.** $\frac{121}{4}$ **10.** $\frac{169}{4}$ **11.** $\frac{225}{4}$
12. $\frac{1}{36}$ **13.** $\frac{25}{144}$ **14.** $\frac{9}{256}$
15. 9 (Hint: $4x^2-12x=4(x^2-3x)$, $4(x^2-3x+\frac{9}{4})=4x^2-12x+9=(2x-3)^2$ **16.** 36
17. $x^2+6x+9=(x+3)^2$ **18.** $x^2-10x+25=(x-5)^2$ **19.** $a^2-3a+\frac{9}{4}=(a-\frac{3}{2})^2$
20. $m^2+\frac{3}{4}m+\frac{9}{64}=(m+\frac{3}{8})^2$ **21.** $m^2-1.8m+0.81=(m-0.9)^2$ **22.** $4x^2+24x+36=(2x+6)^2$
23 ~ 58 Hint: If the answer is not a radical number, we can also solve it by factoring.
23. $x=-8$ or 2 **24.** $x=8$ or -2 **25.** $x=11$ or -1 **26.** $x=-11$ or 1 **27.** $x=11\pm4\sqrt{5}$
28. $x=-11\pm2\sqrt{30}$ **29.** $a=1\pm2\sqrt{3}$ **30.** $x=1\frac{1}{2}$ or $\frac{1}{2}$ **31.** $x=-19\pm6\sqrt{10}$
32. $x=\frac{11}{2}\pm\sqrt{30}$ **33.** $x=-\frac{13}{2}+2\sqrt{10}$ **34.** $a=-1\pm\sqrt{13}$ **35.** $a=-5\pm2\sqrt{7}$
36. No real-number solution **37.** $y=7\pm3\sqrt{2}$ **38.** No real-number solution
39. $p=10$ or -12 **40.** $x=10$ or 6 **41.** $x=-7$ or 1 **42.** $x=2\pm3\sqrt{6}$ **43.** $x=-5\pm2\sqrt{7}$
44. $n=-1\pm2\sqrt{2}$ **45.** $n=7\pm3\sqrt{2}$ **46.** $y=-4\pm2\sqrt{5}$ **47.** $p=-3\pm\sqrt{15}$
48. $x=-\frac{1}{2}\pm\frac{\sqrt{13}}{2}$ **49.** $x=\frac{3}{2}\pm\frac{\sqrt{17}}{2}$ **50.** $k=-4\pm3\sqrt{2}$ **51.** $k=2\pm\sqrt{6}$ **52.** $m=\frac{7}{2}\pm\frac{\sqrt{73}}{2}$
53. $x=-\frac{1}{4}\pm\frac{\sqrt{65}}{4}$ **54.** $x=\frac{1}{3}\pm\frac{\sqrt{10}}{3}$ **55.** $x=-\frac{3}{2}\pm\frac{\sqrt{3}}{2}$ **56.** $x=-1\pm\frac{\sqrt{3}}{3}$ **57.** $x=4\pm2\sqrt{7}$
58. $x=\frac{3}{4}\pm\frac{\sqrt{129}}{4}$ **59.** $(\frac{b}{2})^2$ **60.** $(\frac{b}{2a})^2$ **61.** $\frac{b^2}{4a}$ or $(\frac{b}{2})^2\cdot\frac{1}{a}$
(Hint: $ax^2+bx=a(x^2+\frac{b}{a}x)$, $a(x^2+\frac{b}{a}x+\frac{b^2}{4a^2})=a(x+\frac{b}{2a})^2$, $a(x^2+\frac{b}{a}x+\frac{b^2}{4a^2})=ax^2+bx+\frac{b^2}{4a}$

CHAPTER 12 – 3 The Quadratic Formula Page 250

1. $x=\pm2$ **2.** No real-number solution **3.** $x=0, 4$ **4.** $x=0, -4$ **5.** $x=-2\pm\sqrt{6}$

6. $y=4\pm2\sqrt{3}$ **7.** $x=-5\pm3\sqrt{2}$ **8.** No real-number solution **9.** No real-number solution

10. $x=1\frac{1}{3}, -1$ **11.** $x=\frac{3\pm\sqrt{105}}{12}$ **12.** $x=\frac{3}{4}, -1\frac{2}{3}$ **13.** $x=1\frac{1}{3}, 1$ **14.** $x=-\frac{1}{3}, 1\frac{1}{3}$ **15.** $x=1\pm\sqrt{5}$

16. $y=2, 5$ **17.** No real-number solution . **18.** $y=-1, 8$ **19.** $x=-1\frac{1}{3}, 1$ **20.** $x=\frac{4\pm2\sqrt{13}}{9}$

21. No real number solution **22.** $t=-2\pm\sqrt{13}$ **23.** No real-number solution **24.** $x=\frac{1\pm\sqrt{7}}{3}$

25. $x=-0.3$ or -0.5 **26.** $x=\frac{-0.8\pm\sqrt{1.24}}{2}$ **27.** $x=\frac{0.3\pm\sqrt{2.09}}{4}$

28. $x=1$ or $-\frac{2}{3}$ (Hint: multiply both sides by 3) **29.** $x=\frac{-2\pm\sqrt{22}}{6}$ (Hint: Multiply both sides by 6)

30. $x=\frac{-3\pm\sqrt{105}}{24}$ (Hint: Multiply both sides by 6)

31. $a\approx4.46$ or -2.46 (Hint: $1\pm2\sqrt{3}\approx1\pm2(1.73)\approx1\pm3.46\approx4.46$ or -2.46

32. $x\approx-0.04$ or -37.96 **33.** $y\approx2.61$ or -4.61 **34.** $a\approx0.30$ or -10.30

35. $y\approx11.23$ or 2.77 **36.** $p\approx0.87$ or -6.87

37. $x\approx1.77$ or -2.27 (Hint: $\frac{-1\pm\sqrt{65}}{4}\approx\frac{-1\pm8.06}{4}\approx\frac{-1+8.06}{4}$ or $\frac{-1-8.06}{4}$, 1.77 or -2.27)

38. $x\approx1.39$ or -0.72 **39.** $x\approx-0.64$ or -2.37

40. $r_1+r_2=-\frac{b}{a}$ **41.** $r_1\cdot r_2=\frac{c}{a}$ **42.** $x^2+2x-3=0$ **43.** $x^2-6=0$

CHAPTER 12 – 4 Graphing Quadratic Functions-Parabolas page 253 ~ 254

1. Opens downward **2.** Opens upward **3.** Opens upward **4.** Opens upward
5. Opens downward **6.** Opens upward **7.** Opens upward **8.** Opens downward
9. Open upward

10. One **11.** Two **12.** One **13.** None **14.** Two **15.** One **16.** One **17.** None **18.** Two
19. Two **20.** None **21.** One

22. x – intercepts = 3 and –2 **23.** x – intercepts = –3 and 2 **24.** x – intercepts = 6 and –1
25. x – intercepts = –6 and 1 **26.** x – intercepts = 7 and –4 **27.** x – intercepts = –5 and –7
28. x – intercept = 1 only **29.** x – intercepts = 2.2 and –0.2 **30.** x – intercepts = 2 and –2
31. x – intercepts = none **32.** x – intercepts = 4.6 and –0.6 **33.** x – intercept = 1.9 and –0.2

34. (7, 5) **35.** (4, 1) **36.** (2, –4) **37.** (–4, –1) **38.** (7, –5) **39.** (–2, 4) **40.** (4, –2)
41. (5, –1) **42.** (–4, –2) **43.** (–5, –1) **44.** (–7, 9) **45.** (6, 12) **46.** (–12, 4) **47.** (1, –7)
48. (10, –9) **49.** (0, 12) **50.** (0, –7)

51. $y=(x+2)^2$, vertex (–2, 0) **52.** $y=(x-2)^2$, vertex (2, 0)
53. $y-5=(x+2)^2$, vertex (–2, 5) **54.** $y+5=(x-2)^2$, vertex (2, –5)
55. $y+2=(x-3)^2$, vertex (3, –2) **56.** $y-2=(x+3)^2$, vertex (–3, 2)
57. $y+5=(x-1)^2$, vertex (1, –5) **58.** $y+12=(x-5)^2$, vertex (5, –12)
59. $y-9=(x+7)^2$, vertex (–7, 9) **60.** $y+1=(x+5)^2$, vertex (–5, –1)
61. $y+4=3(x-2)^2$, vertex (2, –4) **62.** $y-2=2(x+4)^2$, vertex (–4, 2)

63. $y=\frac{1}{2}x^2$ **64.** $y=-\frac{1}{2}x^2$ **65.** $y=x^2+1$ **66.** $y=-x^2+2$

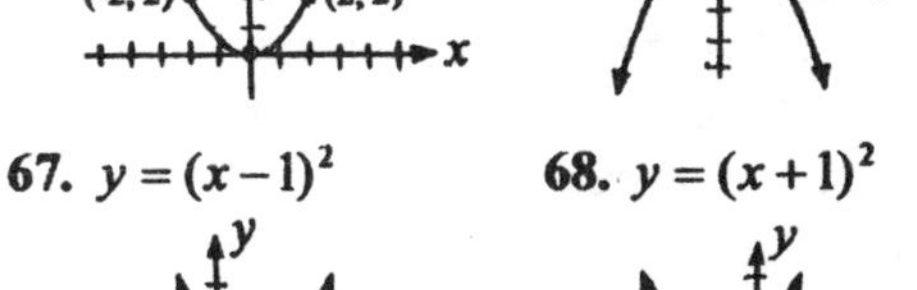

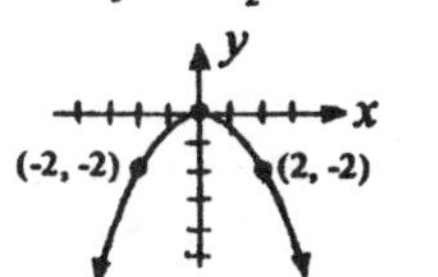

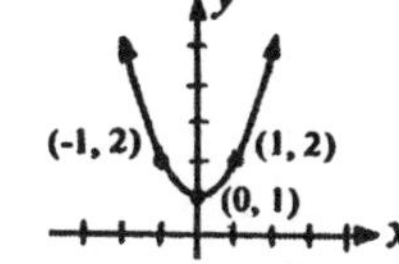

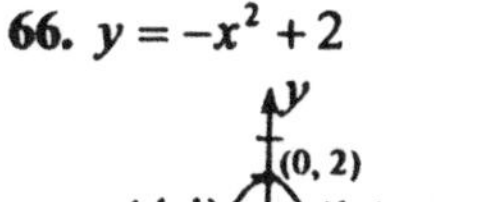

67. $y=(x-1)^2$ **68.** $y=(x+1)^2$ **69.** $y=2(x-2)^2$ **70.** $y=(x-3)^2+1$

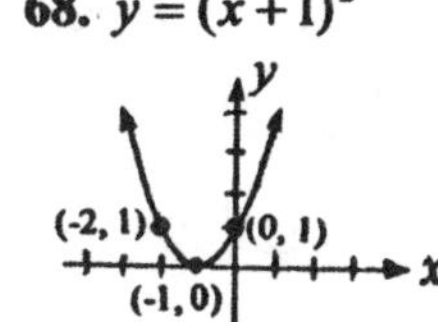

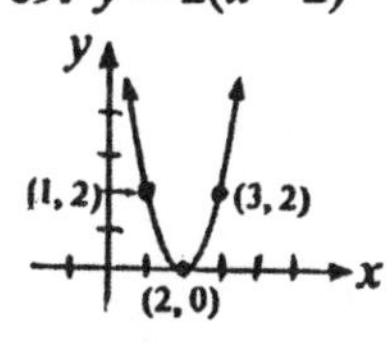

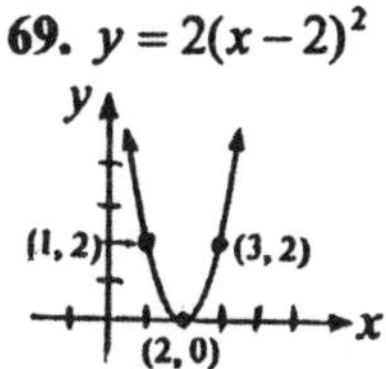

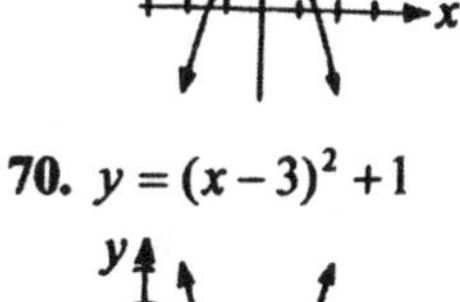

71. $y=2(x-3)^2+1$ **72.** $y=x^2+4x+3$ **73.** $y=x^2+2x-3$ **74.** $y=x^2+2x+2$

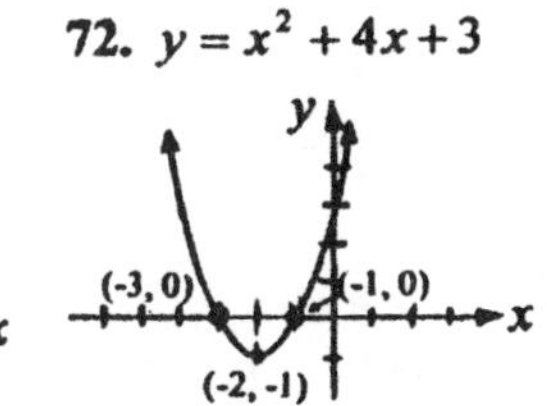

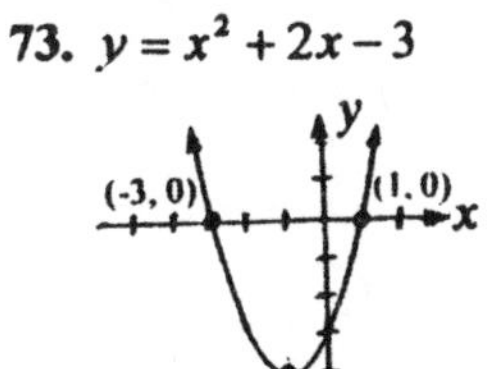

75. $k=25$ **76.** $k>-\frac{9}{20}$ **77.** $k<-\frac{1}{20}$ **78.** It moves the graph of $y=x^2$, one unit upward.
79. It moves the graph of $y=x^2$, 1 unit to the left
80. It moves the graph of $y=x^2$, 2 units to the right and 3 units downward.
81. a. 1 x – intercept **b.** None **c.** 2 x – intercepts **82.** Proof is left to the students.
83. $y=x^2+2x-3$ **84.** $y=-x^2+2$ **85.** $y=x^2-2x+1$

In an English class, the teacher asked the kids to write a letter to his (her) mother regarding what he (she) did today in school. The teacher noticed that John was writing very slowly.
Teacher: John, why do you write so slowly ?
John: My mom could not read fast.

CHAPTER 12 Exercises Quadratic Equations and Functions Page 255 ~ 256

1. $(x-3)^2$ **2.** $(x+5)^2$ **3.** $(y+9)^2$ **4.** $(y-8)^2$ **5.** $(a+10)^2$ **6.** $(a-12)^2$
7. $(x-\frac{1}{3})^2$ **8.** $(x+\frac{1}{6})^2$ **9.** $(2x-5)^2$ **10.** $(3x+6)^2$
11. $x^2-12x+36=(x-6)^2$ **12.** $x^2+14x+49=(x+7)^2$ **13.** $a^2+5a+\frac{25}{4}=(a+\frac{5}{2})^2$
14. $a^2-a+\frac{1}{4}=(a-\frac{1}{2})^2$ **15.** $p^2-2.4p+1.44=(p-1.2)^2$ **16.** $9x^2-6x+1=(3x-1)^2$
17. $x=\pm1$ **18.** $y=\pm1.5$ **19.** $y=\pm2\sqrt{7}$ **20.** $x=\pm4\sqrt{3}$ **21.** $x=\pm1.6$ **22.** $a=\pm\frac{2}{5}$
23. $a=\pm\frac{5}{2}$ **24.** $p=\pm\frac{\sqrt{3}}{2}$ **25.** $p=\pm\frac{2\sqrt{3}}{3}$ **26.** $x=6$ or 0 **27.** No real-number solution
28. No real-number solution **29.** $x=0$ or -8 **30.** $x=8\pm3\sqrt{2}$ **31.** $x=8\pm2\sqrt{5}$
32. $x=1+\sqrt{5}$ **33.** $x=1\pm2\sqrt{2}$ **34.** $x=4\frac{1}{3}$ or $-1\frac{2}{3}$ **35.** $x=-1\pm\frac{\sqrt{6}}{2}$ **36.** $y=1\pm2\sqrt{2}$
37. $x=7\pm2\sqrt{3}$ **38.** $x=\pm\frac{3}{4}$ **39.** $x=\pm\sqrt{6}$ **40.** $x=5\pm\frac{\sqrt{3}}{4}$ **41.** $x=1$ or -4
42. $x=-5$ or 7 **43.** $y=-8$ or 11 **44.** $y=8$ or 10 **45.** $t=9$ or -1 **46.** $x=\frac{1}{2}$ or -4
47. $x=-2\frac{1}{2}$ or 1 **48.** $x=1\pm\sqrt{5}$ **49.** $x=1\pm2\sqrt{2}$ **50.** $x=3\pm2\sqrt{2}$ **51.** $x=-2\pm3\sqrt{2}$
52. $x=\frac{2}{3}$ or -1 **53.** $x=\frac{1}{3}$ or -1 **54.** No real-number solution **55.** $x=\frac{4\pm\sqrt{6}}{5}$
56. $x=\pm7.21$ **57.** $x=10.32$ or -2.32 **58.** $x=1.83$ or -3.83 **59.** $x=-1.54$ or -8.46
60. $x=1.62$ or -0.62 **61.** $y=1.20$ or -4.20 **62.** $y=0.37$ or -5.37 **63.** $a=1.13$ or -7.13
64. $a=1.86$ or -4.86 **65.** $x=2.77$ or -1.27 **66.** $x=0.77$ or -0.44 **67.** $x=0.65$ or -1.15
68. $x=0$ or ±3.46 **69.** $x=0$ or ±4.24 **70.** $x=0, -0.38,$ or -2.62
71. Vertex(0, 4), no x – intercept **72.** Vertex(0, −8), x – intercepts = 2.8 and −2.8
73. Vertex(−2, −9), x – intercepts = 1 and −5 **74.** Vertex(2, 1), no x – intercept
75. Vertex(−7, 10), no x – intercept **76.** Vertex(−6, 3), x – intercepts = −4.3 and −7.7
77. Vertex(4, −1), x – intercepts = 4.5 and 3.6 **78.** Vertex(11, 8), no x – intercept
79. Vertex(−3, −15), x – intercepts = −1.7 and −4.3 **80.** Vertex(6, 0), x – intercept = 6
81. Vertex(0, 6), no x – intercept **82.** Vertex(4, −7), no x – intercept
83. Vertex(−2, −9), x – intercepts = 5 and −1 **84.** Vertex(2, −6), x – intercepts = 4.5 and −0.5
85. Vertex$(-\frac{3}{2}, -\frac{5}{4})$, x – intercepts = −0.38 and −2.62

86. $y=(x+1)^2-1$ **87.** $y=2(x-3)^2+1$ **88.** $y=-(x-2)^2+4$ **89.** $y=x^2+2x-3$

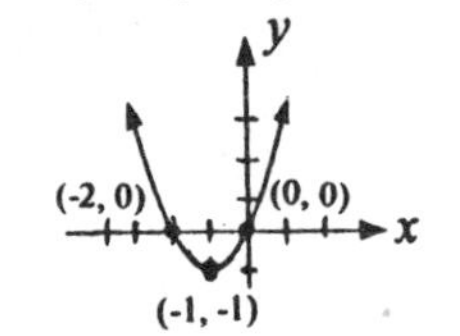

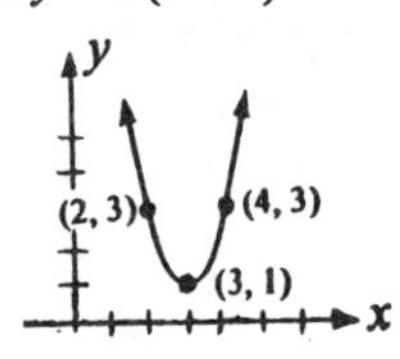

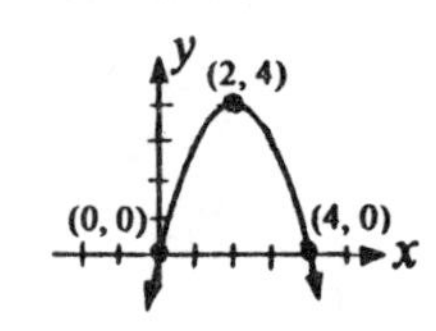

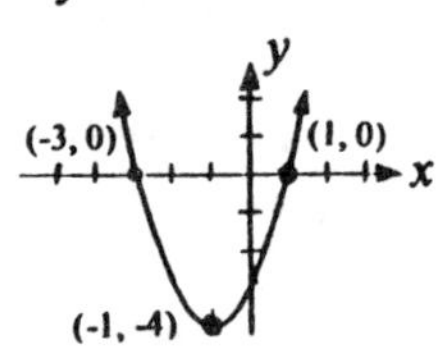

90. 2 x – intercepts **91.** No x – intercept **92.** $k=\pm8$ (Hint: $b^2-4ac=0$)
93. $k>4$ (Hint: $b^2-4ac<0$) **94.** The axis of symmetry is the line $x=-1$.
95. The axis of symmetry is the line $x=3$.
96. It moves the graph of $y=x^2$, 4 units downward. **97.** It moves the graph of $y=x^2$, 3 units to the right and 4 units upward. **98.** Min. $y=17.5$ when $x=2.5$
99. Max. $y=18$ when $x=2$ **100.** Max. Profit $y=90$ when $x=8$

CHAPTER 13 – 1 Simplify Rational Expressions page 260

1. $x=0$ **2.** $x=1$ **3.** $a=5$ **4.** $y=-4$ **5.** $x=3$ **6.** $n=2,-1$ **7.** $n=-5,1$
8. $t=0,-2$ **9.** None **10.** None **11.** $y=5,-1$ **12.** $y=0,2,-1$

13. $3, x\neq 0$ **14.** $3x^3, x\neq 0$ **15.** $\frac{3}{2x^3}, x\neq 0$ **16.** $\frac{3b^3}{5a^4}, a\neq 0$ **17.** $2x+y$ **18.** $\frac{1}{2x+y}, x\neq -\frac{y}{2}$

19. $\frac{2}{y-4}, y\neq 4$ **20.** $\frac{y-4}{2}$ **21.** $\frac{x+y}{2(x-y)}, x\neq y$ **22.** $-1, x\neq 5$ **23.** $-2, a\neq 3$

24. $-(a+2), a\neq 2$ **25.** $-\frac{1}{2+a}, a\neq -2$ **26.** $\frac{1}{x+8}, x\neq -8, 8$ **27.** $\frac{2+x}{x}, x\neq 0$ **28.** $3, y\neq -3$

29. $\frac{2}{x-y}, x\neq -2y, y$ **30.** $\frac{x-y}{3}, x\neq -y$ **31.** $a-2, a\neq 2$ **32.** $\frac{1}{3(a-1)}, a\neq 1$

33. $\frac{a-b}{a+b}, a\neq -b, b$ **34.** $x-4, x\neq 4$ **35.** $x-3, x\neq 5$ **36.** $-\frac{1}{n+1}, n\neq 3, -1$

37. $\frac{2}{a+2}, a\neq 6, -2$ **38.** $\frac{k-3}{5}, k\neq -4$ **39.** $-\frac{2}{p+1}, p\neq 5, -1$ **40.** $\frac{x-2}{x-4}, x\neq -7, 4$

41. It is undefined when $x=6$ or -3 **42.** It equals zero when $x=0$ or -4.
43. $x=a-b, a\neq -b$ **44.** The width is given by $x+3$.

CHAPTER 13 – 2 Adding and Subtracting Rational Expressions Page 262

1. 15 **2.** 36 **3.** 6 **4.** 15 **5.** 30 **6.** 24 **7.** $10x$ **8.** $12x$ **9.** $12x$ **10.** $40x$ **11.** $8ab$
12. $8ab$ **13.** $6x^2$ **14.** xy **15.** $28y^3$ **16.** $(x+3)(x-3)$ **17.** $48x^2y^2$ **18.** a^2-9
19. $30(a-2)^2$ **20.** $x(x^2-1)$
21. 15 **22.** 15 **23.** $10x$ **24.** $40x$ **25.** $8ab$ **26.** $6x^2$ **27.** $(x+3)(x-3)$ **28.** a^2-9
29. x **30.** a **31.** xy **32.** x^2y **33.** $x(x-5)$ **34.** $6c(c-2)$ **35.** $(t+1)^2$ **36.** $(x-2)(x+1)$

37. $\frac{8x}{15}$ **38.** $\frac{11a}{15}$ **39.** $-\frac{1}{10x}, x\neq 0$ **40.** $\frac{41}{40x}, x\neq 0$ **41.** $\frac{6b-1}{8ab}, a\neq 0, b\neq 0$

42. $\frac{4x-5}{6x^2}, x\neq 0$ **43.** $\frac{7x-3}{(x+3)(x-3)}, x\neq -3, 3$ **44.** $\frac{4}{a+3}, a\neq -3, 3$ **45.** $\frac{x^2+2}{x}, x\neq 0$

46. $\frac{1-5a}{a}, a\neq 0$ **47.** $\frac{2y+5x}{xy}, x\neq 0, y\neq 0$ **48.** $\frac{2xy+9}{x^2y}, x\neq 0, y\neq 0$

49. $\frac{x-10}{x(x-5)}, x\neq 0, 5$ **50.** $-\frac{11c+8}{6c(c-2)}, c\neq 0, 2$ **51.** $\frac{3}{(t+1)^2}, t\neq -1$ **52.** $\frac{5}{2(x+1)}, x\neq -1$

53. $\frac{10x-11}{(x-2)(x+1)}, x\neq 2, -1$ **54.** $\frac{6x-1}{3(x-4)}, x\neq 4$ **55.** $-\frac{1}{2(x+3)}, x\neq -3, 3$ **56.** $\frac{x-1}{x-2}, x\neq 2$

57. $2, n\neq \frac{1}{2}$ **58.** $\frac{-3}{(a-1)(a+2)}, a\neq 1, a\neq -2$ **59.** $\frac{1}{x-1}, x\neq -3, 1$ **60.** $\frac{-y}{(y-3)^2}, y\neq 3$

61. $\frac{2b-4}{(b+1)(b+3)(b-5)}, b\neq -3, -1, 5$ **62.** $\frac{13-6c}{9-c^2}, c\neq -3, 3$ **63.** $\frac{-6x+10}{x(x+5)(x-5)}, x\neq 0, -5, 5$

CHAPTER 13 – 3 Multiplying and Dividing Rational Expressions Page 264

1. $\frac{1}{2x}$, $x \neq 0$ 2. $4x^2$, $x \neq 0$ 3. $\frac{9}{y}$, $y \neq 0$ 4. $\frac{5y^3}{2}$, $y \neq 0$ 5. $\frac{3y^3}{x^3}$, $x \neq 0$, $y \neq 0$

6. $\frac{9}{a^4b^2}$, $a \neq 0, b \neq 0$ 7. $\frac{3a^3b^4}{2}$, $a \neq 0, b \neq 0$ 8. $\frac{6}{d^5}$, $c \neq 0, d \neq 0$ 9. $\frac{4x^2}{y^2}$, $x \neq 0$, $y \neq 0$

10. $\frac{8x}{y}$, $x \neq 0$, $y \neq 0$ 11. $\frac{x+2}{x-2}$, $x \neq 2, 5$ 12. $4y$, $y \neq 0$, $x \neq 2y$ 13. -1, $a \neq 3$

14. -5, $a \neq 2b$ 15. $\frac{n-4}{2}$, $n \neq 0, -4$ 16. $\frac{n+1}{n+5}$, $n \neq 5, -5$ 17. $\frac{3}{2x}$, $x \neq 0$

18. $\frac{x^2}{y^2}$, $x \neq 0$, $y \neq 0$ 19. $\frac{10p^2}{n^2}$, $n \neq 0$, $p \neq 0$ 20. $\frac{2b^2}{a^2}$, $a \neq 0$, $b \neq 0$ 21. $\frac{9}{x}$, $x \neq 0$

22. $\frac{9y}{40}$, $y \neq 0$ 23. $\frac{x+2}{x-2}$, $x \neq 2, 5$ 24. $4y$, $y \neq 0$, $x \neq 2y$ 25. $\frac{3a}{2}$, $a \neq 0, -1$

26. $3(x+2)$, $x \neq 2$ 27. $-\frac{1}{2}$, $x \neq y$ 28. $\frac{4}{3}$, $a \neq 5$ 29. $\frac{4(x+3)}{x+4}$ 30. $\frac{2(x-2)}{x+3}$

31. $\frac{x-3}{5}$ 32. $\frac{1}{(x-3)(x-1)}$ 33. $\frac{(m-4)(m-5)}{m-2}$ 34. $\frac{x+y}{4x^2}$ 35. $\frac{a+3}{a+2}$ 36. $\frac{9(a-6)}{5(a-4)}$

37. $x-7+\frac{20}{x+2}$ 38. $x+7+\frac{8}{x-2}$ 39. $2a+3-\frac{9}{2a-1}$ 40. $2a-3+\frac{16}{3a+2}$

41. $x^2-2x+3-\frac{11}{2x+4}$ 42. $x^2+2x+5+\frac{20}{2x-5}$ 43. $a^3+4a^2+16a+64+\frac{252}{a-4}$

44. $2n^2-5n+8-\frac{13n+3}{n^2+2n+1}$

CHAPTER 13 – 4 Solving Rational Equations page 266

1. $x = 2\frac{2}{5}$ 2. $x = 3\frac{3}{5}$ 3. $x = 5\frac{5}{6}$ 4. $x = 1\frac{11}{24}$ 5. $y = 1\frac{3}{4}$ 6. $y = 5\frac{2}{5}$ 7. $x = 5$

8. $x = -14$ 9. $a = 9$ 10. $a = 7$ 11. $n = 53$ 12. $n = 16$ 13. $x = 1\frac{1}{2}$ 14. $n = 2\frac{4}{5}$

15. $x = 1$ 16. $t = 8$ 17. $a = \frac{4}{7}$ 18. No solution 19. $y = -\frac{2}{5}$ 20. $x = -\frac{2}{3}, 1$

21. No solution 22. All real numbers except $x = \frac{3}{2}$ 23. $y = 6\frac{1}{2}$ 24. $y = -1\frac{5}{9}$ 25. $x = 4\frac{4}{9}$

26. $x = -42$ 27. $x = \frac{7}{18}$ 28. $x = \frac{23}{42}$ 29. $x = 8\frac{1}{2}$ 30. $x = 1$ 31. $y = \frac{3}{4}$ 32. $y = -1\frac{1}{3}$

33. $n = 1$ (a double root) 34. $n = \frac{1}{3}, 4$ 35. $a = -6$ 36. $x = -4$ 37. $c = 5$ 38. $c = 3$

39. $m = 20$ 40. No solution 41. All real numbers except 2 42. $a = 4$

43. $\frac{x}{y} = -5$ 44. $\frac{x}{y} = \frac{2}{7}$

45. Slope: $m = -\frac{2}{3}$ (Hint: Simplify the equation in the form $y = mx + b$.)

In the shopping mall, a boy said to his father.
Boy: Why are you always holding mom's hand, dad ?
Father: If I don't hold her hand, she will keep on buying new clothes.

CHAPTER 13 – 5 Solving Word Problems by Rational Equations page 268

1. 12 **2.** 12 **3.** 12 and 32 **4.** 4 and 6 **5.** $\frac{3}{4}$, or $\frac{4}{3}$ **6.** 5 and 7 **7.** \$16.4 **8.** \$8.28
9. 10 gallons **10.** 272 miles **11.** 5000 voters **12.** 300 meters by 200 meters **13.** 1.5 hours
14. 2.4 hours **15.** $2\frac{3}{4}$ hours **16.** $3\frac{1}{3}$ miles/h **17.** 60 miles/h **18.** 50 km/h **19.** 5 miles/h
20. 8 gal. of water **21.** 4.5 gal. of water **22.** 1.125 gal. of pure salt

CHAPTER 13 – 6 Inverse Variations page 270

1. $k=20$; $xy=20$ **2.** $k=50$; $xy=50$ **3.** $k=4000$; $mn=4000$ **4.** $k=2700$; $mn=2700$
5. $k=1500$; $(x+8)y=1500$ **6.** $y=\frac{5}{3}$ or $1\frac{2}{3}$ **7.** $x=\frac{5}{3}$ or $1\frac{2}{3}$ **8.** $y=\frac{5}{3}$ or $1\frac{2}{3}$ **9.** $n=45$
10. $m=45$ **11.** $y_1=36$ **12.** $y_1=106\frac{2}{3}$ **13.** $y_1=12.8$ **14.** $y_2=2\frac{2}{3}$ **15.** $x_1=90$
16. $x_2=0.4$ **17.** $y_1=\frac{1}{20}$ **18.** $y_2=\frac{3}{100}$ **19.** $y_1=\frac{4}{49}$ **20.** $y_2=\frac{1}{5}$
21. Yes, If $xy=5$, then $y=\frac{5}{x}$. y is decreased when x is increased.
22. Yes. 6 × 4 =24, 3 × 8 = 24, 2 × 12 = 24 **23.** 8 workers **24.** 4.8 hours **25.** $36m$
26. 40 miles/h **27.** $z=40.5$ (Hint: $z=k\frac{x}{y}$)

CHAPTER 13 Exercises Rational Expressions page 271 ~ 272

1. $\frac{2}{x}$, $x\neq 0$ **2.** $3x^2$, $x\neq 0$ **3.** $\frac{3a^2}{4b^3}$, $a\neq 0$, $b\neq 0$ **4.** $2x-3$ **5.** -2, $x\neq\frac{1}{2}$ **6.** $n+3$, $n\neq 3$
7. $\frac{8(a+3)^2}{3}$, $a\neq -3$ **8.** $\frac{1}{x+5}$, $x\neq 4,-5$ **9.** $\frac{a+5}{a+2}$, $a\neq 5,-2$ **10.** $\frac{t-2}{t}$, $t\neq 0,-2$
11. $\frac{y+6}{y+7}$, $y\neq 4,-7$ **12.** $\frac{x-1}{x(x+2)}$, $x\neq 0,-3,-2$ **13.** $\frac{9x}{14}$ **14.** $\frac{5y}{14}$ **15.** $\frac{5}{14x}$ **16.** $\frac{9}{14y}$
17. $\frac{2}{a^3}$ **18.** $\frac{3}{x}$ **19.** $\frac{10}{n^3}$ **20.** $4n^3$ **21.** $\frac{x+y}{x^2y^2}$ **22.** $\frac{2(a+2)}{a^2b}$ **23.** $\frac{a}{2}$ **24.** $\frac{y}{2x}$ **25.** $\frac{3a+7}{(a+2)(a+3)}$
26. $\frac{-n+12}{(n-2)(n+3)}$ **27.** $-\frac{1}{2}$ **28.** $-\frac{4}{3}$, or $-1\frac{1}{3}$ **29.** $\frac{7a+6}{a(a+2)}$ **30.** $\frac{-a+14}{a^2-4}$ **31.** $2(a+2)$
32. $2(a-2)$ **33.** $\frac{25-4x}{x^2-25}$ **34.** $\frac{2(x+8)}{x^2-25}$ **35.** $\frac{6}{x+5}$ **36.** $\frac{x+4}{3}$ **37.** x **38.** $x+6$ **39.** $\frac{a^3(a+b)}{2}$
40. $m(m+3)$ **41.** $\frac{3}{c+1}$ **42.** $-\frac{1}{x+3}$ **43.** $\frac{1}{2}$ **44.** $x-6+\frac{7}{x+2}$ **45.** $x^2+2x+12+\frac{42}{x-4}$
46. $2x^2-x-1+\frac{5}{2x+1}$ **47.** $x^2-ax+a^2-\frac{2a^3}{x+a}$ **48.** $x=3\frac{3}{4}$ **49.** $x=22\frac{1}{2}$ **50.** $y=6$
51. $y=6$ **52.** $n=4\frac{6}{7}$ **53.** $n=-\frac{3}{4}$ **54.** $x=5$ **55.** $x=5$ **56.** $k=7$ **57.** ϕ (no solution)
58. ϕ (no solution) **59.** All real numbers **60.** All real numbers **61.** $y=-3$ **62.** $x=-8$
63. $x=10$ **64.** $x=8$ **65.** $y=30$ **66.** $x=-2$ **67.** $x=12$ **68.** $y=\frac{2}{7}$ **69.** $x=3$ and 4
70. $c=5$ **71.** ϕ (no solution) **72.** $x=9$ **73.** $a=6$ **74.** $x=1\frac{2}{3}$ **75.** $x=2$
76. 20 **77.** $\frac{3}{7}$, or $\frac{7}{3}$ **78.** 12 and 36 **79.** \$20.16 **80.** $2\frac{11}{12}$ days **81.** $2\frac{1}{5}$ days
82. 5 gallons of water **83.** 0.61 gallons of pure alcohol **84.** $y=7.5$ **85.** $y_2=10$
86. 12 workers **87.** $z=48$ **88.** $f=k\frac{m_1\cdot m_2}{r^2}$ (Hint: It is the Law of Universal Gravitation.)

CHAPTER 14 – 1 Mean, Median, Mode, and Range Page 274

1. 4 **2.** 7 **3.** 20.5 **4.** 15 **5.** 14.4 **6.** 20.4 **7.** 30.6 **8.** 39.8 **9.** 237.4 **10.** 245.5 **11.** 3.4 **12.** 4.1 **13.** 5.5 **14.** 5.3 **15.** 28.4

16. Median 12, Mode 18, Range 14
17. Median 18, Mode 9, Range 23
18. Median 14, Mode 14, Range 26
19. Median 18, Mode 12 and 18, Range 14
20. Mmedian 10, No mode, Range 22
21. Median 16.5, Mode 10 and 20, Range 10
22. Median 2.2, Mode 1.4, Range 7.3
23. Median 2.5, Mode 1.8, Range 3.3
24. Median 29, Mode 26, Range 34
25. Median 26, Mode 25 and 27, Range 29
26. a) Mean 82.9 **b)** 3 tests **c)** Median 84.5, Mode 85, Range 26 **27.** 89 **28.** Median 12
29. a) Mean 79.85 **b)** Mode 78
30. Mean 12.85 hours, Median 12 hours, Mode 12 hours, Range 16.5 hours
31. Mean, Median, and Mode **32.** Range

CHAPTER 14 – 2 Statistical Graphs page 276

1. Ron **2.** David **3.** 16 hours **4.** John and Jim **5.** 57.1% (Hint: $\frac{22-14}{14}$)
6. 72.7% (Hint: $\frac{22-6}{22}$) **7.** 7,900 points **8.** 7,350 points **9.** 1,074 **10.** 15%
11. Chain stores **12.** College store **13.** 3,000 copies **14.** 225 copies **15.** 825 copies
16. 400 copies

CHAPTER 14 – 3 Histogram and Normal Distribution page 278

1. a. and **b** (The chart and the histogram are left to the students.)
2. a. 68% **b.** 84% **c.** 16% **d.** 95%
3. a. 16% **b.** 84%
4. 97.5%

CHAPTER 14 – 4 Stem-and-Leaf Plots & Box-and-Whisker Plots page 280

1.

Stems	leaves
4	8, 9
5	2, 4
6	3, 6, 8
7	0, 4, 5, 6, 8, 9, 9
8	1, 1, 2, 9
9	0, 4

2.

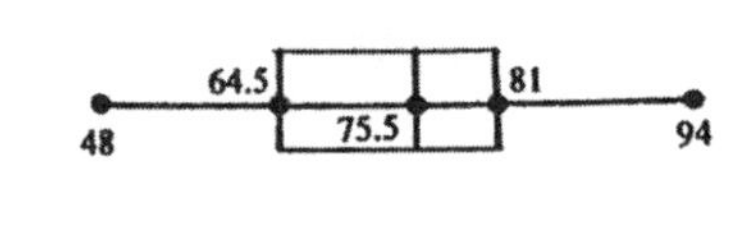

3. Mean = 74, Median = 75.5, Mode = 79 and 81, Range = 46
4. 48 and 94 **5.** $Q_1 = 64.5$, $Q_2 = 75.5$, $Q_3 = 81$ **6.** 81 **7.** Range = 63 – 48 = 15
8. 25% **9.** 75% **10.** The histogram are left to the students.
11. 76 **12.** 25% **13.** 25% **14.** Lower quartile = 66, upper quartile = 82
15. The plot does not show the mean in the data.
16. The plot does not show how many students on the test.
17. 182 copies **18.** None. Each section contains the same 25% of the data.
19. 135 and 210 **20.** Lower quartile = 135 copies, upper quartile = 210 copies

> A young girl walks around in a shopping mall. Her hair and eyes are painted with multiple colors. An old man looks at her frequently.
> Girl: What is wrong with you, old man ? You never got wild when you were young ?
> Man: I am wondering either your father or mother is a peacock. They got wild when they were young.

CHAPTER 14 – 5 Sets and Venn Diagrams page 282

1. $A \cap B = \{7, 9\}$ **2.** $A \cup B = \{1, 2, 3, 4, 5, 7, 9, 10\}$ **3.** $B \cap C = \{9, 10\}$
4. $B \cup C = \{1, 2, 4, 7, 9, 10, 15, 20\}$ **5.** $(A \cup B) \cap C = \{1, 9, 10\}$ **6.** $(A \cap B) \cap C = \{9\}$
7. $A \cap \phi = \phi$ **8.** $A \cup \phi = \{1, 3, 5, 7, 9\}$

9. $\in$ **10.** $\notin$ **11.** $\subset$ **12.** $\subset$ **13.** $=$ **14.** $\not\subset$ **15.** $\subset$ **16.** $\neq$ **17.** $\subset$
18. $\{2, 4, 6, 8\}$ **19.** $\{7, 9, 11, 13\ 15\}$ **20.** $\{1, 2, 3\}$ **21.** ϕ **22.** $\{0\}$ **23.** $\{0, 1, 2, 3, 4, 5\}$
24. ϕ **25.** 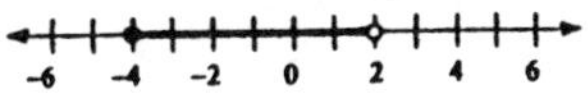

26. $\{1\}, \{2\}, \{3\}, \{1, 2\}, \{2, 3\}, \{1, 3\}, (1, 2, 3\}, \phi$

27. 16 subsets **28.** 32 subsets **29.** 2^n **30.** 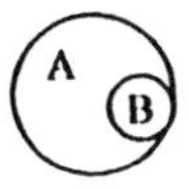

31. 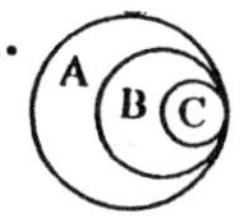

32. M = N

33. 9 students
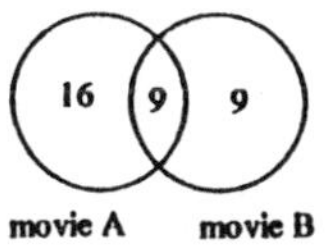

34. 2 students
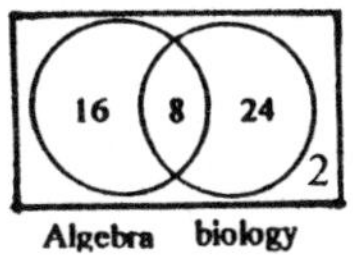

35. a. 5 investors
b. 3 investors
c. 13 investors
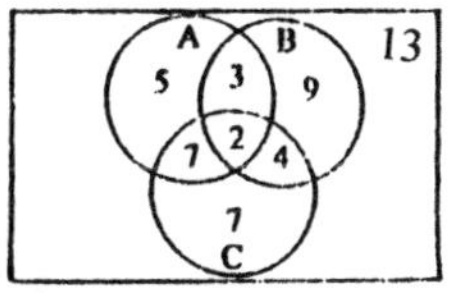

36. {(H, H), (H, T}, (T, H), (T, T)}

37. 10
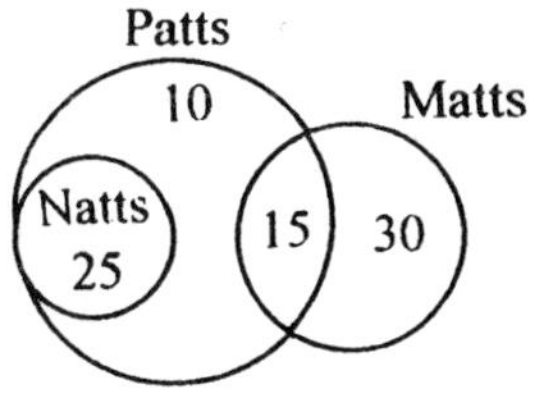

CHAPTER 14 – 6 Probability page 285 ~ 286

1. $\frac{1}{13}$ **2.** $\frac{1}{4}$ **3.** $\frac{1}{52}$ **4.** 0 **5.** $\frac{3}{13}$ **6.** $\frac{1}{2}$ (Hint: hearts and diamonds) **7.** $\frac{2}{13}$ **8.** $\frac{5}{13}$ **9.** $\frac{3}{4}$
10. $\frac{4}{13}$ (Hint: $\frac{4}{52} + \frac{12}{52}$) **11.** $\frac{1}{2}$ **12.** $\frac{1}{5}$ **13.** $\frac{7}{50}$ **14.** $\frac{1}{50}$ **15.** $\frac{17}{50}$ (Hint: $\frac{1}{5} + \frac{7}{50}$) **16.** $\frac{1}{6}$ **17.** 0
18. $\frac{1}{2}$ **19.** $\frac{1}{2}$ **20.** $\frac{1}{3}$ **21.** $\frac{5}{6}$ **22.** $\frac{2}{3}$ **23.** $\frac{1}{2}$ **24.** $\frac{1}{8}$ **25.** $\frac{1}{8}$ **26.** $\frac{1}{4}$ **27.** $\frac{1}{2}$ **28.** $\frac{5}{8}$
29. $\frac{2}{5}$ **30. a.** $\frac{6}{25}$ **b.** $\frac{19}{25}$ **31. a.** $\frac{3}{5}$ **b.** $\frac{1}{2}$ **32. a.** $\frac{2}{5}$ **b.** 2 **c.** $\frac{1}{2}$
33. The tree diagram is left to the students (see example in this section) **34.** 8 **35.** 16
36. 36 **37.** 216 **38.** $\frac{1}{4}$ **39.** $\frac{1}{2}$ **40.** $\frac{7}{8}$ **41.** 75 times **42.** 30 times **43.** $\frac{1}{32}$
44. a. 1,326 voters **b.** 780 voters **c.** 1,794 voters **45.** 60 students **46.** 80 students
47. 2 students **48. a.** $\frac{1}{32}$ **b.** $\frac{1}{16}$ **c.** $\frac{1}{16}$ **49. a.** $\frac{26}{52} \cdot \frac{26}{52} = \frac{1}{4}$ **b.** $\frac{13}{52} \cdot \frac{13}{52} = \frac{1}{16}$ **c.** $\frac{1}{52} \cdot \frac{1}{52} = \frac{1}{2704}$ **d.** $\frac{26}{52} \cdot \frac{26}{52} = \frac{1}{4}$
50. a. $\frac{26}{52} \cdot \frac{25}{51} = \frac{25}{102}$ **b.** $\frac{13}{52} \cdot \frac{12}{51} = \frac{1}{17}$ **c.** 0 (there is only one ace of heart.) **d.** $\frac{26}{52} \cdot \frac{26}{51} = \frac{13}{51}$
51. $\frac{6}{36} = \frac{1}{6}$ (Hint: {(4, 6), (5, 5), (5, 6), (6, 4), (6, 5), (6, 6)} **52.** 2,500 salmons **53.** $\frac{4}{13}$ **54.** $\frac{7}{13}$ **55.** $\frac{1}{26}$

CHAPTER 14 Exercises Statistics and Probability Page 287 ~ 288

1. Mean 5.7, Median 5.5, Mode 5 and 6, Range 6
2. Mean 5.9, Median 7, Mode 9, Range 10
3. Mean 15.1, Median 17, No mode, Range 18
4. Mean 2.8, Median 2.5, Mode 2.4, Range 2.2
5. Mean 2.3, Median 1.5, Mode 1.5, 4.8
6. Mean 31, Median, 31, No mode, Range 44
7. The mean, median, or mode. **8.** The range **9. a.** 77.9 **b.** 5 students **10.** 82

11. a.

Test scores	Tally	Numbers of Students
30-40	///	3
40-50	~~////~~ ///	8
50-60	~~////~~ ~~////~~	10
60-70	~~////~~ ~~////~~ ///	13
70-80	////	4
80-90	//	2

b.
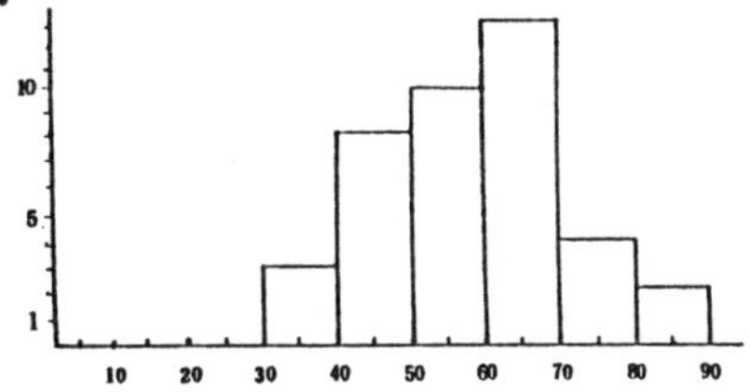

12. a. 60 flights **b.** 40 flights **c.** 40 flights **13. a.** 16% **b.** 50%
14. a. 220 **b.** 30 **c.** May **d.** March **e.** 15.9% **15. a.** 330 people **b.** 170 people
16. a. \$10.13 **b.** \$10.38 **c.** Wednesday **d.** 1.3% **17. a.** 66 **b.** 25% **c.** 25%
18. 35 students **19. a.** $\frac{1}{25}$ **b.** $\frac{3}{25}$ **c.** $\frac{4}{25}$ **d.** $\frac{1}{25}$ **e.** $\frac{4}{25} + \frac{1}{25} = \frac{1}{5}$ **20.** 27.8% **21. a.** $\frac{4}{9}$ **b.** $\frac{14}{33}$

Space for Taking Notes

Sample Test (1)
Numbers

1. **The questions shown here were tested year by year in all kinds of tests, such as PSAT, SAT, ACT, STAR/CAT 6, and High School Exit.**
2. **These two pages cover all possible "basic questions" about arithmetic calculations with integers, decimals, fractions, exponents, scientific notation, percents, absolute values, and square roots.**

Evaluate each expression.

1. $5-9=$

2. $9+(-5)=$

3. $9-(-5)=$

4. $5\times(-9)=$

5. $-9\div(-5)=$

6. $2+3\times4-8\div2=$

7. $3.2+5.3=$

8. $3.2-5.3=$

9. $-3.2-(-5.3)=$

10. $0.4\times(-0.6)=$

11. $-0.48\div0.6=$

12. $0.4\times(-0.6)\div0.3=$

13. $\frac{4}{5}+\frac{3}{5}=$

14. $\frac{3}{4}-\frac{2}{5}=$

15. $\frac{13}{15}-(\frac{1}{3}+\frac{2}{5})=$

16. $\frac{4}{5}\times\frac{3}{5}=$

17. $-\frac{4}{5}\div\frac{3}{5}=$

18. $-2\frac{2}{3}\div(-1\frac{1}{3})=$

19. $5^2=$

20. $5^{-2}=$

21. $\frac{1}{5^{-2}}=$

22. $(-\frac{3}{4})^3=$

23. $5^2\times5^{-2}=$

24. $\frac{2^{-3}}{4^{-2}}\times10^4=$

25. $9^5\times9^3=$

26. $(9^5)^3=$

27. $\frac{9^5}{9^3}=$

28. $9^5\times9^{-3}=$

29. $\frac{9^5}{9^{-3}}=$

30. $(-2)^3\cdot(-2)^{-4}\cdot3^0=$

31. Find the prime factored form of the least common denominator of $\frac{5}{18}+\frac{7}{30}$.
Solution:

32. The distance (d) in feet that an object falls in t seconds is given by the formula $d=16t^2$. Find the distance of an object falls in 8 seconds.
Solution:

------Continued-----

Write each number in scientific notation.

33. 125,000 =

34. 0.000125 =

Write each number in decimal numeral.

35. 9.42×10^{6} =

36. 9.42×10^{-6} =

Convert each fraction to a decimal. Round to the nearest hundredth.

37. $\frac{9}{20}$ = **38.** $\frac{1}{8}$ = **39.** $\frac{3}{7}$ =

Convert each decimal to a percent.

40. 0.45 =

41. 1.125 =

Convert each percent to a decimal.

42. 4% =

43. 45% =

Convert each fraction to a percent. Round to the nearest tenth of a percent.

44. $\frac{9}{20}$ = **45.** $\frac{1}{8}$ = **46.** $2\frac{3}{7}$ =

Convert each percent to a fraction in simplest form.

47. 4% = **48.** 40% = **49.** 4.5% =

50. A suit that is regularly $45 is on sale for 20% discount. What is the sale price of the suit ?.
Solution:

51. Maria earns 4% commission on each car he sells. What is her commission on a car she sells for $19,000 ?
Solution:

52. The price of a shirt was reduced from $20 to $16. What is the percent of decrease ?
Solution:

53. The bill of electricity was increased from $60 to $75. What is the percent of increase ?
Solution:

54. If you deposited $75,000 in a saving account at an annual simple interest rate 6%, how much money will be in the account after one year ?
Solution:

55. If you deposited $75,000 in a saving account at an annual interest rate of 6% compounded twice a year, how much money will be in the account after one year ?
Solution:

Evaluate each expression (absolute values and square roots).

56. $|15|$ = **57.** $|-15|$ = **58.** $-|-15|$ = **59.** $|-3\frac{4}{5}|$ =

60. $\sqrt{64}$ = **61.** $\sqrt{81}$ = **62.** $\sqrt{8}$ = **63.** $\sqrt{48}$ =

64. Between which two integers does $\sqrt{85}$ lie ? Your answer: ___ and ___ .

Sample Test (2)
Quantity Measurements

1. **The questions shown here were tested year by year in all kinds of tests, such as PSAT, SAT, ACT, STAR/CAT 6, and High School Exit.**
2. **These two pages cover all possible "basic questions" about converting units of measurement within and between systems (customary or metric); converting rates and solving direct variation problems; computing perimeter, circumference, area, surface area, and volume; using scale factors (page 99); understanding how and when to apply the Pythagorean Theorem (page 237).**

Comparable Chart of Measures

Length	Weight (Mass)	Volume (Capacity)
1 yard (yd) = 3 feet (ft) 1 foot (ft) = 12 inches (in.) 1 in. = 2.54 centimeters (cm) 1 mile (mi) = 5280 feet (ft) 1 mile (mi) = 1.6 km	1 ton = 2,000 pounds (lb) 1 pound (lb) = 16 ounces (oz) 1 kilogram (kg) = 2.2 lb 1 ounce (oz) = 28 grams (g)	1 gallon (gal) = 4 quarts (qt) 1 quart (qt) = 2 pints (pt) 1 pint (pt) = 2 cups (cu) 1 cup (cu) = 8 fluid ounces (fl.oz) 1 gallon (gal) = 128 fl. oz. = 3.8 liters (ℓ)
1 kilometer(km) = 1000 meters (m) 1 meter(m) = 100 centimeters (cm) 1 meter(m) = 1000 millimeters (mm)	1 kilogram (kg) = 1000 grams (g) 1 gram (g) = 100 centigrams (cg) 1 gram(g) = 1000 milligrams (mg)	1 kiloliter ($k\ell$) = 1000 liters (ℓ) 1 liter (ℓ) = 100 centiliters ($c\ell$) 1 liter(ℓ) = 1000 milliliters ($m\ell$)
Other Measures: 1 cc.(or c.c) = 1 cubic centimeter 1 tablespoon = 3 teaspoons = 15 cc. 1 teaspoon = $\frac{1}{3}$ tablespoon = 5 cc. 1 tablespoon = $\frac{1}{2}$ fluid ounce	1 bushel = 36 liters 1 peck = 9 liters 1 bushel = 4 pecks	" hecto " means " hundred ". " deca " means " ten ". 1 hectometer and 2 decameters = 120 meters

Area of a rectangle: $A = \ell \times w$ where ℓ is the length and w is the width

Volume of a rectangular solid: $V = \ell \times w \times h$ where h is the height

Area of a parallelogram: $A = b\,h$ where b is the base and h is the height

Area of a triangle: $A = \frac{1}{2} b\,h$ where b is the base and h is the height

Area of a circle: $A = \pi\, r^2$ where r is the radius

Circumference of a circle: $C = \pi\, d$ where d is the diameter

Surface area of a sphere: $S = 4\pi\, r^2$ where r is the radius

Volume of a sphere: $V = \frac{4}{3}\pi\, r^3$ Where r is the radius

Area of a right cylinder: $A = B\,h$ Where B is the base area and h is the height

Volume of a right circular cone: $V = \frac{1}{3}\pi\, r^2 h$ where r is the radius and h is the height

Volume of a pyramid: $V = \frac{1}{3} B\,h$ where B is the base area and h is the height

Lateral Area of a right circular cone: $S = \frac{1}{2} c\,\ell$ where c is the circumference and ℓ is the slant height

Complete each statement.

1. 5 yd = _____ ft **2.** 36 ft = _____ yd **3.** 38 ft = _____ yd

4. 4 ft = _____ in. **5.** 48 in. = _____ ft **6.** 55 in. = _____ ft

7. 12 in. = _____ cm **8.** 30.48 cm = _____ in. **9.** 100 in. = _____ m

-----Continued-----

Complete each statement.

10. 3 mi = _____ ft
11. 1760 ft = _____ mi
12. 1.5 km = _____ m
13. 1.5 m = _____ cm
14. 45 mm = _____ m
15. 4 lb = _____ oz
16. 2.4 lb = _____ oz
17. 12, 500 lb = _____ tons
18. 2.4 kg = _____ g
19. 1.5 g = _____ mg
20. 7 gal = _____ qt
21. 3 ft^2 = _____ $in.^2$
22. 648 $in.^2$ = _____ ft^2
23. 648 $in.^3$ = _____ ft^3
24. 5.2 yd^3 = _____ ft^2

Converting rates.

25. 60 miles/hr = _____ feet/hr.
26. 80 miles/hr = _____ feet/sec.
27. 180 yd/hr = _____ yd/min.
28. 120,000 cm/min = _____ m/sec.

Solving direct variation problems.

29. Three pounds of beef cost $9.54. How much will 5 pounds of beef cost ?

30. A bus travels 130 miles in 2 hours. How far can the bus travel in 6 hours ?

31. You earned $75 in 5 hours, How much will you earn in 24 hours ?

32. You typed 320 words in five minutes. H[illegible] long will it take you to type 480 words ?

Find the perimeter and area.

33.

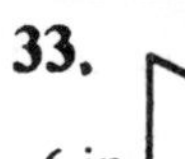

Solution:

34.

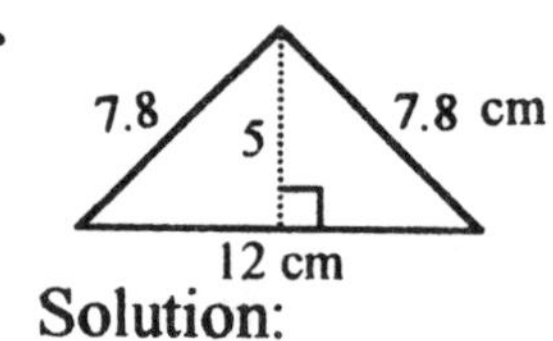

Solution:

35.

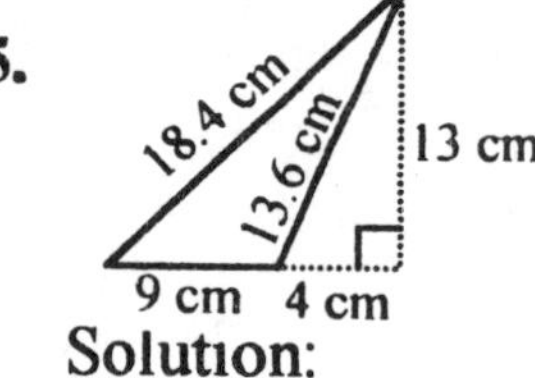

Solution:

36.

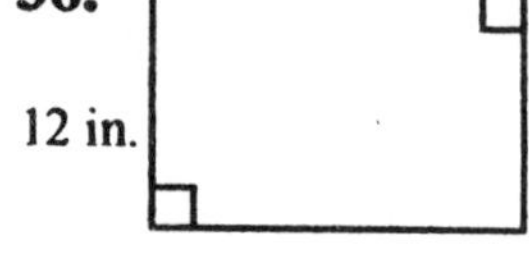

Solution:

37.

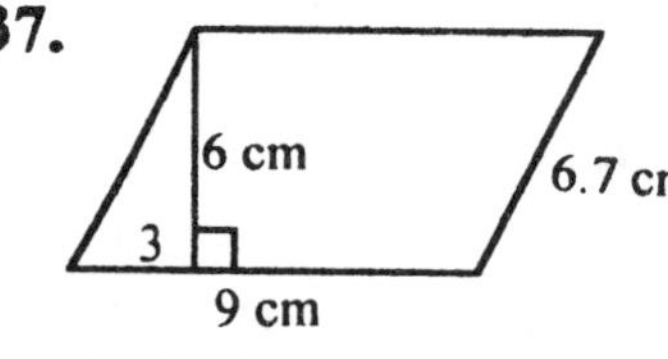

Solution:

38.

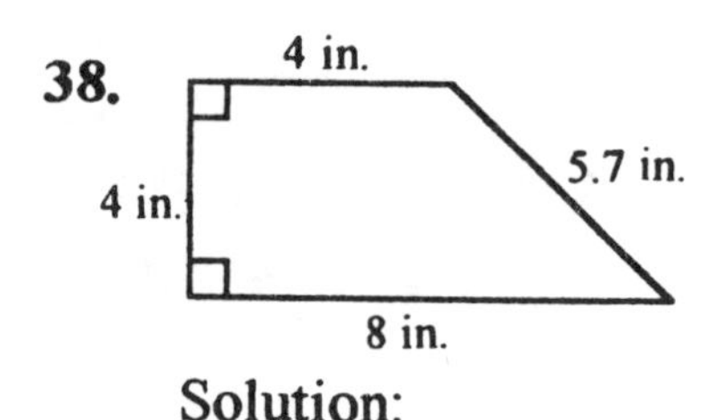

Solution:

39. Find the circumference and area.

Solution:

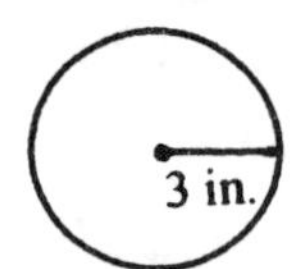

40. Find the value of x in the triangle.

Solution:

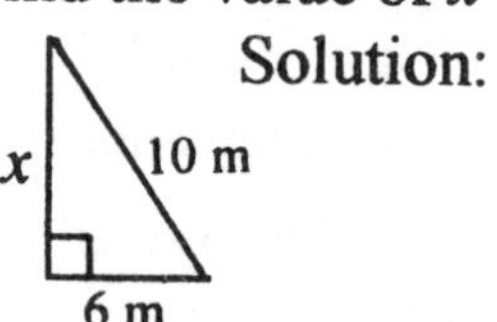

41. The scale drawing shows a rectangle. The actual width is 24 centimeters. What is the actual length ?

Solution:

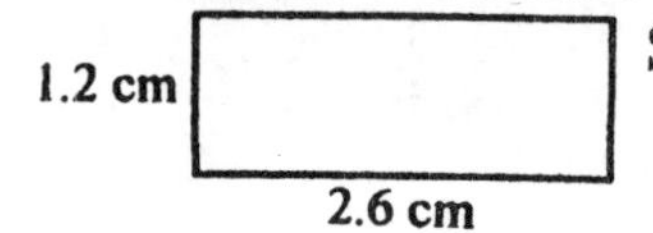

42. Find the volume of the box.

Solution:

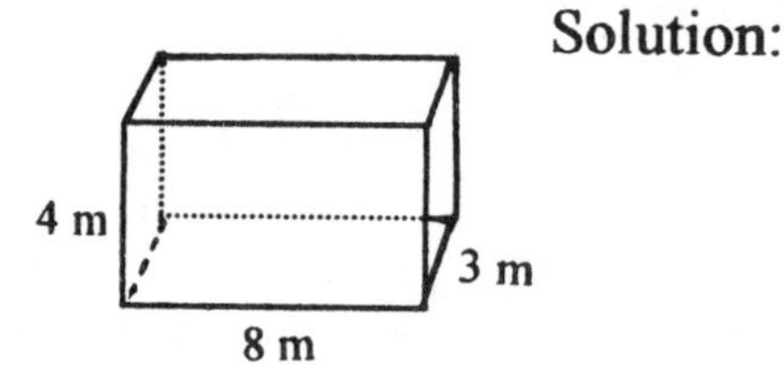

Sample Test (3)
Algebra, Functions, and Modeling

1. **The questions shown here were tested year by year in all kinds of tests, such as PSAT, SAT, ACT, STAR/CAT 6, and High School Exit.**
2. **These four pages cover all possible "basic questions" about the algebraic modeling and symbolic language with expressions, monomials and polynomials, linear equations and inequalities, systems of equations and inequalities, absolute-value equations and inequalities, linear functions and slopes, relation and function, parallel lines, graphs interpreting, and nonlinear equations.**

Write each word phrase to a variable expression.

1. Nine more than x.

2. Nine less than x.

3. x is not equal to nine.

4. 10 more than the sum of x and 5.

5. 9 less than the product of x and 5 is 10.

6. y more than x is 10 and y less than x is 5.

7. x minus 10 is y and x times y is -16.

8. Nine is more than x.

9. Nine is less than x.

10. x is greater than or equal to nine.

11. 10 is more than the product of x and 5.

12. 5 is less than the sum of 2 and twice y.

13. 5 is less than x, which is less than 10.

14. 5 times x plus y is greater than 10, and twice y is less than or equal to 5.

Evaluate each expression.

15. $3x-5$ when $x=4$.

16. $3x-5$ when $x=-4$.

17. $5-3x$ when $x=-4$.

18. $3a^2-2$ when $a=5$.

19. $3a^2-2$ when $a=-5$.

20. $2-3a^2$ when $a=5$.

21. $10-2y^3$ when $y=2$.

22. $10-2y^3$ when $y=-2$.

23. $\dfrac{a+2b}{5}$ when $a=1$, $b=2$

Simplify each expression with exponents.

24. $x^2 \cdot x^6 =$

25. $(x^2)^6 =$

26. $\dfrac{x^6}{x^2} =$

27. $x^2 \cdot x^{-4} =$

28. $(x^{-2})^6 =$

29. $\dfrac{8x^{-6}}{2x^{-2}} =$

30. $(3a^2b^3)(4a^6b^4) =$

31. $(-2a)^3 \cdot a^{-5} =$

32. $\dfrac{(-3a)^2}{18a} =$

-----Continued-----

Simplify each expression (monomials and polynomials).

33. $4x - 9x =$

34. $4x + 4 - 6x =$

35. $(4x^5y^3)(9xy^4) =$

36. $2x(x-4) =$

37. $(x-2)(x+4) =$

38. $(2x+1)(x-4) =$

39. $\frac{4x^5y^3}{9xy^4} =$

40. $\frac{8x-4x}{2} =$

41. $\frac{4n^3+8n^2-12n}{2n} =$

42. Write an expression for the perimeter.

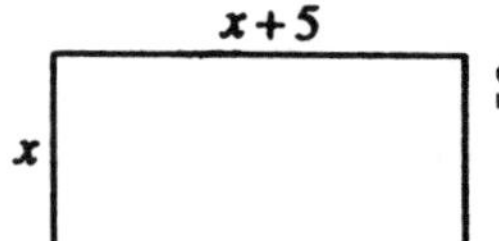

Solution:

43. Write an expression for the area of the shaded region.

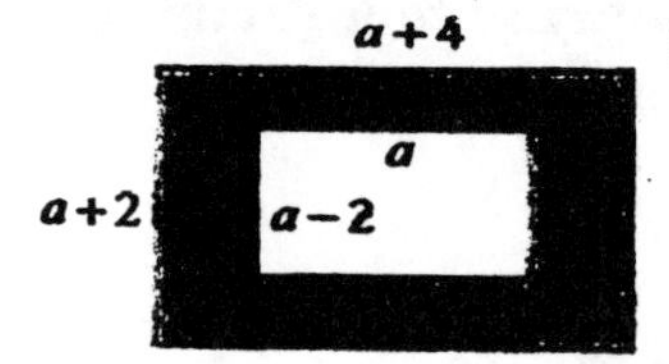

Solution:

Solving linear equations and inequalities.

44. Solve $2x - 3 = 5$.
Solution:

45. Solve $-2x + 3 = 5$.
Solution:

46. Solve $\frac{4}{7}n + 3 = -5$.
Solution:

47. Solve $5(x-2) = 3x + 4$.
Solution:

48. Solve $\frac{5}{2n-1} = \frac{3}{n}$.
Solution:

49. Solve $2x - 3 < 5$
Solution:

50. Solve $2x - 3 \geq 5$.
Solution:

51. Solve $-2x - 3 > 5$.
Solution:

52. $5(x-2) \leq 3x + 4$
Solution:

53. Solve $|x+5| = 2$.
Solution:

54. Solve $|4x-5| = 6$.
Solution:

55. Solve $|x+5| = -2$
Solution:

56. Solve $|x+5| < 2$.
Solution:

57. Solve $|x+5| > 2$
Solution:

58. Solve $7 - |x+5| \geq 2$.
Solution:

-----Continued-----

59. Is the ordered pair (3, –2) a point on the graph $2x - y = 8$?
Solution:

60. Is the ordered pair (–3, 2) in the solution set of $2x - y = 8$?
Solution:

61. The graph of $2x - y = 6$ crosses the x – axis at which point ?
Solution:

62. The graph of $2x - y = 6$ crosses the y – axis at which point ?
Solution:

63. Find the x – intercept and y – intercept of $2x - y = 6$, then graph it.
Solution:

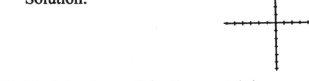

64. Find the slope of the line whose equation is $2x - y = 6$.
Solution:

65. Find the slope of the line containing points (1, –4) and (3, 7).
Solution:

66. Find the slope of the line containing points (–1, 4) and (3, –7).
Solution:

67. Find the equation of the line having slope –3 and y – intercept 5.
Solution:

68. Find the equation of the line having slope –3 and passing through the point (1, 2).
Solution:

69. What is the slope of a line parallel to the line $2x - y = 6$.
Solution:

70. Find the equation of the line parallel to the line $2x - y = 6$ and passing through the point (1, 4).
Solution:

71. The graph below shows the closing price per share of a company's stock. In which day that had the greatest increase in price over the previous day ?
Solution:

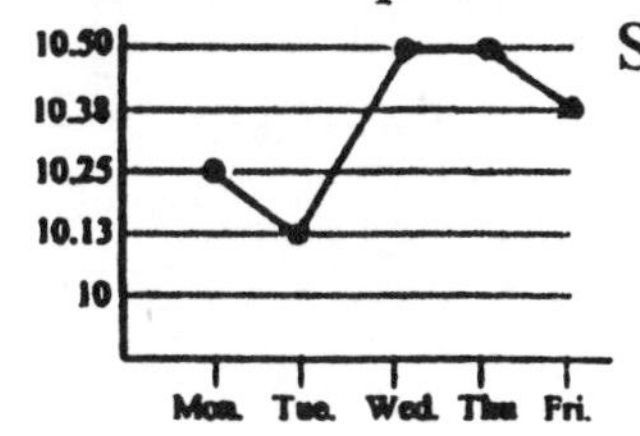

72. The graph below shows the sales of a product. In which month that had the sale twice as many as the sale in February.
Solution:

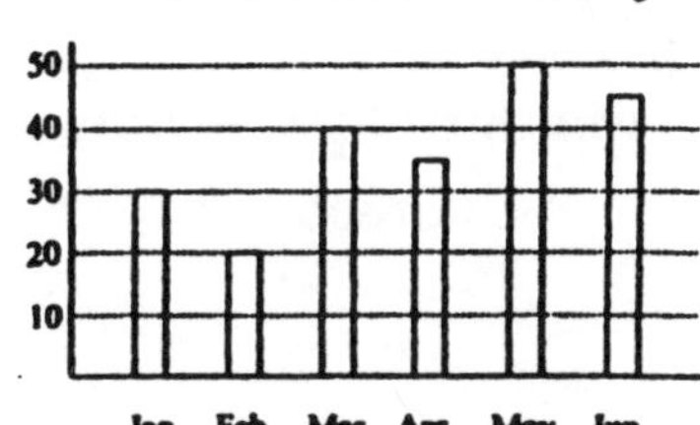

-----Continued-----

State whether each of the following graph shows that y is a function of x ? Explain.

73. 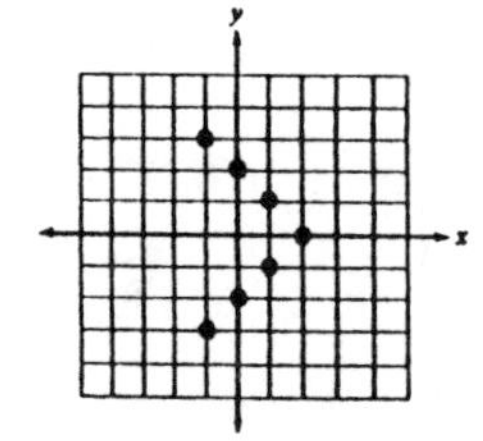
74. 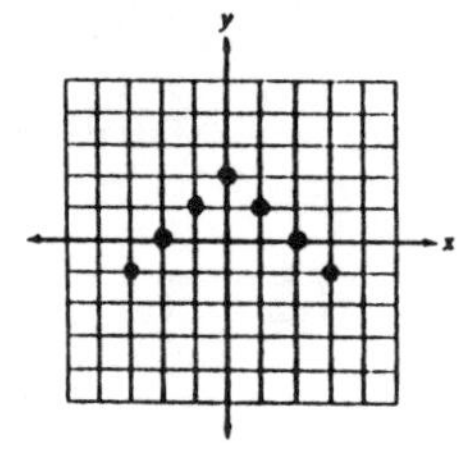
75. 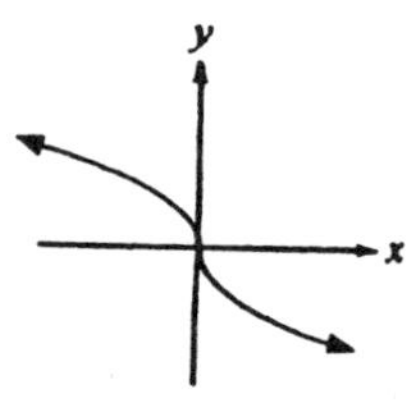
76. 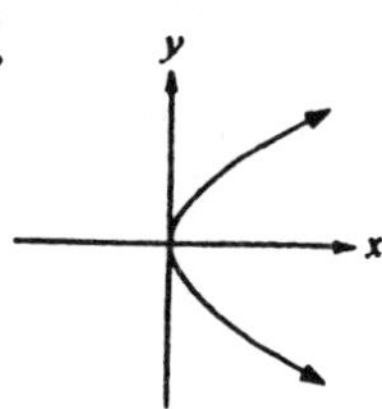

Match the function with its graph.

77. $y = x$ 78. $y = x^2$ 79. $y = x^3$ 80. $y = 3$

a. 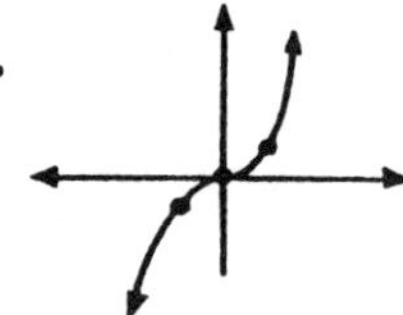
b. 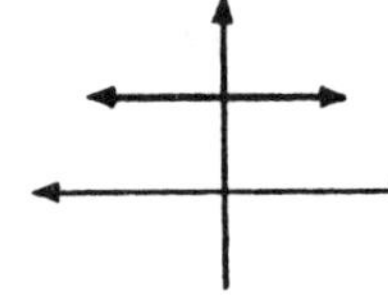
c. 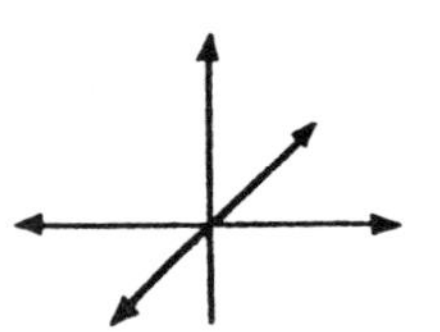
d. 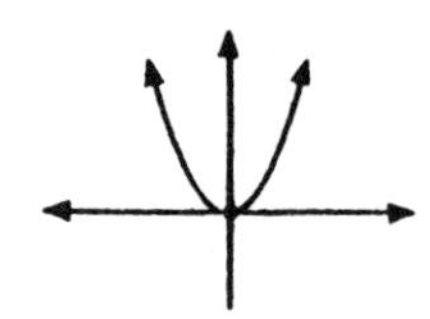

Solving each system of linear equations and inequalities.

81. Solve $\begin{cases} x + y = 5 \\ x - y = 11 \end{cases}$ **82.** Solve $\begin{cases} x - y = 1 \\ 3x - 2y = 7 \end{cases}$ **83.** Solve $\begin{cases} 2x - y = 3 \\ 5x + 3y = 9 \end{cases}$

84. Solve $\begin{cases} x + y = 5 \\ 2x + 2y = 7 \end{cases}$ **85.** Solve $\begin{cases} x - 2y = 5 \\ 3x - 6y = 15 \end{cases}$ **86.** Solve $\begin{cases} x - 2y \geq 4 \\ 2x + y < 6 \end{cases}$

Solving word problems with linear equations and inequalities.

87. John has twice as much money as Carol. Together they have $81. How much money does each have ?

88. A number divided by 3 is equal to the the number decreased by 3. Find the number.

89. A club has 42 members. There are 6 more men than women. How many men and how many women are on the club ?

90. Your scores on your 3 math tests wer 71, 74 and 82. What is the lowest scc you need in your next test in order to have an average score of at lease 80 '

Solving word problems with systems of linear equations.

91. Twice a number is 20 more than another number. One fourth of the first number is 29 less than the second number. Find the two numbers.

92. There are dimes and quarters in a box Dimes are 4 times as many as quarte Their total value is $13. How many Dimes and how many quarters are in in the box ?

Graphing nonlinear equations (parabolas and cubic functions).

93. Graph $y = 2x^2$. **94.** Graph $y = -2x^2$. **95.** Graph $y = 2x^3$. **96.** Graph $y = -2.$

Sample Test (4)
Statistics and Probability

1. **The questions shown here were tested year by year in all kinds of tests, such as PSAT, SAT, ACT, STAR/CAT 6, and High School Exit.**
2. **These two pages cover all possible "basic questions" about finding mean, median, mode, and range; data displays including scatterplots, bar graphs, stem-and-leaf plot, and box-and-whisker plot to compute the minimum, the lower quartile, the median, the upper quartile, and the maximum of a data set; computing the probability of an event as a ratio (fraction), a decimal, or a percent; computing the probability of two events which are independent or dependent.**

Hint: There are three different ways to write a probability: fraction, decimal, and percent (%)

1. Find the mean, median, mode, and range of John's math scores for 9 tests.
73, 84, 84, 82, 73, 85, 85, 85, 93,

2. Find the mean, median, mode, and range of the ages of 10 people.
11, 13, 10, 12, 16, 23, 23, 10, 42, 76.

3. The stem-and-leaf plot below shows the ages of a group of people.

Stems	Leaves
0	6, 8
1	5, 5, 9
2	3, 6, 6, 6, 7
3	2, 4, 6, 8, 8, 9
4	5, 6, 7, 7, 9
5	1, 1, 2, 3, 3
6	8, 9
7	0, 4, 8, 9

1|5 = 15

a. What is the lowest age ?
b. What is the highest age ?
c. What is the median ?
d. What is the mode ?
e. What is the range ?

4. Create a box-and-whisker plot for the ages of 9 people: 15, 12, 17, 20, 20, 32, 46, 35, 20.
a. What is the minimum of the ages ?
b. What is the maximum of the ages ?
c. What is the median of the ages ?
d. What is the lower quartile ?
e. What is the upper quartile ?
f. Use the five numbers (minimum, maximum, median, lower quartile, and upper quartile) to create a box-and-whisker plot.

5. The box-and-whisker plot below shows the monthly sales in copies of a book.

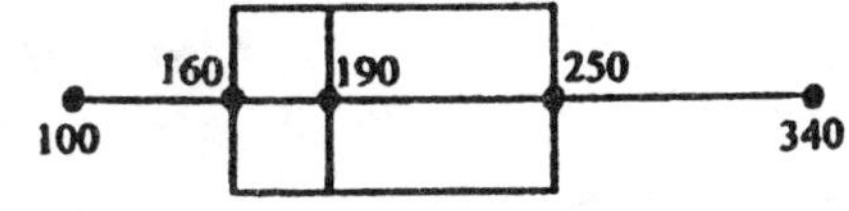

a. What is the minimum of the sales ?
b. What is the maximum of the sales ?
c. What is the median of the sales ?
d. What is the mean of the sales ?
e. What is the lower quartile of the sales ?
f. What is the upper quartile of the sales ?
g. Between what two readings is the middle 50% of the sales ?

-----Continued-----

6. The bar graph below shows the number of a product sold during a 6-month period.
   a. What is the total number of the product sold ?
   b. What is the range of the product sold in 6 months ?
   c. Which month were the month that sold the most number of the product ?
   d. Which month were the month that sold the product two times as many as were sold in February ?
   e. What percent of the product sold during the 6-month period were sold in April ?

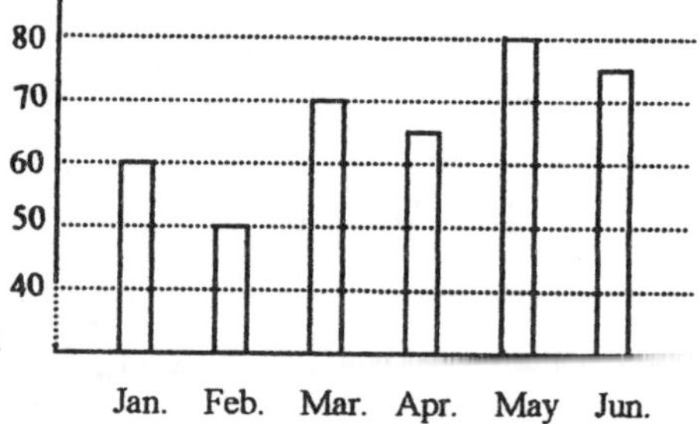

7. Each scatterplot below represents two numerical sets of data on a graph using points. Identify each plot as a correlation of strong positive, strong negative, weak positive, or weak negative.

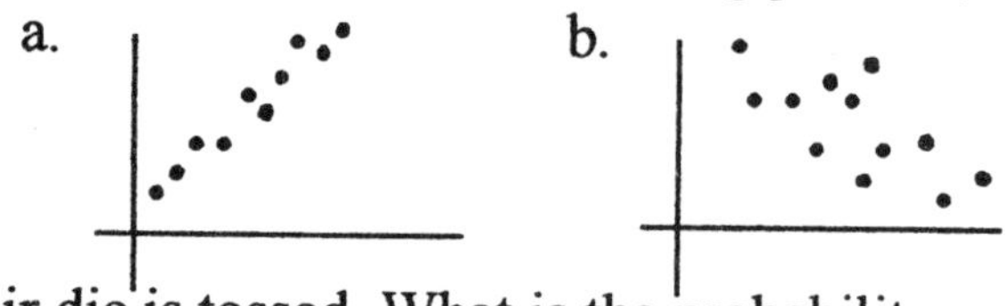

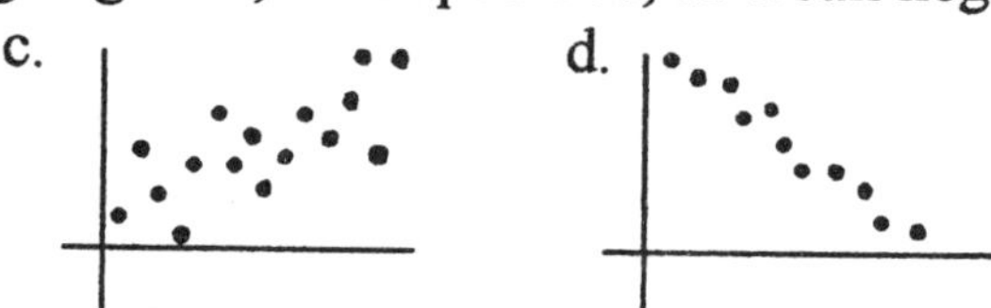

8. A fair die is tossed. What is the probability of getting a 5 ?

9. A cube is numbered from 1 to 6. What is the probability of rolling an even number ?

10. One card is drawn at random from a 52-card bridge deck. What is the probability of the card drawn turned up a 12 ?

11. One card is drawn at random from a 52-card bridge deck. What is the probability of the card drawn turned up a club ?

12. A spinner is numbered from 1 to 8. What is the probability that after a spin the arrow will stop on region 4 ?

13. A spinner is numbered from 1 to 10. What is the probability that after a spin the arrow will stop on region with a number less than 7 ?

14. A box contains 4 red balls, 8 white balls, and 6 green balls. A ball is drawn at random. What is the probability that the ball drawn is red ?

15. A box contains 4 red balls, 8 white balls, and 6 green balls. A ball is drawn at random. What is the probability that the ball drawn is not red ?

16. A box contains 4 red balls, 8 white balls, and 6 green balls. What is the probability of drawing a red ball and then drawing another green ball from the remaining balls ? (without replacement, dependent)

17. A box contains 4 red balls, 8 white balls, and 6 green balls. What is the probability of drawing a red ball and then a green ball. The ball is put back after the first draw. (with replacement, independent)

18. The probability of an event A is $p(A) = \frac{5}{14}$. What is the probability that event A will not occur ?

19. The probability of three even A, B, and C are $p(A) = 0.68$, $p(B) = \frac{2}{3}$, and $p(C) = 70\%$. Which event is the least likely to occur ?

20. Mary tosses a fair coin 50 times and 28 times turned up heads. What is the experimental probability that it will turn up head in the next toss ?

21. Mary tosses a fair coin 50 times and 28 times turned up heads. What is the theoretical probability that it will turn up head in the next toss ?

22. A school with 2,320 students has 58 students who are left-handed this year. What is the probability that a student chosen at random will be left-handed ?

23. The probability of a person being colorblind is $\frac{1}{60}$. There are 1,680 students in a high school. How many students would probably be colorblind ?

Notes

Notes

Additional Example

Amazing Speed-Squared

The square of a two-digit integer with last digit 5 can be found quickly by the following formula:

$$(10a+5)^2 = a \times (a+1) \times 100 + 25$$

Examples:

$$15^2 = 1 \times 2 \times 100 + 25 = 225$$
$$25^2 = 2 \times 3 \times 100 + 25 = 625$$
$$35^2 = 3 \times 4 \times 100 + 25 = 1225$$
$$45^2 = 4 \times 5 \times 100 + 25 = 2025$$
$$55^2 = 5 \times 6 \times 100 + 25 = 3025$$
$$65^2 = 6 \times 7 \times 100 + 25 = 4225$$
$$75^2 = 7 \times 8 \times 100 + 25 = 5625$$
$$85^2 = 8 \times 9 \times 100 + 25 = 7225$$
$$95^2 = 9 \times 10 \times 100 + 25 = 9025$$

We can easily multiply the tenth digit and " tenth digit's next integer ", then put 25 behind it.

Prove the formula: $(10a+5)^2 = a \times (a+1) \times 100 + 25$

Proof:
$$(10a+5)^2 = 100a^2 + 100a + 25$$
$$= 100a(a+1) + 25$$
$$= a \times (a+1) \times 100 + 25$$
The proof is complete.

The above formula can be extended to any number with last digit 5:

$$105^2 = 10 \times 11 \times 100 + 25 = 11025$$
$$205^2 = 20 \times 21 \times 100 + 25 = 42025$$
$$305^2 = 30 \times 31 \times 100 + 25 = 93025$$
$$\vdots$$
$$905^2 = 90 \times 91 \times 100 + 25 = 819025$$
$$1005^2 = 100 \times 101 \times 100 + 25 = 1010025$$

Other formula: $a^2 = (a+b)(a-b) + b^2$

$$98^2 = (98+2)(98-2) + 4 = 100 \times 96 + 4 = 9604$$
$$107^2 = (107+7)(107-7) + 49 = 114 \times 100 + 49 = 11449$$

A teacher is testing a seven-year old boy.
Teacher: What is the answer of 5×6 ?
Boy: 15
Teacher: That is not right.
The answer should be 30.
Boy: At least my answer is half right.

Additional Example

Proof of " Division by zero is undefined "

In mathematics, we say that " Division by zero is undefined. ".
It means " Division by zero is not allowed. ".

Example: Prove the statement " Division by zero is undefined ".

Proof: Let a and b are two real numbers, $a > b$.

We use $a - b = c$

$$a = b + c$$

Multiply each side by $(a - b)$:

$$a(a-b) = (b+c)(a-b)$$

$$a^2 - ab = ab - b^2 + ac - bc$$

$$a^2 - ab - ac = ab - b^2 - bc$$

$$a(a-b-c) = b(a-b-c)$$

Divide each side by $(a - b - c)$

We have $a = b$ (false statement)

Since $a > b$, the above result $a = b$ is a false statement.

The false statement occurred because the division by $(a - b - c)$ is used in the process.

Since $a - b - c = 0$, therefore division by zero is not allowed.

Additional Examples

Finding the reciprocal of a number

Reciprocal: Two numbers are reciprocals of each other if their product equals 1.

Examples: **1.** 2 and $\frac{1}{2}$ are reciprocals of each other because $2 \times \frac{1}{2} = 1$.

2. 5 and $\frac{1}{5}$ are reciprocals of each other because $5 \times \frac{1}{5} = 1$.

3. 2 and 0.5 are reciprocal of each other because $2 \times 0.5 = 1$.

4. 5 and 0.2 are reciprocals of each other because $5 \times 0.2 = 1$.

5. 0.4 and 2.5 are reciprocals of each other because $0.4 \times 2.5 = 1$.

Example 6: Prove $2 + \sqrt{3}$ and $2 - \sqrt{3}$ are reciprocals of each other.

Proof: $(2+\sqrt{3})(2-\sqrt{3}) = 2^2 - (\sqrt{3})^2 = 4 - 3 = 1$. The proof is complete.

Check: $\frac{1}{2-\sqrt{3}} = \frac{2+\sqrt{3}}{(2-\sqrt{3})\cdot(2+\sqrt{3})} = \frac{2+\sqrt{3}}{4-3} = 2 + \sqrt{3}$.

Additional Example

Finding GCF and LCM by Rollabout Divisions

In Chapter 3-4, we have learned how to find the GCF and the LCM of two given numbers. However, it is not easy to find the GCF and the LCM if the numbers are very large.

To find the GCF and the LCM of two large numbers, we use the **Rollabout Divisions** to find the largest number that can divide into all of the given numbers.

The Rollabout Divisions was first invented by Euclid, a famous Greek mathematician who lived about 300 B.C.

Rollabout Divisions

Step 1: Divide the larger number by the smaller number to find the remainder.
Step 2: Divide the smaller number by the remainder in Step 1.
Step 3: Divide the previous remainder by the new remainder.
Step 4: Repeat the Step 3 until no remainder is found.
Step 5: The divisor of the last division is the GCF if the remainder is zero.
Step 6: Find LCM by the same method on page 50.

Example: Find the GCF and the LCM of 3,473 and 1,963.

Solution:

1.

$$1963\overline{)3473}$$ quotient 1
1963
1510 remainder

2.

$$1510\overline{)1963}$$ quotient 1
1510
453 remainder

3.

$$453\overline{)1510}$$ quotient 3
1359
151 remainder

4.

$$151\overline{)453}$$ quotient 3
453
0

We have: GCF =151 Ans.

$$151 \,\underline{\lfloor 3473 \quad 1963}\; ; \quad 3473 \div 151 = 23 \text{ and } 1963 \div 151 = 13$$
$$\quad\;\; 23 \quad\;\; 13$$

We have: LCM = $151 \times 23 \times 13 = 45149$ Ans.

Abraham Lincoln was the 16th President of the United States. One day, as he was giving a speech before the crowd, a pro-slave audience wrote " foolish " on a piece of paper and gave it to Lincoln.

Lincoln: Yesterday, I received a letter from a friend who forgot to sign his name.
Today, I just received a letter from a friend who signed his name only.

Additional Example

Amazing Perfect Square

Examples: $1 \cdot 2 \cdot 3 \cdot 4 + 1 = 25 = 5^2$

$2 \cdot 3 \cdot 4 \cdot 5 + 1 = 121 = 11^2$

$3 \cdot 4 \cdot 5 \cdot 6 + 1 = 361 = 19^2$

We can prove that it is true for all cases as below.

Example: Prove that the sum of the product of four consecutive integers and 1 is a perfect square.

Proof:

$$\begin{aligned} & n(n+1)(n+2)(n+3)+1 \\ &= n(n+3) \times (n+1)(n+2)+1 \\ &= (n^2+3n) \times (n^2+3n+2)+1 \\ &= (n^2+3n)(n^2+3n)+(n^2+3n) \cdot 2+1 \\ &= (n^2+3n)^2+2(n^2+3n)+1 \\ &= (n^2+3n+1)^2 \end{aligned}$$

It is a perfect square. The proof is complete.

Additional Example

Using three of any nonzero real number to obtain 1

Example: Using three 5's to write a math expression which has the result 1.

Solution: $\left(\frac{5}{5}\right)^5 = 1^5 = 1$. Ans.

We can write a math expression using three of any nonzero real number to obtain 1.

Additional Example

Using three 2's to write any integer

Formula: a number $n = -\log_2(\log_2 \sqrt{\sqrt{\sqrt{\cdots\sqrt{2}}}})$

There are n square roots of 2 in the radicals.

To understand this formula, you need to learn advanced algebra.

(Read the proof in the book "A-Plus Notes for algebra", page 255.)

> In a toy store, a five-year old girl wants her father to buy her an expensive toy.
>
> Father: No, it is too expensive.
>
> Girl: Dad, I like it. Honey and Sweetheart.
>
> Father: Why are you calling me " Honey and Sweetheart".
>
> Girl: Because whenever mom wants to buy a new dress, she calls you "Honey" or "Sweetheart".

INDEX and APPENDIX

A

B

C

D

E

F

G

H

R

S

T

U

V

W

Z

APPENDIX

Now, you have completed this book. Are you ready to study the Advanced Algebra ? See you in next book !!

The A-Plus Notes Series:

1. **A-Plus Notes for Beginning Algebra (Pre-Algebra and Algebra 1)**

2. **A-Plus Notes for Algebra (Algebra 2 and Pre-Calculus)**
   (A Graphing Calculator Approach)

3. **A-Plus Notes for SAT Math (SAT/PSAT and Subject Tests)**
   (Covers Geometry and Trigonometry)

4. **A-Plus Notes for GRE/GMAT (General and Subject Tests)**
   (GRE revised General Test)

If you are an educator and want to write your own book (all subjects) to improve the education in schools, please contact us.

Notes

Notes